AF389353

Integration of Renewables in Power Systems by Multi-Energy System Interaction

Integration of Renewables in Power Systems by Multi-Energy System Interaction

Editors

Birgitte Bak-Jensen
Jayakrishnan Radhakrishna Pillai

MDPI • Basel • Beijing • Wuhan • Barcelona • Belgrade • Manchester • Tokyo • Cluj • Tianjin

Editors
Birgitte Bak-Jensen
Department of Energy
Technology, Aalborg University
Denmark

Jayakrishnan Radhakrishna Pillai
Department of Energy
Engineering, Aalborg University
Denmark

Editorial Office
MDPI
St. Alban-Anlage 66
4052 Basel, Switzerland

This is a reprint of articles from the Special Issue published online in the open access journal *Energies* (ISSN 1996-1073) (available at: https://www.mdpi.com/journal/energies/special_issues/Multi-Energy_System_Interaction).

For citation purposes, cite each article independently as indicated on the article page online and as indicated below:

LastName, A.A.; LastName, B.B.; LastName, C.C. Article Title. *Journal Name* **Year**, *Volume Number*, Page Range.

ISBN 978-3-0365-0342-4 (Hbk)
ISBN 978-3-0365-0343-1 (PDF)

Contents

About the Editors

Birgitte Bak-Jensen professor at Aalborg University, Department of Energy Technology (senior IEEE member 2012), received her M.Sc. degree in electrical engineering in 1986 and a Ph.D. degree in "Modelling of High Voltage Components" in 1992, both degrees from the Department of Energy Technology, Aalborg University, Denmark. In 1986-1988, she was with Electrolux Elmotor A/S, Aalborg, Denmark as an electrical design engineer. She is now professor of intelligent control of the power sistribution system at the Department of Energy Technology, Aalborg University, where she has worked since August 1988. Her fields of interest are mainly related to the operation and control of the distribution network grid, including power quality and stability in power systems, taking the integration of dispersed generation and smart grid issues like demand response into account. Additionally, multienergy systems including the interaction between the electrical grid and, e.g., the heating and transport sector, is a key area of interest. She has participated in many projects concerning the control and operation of small dispersed generation units in distribution networks operated in connected and islanded mode, and the utilization of demand-side management, for instance, using electrical vehicles, electrical boilers, or heat pumps as energy storages used for leveling out fluctuations from renewable power units. She is now convenor of Cigre WG C6/1.33 on multienergy systems.

Jayakrishnan Radhakrishna Pillai associate professor, Department of Energy Technology, Aalborg University, received an M.Tech. degree in power systems from the National Institute of Technology, Calicut, India, in 2005, an M.Sc. degree in sustainable energy systems from the University of Edinburgh, Edinburgh, U.K., in 2007, and a Ph.D. degree in power systems from Aalborg University, Aalborg, Denmark, in 2011. His current research interests include distribution system analysis, grid integration of electric vehicles and distributed energy resources, multicarrier energy systems, and smart grids. He has participated in many projects related to the smartening of electricity grids and smart energy systems. He is active in IEEE and CIGRE C6 Working Groups on active distribution systems and distributed energy resources. He has also served as an electrical engineer in the petrochemical industry and as a software engineer in systems software development and mainframe technologies.

Preface to "Integration of Renewables in Power Systems by Multi-Energy System Interaction"

This Special Issue will focus on the interactions between different energy vectors, that is, between electrical, thermal, gas, and transportation systems, with the purpose of optimizing the future energy system. More and more renewable energy is integrated into the electrical system, and to optimize the usage and ensure that full production can be hosted and utilized, the power system has to be controlled in a more flexible manner—as an example, using excess electricity in the thermal system, using heat pumps or electrical boilers, and storing energy as thermal energy in storage tanks or in the district heating system. Another solution is to use and store electrical energy in the batteries of electrical vehicles, either to be used for transport or to be fed back to the power system again (V2G principle). The gas system can also be involved, using electrolyzers and storing hydrogen. In order not to overload the electrical distribution grid, the new large loads have to be controlled using demand response, perchance through a hierarchal control set-up where some controls are dependent on price signals from the spot and balancing market, but also where local real-time control and coordination are performed based on local voltage or frequency measurements where grid hosting limits are not violated. We welcome contributions on multienergy systems to explore the different possibilities for future smart energy systems with a high level of interaction among the different energy systems. The topics of interest include, but are not limited to:

- Modeling, optimization, and analysis of multienergy systems;

- Planning, operation, and control;

- Interaction and coupling between different energy supply systems and networks;

- Flexible demand and energy storages;

- Energy efficiency and management;

- Reliability and security of multienergy systems;

- Cyberphysical systems, information and communication infrastructure, and data analytics;

- Market, social, regulatory frameworks and policies for multienergy systems.

Birgitte Bak-Jensen, Jayakrishnan Radhakrishna Pillai
Editors

Article

Flexibility from Electric Boiler and Thermal Storage for Multi Energy System Interaction

Rakesh Sinha [1,*], Birgitte Bak-Jensen [1], Jayakrishnan Radhakrishna Pillai [1] and Hamidreza Zareipour [2]

[1] Department of Energy Technology, Aalborg University, Fredrik Bajers Vej 5, 9100 Aalborg, Denmark; bbj@et.aau.dk (B.B.-J.); jrp@et.aau.dk (J.R.P.)
[2] Department of Electrical and Computer Engineering, Schulich School of Engineering, University of Calgary, 2500 University Dr NW, Calgary, AB T2N 1N4, Canada; hzareipo@ucalgary.ca
* Correspondence: ras@et.aau.dk

Received: 7 November 2019; Accepted: 20 December 2019; Published: 24 December 2019

Abstract: Active use of heat accumulators in the thermal system has the potential for achieving flexibility in district heating with the power to heat (P2H) units, such as electric boilers (EB) and heat pumps. Thermal storage tanks can decouple demand and generation, enhancing accommodation of sustainable energy sources such as solar and wind. The overview of flexibility, using EB and storage, supported by investigating the nature of thermal demand in a Danish residential area, is presented in this paper. Based on the analysis, curve-fitting tools, such as neural net and similar day method, are trained to estimate the residential thermal demand. Utilizing the estimated demand and hourly market spot price of electricity, the operation of the EB is scheduled for storing and fulfilling demand and minimizing energy cost simultaneously. This demonstrates flexibility and controlling the EB integrated into a multi-energy system framework. Results show that the curve fitting tool is effectively suitable to acknowledge thermal demands of residential area based on the environmental factor as well as user behaviour. The thermal storage has the capability of operating as a flexible load to support P2H system as well as minimize the effect of estimation error in fulfilling actual thermal demand simultaneously.

Keywords: energy flexibility; power-to-heat; multi energy system; flexible demand; thermal storage; electric boiler; estimation of thermal demand

---

1. Introduction

District heating (DH) supplied hot water to 63% of the private Danish houses in 2015 [1]. The concept of 4th generation district heating/cooling system, supported by renewable, is presented in [2]. With the goal to become carbon neutral in the heating sector by 2030, renewables need to contribute all the heating demands. Thus, there is a possibility to integrate the thermal and electric networks to support grid ancillary services by the flexible electrical loads, such as electric boilers (EBs) and heat pumps (HPs), supporting the thermal system [2,3]. The electricity and heating network are coupled together as power-to-heat (P2H) to utilize renewable electricity for district heating. Integrated heat storage decouples demand and generation, to enhance flexibility for a better adaptation of energy requirements. The concept of P2H in the multi-energy system requires minor expansion of grid and storage [4].

The objective of this paper is to acknowledge flexible operation of the thermal unit consisting of an electric boiler (EB) and a storage tank modelled with stratified layers, as a part of P2H system. This is primarily realized through analysis of metered thermal consumption data from the residential area and estimating thermal demand using curve fitting, followed by an optimal schedule of the EB based on the spot price. The multilayer stratified thermal storage tank model is suitably identified for electric grid integration and flexible operation to compensate the error in estimation of thermal demand. The method could as well be applied for a heat-pump system. However, the application of EB is quite significant nowadays in providing energy flexiblity as well as system frequency services [5]. As an example, an EB of 50 kW is used as a flexible load in LIVØ island, Denmark to increase the self-consumption from wind and PV units installed in the island [6].

Advantages of centralized thermal storage in terms of operational flexibility of CHP (combined heat and Power) for district heating is well-explored in [7]. The flexibility of a district heating network for automatic frequency restoration reserve market is studied in [8]. The balancing markets provide an opportunity for introducing more EBs into DH and increase its contribution to flexibility [9]. A crucial aspect here is how system deployment can be realized effectively. Ref. [7] addresses the flexible operation of heat pumps using predictive control strategy, neglecting consumption of hot water for its highly randomized and hardly predictable nature. The predictive control of the heat pump by estimating only outdoor temperature has been studied in [10]. Thus, there is a necessity to investigate simple and effective methods to determine the influencing parameters for thermal demand prediction to manage the flexible operation of the thermal units in P2H technology.

The perspective of heat electrification in a wind dominated market using resistive heating and storage is the most carbon-intensive method [11] with lower investment cost compared to HP [9,12]. Further, large HPs take a long time from a cold start until they reach their optimal efficiency. Thus, they are not very active at balancing markets between hours, due to short start–stop intervals. Rather, they are mainly used as base load [9]. Hence, the flexibility in easy start–stop in balancing services is the main driver for introducing more EBs into the system. EBs in district heating have the potential for negative secondary control power by increasing consumption and supporting grid balance [13]. Reference [14] realized the benefit of demand-side management and the ability of demand response to improve power system efficiency with integrated wind power and electric heating devices considering constant heat load through the day. Higher potential of HPs in DH systems in the future is realized in [15]. Integration of EB with storage in low voltage residential grid as flexible consumer load has been presented in [16]. Hence, there is the potential of good harmony and flexibility between electrical and thermal energy sector supporting each other in multi-energy systems.

The investigation of space heating and domestic hot water needs is presented in [17] based on curve fitting and distribution functions. In [18] peak load ratio index of buildings are used for determining the diversity in thermal loads to generate the thermal profile for residential buildings. Reference [19] calculates the probability of domestic hot water drawn at a time(t) which depends upon probability during the day, weekday, season and holiday as a function of time(t). Higher probability step functions for weekends in comparison to weekdays are used to indicate higher consumption of domestic hot water on weekends. The thermal demand for space heating in a typical winter day is explored in [20]. However, the pattern of usage for the combined effect of space heating (SH) and domestic hot water (*DHW*) still remains unrealized. Proper knowledge of demand pattern for space heating and domestic use, as presented in this paper, is the key factor for developing a good and applicable estimating tool for thermal demand. This is italic in the main text and equations. For the consistence in the paper, please carefully check and change them to italic.

The possibility of estimating heat demand for space heating just a few hours in advance using neural network based on heat consumption in Polish buildings is matched against weather conditions over a 10-year period in [21]. In [21], the forecasting method is based on time series neural network with temperature and thermal consumption at a particular hour, day and previous history are taken into consideration. One month data from a DH network in Riga has been analysed for forecasting in [22] with the comparison between methods using an artificial neural network, polynomial regression model and the combination of both. With these methods, forecasts are performed by updating the statistics of actual load and temperature of the previous measurement. DH from Czechia has been analysed in [23] in a forecast model based on time series of outdoor temperature and time-dependent social components, which may vary for different weekdays and seasons. The Box–Jenkins method is used to realize the forecast of the social component. Reference [24] addresses issues on the selection of appropriate input variables from building energy management systems sensors. Ambient temperature and relative humidity along with solar radiation are the predominant factors for the predictive model [24,25]. In [26], forecasting based on similar day method is well presented for day-ahead power output for small scale solar PV system. However, none of the literature discussed regarding district heating in both summer and winter, as well as thermal demand prediction based on a combined effect of the time factor and environment variables (such as outdoor temperature, humidity, and wind velocity) together. These aspects are significant to be studied in an integrated framework to clearly understand the effective potential of thermal units like EBs. In this way, it is possible for such flexible units to render energy flexibility necessary to support integration of renewable energy in future energy systems.

In this paper, the proposed methodology to obtain flexibility with EB in P2H is summarised in the block diagram as shown in Figure 1. The significant contributions in this paper are the identification of thermal demand pattern, estimation of thermal demand using curve fitting tool, and use of stratified storage tank to verify flexible operation of EB. Actual thermal data from DH operator are analysed to unleash the specific consumption pattern of residential areas associated with usage based on different time factors such as hourly, weekdays, weekends, and seasonal. This information is useful while training the curve fitting tool to estimate thermal demand. With reference to [21–23], thermal demand estimation is based on the past and its current state for winter. A simple, yet effective curve fitting technique for estimating the thermal demand in the residential area, based on dependent parameters such as time factor (based on consumption profile) and environment variables (apparent temperature), has been investigated and compared with actual data as well as results from existing literature. The analysis is performed for thermal demand estimation during winter as well as summer. The curve fitting is simple and overcomes the problem encountered with the update of measured data (due to the failure of measuring equipment) as in time series estimation. The estimated demand is used to determine the optimal schedule of the EB operation in P2H, for the planning of capacities to store and fulfil thermal energy demand simultaneously, based on the spot price of electricity. The use of stratified storage tank in combination with EB emulates the real operating condition where the temperature of hot water being delivered is more realistic compared to ones from an average model of the storage tank, where hot water temperature decreases gradually. The outcome is verified with actual thermal demand to illustrate how the thermal storage copes up with the error in forecasting and contribute as an example of a flexible load in the P2H concept.

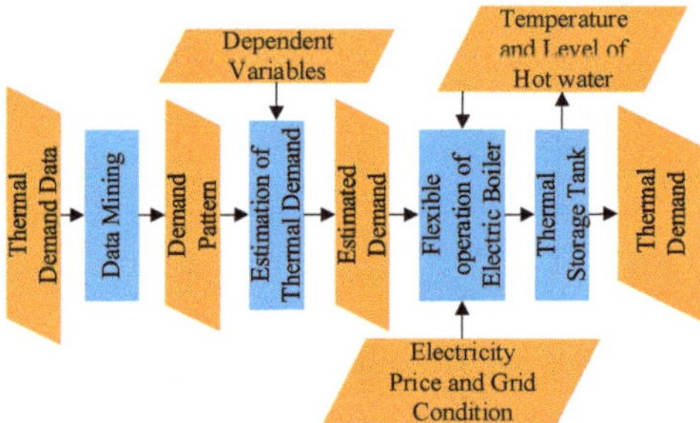

Figure 1. Block diagram of proposed system for flexible operation of electric boiler.

The paper is structured as follows. Analysis of thermal load consumption based on actual measurements at one particular residential site in Denmark supplied with five feeders is analysed to unleash the specific usage pattern and is identified in Section 2. Selection of parameters for effective estimation of thermal demand using various tools such as neural net fitting, and similar day method are discussed in Section 3. Overview of the modelling approach of the stratified hot water storage tank and EB is presented in Section 4 along with validation of the model. In Section 5, the methodology for optimized operation schedule of EB is presented along with EB ON/OFF control strategy. Results of the estimated demand are discussed in Section 6, followed with its application in flexible scheduling of the EB for demand response. Finally, the paper is concluded with the outcome of the research work in Section 7.

2. Analysis of Thermal Data

Thermal data measured at the terminal of five thermal distribution feeders ($F_1 - F_5$) supplying a number of residential buildings, in one particular residential area of Aalborg, Denmark, are used for analysis. Available measured data of hourly thermal consumption, from the period of 21 December 2015 to 4 December 2016 are analysed. Figure 2 shows the total annual consumption of thermal demand (Q_{DHW}) for residences in feeders ($F_1 - F_5$) supplying residential buildings. The annual consumption varies from 723.7 MWh as lowest consumption for F_1 to 1278.5 MWh as the highest consumption in F_4. This variation is due to the different number of residents in the area and their level of comfort. The total annual consumption was 5195.7 MWh. Figure 3a,b shows the plot of hourly consumption of Q_{DHW} for feeders ($F_1 - F_5$) and their total consumptions respectively, throughout the year. Figure 3a,b clearly shows that there is seasonal variation.

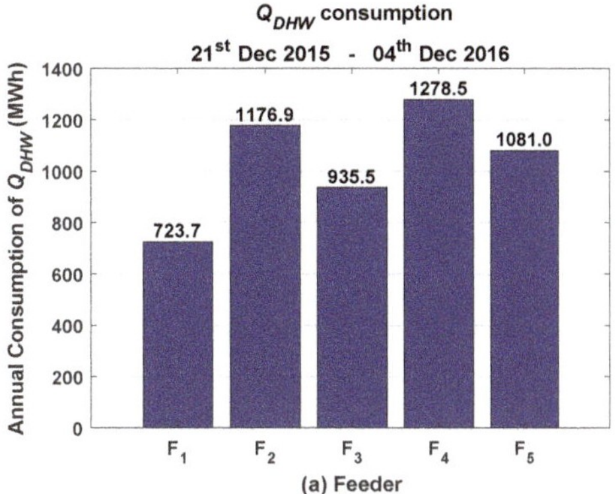

Figure 2. Yearly consumption of Q_{DHW} in different feeders.

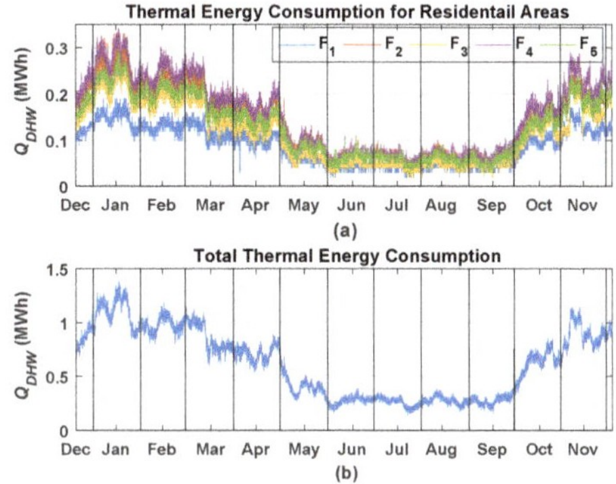

Figure 3. (a) Yearly Q_{DHW} consumption pattern of all feeders. (b) Total yearly Q_{DHW} consumption pattern of all feeders.

Figure 3b shows that there is a sudden transition in thermal consumption at certain time period such as towards the end of January, mid of March, and beginning of May. However, there is a significant difference in thermal consumption between mid-May to September end which is less than 35% of the peak winter consumption. Thus, to simplify the further analysis, the trend of thermal consumption is roughly divided into two seasons, winter and summer, irrespective of autumn and spring. Hence, October to April is considered as winter season and May to September is considered as summer season. The transition period at the beginning of May and October is not considered in this analysis. It seems that there is slightly more thermal demand in May than in September, due to transition from winter to summer and is around $30 \pm 5\%$ of the peak winter consumption. It is interesting to see the analysis of data from seasonal perspectives: winter and summer consumption. In the rest of the paper, analysis is done taking the combined effect of all feeders. As a result, the maximum heat demand is likely to be less than the sum of the individual feeder's peak load. This also reduces the intermittent variation in demand for individual feeders.

The average consumption per hour of Q_{DHW} for all the feeders, considering yearly consumption, is 618.5 kWh. During winter, it is 881.8 kWh, which is 205.8% more than summer consumption of 288.4 kWh.

Figure 4a,c shows the graph of hourly average thermal consumption pattern of different days of the week during winter and summer respectively. It is clearly seen that there exist a unique pattern of average thermal consumption with peaks. The pattern is different on weekends (Saturday and Sunday) in comparison to weekdays (Monday–Friday). To simplify the graphs shown in Figure 4a,c, graphs with an average consumption of thermal energy during the week, weekdays and weekend has been plotted in Figure 4b,d for winter and summer respectively. It is observed that there are some definite patterns of hourly usage of an average Q_{DHW}. There are two peaks and two valleys. It is clear that the amount of variation in thermal consumption with respect to minimum consumption is higher for weekends than for weekdays indicating higher consumption of domestic hot water as mentioned in [19].

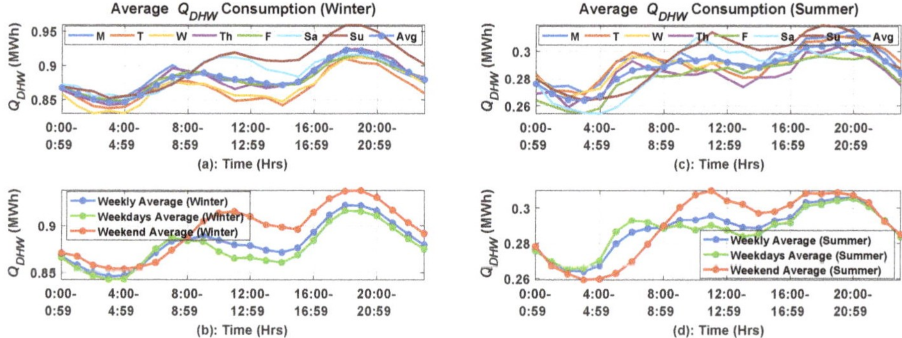

Figure 4. Analysis of Q_{DHW} December 2015–December 2016. (**a,c**) Analysis of average thermal consumption in hourly basis for different days of week for summer and winter respectively. (**b,d**) Analysis of average thermal consumption in hourly basis for a week, weekdays and weekends.

Figure 5 shows the consumption pattern for the week, weekdays and weekends for the period of Dec 2016 to Aug 2017 for winter and summer respectively.

Unlike in Figure 4b,d the total consumption at weekends are lower than weekdays. Thus, the amount of thermal consumption based on weekend and weekdays are not much relevant. However, the hourly pattern of consumption for weekdays and weekends are comparable with similar peaks and valley at particular hours seen in Figure 4b,d. Hence, knowledge of these patterns of thermal consumption during weekdays and weekend is much helpful to train the estimation tool to compensate for the error due to temperature independent factors such as user behaviour. The lowest consumption is during the period 03:00–04:59 h which rises gradually until 07:00–07:59 h during normal weekdays when people get ready for their job (Figure 4b,d). On the weekend there is a shift in this peak which is around 10:00–12:59 h. The shift in peak could be because people prefer to wake up late on the weekend.

After the morning peak, there is decrement of thermal consumption until 2:00–3:59 h when people are at work during weekdays. Throughout the week, the evening peak is around 18:00–20:59 which gradually decreases to 4:59 h in the early morning. However, in summer there is a shift in evening peak compared to that in winter. This analysis shows the relevance of time, day and season to determine the usage pattern of thermal consumption and that it is significant for forecasting as seen in [21] for thermal load similar to the forecasting of electrical load [27].

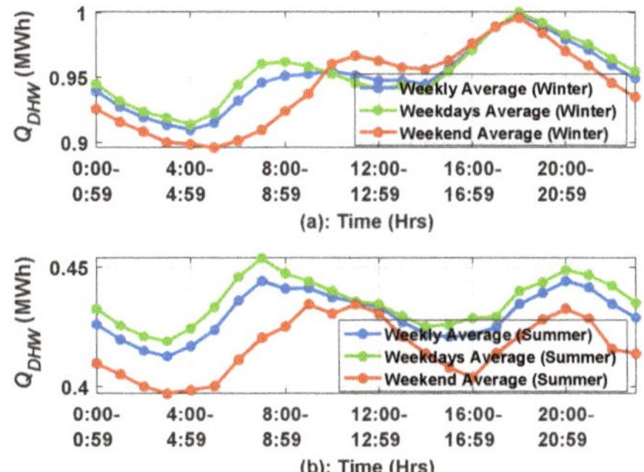

Figure 5. Average hourly thermal consumption of week, weekdays and weekends from period (December 2016–August 2017) (**a**) Winter (**b**) Summer.

3. Thermal Demand Estimation

It is difficult to estimate thermal demand for the residential area, as it is not only largely depending on the environmental variables (weather), but also on the user behaviour and building geometry. In reality, analysis for occupancy and user level comfort is difficult and leads to challenges incorporated with privacy issues of the individuals. This leads to a significant effort to compromise between errors in estimated variable and dependent parameters. Analysis of thermal data from residential areas gives remarkable information on the pattern of thermal demand, without compromising the privacy issue of individuals. These informations are helpful in selecting effective variables for the estimation of thermal demand from the perspective of user behaviour, which defines the pattern of demand. Time of day and days of the week (weekdays or weekends) are the two major parameters associated with the pattern of thermal consumption based on user level comfort.

The estimated parameters are subjected to identify the flexible operation of the thermal system based on demand, supply, capacity, and energy prices. In this paper, for the estimation of thermal consumption in the residential area, thermal data associated with Figure 5 are used.

3.1. Dependent Variables for Thermal Demand Estimation

Thermal demand is highly influenced by the environmental variable such as air temperature. Figure 6a shows the hourly value of thermal demand and corresponding average external temperature of the environment. It shows that decrease in temperature increases thermal demand. Beside air temperature, cold air with high relative humidity increases the conduction of heat from the body in comparison to dry air with the same temperature. In order to incorporate the combined effect of relative humidity, wind and air temperature together, responsible for heat loss from a body, apparent temperature is considered. The apparent temperature is calculated using (1) and (2) [28]. Figure 6b shows the hourly value of thermal demand and corresponding apparent temperature. The correlation coefficient of thermal demand with respect to external ambient temperature and apparent temperature, is −0.88 and −0.89 respectively.

$$AT = T_a + 0.33e - 0.7v - 4.00 \tag{1}$$

$$e = \frac{RH}{100} 6.105 \exp\left(\frac{17.27 T_a}{237.7 + T_a}\right) \tag{2}$$

where, AT = Apparent Temperature [°C]. T_a = Dry bulb temperature of external environment [°C]. e = water vapour pressure [hpa]. v = wind speed [m/s]. RH = Relative humidity [%].

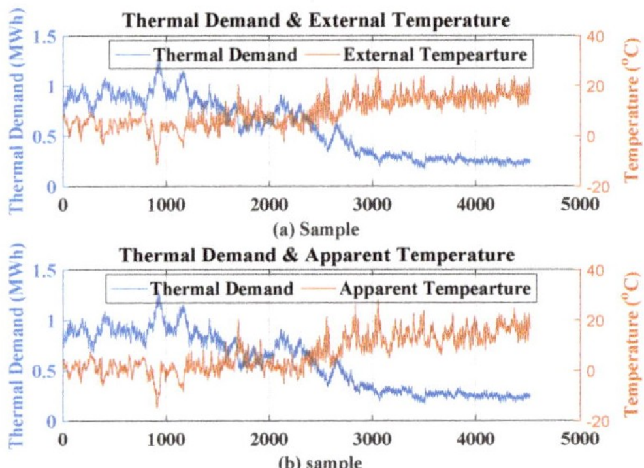

Figure 6. (**a**) Graph of thermal demand and external temperature; (**b**) thermal demand and apparent temperature.

Figure 7a shows the graph of apparent temperature vs. thermal demand throughout the period of December 2016 to August 2017. Figure 7b shows the distribution of thermal demand with respect to apparent temperature during summer and winter only. It is clear from Figure 7b that thermal demand during winter is inversely proportional to the apparent temperature. Whereas, during summer, proportional relationship between each other is very small. This could be due to the reason that apart from external temperature, thermal consumption is mostly for domestic purposes such as bathing, washing, space heating for toilet/bathroom, and transmission losses. Thus, it is logical to conclude that seasonal effect needs to be considered as input variable in the model for estimation.

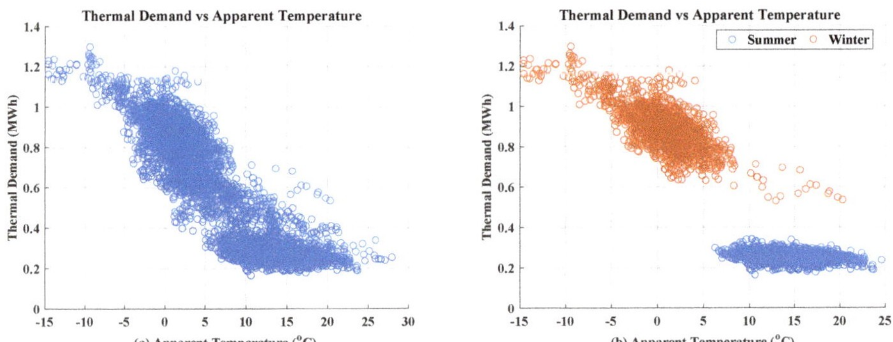

Figure 7. (**a**) Thermal demand vs. apparent temperature throughout the period. (**b**) Thermal demand vs. apparent temperature during winter and summer.

The parameters for estimation of thermal loads in residential areas are based on factors such as user behaviour (hour, weekdays, and weekends), and environmental condition (apparent temperature and season).

3.2. Estimation Technique of Thermal Demand

Different approaches to estimating thermal demand based on curve fitting technique such as the neural net fitting, and similar day method are considered as they are widely used. MATLAB inbuilt tools and functions are used for developing the estimation model using the neural net tool. Different scenarios based on the seasonal variations (summer and winter) are analysed.

For the neural net fitting tool, 50% of the seasonal data set are used for training, 25% for validating, and 25% for testing to develop the model. The datasets are divided randomly for training, testing, and validation of the model. After developing the model, 50% of the remaining seasonal data set are used in estimation.

For the similar day approach, the hourly data of a day is arranged according to season (summer and winter) and weekdays and weekends as shown in Figure 8.

Data Set								
	Summer				Winter			
Hour	Weekdays		Weekend		Weekdays		Weekend	
1	$[AT]$	$[Q_{DHW}]$	$[AT]$	$[Q_{DHW}]$	$[AT]$	$[Q_{DHW}]$	$[AT]$	$[Q_{DHW}]$
2	$[AT]$	$[Q_{DHW}]$	$[AT]$	$[Q_{DHW}]$	$[AT]$	$[Q_{DHW}]$	$[AT]$	$[Q_{DHW}]$
3	$[AT]$	$[Q_{DHW}]$	$[AT]$	$[Q_{DHW}]$	$[AT]$	$[Q_{DHW}]$	$[AT]$	$[Q_{DHW}]$
.	.	.	.	.	.	.	.	.
.	.	.	.	.	.	.	.	.
.	.	.	.	.	.	.	.	.
.	.	.	.	.	.	.	.	.
.	.	.	.	.	.	.	.	.
24	$[AT]$	$[Q_{DHW}]$	$[AT]$	$[Q_{DHW}]$	$[AT]$	$[Q_{DHW}]$	$[AT]$	$[Q_{DHW}]$

Figure 8. Data set.

50% of each dataset (weekdays and weekends for summer and winter) are used as the historical data to build a Euclidean distance (ED) for measure of similarity. In similar day method, it is assumed that the thermal demand is associated with apparent temperature (AT) for similar day (weekdays and weekends for summer or winter), and will result into similar thermal demand. EDs value based on recorded normalised AT ($\tilde{AT}$) values at particular hour(h) of the day (d) are calculated for each and every historical similar days(d^i) using (3) [26]

$$ED(\tilde{AT}, d, d^i) = \sum_{h=1}^{24}(\tilde{AT}_h^{(d)} - \tilde{AT}_h^{(d^i)})^2 \tag{3}$$

where, $ED(\tilde{AT}, d, d^i)$ is the ED between day d and historical days d^i with respect to the value of $\tilde{AT}$. Days with similar pattern of AT will have very small values of ED, hence corresponding value of thermal demand is selected as the estimated value.The parameters for AT can be achieved from the forecasted meteorology data.

4. Electric Boiler and Stratified Storage Tank

Modelling of the hot water storage tank for the electric boiler (EB) is equally important to be able to analyze the flexibility in power to heat conversion with effective thermal demand and supply. The EB and a storage tank with stratified layers as shown in Figure 9, has been well-defined and theoretically verified in [29] based on the principle of conservation of energy in a control volume and surface. The detailed derivation of EB and storage tank presented here is suitably adapted to utilize flexibility in its synergy operation with the electricity network. Also, the single general equation is derived which is suitable for charging and discharging of the storage tank, with and without discharge of hot water from the tank.

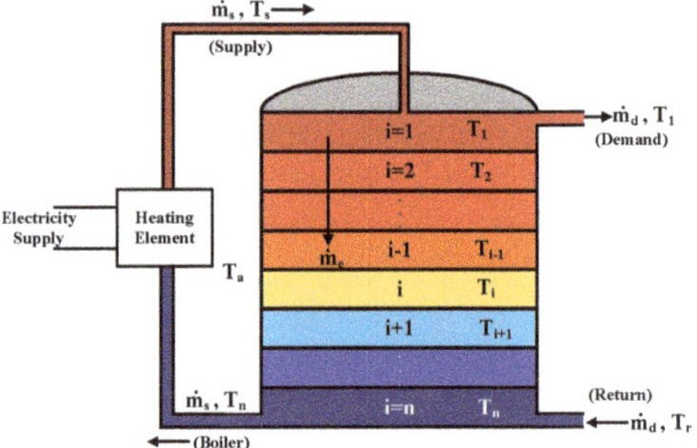

Figure 9. Stratification in hot water storage tank (b) energy flow in stratified layers.

In Figure 9, T_s = temperature of supply hot water in the tank [K], T_r = temperature of return water in the tank [K], T_a = temperature of ambient environment [K], T_i = temperature of ith stratified layers, ($i = 1, 2, ...n$) [K], $\dot{m}_e$= effective mass flow between the stratified layer [kg/s], $\dot{m}_s$ and $\dot{m}_d$ are the inlet and demand mass flow [kg/s] of water from the heating source, and out of storage tank respectively.

4.1. Modelling of EB and Storage Tank

The EB is considered to be a constant impedance load [30] and is operated with constant power ($P_{r,b}$ = rated power of boiler [W]) as shown in (4). Here, $V_{r,b}$ and V_{poc} are rated voltage of boiler [V] and voltage at point of coupling of boiler into the grid [V] respectively. η is the efficiency of boiler [%] and $\dot{Q}_{heat}$ is the heat flow rate of heating element [J/s]. C_w is the specific heat capacity of water [J/kg·K]. As per the manufacturer, the recommended flow of $\dot{m}_s$ is produced at a $\Delta T = 10\,°C$ ($\Delta T = T_s - T_n$) with the EB on full power. However, the allowable maximum flow of $\dot{m}_s$ is produced at $\Delta T = 5\,°C$.

$$\dot{Q}_{heat} = \frac{\eta}{100}\left(\frac{V_{poc}}{V_{r,b}}\right)^2 P_{r,b} = \dot{m}_s C_w (T_s - T_n) \qquad [W] \qquad (4)$$

The energy flow in different layers of stratification is shown in Figure 10. The amount of heat exchanged by the layer i of thickness z with the surrounding layers ($i - 1$) and ($i + 1$) of same thickness (z) and area (A_q) due to thermal conduction ($\dot{Q}_{exc}$) in vertical direction of the storage tank is calculated using (5) [29].

$$\begin{aligned}\dot{Q}_{exc,i} &= \dot{Q}_{exc,i-1\to i} - \dot{Q}_{exc,i\to i+1}\\ &= \frac{A_q \lambda_w}{z}(T_{i-1} - T_i) - \frac{A_q \lambda_w}{z}(T_i - T_{i+1})\\ &= \frac{A_q \lambda_w}{z}(T_{i-1} + T_{i+1} - 2T_i)\end{aligned} \qquad (5)$$

where, λ_w = effective vertical heat conductivity of water ($1-1.5$ W/mK) [29]. $\dot{Q}_{exc}$ is heat exchange due to natural convection and thermal conduction [W]. The effective mass flow of water between the stratified layer ($\dot{m}_e$) is given by (6).

$$\dot{m}_e = \dot{m}_s - \dot{m}_d \qquad [kg/s] \qquad (6)$$

Let, δ^+ (7) and δ^- (8) indicate an effective flow of water inside the storage tank from top to bottom or bottom to top respectively. $\delta^+ = 1$ represents that the heat transfer inside the storage tank due to the water mixing is from top to bottom (downwards). Similarly, $\delta^- = 1$ indicates that the heat transfer inside the storage tank due to the mixing of water is from bottom to top (upward).

$$\delta^+ = 1 \quad \text{if } \dot{m}_e > 0$$
$$\quad\;\; = 0 \quad \text{if } \dot{m}_e \leq 0 \tag{7}$$

$$\delta^- = 1 \quad \text{if } \dot{m}_e < 0$$
$$\quad\;\; = 0 \quad \text{if } \dot{m}_e \geq 0 \tag{8}$$

Let, $\delta_{(X)}$ be the conditional operator as defined in (9). $\delta_{(X)} = 1$ when conditions defined by X is true, else, $\delta_{(X)} = 0$

$$\delta_{(X)} = 1 \quad \text{if } X = \text{True}$$
$$\quad\;\; = 0 \quad \text{if } X = \text{False} \tag{9}$$

A general equation for n stratified layers, considering $i = 1$ as the top layer and $i = n$ as the bottom layer, is derived and given by (10).

$$mC_w \frac{dT_i}{dt} = \dot{m}_s C_w (T_s - T_1)\delta_{(i=1)} + \dot{m}_d C_w (T_r - T_n)\delta_{(i=n)}$$
$$+ \dot{m}_e C_w (T_{i-1} - T_i)\delta^+ \delta_{(i \neq 1)} + \dot{m}_e C_w (T_i - T_{i+1})\delta^- \delta_{(i \neq n)} \tag{10}$$
$$- UA_s(T_i - T_a) + \frac{A_q \lambda_w}{z}[(T_{i-1} - T_i)\delta_{(i \neq 1)} - (T_i - T_{i+1})\delta_{(i \neq n)}]$$

Here, U = overall heat transfer coefficient [W/m^2·K].

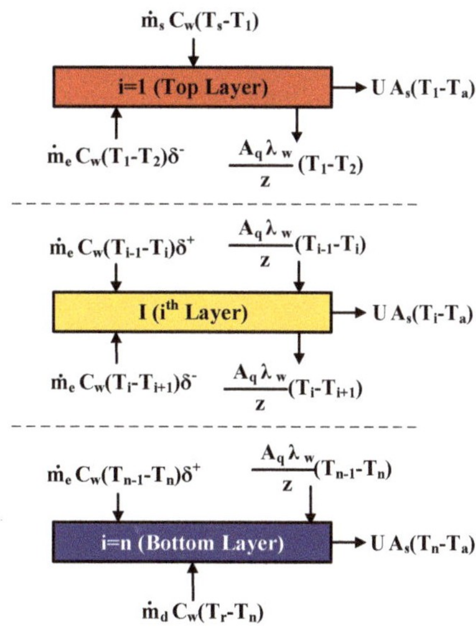

Figure 10. Energy flow in stratified layers.

4.2. Model Verification

The dynamic EB storage was verified based on three different scenarios as classified in Table 1. These values were taken just to verify the model.

Table 1. Electric boiler (EB) storage model verification scenarios.

Scenario	Situation	$\dot{m}_d$	T_s	T_r
Scenario 1	Discharging	10-L/s	–	40 °C
Scenario 2	Charging	0-L/s	80 °C	-
Scenario 2	Charging and Discharging	10-L/s	80 °C	40 °C

The technical parameters of EB and storage is presented in Table 2. The temperature distribution for a 200-m^3 hot water storage tank with 10 stratified layers are shown in Figure 11a for all three scenarios mentioned in Table 1. Figure 11b represents the EB status (ON/OFF) and discharge of hot water from the storage tank (L/s) corresponding to Figure 11a.

Table 2. Technical parameters of EB and storage.

Parameters	Definition	Value	Units
V	storage volume	200	[m^3]
n	number of stratified layers in storage tank	10	-
λ_w	effective heat conductivity of water	0.644	[W/mK]
U	heat transfer coeff. of the storage walls	0.12	[W/m^{2}K]
x	diameter to height ratio of storage	2.24	-
T_a	ambient Temperature	10	[°C]
x T_s	supply Temperature	80	[°C]
T_r	return Temperature	40	[°C]
P_b	Rated Power of EB	2.4	[MW]
C_w	specific heat capacity of water	4190	[J/kg·K]

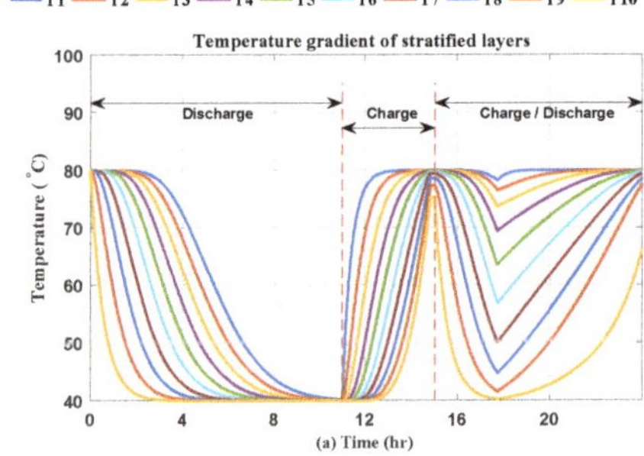

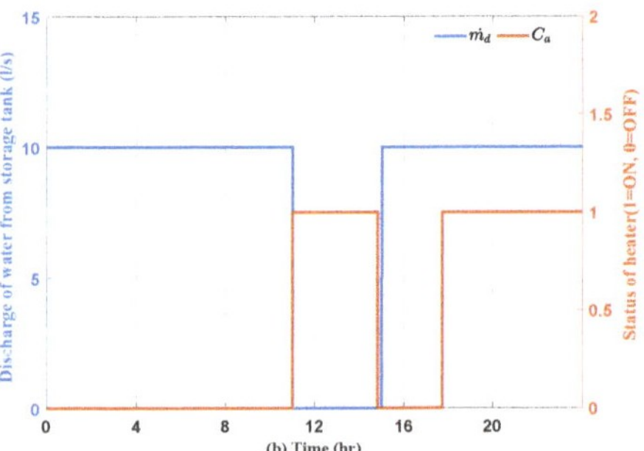

Figure 11. Temperature gradient of stratified layers (**a**) during discharge, charge, and charge-discharge of storage tank (**b**) discharge of water from the storage tank and status of the heater (ON/OFF).

Scenario 1 represents the temperature dynamic in the storage tank subjected to discharge of hot water at a constant rate of 10-L/s and the constant return temperature of cold water ($T_r = 40$ °C). Initially, the storage tank is fully charged and is filled with hot water with an initial temperature of all layers ($T_i(ini)$) equal to 80 °C. The temperature gradient of each layer is different and falls gradually until it reaches 40 °C and this situation is considered as a fully discharged storage tank. The temperature of the bottom layer drops rapidly compared to the adjacent top layer.

Scenario 2 represents the temperature dynamic in the storage tank subjected to charging with hot water from EB of power rating 2.4 MW based on Equation (4). The temperature of supply hot water is maintained constant at $T_s = 80$ °C. The discharge from the storage tank during this period is 0-L/s. The temperature of each layer in the storage tank gradually increases until $T_{10} = 75$ °C, so that the allowable maximum flow from EB is not exceeded as discussed in Section 4.1. The temperature of the top layer increases quickly compared to the adjacent bottom layer.

Scenario 3 represents the temperature dynamic in the storage tank subjected to both charging and discharging process to resemble an actual scenario. Initially, the storage is fully charged. It is then discharged at the rate of 10-L/s with the heater turned off until the temperature of 7th layer is below 50 °C. Now, the heater is turned ON with discharge remaining 10-L/s from the storage tank. During this period, the temperature of water in different layers rises at a slower rate compared to charging without any discharge. The temperature of the bottom layer increases at a much slower rate compared to its adjacent top layer. Hence, the mode of EB storage tank defined by (10) is verified using these different scenarios.

5. Operation Schedule of EB for Flexibility

In order to schedule the time of operation of the EB to charge the hot water storage tank, the optimization procedure given by (11) and (12) are followed. The objective function is to minimize the cost of electricity for production of hot water to meet the demand and storage needs. The constraints calculate the energy stored in storage tank and does not allow the storage tank to charge more than its allowable maximum and minimum limit. The energy extracted from the grid is either 0 (when EB is turned OFF) or is equal to the rated power of EB heater (P_b, when EB is turned ON). The energy extracted from the grid must be able to charge the storage as well as fulfil the demand. Although there are possibilities to control the EB power in several stages, the problem here is simplified with just ON and OFF in order to demonstrate the flexibility in operation of EBs under dynamic tariff conditions, with the help of estimated demand. Also, the operation of EB during peak hours in evening are restricted to minimise problems related to grid congestion and under voltage in Danish low voltage residential grid, due to integration and operation of electric boilers (EBs) [6]. The thermal energy stored in the tank at the end of the day is maximized to illustrate that storage tank is not only providing flexibility by supplying the thermal demand at the time of high electricity price and peak electricity demand, but also stores energy during the period of low electricity price during the 24 hour period of spot price in the electricity market.

$$\text{Minimize} \qquad \sum_{t=1}^{24} C_t P_{g,t} \tag{11}$$

Constraints

$$
\begin{aligned}
&S_{t+1} = S_t - Q_{DHW,t} + P_{g,t} \\
&S_{\min} \leq S_t \leq S_{\max} \\
&P_{g,t} \in [0, P_b \Delta t] \\
&P_{g,t} = 0 \ \ for \ \ 17 \leq t \leq 20 \\
&(S_{\max} - P_b \Delta t) \leq S_t \leq S_{\max} \ for \ \ t = 24
\end{aligned}
\tag{12}
$$

Here, C = energy price[EUR/MWh]. P_g= energy extracted from the grid [MWh]. S = energy that can be extracted from storage [MWh]. Q_{DHW} = thermal demand [MWh]. P_b = rated power of EB [2.4 MW]. Subscripts: t = time [h], min = minimum, max = maximum, ini = initial value. The maximum energy that can be stored in hot water storage tank is given by (13)

$$S_{\max} = M_b C_w (T_s - T_r)/(3600 \times 10^6) \qquad \text{[MWh]} \tag{13}$$

$$S_{\min} = 0.4 \times S_{\max} \qquad \text{[MWh]} \tag{14}$$

Here, M_b = Mass of water in storage [2×10^5kg]. T_s = temperature of supply hot water in the tank [80 °C]. T_r = temperature of return water in the tank [40 °C]. C_w = specific heat capacity of water [4190 J/kg·K].

The optimization problem was solved by minimizing the cost function using brute force optimization in MATLAB. All possible candidates for the solutions are generated and then checked against the satisfaction of problem statement as given in (11) and (12). For more than one solutions, the solution with less ON/OFF operation of EB is selected. The solutions were verified using "PuLP", linear programming modeller written in python.

Control of EB

The optimized schedule for operation of the EB is determined based on the estimated thermal demand. On the other hand, the actual thermal demand would vary to some extent compared to the estimated value. This leads to an estimation error. In case the error is large, it can lead storage tank temperature to be away from the specified limit ($T_{10} \leq 75\,°\text{C}$ when storage is charged and $T_7 \geq 46\,°\text{C}$ to limit storage discharge up to 70% of its capacity). Thus, in order to compensate for the large error in estimated demand with respect to the actual value, the optimized schedule for operation of EB is reinforced with limit controllers based on hysteresis control, realized with RS flipflop, to turn ON/OFF EB as shown in Figure 12. This ensures that the temperature of hot water in the storage tank is within the specified limit.

Figure 12a shows that, when the temperature of bottom layer $T_{10} \geq 75\,°\text{C}$ the EB needs to be turned OFF as discussed in Section 4.1. It is turned OFF only for a short period until the temperature of the seventh layer (T_7) is less than 78 °C, so that it can further follow the schedule. Figure 12b ensures that if $T_7 < 46\,°\text{C}$ (storage is discharged more than 70% of its capacity), EB is turned ON until it is fully charged (i.e., $T_{10} \geq 75\,°\text{C}$). Apart from these two conditions, the EB is operated as per the determined schedule. The overall control strategy is shown in Table 3, where C_a is the Control signal for turning ON and OFF of EB and C_{a1} is the signal from scheduled ON/OFF of EB.

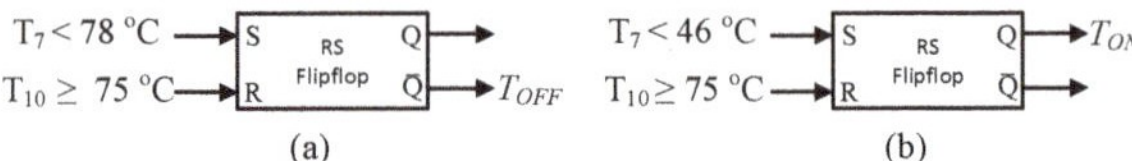

Figure 12. Limit controller schematic for ON/OFF operation of EB (a) upper temperature limit (b) lower temperature limit .

Table 3. Truth table for control of EB based on outputs of Figure 12.

T_{OFF}	T_{ON}	C_a
1	x	0
x	1	1
0	0	C_{a1}

6. Result and Discussion

6.1. Thermal Demand Estimation

Figure 13a,b shows the estimated vs. actual demand and associated %error in estimation using the neural net fitting tool are presented respectively. The range of %error associated with the neural net tool is lower than for the similar day method. Figure 13c presents the histogram of %error normalized with the frequency of occurrence as well as the probability density function. The normal distribution of %error is well-defined with the mean and standard deviation of error as well as its deviation from the mean. Finally, the average value of estimated data and actual data are compared with respect to time of day, weekdays and weekend in Figure 13d. The estimated data closely follow the trend and pattern associated with thermal demand.

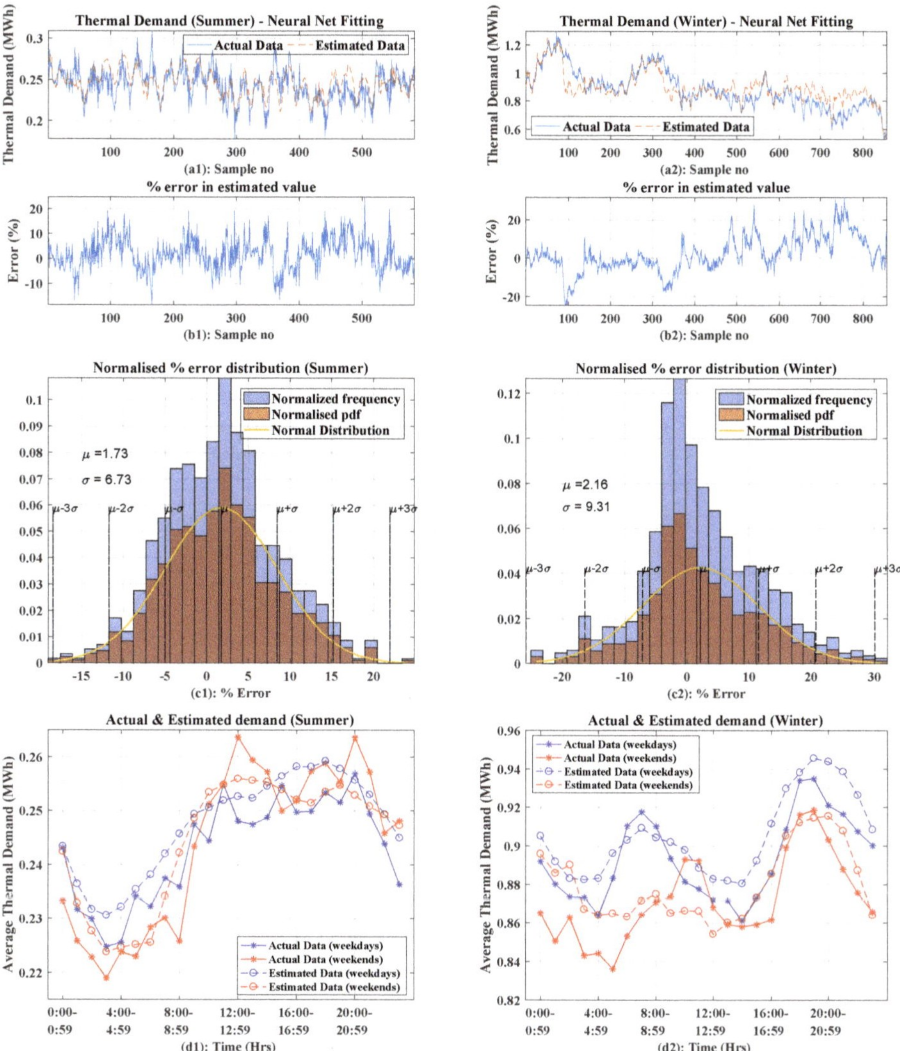

Figure 13. Forecasting of thermal demand using neural net fitting tool for summer and winter. Row (**a**): actual data and estimated data. Row (**b**): % error associated with estimation of thermal demand. Row (**c**): histogram of %error and its normal distribution. Row (**d**): Analysis of average thermal demand (actual and estimated) in hourly basis for different weekdays and weekends.

Similar analysis has been conducted for a similar day method and is presented in Figure 14. The range of percentage error between estimated and actual value(Figure 14(b1,b2)) is higher than with the neural net fitting tool. However, with less number of the dataset, similar day approach is equally reliable as a neural net fitting tool.

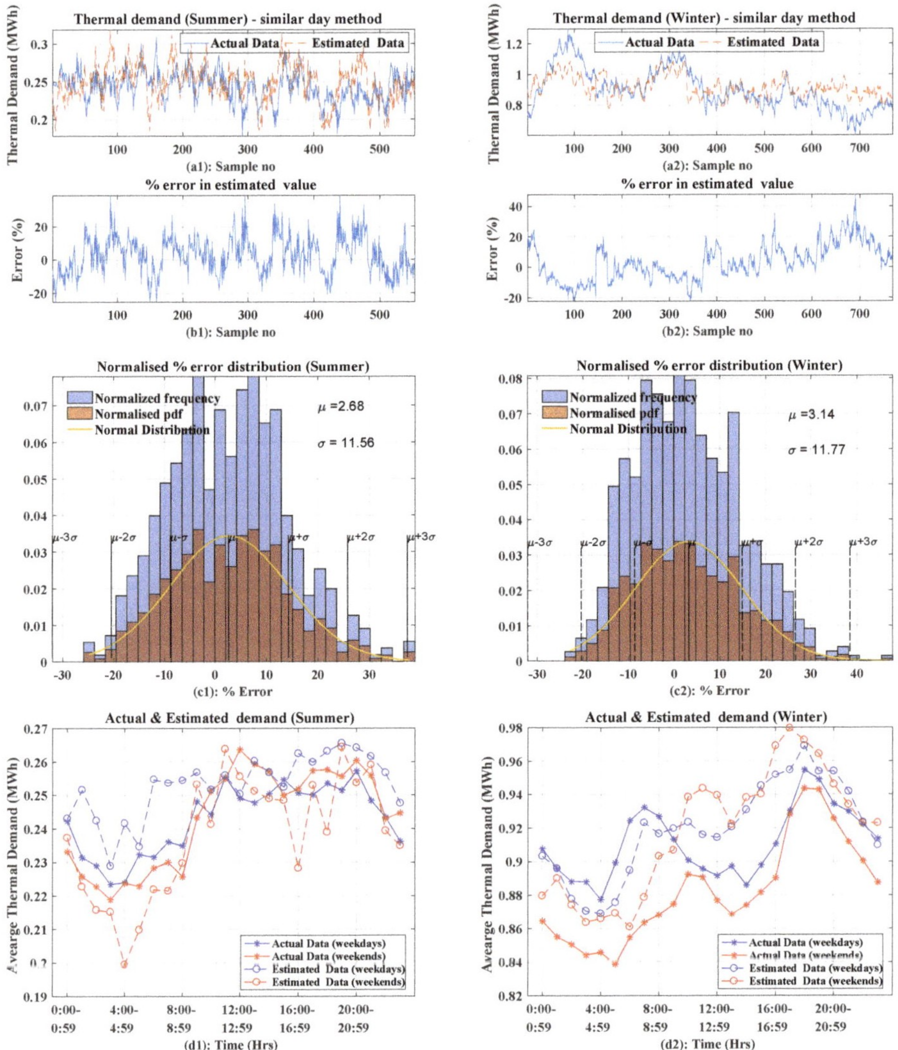

Figure 14. EStimation of thermal demand using similar day method for summer and winter. Row (**a**): actual data and estimated data. Row (**b**): % error associated with prediction of thermal demand. Row (**c**): histogram of %error and its normal distribution. Row (**d**): Analysis of average thermal demand (actual and estimated) in hourly basis for different weekdays and weekends.

Table 4 summarises the results of different methods of estimation. Root mean square error (RMSE) is used to measure the difference between the estimated value by a model and the actual values observed. Mean absolute percentage error (MAPE) estimates how close estimated values are to actual values in percentage.

Table 4. Errors from thermal demand estimations.

	Neural Network		Similar Day	
	Summer	Winter	Summer	Winter
MAPE	5.459	7.30	9.57	9.72
RMSE	0.016	0.08	0.028	0.103
Mean (μ) %error	1.73	2.16	2.68	3.14
Std.Dev (σ)	6.73	9.31	11.56	11.77

Results in Figures 13 and 14 shows that the knowledge of apparent temperature, (which includes the effect of wind, relative humidity, water vapour pressure, and ambient temperature) along with the user pattern behaviour was good enough for the estimation of thermal demand, for both summer and winter season, without considering user behaviour and geometry of the building. The consumption pattern of thermal demand with peaks and valleys were well preserved with estimated demand as seen in Figure 13d1,d2 and Figure 14d1,d2. This pattern is due to training of estimation tool based on parameters such as time, day and season as well. However, the error in the magnitude of estimated demand was expected from the thermal components that were not much dependent on the external temperature, e.g., domestic hot water usage. Using a thermal storage tank, error in estimated demand wes well compensated.

Results of thermal demand estimation using a curve fitting technique based on neural net fitting were compared with the results from time series estimation, in existing literature, based on estimation errors. Errors in the estimated value using the neural net fitting tool for winter season ranged between -23 and 31 whereas, in [22] it ranged from -33 to 15 (with 10% less range) with use of time series ANN. Further, the MAPE for 24 h ahead forecast according to [23] was 7.28% for winter which is similar to 7.3% during winter using neural net fitting as discussed here. Also, in [25] the MAPE errors for different machine learning technique and for different area varies between 5–27%. Hence, the curve fitting techniques discussed here is a well effective and simple estimation tool for thermal demand estimation using AT and hourly pattern with respect to weekdays and seasons.

6.2. Flexible Operation of Electric Boiler

One of the important applications of deploying thermal demand estimation is to prepare an algorithm for energy management, to support the future smart grid system. An example of scheduling thermal storage under dynamic tariff condition is introduced in [31] with an average model of the storage tank. Here, a time series based concept for thermal demand forecast and the schedule is proposed for heat pumps without considering ON/OFF delay for thermal production. It usually takes around 10–15 min to achieve a steady state condition [32]. In [6] the models of EB and HP with a storage tank having only two stratified layers, are briefly discussed in relation to the actual control and flexibility based on grid condition and status of storage tank temperature or position of the stratified layer.

The EB with an n-stratified layered storage tank as modelled in Section 4 is used to acknowledge its operational flexibility. The optimal schedule of the EB is based on estimated thermal demand using the curve fitting technique. The temperature of hot water inside the storage tank at different layers (1, 7, and 10 as in Figure 9) gives the indication of flexibility performance of the storage tank to be able to supply the demand. The operational flexibility in terms of ON and OFF of the EB under dynamic tariff conditions based on estimated thermal demand is presented in this section. The storage is never scheduled to discharge more than up to 40% of its maximum capacity in order to accommodate substantial error in the estimation of thermal demand. Figure 15 shows the complete system methodology implemented for flexible operation of the EB. Thermal demand estimation and optimized schedule for operation of EB are determined using Matlab. Then the flexibility in the operation of the EB is verified using DigSILENT power factory simulation tool.

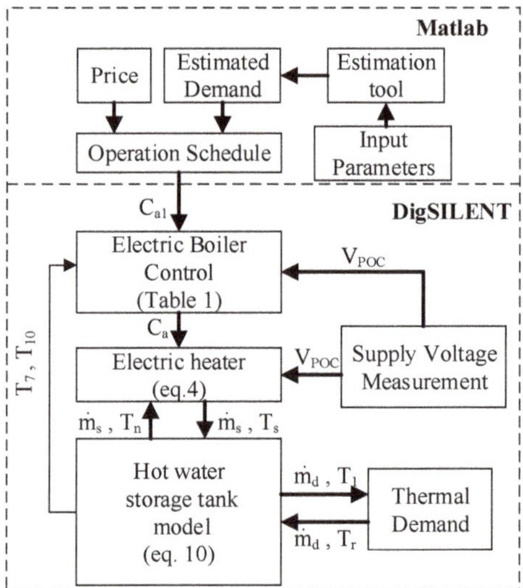

Figure 15. System implementation for flexible operation of EB.

Figure 16a shows the day-ahead hourly spot price of Denmark's northern region for case 1: 6 January 2017 [33]. On this particular day, the price seems to be decreasing gradually according to the time of the day. Sometimes the prices even go to negative as on case 2: 24 December 2016 [33]. This is because of the high electricity generation from renewable sources such as wind or solar. With an optimization problem to reduce energy cost, it is obviously beneficial to turn ON the boiler during low electricity price to store thermal energy.

Figure 16b shows the estimated thermal demand based on the neural network on 6 January 2017 and is compared with actual demand on the same day. Here, the MAPE error is significantly less compared to other days estimation. Based on the estimated thermal demand, the optimal schedule of EB, as well as thermal energy storage status, is determined (0 = fully discharged, 1 = fully charged) to avoid higher price and time of peak electrical demand as shown in Figure 16c for case 1. Using this schedule for operation of the EB with actual demand, the storage temperature is monitored in Figure 16d. It is observed from Figure 16c that the controller (C_a) follows the estimated schedule (C_{a1}). The actual EB storage status (AS_{status}) is also closer to estimated value (PS_{status}). A similar observation is made for case 2 as well. The storage temperature of the top layer confirms that demand has been fulfilled taking care of grid flexibility and end-user satisfaction. The storage tanks are charged when the electricity price is lower as seen from case 1 and case 2, where spot pricing is different whereas demand remains the same for both cases.

Similarly, Figure 17a shows the electricity price of 23 March 2017. Here, the estimated value of thermal demand has the maximum forecasting error as seen in Figure 17b. Estimated EB schedule (C_{a1}) along with the controller action (C_a) is shown in Figure 17c. The controller turns OFF the EB twice during, 6th and 24th hour as the storage tank is fully charged as seen in Figure 17d. This is due to the over estimation of thermal demand. The temperature of the hot water as seen from Figure 17d indicates that storage has been fully charged and heating element needs to be shut down.

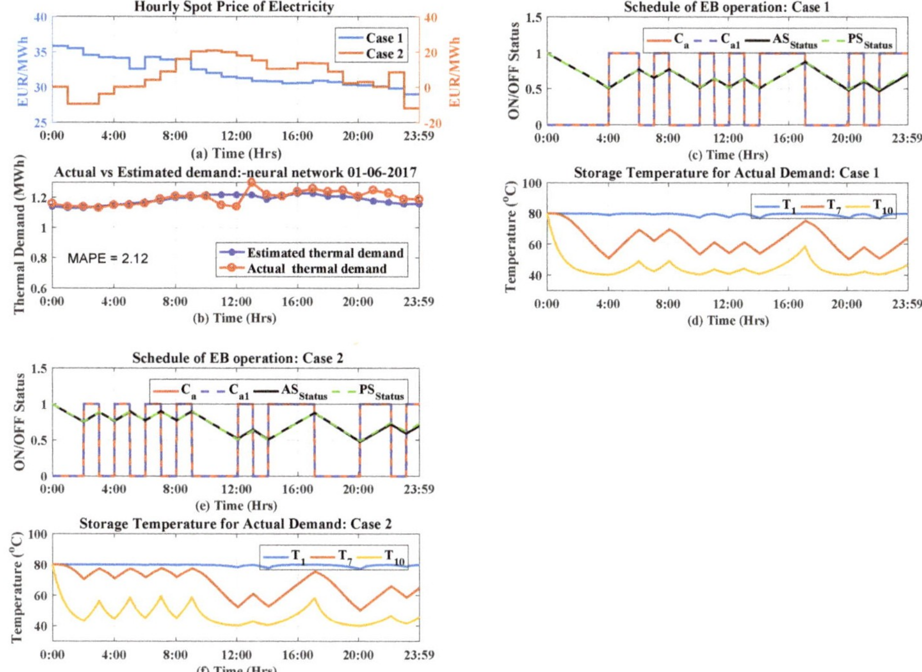

Figure 16. (**a**) Hourly spot prices of electricity in Denmark region DK1 case 1: 6 January 2017 and case 2: 24 December 2016 (**b**) Actual and estimated thermal demand on 6 January 2017 (**c**) estimated EB schedule based on estimated demand (C_{a1}), estimated EB storage status (PS_{status}), and actual schedule of EB based on actual demand (C_a), and actual EB storage status (AS_{status}) (case 1). (**d**) Storage temperature of various layers for actual demand and schedule (case 1). (**e**) estimated EB schedule based on estimated demand (C_{a1}), estimated EB storage status (PS_{status}), and actual schedule of EB based on actual demand (C_a), and actual EB storage status (AS_{status}) (case 2). (**f**) Storage temperature of various layers for actual demand and schedule (case 2).

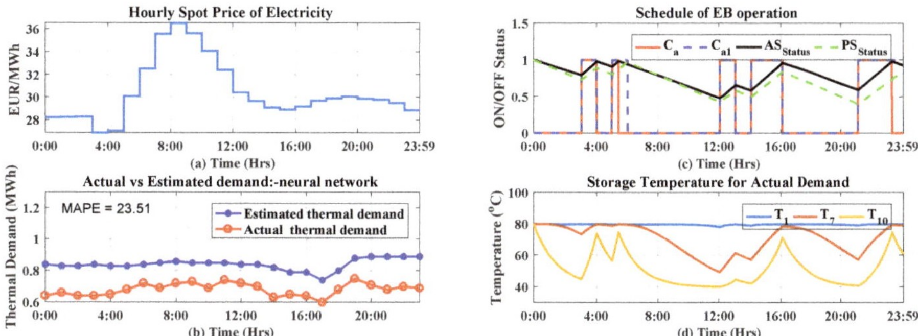

Figure 17. (**a**) Spot price of electricity in Denmark region DK1 on 23 March 2017; (**b**) actual and estimated thermal demand on 23 March; (**c**) estimated EB schedule based on estimated demand (C_{a1}), estimated EB storage status (PS_{status}), and actual schedule of EB based on actual demand (C_a), and actual EB storage status (AS_{status}); (**d**) storage temperature of various layers for actual demand and schedule.

The knowledge of apparent temperature, and hourly pattern of thermal consumption during weekdays or weekends has helped the estimation tool to estimate the thermal demand effectively. The estimation of thermal demand has supported immensely in the decision-making process to schedule flexible operation of the EB in the multi-energy system, where energy prices are lower during high generation from renewable sources such as solar and wind. Despite a substantial error in estimation, the thermal storage has the capability of operating as a flexible load by decoupling demand and generation, and enhancing accommodation of renewable energy. The energy in the storage tank tries to attain at its best status to fulfil evening thermal peak demand, as well as avoid EB operation to support peak shaping during high electricity demand in the evening. The temperature of the top bottom and middle layer of stratified hot water storage tank indicates that the thermal demand is fulfilled by maintaining operational flexibility. The use of stratified layered storage tank has the advantage over average temperature model of storage tank, as an average model is not able to illustrate the actual condition of supply temperature which regulates the flow of hot water from the EB to storage tank while charging process and monitor the status of storage tank as realized in practice.

7. Conclusions

This paper shows insight on the daily usage pattern of the thermal energy, during summer and winter, in a residential area along with the factors influencing the estimation of thermal demand such as user behaviour parameters and external environment parameters. Based on these factors, the neural network model and the similar day method for estimation has been implemented. Using this model, estimation of thermal usage for the same area can be achieved, but not for other areas. So, an already available model is unlikely to be used for new subjects. However, the findings of this paper on the use of input parameters for determining the thermal demand of a particular area and its influence on the pattern of usage has been justified.

The findings of the data analysis of thermal consumption (Q_{DHW}) yields some important conclusions on the pattern of energy demand based on time and day of usage reflecting user behaviour without compromising the privacy issue of individuals. This valuable information is useful for determining the generation of thermal demand and need for storage. When large CHP units are replaced by small heat pumps or electric boilers and integrated to the electric grid network, it will increase the electricity demand with the profile discussed in Figure 4. Thus, during off-peak hours, when electricity demand is low, the thermal storage unit can be used to store the surplus electricity generated by wind turbines and other renewable generation. This storage of thermal power can be utilized during its peak hours reducing greenhouse emission on the production of hot water. Further, the estimated value of thermal demand helps in determining the range of requirement of thermal storage to meet the consumer demand as well as demand response to utilizing thermal storage unit as flexible consumer load in the multi-energy system.

Author Contributions: R.S. developed the methodology, corresponding models and wrote the paper. B.B.-J., J.R.P., and H.Z. supervised the work and reviewed the paper. All authors have read and agreed to the published version of the manuscript.

Funding: This work is supported by the Innovation fund Denmark through the Dicyps project. The data for the analysis has been provided by Aalborg Energi Holding A/S and Aalborg University, Denmark, as per the agreement between project partners for research.

Acknowledgments: The data for the analysis has been provided by Aalborg Energi Holding A/S and Aalborg University, Denmark, as per the agreement between project partners for research.

Conflicts of Interest: The authors declare no conflict of interest.

References

1. Danish Energy Agency (DEA). *Regulation and Planning of District Heating in Denmark*; 2017; pp. 1–24. Available online: https://ens.dk/sites/ens.dk/files/Globalcooperation/regulation_and_planning_of_district_heating_in_denmark.pdf (accessed on 6 November 2019).
2. Lund, H.; Werner, S.; Wiltshire, R.; Svendsen, S; Thorsen, J.E.; Hvelplund, F.; Mathiesen, B.V. 4th Generation District Heating (4GDH): Integrating smart thermal grids into future sustainable energy systems. *Energy* **2014**, *68*, 1–11. [CrossRef]
3. Danish Ministry of Climate, Energy and Building. *Smart Grid Strategy-The intelligent Energy System of the Future*; Danish Ministry of Climate, Energy and Building: Copenhagen, Denmark, 2013; pp. 1–23. Available online: https://ens.dk/sites/ens.dk/files/Globalcooperation/smart_grid_strategy_eng.pdf (accessed on 6 November 2019).
4. Lund, H. Renewable heating strategies and their consequences for storage and grid infrastructures comparing a smart grid to a smart energy systems approach. *Energy* **2018**, *151*, 94–102. [CrossRef]
5. Energinet. Available online: https://www.energidataservice.dk/en/dataset/electricitybalance/resource_extract/498c68e3-d248-4965-b36f-3aa738130adc (accessed on 6 November 2019).
6. Sinha, R.; Jensen, B.B.; Pillai, J.R.; Moller-Jensen, B. Unleashing Flexibility from Electric Boilers and Heat Pumps in Danish Residential Distribution Network. In Proceedings of the CIGRE 2018, Paris, France, 26–31 August 2018.
7. Nuytten, T.; Claessens, B.; Paredis, K.; Bael, J.V.; Six D. Flexibility of a combined heat and power system with thermal energy storage for district heating. *Appl. Energy* **2013**, *104*, 583–591. [CrossRef]
8. Korpela, T.; Kaivosoja, J.; Majanne, Y.; Laakkonen, L.; Nurmoranta, M.; Vilkko, M. Utilization of District Heating Networks to Provide Flexibility in CHP Production. *Energy Procedia* **2017**, *116*, 310–319. [CrossRef]
9. Skytte, K.; Olsen, O.J. Regulatory barriers for flexible coupling of the Nordic power and district heating markets. In Proceedings of the 13th European Energy Market Conference-EEM 2016, Porto, Portugal, 6–9 June 2016; pp. 1–5.
10. Pospíšil, J.; Špiláček, M.; Kudela, L. Potential of predictive control for improvement of seasonal coefficient of performance of air source heat pump in Central European climate zone. *Energy* **2018**, *154*, 415–423. [CrossRef]
11. Vorushylo, I.; Keatley, P.; Shah, N.; Green, R.; Hewitt, N. How heat pumps and thermal energy storage can be used to manage wind power: A study of Ireland. *Energy* **2018**, *157*, 539–549. [CrossRef]
12. Schweiger, G.; Rantzer, J.; Ericsson, K.; Patrick, L. The potential of power-to-heat in Swedish district heating systems. *Energy* **2017**, *137*, 661–669. [CrossRef]
13. Böttger, D.; Götz, M.; Theofilidi, M.; Bruckner, T. Control power provision with power-to-heat plants in systems with high shares of renewable energy sources—An illustrative analysis for Germany based on the use of electric boilers in district heating grids. *Energy* **2015**, *82*, 157–167. [CrossRef]
14. Fitzgerald, N.; Aoife, M.F.; McKeogh, E. Integrating wind power using intelligent electric water heating newblock. *Energy* **2012**, *48*, 135–143. [CrossRef]
15. Kontu, K.; Rinne, S; Junnila, S. Introducing modern heat pumps to existing district heating systems—Global lessons from viable decarbonizing of district heating in Finland. *Energy* **2018**, *166*, 862–870. [CrossRef]
16. Sinha, R.; Jensen, B.B.; Radhakrishnan Pillai, J. Impact Assessment of Electric Boilers in Low Voltage Distribution Network. In Proceedings of the 2018 IEEE Power Energy Society General Meeting (PESGM), Portland, OR, USA, 5–9 August 2018; pp. 1–5. [CrossRef]
17. Mendaza, I.D.C.; Pigazo, A.; Jensen, B.B.; Chen, Z. Generation of domestic hot water, space heating and driving pattern profiles for integration analysis of active loads in low voltage grids. In Proceedings of the IEEE PES ISGT Europe 2013, Lyngby, Denmark, 6–9 October 2013; pp. 1–5.
18. Weissmann, C.; Hong, T.; Graubner, C.A. Analysis of heating load diversity in German residential districts and implications for the application in district heating systems. *Energy Build.* **2017**, *139*, 302–313. [CrossRef]
19. Jordan, U.; Vajen, K. DHWcalc: Program to generate domestic hot water profiles with statistical means for user defined conditions. In Proceedings of the ISES Solar World Congress, Orlando, FL, USA, 8–12 August 2005; pp. 1–6.

20. Clegg, S.; Mancarella, P. Integrated electricity-heat-gas modelling and assessment, with applications to the Great Britain system. Part I: High-resolution spatial and temporal heat demand modelling. *Energy* **2019**, *184*, 180–190. [CrossRef]

21. Wojdyga, K. Predicting Heat Demand for a District Heating Systems. *Int.J. Energy Power Eng.* **2014**, *3*, 237–244.[CrossRef]

22. Petrichenko, R.; Baltputnis, K.; Sauhats, A.; Sobolevsky, D. District heating demand short-term forecasting. In Proceedings of the 2017 IEEE International Conference on Environment and Electrical Engineering and 2017 IEEE Industrial and Commercial Power Systems Europe (EEEIC/I CPS Europe), Milan, Italy, 6–9 June 2017; pp. 1–5.

23. Chramcov, B. Heat demand forecasting for concrete district heating system *Int. J. Math. Models Methods Appl. Sci.* **2010**, *4*, 231–239.

24. Kapetanakis, D.S.; Mangina, E.; Ridouane, E.H.; Kouramas, K.; Finn, D. Selection of Input Variables for a Thermal Load Prediction Model. *Energy Procedia* **2015**, *78*, 3001–3006. [CrossRef]

25. Idowu, S.; Saguna, S.; Åhlund, C.; Schelén, O. Forecasting heat load for smart district heating systems: A machine learning approach In Proceedings of the 2014 IEEE International Conference on Smart Grid Communications (SmartGridComm), Venice, Italy, 3–6 November 2014; pp. 554–559.

26. Zhang, Y.; Beaudin, M.; Taheri, R.; Zareipour, H.; Wood, D. Day-Ahead Power Output Forecasting for Small-Scale Solar Photovoltaic Electricity Generators. *IEEE Trans. Smart Grid* **2015**, *6*, 2253–2262. [CrossRef]

27. Chitsaz, H.; Shaker, H.; Zareipour, H.; Wood, D.; Amjady, N. Short-term electricity load forecasting of buildings in microgrids. *Energy Build.* **2015**, *99*, 50–60. [CrossRef]

28. Australian Government Bureau of Meteorology. Available online: http://www.bom.gov.au/info/thermal_stress/ (accessed on 15 February 2018).

29. Eicker, U. *Solar Technologies for Buildings*; John Wiley & Sons: Hoboken, NJ, USA, 2006.

30. Sinha, R.; Jensen, B.B.; Pillai, J.R.; Bojesen, C.; Moller-Jensen, B. Modelling of hot water storage tank for electric grid integration and demand response control. In Proceedings of the 2017 52nd International Universities Power Engineering Conference (UPEC), Crete, Greece, 28–31 August 2017; pp. 1–6.

31. Harb, H.; Schutz, T.; Streblow, R.; Muller, D. Adaptive model for thermal demand forecast in residential buildings. In Proceedings of the WSB, Barcelona, Spain, 28–30 Otober 2014.

32. Masuta, T.; Yokoyama, A.; Tada, Y. Modeling of a number of Heat Pump Water Heaters as control equipment for load frequency control in power systems. In Proceedings of the 2011 IEEE Trondheim PowerTech, Trondheim, Norway, 19–23 June 2011; pp. 1–7.

33. NordPool. Available online: https://www.nordpoolgroup.com/Market-data1/Dayahead/Area-Prices/DK/Hourly/?view=table (accessed on 6 June 2017).

Article

A GIS-Based Planning Approach for Urban Power and Natural Gas Distribution Grids with Different Heat Pump Scenarios

Jolando M. Kisse [1,2,*], Martin Braun [1,2], Simon Letzgus [3] and Tanja M. Kneiske [2]

[1]   Department of Energy Management and Power System Operation, University of Kassel, Wilhelmshöher Allee 73, 34121 Kassel, Germany; martin.braun@uni-kassel.de
[2]   Fraunhofer Institute for Energy Economics and Energy System Technology IEE, Department of Grid Planning and Operation, Königstor 59, 34119 Kassel, Germany; tanja.kneiske@iee.fraunhofer.de
[3]   Department of Energy Systems, Technische Universität Berlin, Straße des 17. Juni 135, 10623 Berlin, Germany; simon.letzgus@tu-berlin.de
*   Correspondence: jolando.kisse@uni-kassel.de

Received: 30 June 2020; Accepted: 31 July 2020; Published: 5 August 2020

Abstract: Next to building insulation, heat pumps driven by electrical compressors (eHPs) or by gas engines (geHPs) can be used to reduce primary energy demand for heating. They come with different investment requirements, operating costs and emissions caused. In addition, they affect both the power and gas grids, which necessitates the assessment of both infrastructures regarding grid expansion planning. To calculate costs and CO_2 emissions, 2000 electrical load profiles and 180 different heat demand profiles for single-family homes were simulated and heat pump models were applied. In a case study for a neighborhood energy model, the load profiles were assigned to buildings in an example town using public data on locations, building age and energetic refurbishment variants. In addition, the town's gas distribution network and low voltage grid were modeled. Power and gas flows were simulated and costs for required grid extensions were calculated for 11% and 16% heat pump penetration. It was found that eHPs have the highest energy costs but will also have the lowest CO_2 emissions by 2030 and 2050. For the investigated case, power grid investments of 11,800 euros/year are relatively low compared to gas grid connection costs of 70,400 euros/year. If eHPs and geHPs are combined, a slight reduction of overall costs is possible, but emissions would rise strongly compared to the all-electric case.

Keywords: heat pumps; power grid; gas distribution; grid expansion planning; load-profiles

---

1. Introduction

One important aspect of mitigating climate change is the increase in energy efficiency, particularly in the building sector. The European Union's amended Directive on Energy Efficiency (2018/2002) sets an energy efficiency target for 2030 of at least 32.5% improvement compared to the 2007 business-as-usual scenario [1,2]. The German government has set a goal of an 80% reduction in primary energy consumption by 2050 compared to 2008 for the building sector [3]. The primary energy consumption of the German building sector decreased by 18.8% from 2008 to 2017 while the overall German primary energy consumption sunk by 5.5%. Space heating and domestic hot water provision in private households accounted for 21.9% of the total final energy consumption in Germany in 2017 [4].

The research field of future energy-efficient heat supply is manifold and consists of various sub-research areas that overlap with other energy research questions. There are three main areas of particular relevance for the present paper that are elaborated on later in this section. Firstly, heat pumps are an important technology with which primary energy consumption can be reduced [5]. They can be evaluated in the light of energy management [6] and regarding their impact on grid infrastructure [7]. Secondly, this impact of heat pumps and other grid-connected heat generators can, in general, lead to violations of operating limits in the respective infrastructure. Thus, grid expansion and reinforcement planning is another relevant research area [8]. While past infrastructure was often planned separately for different energy carriers, e.g., electricity and natural gas, integrated planning of energy infrastructure has become increasingly important in recent publications [9]. Finally, an energy system can be modeled with different spatial resolutions and levels of abstractions [10,11]. On a regional level, geographical information (GIS) on buildings, streets, and existing infrastructure can be included to find site-specific solutions and account for local characteristics.

One of the main instruments with which to reduce CO_2 emissions is better insulation of buildings, thereby leading to lower thermal demand for each building that can be coupled with an efficient thermal energy system such as a heat pump (HP) [5]. Innovative concepts exist to build low energy, zero energy, or even plus energy buildings [12]. For existing buildings, however, energetic refurbishment measures have to be applied to achieve lower heat demand and low flow temperatures, which are suitable for efficient heat pump operation [13]. For some historic buildings, particular regulations on conservation principles limit the options for energetic refurbishments, such as insulation measures [14]. For buildings with higher energy demands, efficient gas-based heating systems with innovative energy management are discussed to help to reduce CO_2 emissions at reasonable costs [15]. For instance, in [16], a combined control for hybrid systems consisting of photovoltaic solar panels (PV) and combined heat and power (CHP) was introduced based on a model predictive control for the electrical and thermal components and a short term, rule-based control for each component. The new control can manage the PV-CHP system with higher efficiency, lower CO_2 emissions, and lower operational costs. The model predictive control was successfully tested in the laboratory [17]. In [18], the authors also showed the flexibility potential of thermal-electrical systems using a model predictive control, which could be used for market or grid-friendly behavior. Nonetheless, even with highly innovative optimized control algorithms, natural gas-fired CHP plants have unavoidable CO_2 emissions. To improve the primary energy factor, ambient heat can be used by heat pumps. Usually, the heat pumps' refrigerant circuit is driven by electric motors, and auxiliary heating coils might be used. Besides that, heat pump systems are also available with gas-powered engines and additional waste heat recovery. The emission and combustion characteristics of gas engine heat pumps (geHPs) are shown in [19].

Various studies on geHPs for industrial and residential use were reviewed in [20], which concluded that efficiency gains can be reached by using geHPs not only for space heating but also for hot water generation. In-depth energy efficiency analysis of a geHP was conducted in [21,22]; a primary energy ratio up to 1.83 was reported.

In the present study, the idea is to use different heat pump systems as a flexibility option in grid planning, not in energy management. The heat pump hot water storage systems are assumed to have rule-based controls based on leveling the storage temperature to supply the necessary space heating and domestic hot water during a year.

The impact of heat pumps on an electrical distribution grid has been studied in great detail in [23] and corrected in [24] as a function of building type and district properties. It was found that cable overloading can be expected for large rural feeders at heat pump penetrations as low as 30%, depending on the cable, while voltage problems start usually at slightly higher percentages. Additionally, building characteristics show high correlations with the examined grid performance indicators, revealing a promising potential for statistical modeling of the studied indicators. Electrical heat pumps and grid integration were also discussed in the context of flexibility options and demand-side management, e.g., in [25]. The authors show that ground source heat pumps are a very high-efficiency technology for space conditioning in buildings, and present a high potential for electric load management as a flexible load when combined with the thermal storage capacity of the building. In addition, the authors in [26] demonstrate that the combined use of a grid-connected PV system for heat pump water heaters has a larger economic benefit than solar thermal heaters when combined with optimal scheduling and therefore helps with grid integration of PV systems. An energy index was introduced in [27] to assess plants, such as a PV system and heat pump, capable of producing electricity from a renewable source. The index evaluates the bidirectional energy flows on the external power grid, in comparison to the electricity demand of a building. None of these studies considered the costs of a gas distribution grid and the effects of the gas heating systems when building owners changed their heating systems to electrical heat pumps. Such effects can include costs for dismantling existing gas network elements.

Even if electrical heat pump systems based on renewable electricity will be a zero-emission solution in the future, a high number of newly installed electrical heat pumps will lead to problems with the electrical power grid, such as undervoltage and line overloading [28]. As a result, power grid reinforcement measures are required to stabilize the grid. In addition, the start-up characteristics of heat pumps can affect the power grid on short time scales [29]. This can also result in need for expensive grid expansion measures [30]. Frameworks for distribution grid planning are introduced in [31,32]. The latter can be also used for large-scale networks, as shown in [33]. In the past, several aspects of distribution grid planning have been studied. A comparison of meta-heuristics for meshed power grids is elaborated in [34]. Optimizations of low and medium voltage grids have been done [35,36]. A novel approach wherein grid expansion is modeled together with asset management is shown in [37]. A geographic information system (GIS)-based approach was developed by the authors of [38]. In the present study, we also used a GIS-based approach, which includes not only the electrical but also the thermal characteristic of energy demand and its influence on an existing gas distribution grid.

A comparison of optimization methods of gas distribution grids is presented in [39]. The uncertain necessity of dismantling parts of the distribution grid in different ways due to a possibly declining natural gas demand has been recently addressed in [40]. The effects of increasing grid charges for natural gas that incentivize energetic refurbishment have been further investigated in [41].

More research has been done on coupled gas and power grids. A new model for optimal joint scheduling of power-to-gas and gas-fired generation units in a power-gas embedded grid was studied in [42]. This work dealt with operational aspects of a combined infrastructure, not with strategic planning. The authors in [43] analyzed combined planning for enhancing the power grid resilience. A co-simulation approach of gas and power grids was used in [44]. The possibility of using the co-simulation for grid planning was mentioned but not run in detail. A simulation tool for combined power and gas infrastructure (*SAInt*) was developed in [45] and applied on a 158-bus power grid and a 352-node gas network. However, the case study focused on security-related events and planning was a minor aspect.

A survey on models of integrated power and natural gas grid coordination is presented in [46]. A model for integrated generation, transmission, and gas expansion planning is shown in [47]. In [48], a co-optimization using mixed-integer non-linear programming for modeling a 6-bus power system with a 6-node gas network was shown to be an effective tool for a small number of nodes. Additionally, a mixed-integer linear programming model for optimized integrated planning of power and gas networks was developed in [49] and applied in a case study with an 18 node-power distribution grid that was interconnected with an 18 node-gas network. In the present study, however, no co-optimization approach was used, due to the larger size of the low voltage grid with around 3000 nodes. The focus here is on a techno-economic analysis using a high number of electrical and thermal household loads in a German town for three different cases. A heuristic optimization was used for cost minimal expansion planning of the power grid.

A review of GIS-based modeling of urban energy systems and the FlexiGIS platform is described in [50]. FlexiGIS is an open-source GIS-based platform for modeling urban energy systems. The framework relies mainly on spatial features of urban objects extracted from open databases such as OpenStreetMap. GIS-based studies in [51,52] use the platform to allocate distributed battery storage optimally in urban areas. The simulation and planning of the coupled grid infrastructure is not part of the FlexiGIS model and not mentioned in the literature review. The platform was developed for the optimization of generation and demand of flexibilities in urban regions. In the present paper, we go a step further and add detailed modeling of the gas and power grid infrastructure to the neighborhood energy model of an example town.

In the following, we model both topics in great detail so that it is possible to derive the total costs for a future energy infrastructure that is able to supply electricity and heat in a German town for the years 2030 and 2050. On the one hand, the building owners' perspective regarding deciding on insulation and the type of heating system is modeled. On the other hand, the grid operator's perspective toward investing in the power and gas distribution grid to assure a reliable and efficient network infrastructure is modeled as well. The innovation in this paper is to analyze the costs and emissions of heat generation under consideration of different building ages and energetic refurbishment variants in combination with the related costs for grid reinforcement in both the natural gas and the low voltage network of a German town. Using this approach, in contrast to other studies, a more comprehensive assessment of different heating technologies can be conducted, including the cost for heat pump systems and infrastructure costs together. The study shows where a significant number of costs are located and which type of heat pump could be used to minimize the costs but also to reduce CO_2 emissions.

In the context of grid expansion planning, this study presents the following main innovations:

- The mutual investigation of power and natural gas distribution infrastructure for a whole town using a pipe and power-flow grid analysis.
- Deriving open models from a large number and different types of public data only, creating a highly diversified spatial and temporal resolution.
- Using a multi-perspective approach that considers electricity and natural gas grid investments, heat pump costs, and CO_2 emissions for three cases.

The structure of the paper is as follows: First, the steps to model the existing building stock in the town and electric and gas grid infrastructure are described in Section 2. Second, fixed numbers of heat pumps are assumed to be added to the infrastructure by 2030 and 2050. Three cases are studied, including (1) only electric heat pump systems being installed, (2) only gas engine heat pump systems being installed, and (3) a mix of electric and gas engine heat pump systems being installed. Heat pump related costs and emissions are presented in Section 3.1 for individual building types with different building ages and three energetic refurbishment variants. In Section 3.2, the costs for all heat pumps that are assumed to be installed in the town are summarized. The costs for grid expansion of the power and gas grid depending on the number of electric and gas engine heat pumps are presented in Section 3.3. In Section 4, the results are discussed for the three cases, and the conclusion on the costs and CO_2 emissions of scenarios with different numbers of heat pump types are drawn.

2. Materials and Methods

The approach of evaluating the total costs and CO_2 emissions of planning an urban power and gas grid is based on detailed geographic information system (GIS) data. An overview of the different steps is shown in Figure 1. The data used in this study are all publicly available on the Internet as open-data or derived from assumptions based on studies mentioned in the following paragraphs.

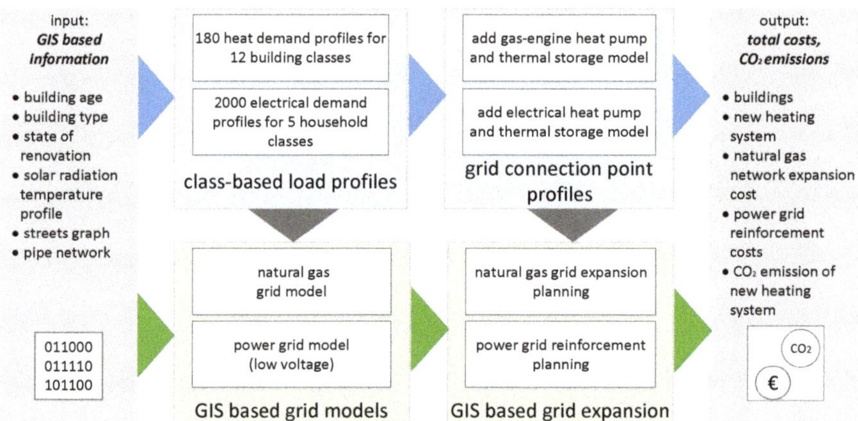

Figure 1. GIS-based approach for evaluating the total costs for an overall future infrastructure plan.

The analysis was performed for the German town "Schutterwald" in the geographical region of "Oberrhein" in southern Germany. The basic model of the studied town was built from GIS data of buildings and streets derived from OpenStreetMap (OSM) [53]. Additional information from public sources [54,55] on building age and energetic refurbishment was connected to the buildings of the town. From this geographical connected information, the load profiles and the grid models were derived. All buildings were treated as residential buildings, since commercial buildings are rather rare in the focus area. Most of the town's commercial and industrial area is located in the northern part of the town, which is not considered in this study. The load profiles cover electrical and thermal profiles. While the electrical load profiles were modeled for each building, thermal profiles were modeled for 12 different building classes with three different renovation standards and five different household types. The electrical load profiles were then geographically connected to the power grid model. The power grid and the gas grid model were also based on GIS data and were modeled based on publicly available information [53–57]. However, assumptions had to be made if the level of detail from the sources was not sufficient. The grids in this study can therefore differ from the actual grid in features such as cable type and exact routing or transformer parameters and location. The thermal profiles served as input for the gas consumption profiles and the future electrical load profiles, which were modeled for the grid planning in 2030 and 2050. The gas and power grid connection point profiles were based on a model of heat pump and storage systems including a rule-based control to apply the thermal energy for space heating and domestic hot water for a household. In future scenarios, additional loads may occur in the grid, e.g., more air conditioning units due to global warming, and electric vehicles. However, these effects are out of the scope of the present study and require further research. From the resulting time-series, the operating costs and CO$_2$ emissions for the buildings and thermal energy systems on the one hand, and required grid infrastructure investments, on the other hand, were calculated.

2.1. GIS-Based Information

The GIS coordinates of the building were taken from OSM data. The buildings were classified in accordance with the TABULA-building topology for Germany [58–60]. Twelve different periods were classified according to their year of construction, as shown in Table 1 ("construction year classes"). The state of energetic refurbishment was modeled by three different states:

- Variant 1: "standard (no refurbishment)."
- Variant 2: "moderate refurbishment."
- Variant 3: "advanced refurbishment."

Table 1. Different construction year classes for Germany as defined in [59].

Name	Years
A	1859 and earlier
B	1860–1918
C	1919–1948
D	1949–1957
E	1958–1968
F	1969–1978
G	1979–1983
H	1984–1994
I	1995–2001
J	2002–2009
K	2010–2015
L	2016 and later

Variant 1 refers to the requirements according to the applicable energy efficiency guideline for buildings in the respective construction year. Moderate and advanced refurbishments differ in thickness of wall insulation, number of insulated walls and other refurbishment measures. A detailed overview is given in [58,59].

The houses' construction year periods in the example town were derived from the technical report [54]. The result is shown in Figure 2.

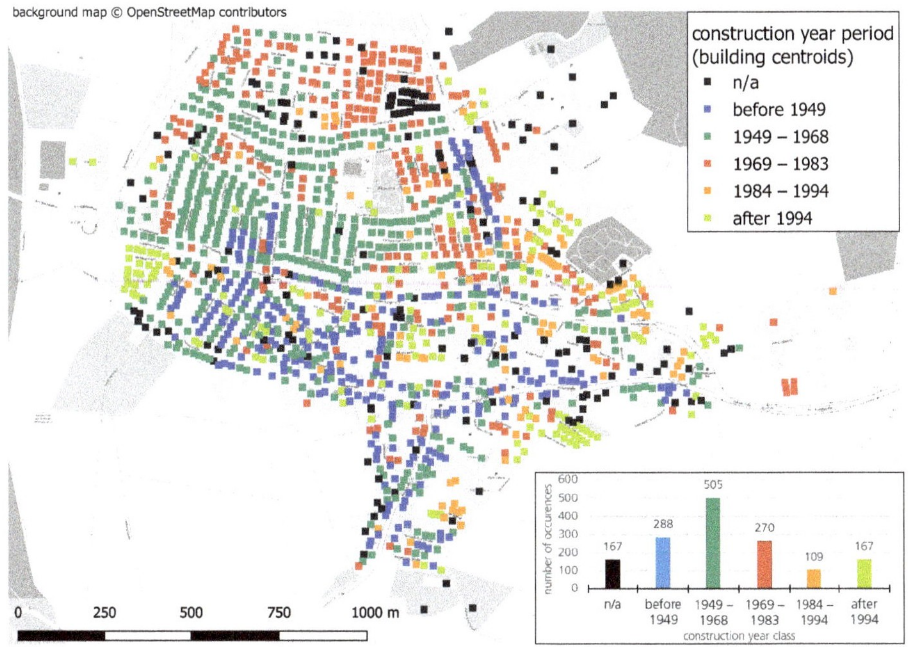

Figure 2. The region of the investigated town Schutterwald. Each colored square depicts a household. The color code describes the construction year class of each building (see legend in the right upper corner and, in the right lower corner, the inset for distribution of the number of houses for each period). Classification of periods based on [54].

The refurbishment variants were randomly assigned to the houses, based on a probabilistic distribution. The distribution of the three variants within each construction year class was based on statistics of German stock of buildings [55] as depicted in Figure 3. Different household types with one to four residents were randomly assigned to the houses, weighted by the census data for the municipality [61]; see Table 2. The household type is needed to calculate the electrical and domestic hot water load profiles.

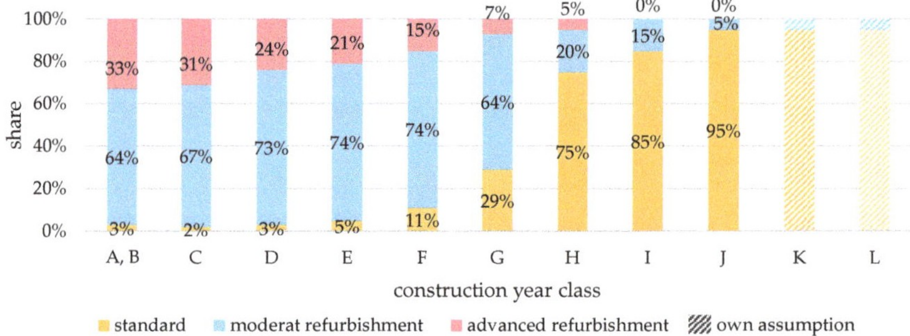

Figure 3. Share of refurbished single-family and two-family homes of German stock of buildings in 2009 per construction year class, adapted from [55]. Class A and B were not differentiated in the reference. As no data for classes K and L were given in the reference, equal distributions to class J were assumed.

Table 2. Shares of different household types in the investigated municipality, based on [61] (summarized).

Household Type	Code	Number	Share
single, retired	SRa	172	6%
single, employed	SOa	537	18%
couple, employed, 0 children	POa	1030	35%
couple, employed, 1 child	P1a	520	18%
couple, employed, 2 children or more *	P2a	686	23%

* Other, bigger households were assigned to household type P2a, too.

In total, the model included 1506 houses; twelve construction year classes with three variants of refurbishment, each completed by five household types.

The temperature and solar radiation data were taken from the weather year 2009.

2.2. Class-Based Electrical and Thermal Load Profiles

To reduce complexity, all buildings were assumed single-family homes. The electrical load profiles without heat generators were generated using a bottom up load profile generator, which is presented in [62]. The profile generator was based on adding up single household devices, such as a television, coffee machine, and washing machine. The devices' operating times were based on probability functions so that each resulting profile was different from the others. The probability functions were specific for different types of household members, e.g., children use the bathroom lightning earlier in the evening than adults. To account for different usage patterns of electrical appliances and different household sizes, five main household types were implemented representing the statistical distribution of singles (employed/retired) and couples with one, two, or without children in the municipality (Table 2). In total, 2000 different electrical load profiles for one year with a resolution of 10 minutes have been computed using the load profile generator.

Space heating (SH) demand and domestic hot water (DHW) demand were computed using another load profile generator that was developed in [63], similar to the method in [64]. It uses occupation models for the five different household types for each of the 36 building types. The profiles for space heating depending on the construction year class and the domestic hot water demand with a

10 min resolution were also computed for one year. Based on those profiles, models of heat pumps and thermal storage systems were applied to derive "grid connection point profiles", i.e., electricity and gas load profiles from the heat demand time-series that affect the power grid and natural gas network (see Figure 1).

2.3. GIS-Based Grid Connection Point Profiles

A heating system was modeled by a heat generator unit and two insulated hot water tanks, one for space heating and one for domestic hot water. The distribution of the heat within the house, including radiators, in-house pipes, and ventilation systems, was not modeled explicitly in this study. Distribution losses were taken into account by a loss factor. Three heat generator options were considered to cover the heat demand

- Natural gas-fired boiler;
- Electric heat pumps with auxiliary heating coils (eHPs);
- Gas engine heat pumps (geHPs).

The gas boiler models were based on Equations (1)–(4).

$$\eta_{Pn} = (A + B \cdot \log_{10}(P_n))/100 \tag{1}$$
$$\eta_{Ppart} = (C + D \cdot \log_{10}(P_n))/100 \tag{2}$$
$$q_{P0} = (E \cdot P_n^F)/100 \tag{3}$$
$$P_{aux} = (G + H \cdot P_n^K)/1000 \tag{4}$$

with rated power P_n; efficiency at rated and partial power η_{Pn} and η_{Ppart}; standby heat loss q_{P0}; electric auxiliary power P_{aux}; and factors A–H and K according to standard DIN V 18599:5.

The coefficient of performance (COP) for electric heat pumps ϵ_{eHP} was calculated from the ideal COP and a system efficiency factor $\eta_{HP} = 0.36$ using Equation (5).

$$\epsilon_{eHP} = \frac{T_c}{T_c - T_e} \cdot \eta_{HP} \tag{5}$$

with temperatures at the condenser T_c and evaporator T_e in Kelvin.

Like the electric heat pumps, gas engine heat pumps use a refrigerant cycle too. The compressor is driven by a gas engine and the heat from the engine cooling cycle and exhaust gas is utilized by additional heat exchangers. Thus, the overall efficiency was estimated according to Equation (6) with relative exhaust heat loss $q_{ex,loss} = 0.25$ [21], heat pump system efficiency $\eta_{HP} = 0.36$ and gas engine efficiency $\eta_{engine} = 0.307$ (assumption: full load, 1400 rpm) [65]. Currently, commercial gas engine heat pumps are only available for rated power of 25 kW upwards but market availability of small-scale geHPs could be reached within a few years [66].

$$\epsilon_{geHP} = 1 - q_{ex,loss} + \eta_{engine} \left(\frac{T_c}{T_c - T_e} \eta_{HP} - 1 \right) \tag{6}$$

The temperatures of the condensers in houses with regular, advanced, and ambitious energetic standards are assumed to be 55 °C, 45 °C, and 35 °C, respectively. Ambient air temperature records of a nearby weather station were used as evaporator temperature [67]. Since the lowest air temperature in 2007–2014 was recorded for 2009 (-15.2 °C), the weather year 2009 was used for the simulations in this study. Extreme winters may also occur in the future due to changes in atmospheric circulation caused by climate change [68]. Thus, the risk of over-sizing the heating systems appears to be limited but cannot be negated.

The losses of the hot water tanks are calculated with Equation (7) (according to DIN V 18599:5) and the stored heat energy is a function of the tank's volume and temperature, see Equation (8).

$$Q_{\text{loss,d}} = 0.4\frac{\text{kWh}}{\text{day}} + 0.14\frac{\text{kWh}}{0.001\,\text{m}^3\,\text{day}} \cdot \sqrt{V_{\text{stor}}} \tag{7}$$

$$Q_{\text{stor}} = V_{\text{stor}} \cdot 1.163\frac{\text{Wh}}{0.001\,\text{m}^3\,\text{K}} \cdot \Delta T \tag{8}$$

with daily heat loss $Q_{\text{loss,d}}$, storage volume V_{stor} in liters, thermal storage capacity Q_{stor} and difference between the maximum and minimum storage temperature ΔT

The investments for heat generators and hot water tanks were estimated according to Equation (9). All price parameters were assumed to stay constant for all investigated scenario years.

$$C_{\text{inv}} = a \cdot X + b \tag{9}$$

where a and b are factors shown in Table 3 and X is rated thermal power in kilowatts or storage volume in liters [69].

Table 3. Assumed price parameters for heat generators [69].

Heat Generator	Investments		Maintenance Costs	Depreciation Period
	a	b	[% of Investment/Year]	
gas condensing boiler	61 EUR/kW	4794 EUR	3.0	20 years
gas engine heat pump	163 EUR/kW	14797 EUR	4.5	20 years
electric air water heat pump	488 EUR/kW	7461 EUR	2.5	20 years
supplementary heating coil	100 EUR/kW	0 EUR	0.0	20 years
hot water storage tank	1120 EUR/m^3	806 EUR	0.0	20 years

Initial costs were converted to annuities, using an interest rate of $i_{\text{hh}} = 2.7\,\%$ for households (average of [70–72]) and $i_{\text{DSO}} = 4.27\,\%$ for distribution system operators (see Table A9).

The energy carrier rates given in Table 4 were applied to calculate the heating systems' fuel or electricity costs. Emissions were calculated based on the emission factors given in Table 5 [73,74]. It has to be noted that only scope 1 and 2 emissions were taken into account; no scope 3 emissions (production and mining). Possible emission changes for gas that may come from changing gas mix were neglected.

Table 4. Prices for energy usage (all prices including VAT).

Energy Carrier	Tariff	Variable [EUR/kWh]	+ Fix [EUR/Year]	Ref.
electricity	standard	0.283	96.39	[75]
	heat pump tariff, 3 × 2 h blocking time	0.231 (high load time) 0.196 (low load time)	71.40	[75]
natural gas	standard (18–50 MWh/year)	0.058	122.40	[76]

Table 5. Emission factors (scope 1 and 2).

Energy Carrier	Scope	Emission Factor	Reference
electricity (2017, domestic cons.)	2	537 g$_{CO2\text{-eq}}$/kWh	[73]
electricity (scenario 2030)	2	217 g$_{CO2\text{-eq}}$/kWh	calculated based on [77]
electricity (scenario 2050)	2	66 g$_{CO2\text{-eq}}$/kWh	calculated based on [77]
natural gas	1	202 g$_{CO2\text{-eq}}$/kWh	[78]

2.4. GIS-Based Grid Models

Based on the existing building stock model, the town's low voltage grid and gas distribution network were modeled.

2.4.1. Low Voltage Power Grid Model

As the real low voltage grid of the town is unknown, a new grid model was created. According to the classification presented in [79], three categories are applicable:

- Synthetic, because it is not based on a real DSO grid model;
- Example, because the main purpose is to illustrate different scenarios;
- Benchmark, because it is used to compare different scenarios by derived grid expansion costs.

The grid model was derived from an open-source 10 kV medium voltage (MV) grid model called "MV Oberrhein" that is provided as a synthetic grid in the Python package *pandapower* [32,80]. The low voltage grids for each of the 14 transformers have been modeled using a semi-automated method that includes the following steps:

1. The street lines downloaded from OpenStreetMap are segmented into sets of points with 1 m distance and an individual ID, called grid-points, using the QGIS-plugin QChainage [81]. It is assumed that cables are routed along the streets and each house is assigned to its nearest grid-point using the extension NNJoin [82]. All grid-points that have no house assigned are deleted.
2. The remaining grid-points are connected to their assigned houses by cables of type NAYY 4x50. To derive a preliminary grid structure, grid-points are connected to their nearest neighbor in the same street with less than 40 m distance by cables of the type NAYY 4x150. The parameters and locations of the MV/LV transformer stations are taken from the pandapower grid "MV Oberrhein". All information is imported to PSS® Sincal to proceed with a graphical interface.
3. The imported information is validated and existing errors due to the automated approach of grid generation are corrected manually. For each transformer, a supply area is chosen and the respective branches are connected by cables of type NAYY 4x150 to the LV-busbar of the transformer. Crossing points of many cables are equipped with switch cabinets. A radial topology without any galvanic connections between transformers is ensured by appropriate switch configuration. The result is shown in Figure 4.
4. For each house, a peak load of $P = 2$ kW and $Q = 0.1$ kVar is assumed, which represents a $cos(\phi)$ of 0.96. With these values, a load flow calculation is performed to make sure that the grid model is valid and the voltage and current of each cable, bus, and transformer stay within given limits. Limits are chosen to be 0.9–1.1 p.u. for bus voltages, 60% capacity for cables, and 130% for transformers (oil insulated) [83].
5. If violations of given voltage and capacity limits are found at this stage, switch measure, direct connection to bus bars; or new, parallel cables are added until all restrictions are met.
6. The Sincal-grid model is imported to pandapower and by using the integrated converter of pandapower-pro.
7. The loads in pandapower are connected with the house data (construction year classes, the status of energetic refurbishment, household type) from Figure 2.
8. Time-series of load profiles for each of the 1506 household loads are matched using one of the 2000 generated load profiles.
9. A final load flow calculation for a whole year (all time-steps) is done to validate the grid and make sure all the voltage and currents are within the given limits.

This electrical distribution grid is the benchmark grid for today and was used for the applications of the different scenarios. All transformer tap positions were set to the second-lowest position (-1). The power flow was calculated using the open-source Python package pandapower [32,80]. A version of the synthetic low voltage grid can be found in the Supplementary Materials as a pandapowerNet-file.

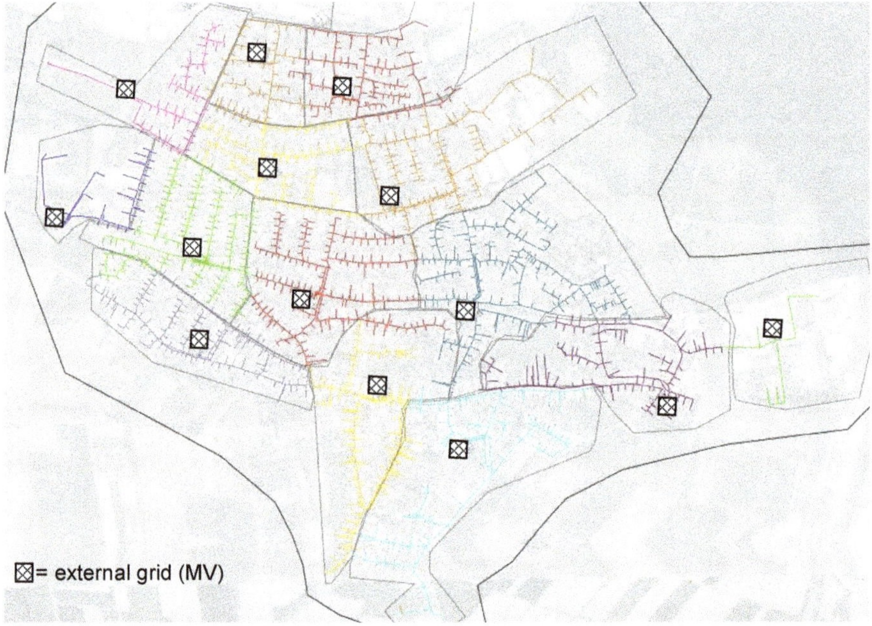

Figure 4. Synthetic electrical distribution grid of Schutterwald derived from open-source data. Each color represents a galvanically connected low voltage grid (grid group) that belongs to one transformer/external grid connection (squares). Due to the status of the switches, the different grid groups are not galvanically connected on the low voltage side.

2.4.2. Natural Gas Grid Model

The gas distribution grid was modeled in STANET® [84] by applying the following steps:

1. The raw network topology was derived from a map presented in [54].
2. Detailed information such as pipe diameters and types was derived from the gas network operator's online planning information platform [57] and was set in STANET® accordingly. The backbone of the gas system is made of pipes of the type 180 PE 100, all other pipes are of type 125 PE 100.
3. The location of the city gate station (pressure regulator station) was taken from the route depicted in the land utilization plan [56] and assumed to provide a constant pressure of 1 bar. It was implemented as a constant pressure node in STANET®.
4. The buildings and their types that were set in the electrical distribution grid model were imported to the gas distribution grid model.
5. Linear connection pipes from houses to the nearest natural gas pipeline were created by using the STANET® function "Create house connection pipes".
6. The STANET® grid model was exported as a CSV-file and imported into pandapipes, an open-source Python package for pipe flow and network simulation [85], for further analysis, e.g., on different lengths of house connection pipes.
7. Gas network capacity tests were conducted to find potential violations of the operation limits (flow velocity and nodal pressure). For these tests, it was assumed that all houses in the model were heated by gas boilers, except for those with an assigned heat pump. Then, time-series simulations were conducted in pandapipes. The highest gas flow velocity and the lowest nodal pressure per time step were logged.

The resulting gas distribution grid is shown in Figure 5 and can be found in the Supplementary Materials in the STANET®-CSV and pandapipesNet-file format.

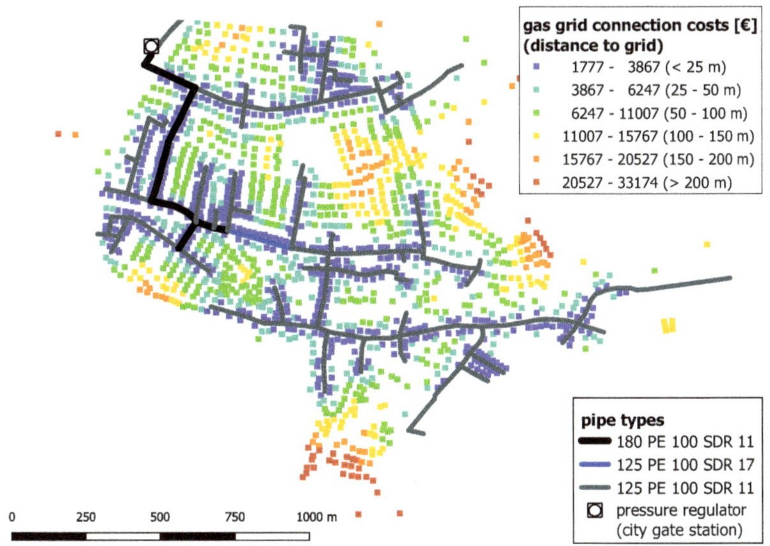

Figure 5. Assumed connection costs for houses based on their linear distance to the natural gas grid and the specific costs given in Table 6.

2.5. GIS-Based Grid Expansion Planning

Based on a scenario data of the INTEEVER project [77] scaled by the number of residents, the numbers of heat pumps in the years 2030 and 2050 were set to 164 and 247, respectively. These heat pumps can be realized as eHPs or geHPs. In the following, three cases are studied and compared for calculating the costs for grid expansion planning of electrical and gas distribution systems.

- Case 1 "electric": All heat pumps are realized as eHPs.
- Case 2 "natural gas": All heat pumps are realized as geHPs.
- Case 3 "mixed": It is assumed that heat pumps are driven by gas engines for houses that are close to the gas grid (less than 67 m linear distance). Heat pumps in other houses are implemented as eHPs.

2.5.1. Allocation of Heat Pumps

Electric heat pumps are connected at the same bus as the household load. For a worst-case scenario, it is assumed that no gas house connections exist or existing house connections are not used. Thus, each geHP is connected to the closest pipe of the gas grid by a linear house connection pipe and pipe investments are required.

The allocation of new heat pumps was not evenly randomly distributed like in other studies but was based on the construction year class. It should also be possible to allocate new heat pumps for a large range of numbers automatically. It should be based on given GIS information and include a sufficient degree of variety to create sets of different samples for the same given number of heat pumps. To achieve this, it was assumed that the likelihood of a house to be equipped with a heat pump depended mainly on the age and degree of refurbishment. For the given stock of buildings, likelihood points were assigned. As heat pumps are especially efficient with modern heating systems that require low-temperature heat, houses with the energetic refurbishment variants "standard", "moderate" and "advanced", received 1, 5, and 10 points, respectively. In a second step, today's shares of heat pumps in each construction year class in Germany were applied and the respective number of houses received 100 points, starting from advanced refurbished houses. For example, 599,201 residential buildings were completed in Germany between 2010 and 2015 (construction year class K), and for 184,110 (30.7%) of those houses, heat pumps were the main heat generator [86]. In the investigated area, around 50 houses got the construction year class K assigned. Thus, 15 of these houses ($\approx$30.7%) received 100 points for the heat pump allocation. The remaining 35 houses stayed at their initial 1, 5, or 10 points. Twenty random allocations were computed for the year 2030 and the remaining heat pumps for 2050 were added subsequently. For replicability, the heuristic's random number generator was initialized with different seeds (0–19) for each allocation. In each case, the same 20 allocations of heat pumps are analyzed and either electric motors or gas engines were assumed.

2.5.2. Grid Analysis

For each sample grid, power flow and pipe flow calculations were conducted in pandapower and pandapipes/STANET®, respectively. To limit the calculation time, the time-series was reduced to time steps with a cumulative load of 90% or more of the annual peak load. From these time steps, the one with the highest voltage band violation (i.e., lowest bus voltage) was used for the following grid extension study.

2.5.3. Grid Reinforcement

Prior to the low voltage grid reinforcement study, switch measures were applied (i.e., opening and closing of switches) by a hill-climbing heuristic for 5 min to balance line loadings and bus voltages among the network. For this sectioning point optimization (SPO) heuristic, the weighted sum of voltage violations and line overloadings was considered. The weighting factor for voltage band violations was set to 15 and line loading violations were weighted by a factor of 1.5. Then, grid reinforcement and extension measures were applied using hill-climbing and iterated local search algorithms from [33]. The allowed voltage band was 0.9 p.u. $\leq u_{bus} \leq$ 1.1 p.u. and the maximum line loading was 60%. The following measures were allowed:

- Replacing overloaded cables or cables that were upstream of voltage band violations by cables with increased diameter (NAYY 4x240).
- Adding parallel cables (NAYY 4x240) to replaced cables.

Subsets of those measures were evaluated by the heuristic optimizer to find a feasible solution and improve it further towards a cost minimum. However, the solutions might not represent the global minima due to a limited computation time of 30 min. Extension costs for the low voltage grid were estimated based on Table 6.

Table 6. Assumed costs for low voltage (LV) grid extension and construction of natural gas house connections.

Conductor	Costs	Reference	Depreciation Period
LV cable, NAYY 4x150 mm^2	95,000 EUR/km	[87]	40 years
LV cable, NAYY 4x240 mm^2	114,000 EUR/km	calc. from [87,88]	40 years
house connection gas pipe, DN 50	1488 EUR + 95 EUR/m	[89]	45 years

In the gas grid, the operational limits are specified as minimum node pressure of $p_{min} = 20$ mbar and maximum gas velocity $v_{gas,max} = 18$ m/s [90]. If these boundaries are violated, additional pipes have to be considered.

3. Results

In the following, the results for different stages of the bottom-up approach are explained. First, the results for the costs and emissions of single buildings are shown. As the effect of different household types is relatively small compared to the influence of building age and energetic refurbishment, only average values for all five household types are presented. Second, the costs and emissions that were specified for each house type are summarized according to the assumed distribution of buildings with heat pumps. Third, heat maps of bus voltages for worst-case time steps are shown and the calculated required grid investments are given. Finally, the combined costs of heat pumps and grid investments are compared for different cases.

3.1. Costs and CO_2 Emissions for Single Buildings and New Heating Systems

The calculated annual demands for space heating, hot water, and electricity (for household devices, not for eHP) are summarized by household type in the Appendix A, Table A1. The electricity consumption increases with increasing household size. In turn, the space heating demands decrease slightly, due to the waste heat provided by the electrical appliances. The DHW demand is almost proportional to the number of household members. The simulated annual heat production and efficiency values for eHPs and geHPs are listed for each construction year class and energetic refurbishment variant in Tables A2 and A3. Figure 6 displays specific costs of heat generation for different combinations of heating technologies and years of construction as well as their state of refurbishment. Namely, electric heat pumps with an auxiliary heating coil, gas engine heat pumps, and gas boilers are compared for the construction year classes E (1958–1968) and L (2016 and later). For all heat generator options and years of construction, the required rated heat generator power—and thus the investments—decrease with further energetic refurbishment. Gas boilers have the lowest overall costs of generated heat, as they require the least investments. Required investments for geHPs are 3.18–3.34 times higher than for gas boilers. For eHP, 1.95–3.72 times the investments for respective gas boilers are necessary. Regarding energy costs, geHP are the most cost-efficient, particularly for well-insulated buildings and buildings according to the latest energy efficiency guidelines.

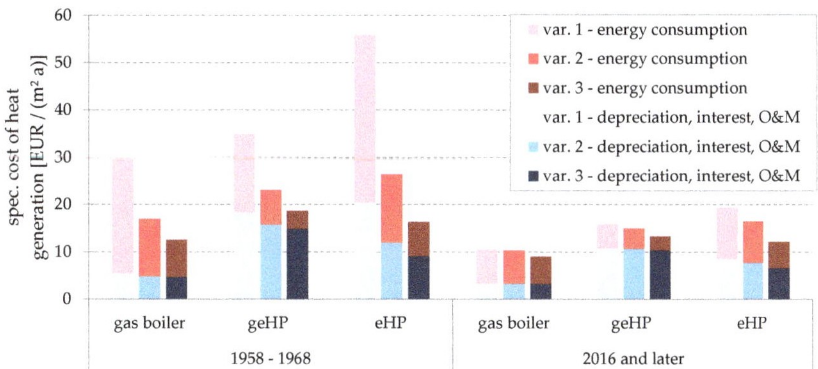

Figure 6. Costs for heat generation for sample buildings of two construction year classes and three energetic insulation variants. Average of five simulated household types.

Total annual costs (depreciation, energy consumption, maintenance, and interest) and CO_2 emissions of different heat generators are shown in Figure 7 for the construction year classes E (1958–1968) and L (2016 and later). In all cases, gas boilers cause the highest annual emissions. The replacement of gas boilers by geHPs reduces emissions by 27–30%, 35–37% and 46–49% (for variants 1, 2 and 3, respectively). For eHPs, the assumed electricity mix is essential. Using the electricity mix of 2017, an eHP in a house constructed between 1958 and 1968 would eliminate between 6% (variant 1) and 40% (variant 3) of CO_2 emissions. Assuming a further decline of fossil–fired power plants (see Table 5), the savings rise to 62–76% in 2030 and 89–93% in 2050.

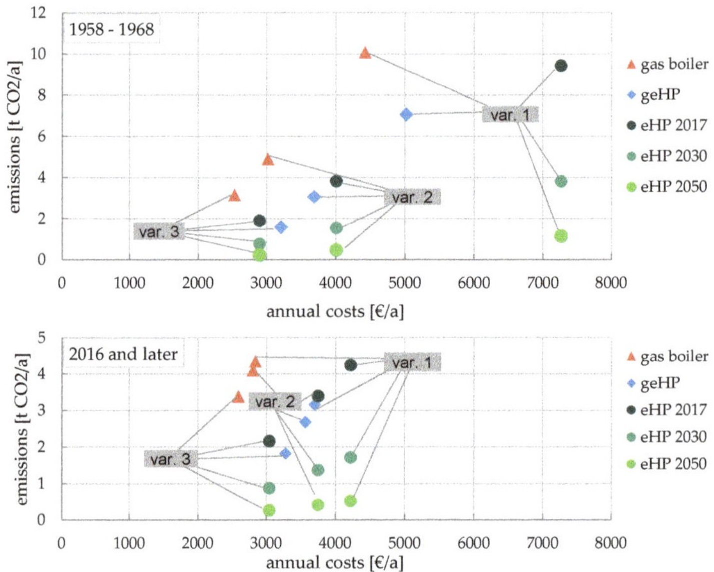

Figure 7. Calculated annual costs (depreciation, energy consumption, maintenance and interests) and emissions of different heating options for single-family homes (construction year period 1958 to 1968 and 2016 onwards). Average of five simulated household types. Underlying numbers can be found in Tables A4 and A5. For electric heat pumps (eHPs), the emissions are given for different years (i.e., decreasing carbon intensity of electricity generation; see Table 5). Costs for energy and heating systems were assumed to stay constant for all years. Possible emission changes for gas that may come from changing gas mix are neglected.

3.2. Cost and Emissions for Investigated Buildings in Schutterwald

The GIS information of the buildings in Schutterwald and the method to allocate new heat pump systems made it possible to calculate the costs and CO_2 emissions for all analyzed buildings in Schutterwald together.

In the investigated cases, 164 and 247 heat pumps were allocated among the 1506 houses of the town. Due to the partially randomized allocation heuristic, the distributions vary among the construction year classes, energetic insulation standards, and household types. The distributions are shown in Figure 8. Figure 8a shows the share of each construction year class and energy refurbishment variant for one exemplary seed. Figure 8b,c present the distributions of heat pumps for 2030 and 2050 respectively.

The average total annual costs for 164 heat pumps are 494,000 and 416,000 euros in case 1—electric and case 2—natural gas. For 247 heat pumps, these numbers rise to 736,000 and 624,000 euros. In case 3—mixed, on average 67% of the heat pumps are implemented as geHPs and 33% as eHPs. This is reflected proportionally in the heat pump investments and energy costs of 443,000 and 662,000 euros for 164 and 247 heat pumps in case 3. In Figure 9, the cost sensitivity to varied energy prices and investment costs is depicted. In case 1, an electricity price increase of 20% leads to raised overall costs of +11%. If the electricity price is lowered by 20%, case 1 has lower heat pump costs than case 3 with default parameters (−2.2%). If required heat pump investments change by 20%, the overall annuities change by 9%, 12%, and 14% in cases 1, 2, and 3, respectively.

The heat pumps' cost difference is relatively small compared to the difference in annual CO_2 emissions. In the 2030 scenario, the heat pumps emit on average 505 $t_{CO2\text{-}eq}$/a if they are implemented as geHPs. In contrast, the same number of eHPs in 2030 causes on average 66% less

CO_2. With decreasing carbon intensity of the electricity mix, this difference rises to 84% less CO_2 from eHPs than from geHPs in 2050. The emissions in case 3 scale proportionally to the respective eHP/geHP ratio.

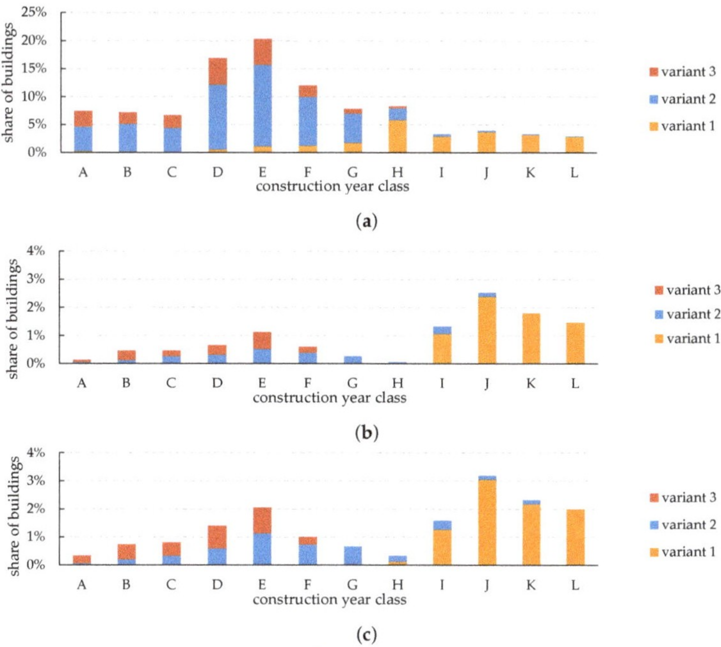

Figure 8. Resulting building stock and assigned heat pumps (example, for the seed "0"; variant 1/2/3 = no/moderate/advanced energetic refurbishment). (**a**) Derived distribution of total housing stock in the town per construction year class and energy variant. (**b**) Share of selected buildings to be equipped with a heat pump. Scenario with 164 heat pumps. (**c**) Share of selected buildings to be equipped with a heat pump. Scenario with 247 heat pumps.

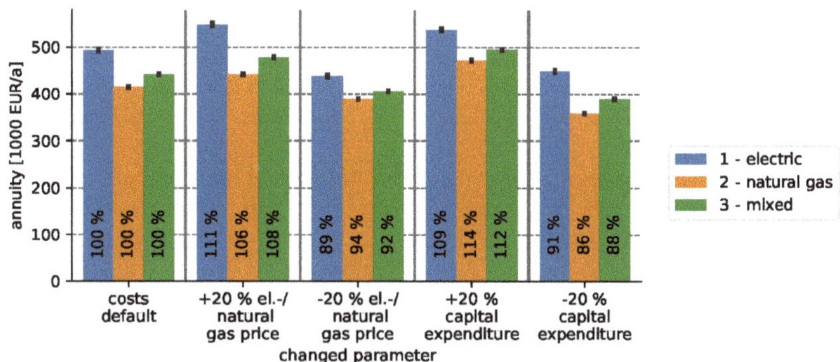

Figure 9. Sensitivity of overall cost for heat pumps in the cases in year 2030/164 heat pumps. Averages of all 20 seeds.

3.3. Required Grid Investments in the Gas and Power Grids

For each allocation of heat pumps, sectioning point optimizations were conducted in the electric and mixed cases. If voltage band violations or line loading violations occurred, grid reinforcement measures were added.

3.3.1. Case 1—Electric (eHPs)

From the sample allocation of 247 heat pumps shown in Figure 10, it can be confirmed that the distribution among the town's houses is not equal, due to the weighted random heuristic. Thus, clustering can be observed in areas with younger buildings, in the western and southeastern part of the town. In these areas, the grid load is particularly high. Table 7 gives an overview of lowest bus voltages and maximum line loads for an exemplary allocation (for the seed "15"). Undervoltage occurs at 54% of the busses but can be reduced to 4.8% of the buses by SPO; 68 lines (1.66 km) are overloaded; that is, 5.0% of the total low voltage line length without house connection cables. The results show that the violations can be significantly lowered by SPO. If the solutions of the automated grid reinforcement planning are applied, all voltage and line loading limits are met.

The required grid extension measures differ highly between different seeds and the success of the automated grid extension hill-climbing heuristic.

For some seeds, very few measures are required or sectioning point optimization was already sufficient to meet all operational restrictions. Overall, the maximum required grid reinforcement investments summed up to 161,994 euros in 2030 and 296,419 euros in 2050 with average costs per allocation of 76,110 euros (2030) and 223,488 euros (2050).

Figure 10. Synthetic low voltage grid of the town with 247 electric heat pumps (case 1—electric, for the seed "15"). Line overloads and bus voltages at time step with lowest bus voltage, prior to sectioning point optimization and grid extension.

Table 7. Exemplary simulation results from pandapower (for the seed "15"): load at worst time step (occurrence of lowest bus voltage) and lowest bus voltage and maximum line loading before and after sectioning point optimization (SPO) and after automated grid reinforcement planning (with cost-optimal cable reinforcement and parallel lines applied).

		Case 1		Case 3	
number of heat pumps		164	247	164 (47 eHP)	247 (76 eHP)
total load at worst time step [MW]		3.297	3.556	2.595	2.687
load caused by eHP [MW]		0.839	1.787	0.225	0.317
	before SPO	0.832	0.796	0.847	0.833
lowest bus voltage [p.u.]	after SPO	0.895	0.863	0.908	0.896
	after grid reinforcement	0.901	0.909	0.919	0.900
	before SPO	99.1	133.4	79.1	90.3
highest line loading [%]	after SPO	92.4	123.4	73.9	84.1
	after grid reinforcement	58.8	59.27	59.3	59.8

3.3.2. Case 2—Natural Gas (geHP)

In this case, the same heat pump allocations as in case 1 were evaluated, but gas engines were assumed instead of electric motors. The required gas connection pipes differ in length and were not limited in this case. As shown in Figure 5, the majority of the town's houses are closer than 100 m to the natural gas grid. However, some areas are further away and the longest required connection pipe has a length of 332.8 m.

This leads to higher grid investments than in case 1 with on average 1,237,127 euros in 2030 and 1,829,862 euros in 2050. These values are theoretical values and indicate an upper limit, as all pipes are implemented as connection pipes. In a real system, the main distribution system may be expanded to supply new demand areas. Thus, shorter house connection pipes would be necessary.

The capacity test shows maximum gas velocities of 13.5 m/s and nodal pressures between 0.80 bar and 1.00 bar in case 2 with 247 geHPs and gas boilers for the remaining 1259 houses. Thus, no predefined operating limits are violated and no reinforcement or extension of the natural gas network is required apart from house connection pipes.

3.3.3. Case 3—Mixed (eHPs and geHPs)

In case 3, both electric and gas engine heat pumps should be deployed. As a reasonable indicator to estimate the house owners' preference for one technology or the other, the house's distance to the gas grid and the respective connection costs were used. For young houses (construction year class L-2016 and later, variant 1) the net present value for heat supply over 20 years is −64,706 euros for eHP and −56,819 euros for geHP, not including connection pipelines. The difference, 7887 euros, equals the costs for a natural gas house connection pipe of 67.2 m length. Thus, this distance is considered as a threshold in case 3.

As in cases 1 and 2, the same allocations of heat pumps were used in case 3. It was assumed that all heat pumps that were closer to the gas grid than 67 m were driven by gas engines (geHPs). Those heat pumps that were further than 67 m away from the existing natural gas network were assumed electrically driven (eHPs). Since the electrical load was reduced (compared to case 1), large parts of the power grid were without violations. However, some load clusters remained and required grid reinforcement (Figure 11). In general, the violations of the admissible voltage range and the maximum admissible line loads were lower than in case 1. In addition, sectioning point optimization can reduce maximum voltage violations more effectively than in case 1 (see Table 7). On average, this leads to reduced low voltage grid investments of 45.4% (2030) and 33.8% (2050) compared to case 1. At the same time, 35.4% (2030) and 37.2% (2050) of the investments in gas grid connection pipes calculated in case 2 are required.

Figure 11. Synthetic low voltage grid of the town with 171 geHPs and 76 eHPs (case 3—mixed). Line overloads and bus voltages at time step with the lowest bus voltage, prior to sectioning point optimization and grid extension. geHPs are not represented in the figure.

3.4. Combined Costs of Heat Supply and Grid Investments

All grid extension and grid connection costs are summarized in Table A7 and shown in Figure 12.

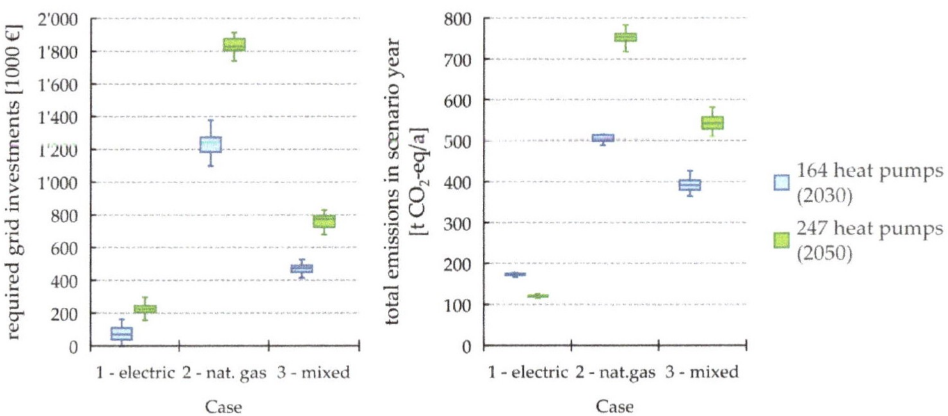

Figure 12. Box plots (n = 20) of calculated required grid investments for 164 and 247 deployed heat pumps (**left**) and emissions caused by deployed heat pumps in 2030 and 2050 (**right**). See also Tables A7 and A8.

The combined costs of required grid investments per case and heat pump investments, maintenance, and operation are shown with the average emissions in 2030/2050 in Figure 13. For comparison, all costs and investments have been converted to annuities, as installed cables and pipes have much longer lifetimes than heat pumps (see Tables 3 and 6). In all three cases, the required grid investments are low compared to the heat pump related costs. In the scenario with 247 heat pumps, the highest shares of grid investments are 2.1%, 10.6%, and 4.7% for cases 1, 2, and 3, respectively. The average grid investments and other cost components for each case are shown in Figure 14.

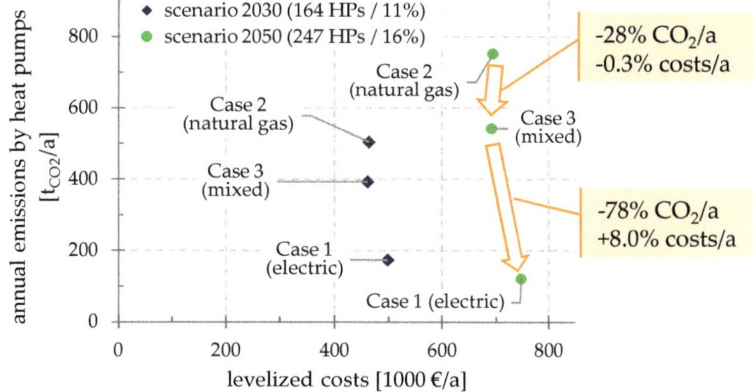

Figure 13. Average annualized costs (grid investment, heat pump investment, and operation) and annual emissions in 2030 and 2050 by heat pumps for cases 1—electric, 2—natural gas, and 3—mixed respectively.

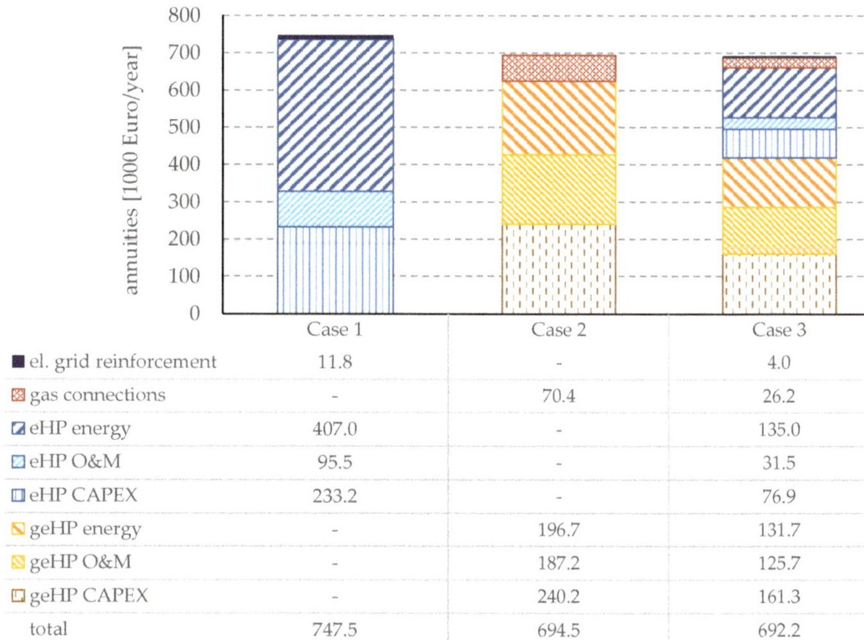

	Case 1	Case 2	Case 3
■ el. grid reinforcement	11.8	–	4.0
⊠ gas connections	–	70.4	26.2
◪ eHP energy	407.0	–	135.0
◩ eHP O&M	95.5	–	31.5
▯ eHP CAPEX	233.2	–	76.9
◥ geHP energy	–	196.7	131.7
◩ geHP O&M	–	187.2	125.7
▤ geHP CAPEX	–	240.2	161.3
total	747.5	694.5	692.2

Figure 14. Annual costs in 1000 euros/year for heat supply of 247 houses with heat pumps and connected required grid investments (annualized). Average of 20 seeds for each case.

In case 2, the cost advantage of geHPs (Table A6) caused by lower energy costs is to a large extent compensated by the higher grid investments. On average, the overall annual costs of case 2 (gas) are 7% lower than in case 1 (electric), both in 2030 and 2050. In case 3 (mixed) a decrease in costs of 0.5%

occurs, compared to case 2. The CO_2-emissions of case 2 are by a factor of 2.9 (2030) and 6.3 (2050) higher than in case 1.

4. Summary

4.1. Conclusions

Amongst others, three innovation highlights were addressed in this paper:

- On the basis of a large number and different types of public data only, a low voltage and gas grid model with a highly diversified spatial resolution has been created for an example town and made available in the Supplementary Materials.
- We did a mutual investigation of power and natural gas distribution infrastructure for a whole town using a pipe and power-flow grid analysis.
- For all three cases, we investigated grid investments, heat pump costs, and CO_2 emissions for a multi-perspective approach.

A low voltage network for a town with around 1500 houses has been modeled in pandapower. In addition, a simplified but realistic natural gas distribution network model has been developed in STANET® for the same town. A model of the existing building stock of the town has been derived from OpenStreetMap coordinates and has been enriched with realistic data on construction year classes. Furthermore, information on the houses' energetic refurbishment and household types has been estimated based on detailed statistics. For each house type, different heat demand profiles were simulated and electric load profiles for heat pumps as well as natural gas load profiles for gas boilers and gas engine heat pumps have been derived. Overall, the models can be seen as synthetic with realistic assumptions. They have been compared in terms of costs and emissions. For the majority of the investigated buildings, it was found that eHPs caused higher specific CO_2-emissions than geHP in 2017. In 2030 and 2050 scenarios, however, the eHP emissions were around 50% and more than 80% below the geHP emission levels, respectively. The specific heat generation costs for eHP decrease strongly with the increasing energetic refurbishment of the buildings. For buildings with little to no energetic refurbishment, a cost advantage for geHP was observed.

Based of the load profiles, the effects of heat pump deployment on either energy infrastructure have been analyzed regarding required grid reinforcement and extension measures. In the power grid, a large share of bus voltage violations could be solved by SPO, which switches lines from one branch to another. If around 11% of the buildings are equipped with electric heat pumps, little to no grid reinforcement is required in the investigated model. For gas engine heat pumps, however, significant grid connection costs can occur if the maximum connection length is not limited. If the share of heat pumps rises to 16%, bigger clusters occur, and average required grid investments in the low voltage power grid increase. Nonetheless, the electric heat pump case (case 1) requires just 12% of the grid investments in the gas engine heat pump case (case 2) and causes 84% less CO_2 emissions (in 2050). If heat pump investments and operating costs are considered as well, case 3 comes on average with 0.3% less annual costs and -28% CO_2 emissions in 2050 compared to case 2. If all remaining geHPs in case 3 are replaced by eHPs, overall annual costs increase on average by 8.0% and CO_2 emissions decrease by 78% (see Figure 13). The observed load clusters indicate the importance of a neighborhood's building characteristics (e.g., energetic refurbishment and age) for grid planning and energy system modeling.

4.2. Discussion and Limitations

In the proposed method, multiple building characteristics (construction year, energetic refurbishment, household type) are taken into account. The grid load analysis indicates that the degree of energetic refurbishment has the strongest effect of grid stress. This is partly related to the assumption of decreasing heating temperatures with higher insulation but also shows the importance of energy efficiency.

In the case study, some simplifications and assumptions were made that may have affected the results. In particular, constant cost parameters due to neglected learning curves may have led to overestimated investment costs for heat pumps. Additionally, the fact that present heat generators and existing gas grid connections were not taken into account also led to higher cost estimations. Instead of adding multiple individual house connection gas pipes, synergies could be used by extending the main natural gas grid to reach new demand clusters. For this, the gas grid could be modeled in pandapipes and an algorithm similar to the automated grid planning in pandapower [33] could be applied.

The buildings' heat demand was likely overestimated as well since the building stock was assumed to stay constant and no refurbishments were taken into account. The distribution of the heat within the house included radiators, in-house pipes, and ventilation systems. These systems were not modeled explicitly in this study, but are important within the context of renovation and usability of thermal systems with different temperature levels. Therefore, it is important to distinguish between old buildings with low refurbishment standards and new buildings with low heat demand. Enabling older buildings to use electrical heat pump systems could create much higher costs than in younger buildings.

This work focuses on single-family homes. However, commercial, non-residential consumers will likely have a strong impact on load in the power and natural gas grids. There are commercial consumers, such as restaurants, cooking with natural gas, who may not be willing to change to electrical systems. Solutions for such individual demands have to be taken into account in real urban systems. As the natural gas composition was assumed to stay constant, the emissions were higher than in a scenario with an increased share of green hydrogen, bio-methane, or synthetic fuel usage. Nevertheless, for a comprehensive assessment of the emissions, supply chain emissions of electricity generation and natural gas supply have to be taken into account as well. This could lead to remarkably higher indirect emissions. Traber and Fell [91] reported a global warming potential (20-year horizon) of the natural gas supply chain of 170–337 $g_{CO2\text{-}eq}/kWh_{fuel}$. Compared to scope 1 emissions of 202 g_{CO2}/kWh_{fuel}, this implies an increase of emissions by 84–266%.

In addition, the investigated cases did not consider other developments that may increase the need for grid reinforcement. In particular, increasing installations of solar PV, charging points for electric vehicles, and the need for air conditioning units can be considered as potential drivers for additional power grid investments.

The grid extension analysis was conducted for a dedicated town with specific infrastructure and building stock characteristics. Therefore, the applied method will likely lead to different results in other towns and further research is required to derive more general results.

4.3. Further Research

The calculations were done for a synthetic German town example and the numbers of new heat pumps were taken from a national energy optimization model. However, the number was rather low, so that 11% and 16% of the buildings were to be equipped with heat pumps in 2030 and 2050, respectively. This leaves questions regarding efficient, renewable heating systems for the rest of the buildings in the town. In Germany, the new installation of decentralized oil heating systems will be forbidden, starting in the year 2025. Currently, more than 60% of the buildings in the example town are equipped with such systems. An evaluation of cost and feasibility needs to be done for future scenarios, including the change of all thermal supply systems in the town to a CO_2 neutral solution.

In this study, a simple control algorithm for heating systems has been used and static blocking time slots set by the DSO were assumed. Advanced algorithms may lead to further variation in load profiles and thus reduce simultaneity. This would lead to lower peak load and reduce required grid extension measures. Furthermore, future research may check the usage of small-scale heat pumps for each building against the possibility of installing district heating grids.

The matching power and gas grid models that were created in this work can be used for future research, in particular in the field of coupled power and gas distribution grids.

For municipalities with increasing numbers of heat pumps being installed, the case study provides an initial impression in which operating costs, capital expenditures for households, and grid operators arise and how they are distributed among the stakeholders. As a next step, the individual stakeholders' investment decisions could be investigated in more detail and policy measures (e.g., regional incentive programs) could be coordinated accordingly. However, the results of the case study cannot be generalized without further research.

Supplementary Materials: The following network models are available online at http://www.mdpi.com/1996-1073/13/16/4052/s1: natural gas network in pandapipes-JSON format (pandapipes version 0.1.2), natural gas network in STANET-CSV format, and low voltage power grid in pandapower-JSON format (pandapower version 2.2.2).

Author Contributions: Conceptualization, T.M.K. and J.M.K.; methodology, J.M.K.; software, J.M.K.; formal analysis, J.M.K.; investigation, J.M.K.; data curation, J.M.K.; writing—original draft preparation, T.M.K. and J.M.K.; writing—review and editing, M.B. and S.L.; visualization, J.M.K. and T.M.K.; supervision, T.M.K. and S.L.; project administration, T.M.K.; funding acquisition, T.M.K. All authors have read and agreed to the published version of the manuscript.

Funding: This research was funded by the Bundesministerium für Wirtschaft und Energie within the projects INTEEVER and INTEEVER II (03ET4020C and 03ET4069C).

Acknowledgments: J.M.K. would like to thank Daniel Then (Fraunhofer IEE, Stadtwerke Bamberg GmbH) for his helpful advice on low voltage network planning. The authors thank bnNETZE for permission to use the natural gas grid data.

Conflicts of Interest: The authors declare no conflict of interest. The funders had no role in the design of the study; in the collection, analyses or interpretation of data; in the writing of the manuscript, or in the decision to publish the results.

Abbreviations

The following abbreviations are used in this manuscript:

aux.	auxiliary
CAPEX	capital expenditure
CHP	combined heat and power
cons.	consumption
constr.	construction
COP	coefficient of performance
DHW	domestic hot water
DSO	distribution system operator
eHP, eHPs	electric heat pump, electric heat pumps
el.	electricity
EMF	emission factor
geHP, geHPs	gas engine heat pump, gas engine heat pumps
Hh., hh	household
NZEB	nearly zero-energy building
O&M	operation and maintenance costs
OSM	OpenStreetMap
prod.	production
PV	photovoltaic
SH	space heating
WACC	weighted average cost of capital
YTM	yield to maturity

Appendix A. Supplementary Tables

Table A1. Simulation results for average (by household type) annual space heating demand, domestic hot water demand, and electricity consumption of household devices. For each household type, 12 construction year classes with 3 energy refurbishment variants each were considered.

Household Type	Space Heating [kWh$_{th}$]	DHW [kWh$_{th}$]	El. Hh. Devices [kWh$_{el}$]
single, employed	19,618	847	2263
single, retired	19,579	869	1912
couple, employed, no children	18,620	1707	3281
couple, employed, 1 child	17,827	2578	4207
couple, employed, 2 children	16,923	3455	4842

Table A2. Calculated electricity consumption, heat generation, and efficiency of eHP systems for different types of single-family homes (average of the five household types).

Constr. Year Class	Energy Refurb. Variant	Heat Demand [kWh/a]	El. Cons. eHP [kWh$_{el}$/a]	Heat Prod. eHP [kWh$_{th}$/a]	Annual COP (eHP Only)	El. Cons. Aux. Heating Coil [kWh$_{el}$/a]	Annual COP (System of eHP + Coil)
A	1	66,975	30,725	73,422	2.39	1856	2.31
	2	22,812	8847	26,082	2.95	739	2.80
	3	13,557	4116	16,051	3.90	488	3.59
B	1	40,965	19,244	46,207	2.40	1164	2.32
	2	15,088	6035	17,938	2.97	499	2.82
	3	9667	3030	12,027	3.97	372	3.64
C	1	59,076	27,184	65,014	2.39	1679	2.31
	2	29,102	11,287	33,219	2.94	922	2.80
	3	19,762	5857	22,555	3.85	712	3.54
D	1	32,674	15,403	37,145	2.41	884	2.33
	2	14,130	5678	16,942	2.98	426	2.85
	3	7991	2577	10,311	4.00	287	3.70
E	1	35,026	16,582	39,929	2.41	976	2.33
	2	16,639	6577	19,554	2.97	549	2.82
	3	10,125	3159	12,517	3.96	384	3.64
F	1	36,594	17,142	41,238	2.41	1043	2.33
	2	17,809	7041	20,796	2.95	568	2.81
	3	12,543	3843	15,033	3.91	491	3.58
G	1	29,902	14,126	34,050	2.41	853	2.33
	2	17,219	6803	20,166	2.96	572	2.81
	3	12,327	3782	14,843	3.92	474	3.60
H	1	24,728	11,810	28,476	2.41	674	2.34
	2	16,176	6428	19,071	2.97	528	2.82
	3	10,589	3298	12,964	3.93	393	3.62
I	1	15,938	7821	18,997	2.43	460	2.35
	2	13,400	5453	16,252	2.98	413	2.84
	3	9116	2865	11,396	3.98	367	3.64
J	1	13,988	6932	16,858	2.43	436	2.35
	2	12,678	5168	15,487	3.00	404	2.85
	3	10,318	3193	12,683	3.97	386	3.65
K	1	18,232	8839	21,419	2.42	528	2.34
	2	16,358	6513	19,321	2.97	519	2.82
	3	11,788	3631	14,335	3.95	442	3.63
L	1	15,178	7448	18,140	2.44	444	2.35
	2	14,490	5826	17,331	2.97	482	2.82
	3	11,611	3602	14,166	3.93	423	3.62

Table A3. Calculated natural gas consumption, heat generation and primary energy ratio (ratio heat output/fuel input) of geHP systems for different types of single-family homes (average of the five household types).

Constr. Year Class	Energy Refurb. Variant	Heat Demand [kWh/a]	Fuel Cons. [kWh$_{fuel}$/a]	Heat Prod. [kWh$_{th}$/a]	Primary Energy Ratio
A	1	66,975	60,189	75,524	1.25
A	2	22,812	18,575	26,478	1.43
A	3	13,557	9514	16,451	1.73
B	1	40,965	37,224	46,770	1.26
B	2	15,088	12,633	18,148	1.44
B	3	9667	6893	12,007	1.74
C	1	59,076	53,214	66,817	1.26
C	2	29,102	23,615	33,590	1.42
C	3	19,762	13,692	23,420	1.71
D	1	32,674	29,579	37,250	1.26
D	2	14,130	11,768	16,942	1.44
D	3	7991	5715	9970	1.74
E	1	35,026	31,760	39,894	1.26
E	2	16,639	13,787	19,799	1.44
E	3	10,125	7205	12,531	1.74
F	1	36,594	33,183	41,760	1.26
F	2	17,809	14,788	21,165	1.43
F	3	12,543	8878	15,352	1.73
G	1	29,902	27,105	34,096	1.26
G	2	17,219	14,342	20,591	1.44
G	3	12,327	8686	15,003	1.73
H	1	24,728	22,614	28,495	1.26
H	2	16,176	13,433	19,252	1.43
H	3	10,589	7501	12,983	1.73
I	1	15,938	14,928	18,894	1.27
I	2	13,400	11,237	16,138	1.44
I	3	9116	6553	11,470	1.75
J	1	13,988	13,243	16,784	1.27
J	2	12,678	10,661	15,353	1.44
J	3	10,318	7275	12,685	1.74
K	1	18,232	16,937	21,404	1.26
K	2	16,358	13,608	19,523	1.43
K	3	11,788	8287	14,389	1.74
L	1	15,178	14,219	18,039	1.27
L	2	14,490	12,069	17,306	1.43
L	3	11,611	8209	14,224	1.73

Table A4. Annual costs (CAPEX, O&M, fuel) in euros/year and emissions in t $_{CO2\text{-}eq}$/year for different heat generators in a single-family home, construction year period E (built between 1958 and 1968).

	Existing State (Var. 1)		Usual Refurbishment (Var. 2)		Adv. Refurbishment (Var. 3)		Information: EMF [g$_{CO2}$/kWh$_{fuel}$]
	Costs	Emissions	Costs	Emissions	Costs	Emissions	
oil boiler	5311	13.7	3386	6.6	2714	4.1	266
gas boiler	4873	10.8	3197	5.0	2625	3.2	202
geHP	5320	7.5	3798	3.1	3263	1.6	202
eHP 2017	7809	10.9	4196	4.3	2983	2.1	537
eHP 2030	7809	4.4	4196	1.7	2983	0.9	141
eHP 2050	7809	1.3	4196	0.5	2983	0.3	66

Table A5. Annual costs (CAPEX, O&M, fuel) in euros/year and emissions in t $_{CO2\text{-}eq}$/year for different heat generators in a single-family home, construction year period L (built in 2016 or later).

	National Minimum Requirement (Var. 1)		Ambitious Standard/ NZEB (Var. 2)		Advanced Refurbishment (Var. 3)		Information: EMF [g_{CO2}/kWh$_{fuel}$]
	Costs	Emissions	Costs	Emissions	Costs	Emissions	
oil boiler	3146	5.7	3114	5.6	2818	4.5	266
gas boiler	2987	4.5	2962	4.3	2713	3.5	202
geHP	3809	3.2	3669	2.8	3345	1.9	202
eHP 2017	4412	4.7	3926	3.8	3146	2.4	537
eHP 2030	4412	1.9	3926	1.6	3146	1.0	141
eHP 2050	4412	0.6	3926	0.5	3146	0.3	66

Table A6. Statistical description of overall annualized heat pump investments and yearly energy costs in 1000 euros/a for each case (20 allocations per case). Q_1, Q_2, and Q_3 represent 25%, 50%, and 75% quartiles.

Year	Number of Heat Pumps	Case	Mean	σ	Min	Q_1	Q_2	Q_3	Max
2030	164	1 - electric	494	6	480	491	494	500	502
		2 - natural gas	416	3	411	415	416	419	419
		3 - mixed	443	3	438	440	442	446	448
2050	247	1 - electric	736	11	707	730	737	741	759
		2 - natural gas	624	5	613	622	624	627	634
		3 - mixed	662	6	646	659	663	667	672

Table A7. Statistical description of required natural gas and low voltage grid investments in 1000 euros for each case (20 different allocations per case).

Year	Number of Heat Pumps	Case	Mean	σ	Min	Q_1	Q_2	Q_3	Max
2030	164	1 - electric	76	46	0	38	69	109	162
		2 - natural gas	1237	74	1097	1184	1238	1273	1376
		3 - mixed	473	30	415	450	472	489	526
2050	247	1 - electric	223	36	157	205	225	244	296
		2 - natural gas	1830	46	1740	1805	1828	1,873	1912
		3 - mixed	757	44	679	725	775	794	829

Table A8. Statistical description of annual CO_2 emission caused by heat pumps in the scenarios in t$_{CO2\text{-}eq}$/a.

Year	Number of Heat Pumps	Case	Mean	σ	Min	Q_1	Q_2	Q_3	Max
2030	164	1 - electric	173	3	167	172	173	176	177
		2 - natural gas	505	8	489	501	504	513	515
		3 - mixed	392	16	365	380	392	403	427
2050	247	1 - electric	120	2	114	119	121	122	125
		2 - natural gas	753	14	718	745	753	761	783
		3 - mixed	543	19	510	530	542	557	582

Table A9. Discount rate assumptions.

Parameter	Value	Reference
discount rate households	2.67 %	average of [13,71,72]
equity interest rate DSO	6.91 %	[92]
debt interest rate DSO	1.33 %	10 year avg. of YTM on German bearer debentures (2009–'18) [93]
equity ratio DSO	52.73 %	[94]
discount rate DSO (WACC)	4.27 %	own calculation

References

1. European Parliament; Council of the European Union. Directive (EU) 2018/2002 of the European Parliament and of the Council of 11 December 2018 amending Directive 2012/27/EU on energy efficiency. *Off. J. Eur. Union* **2018**, *328*, 210–230.
2. Erbach, G. *Understanding Energy Efficiency*; European Parliamentary Research Service: Brussels, Belgium, 2015.
3. German Federal Ministry for the Environment, Nature Conservation, Building and Nuclear Safety (BMUB). *Climate Action Plan 2050: Principles and Goals of the German Government's Climate Policy*; BMUB, Division KI I 1: Berlin, Germany, 2016.
4. Federal Ministry for Economic Affairs and Energy (BMWi). *The Energy of the Future: Second Progress Report on the Energy Transition. Reporting Year 2017*; BMWi, Public Relations Division: Berlin, Germany, 2019.
5. Fraunhofer IWES; Fraunhofer IBP. *Heat Transition 2030*; Agora Energiewende: Berlin, Germany, 2017.
6. von Appen, J. Sizing and Operation of Residential Photovoltaic Systems in Combination with Battery Storage Systems and Heat Pumps: Multi-Actor Optimization Models and Case Studies. Ph.D. Thesis, University Kassel, Kassel, Germany, 2018.
7. von Appen, J.; Braun, M. Sizing and improved grid integration of residential PV systems with heat pumps and battery storage systems. *IEEE Trans. Energy Convers.* **2019**, *34*, 562–571. [CrossRef]
8. Scheidler, A.; Bolgaryn, R.; Ulffers, J.; Dasenbrock, J.; Horst, D.; Gauglitz, P.; Pape, C.; Becker, H. DER Integration Study for the German State of Hesse—Methodology and Key Results. In Proceedings of the 25th International Conference on Electricity Distribution, Madrid, Spain, 3–6 June 2019; CIRED: Madrid, Spain, 2019; pp. 1–5. [CrossRef]
9. Guelpa, E.; Bischi, A.; Verda, V.; Chertkov, M.; Lund, H. Towards future infrastructures for sustainable multi-energy systems: A review. *Energy* **2019**, *184*, 2–21. [CrossRef]
10. Mancarella, P. MES (multi-energy systems): An overview of concepts and evaluation models. *Energy* **2014**, *65*, 1–17. [CrossRef]
11. Prina, M.G.; Manzolini, G.; Moser, D.; Nastasi, B.; Sparber, W. Classification and challenges of bottom-up energy system models—A review. *Renew. Sust. Energ. Rev.* **2020**, *129*, 109917. [CrossRef]
12. D'Agostino, D.; Mazzarella, L. What is a Nearly zero energy building? Overview, implementation and comparison of definitions. *J. Build. Eng.* **2019**, *21*, 200–212. [CrossRef]
13. Ifeu, Fraunhofer IEE, Consentec. *Building Sector Efficiency: A Crucial Component of the Energy Transition*; Agora Energiewende: Berlin, Germany, 2018.
14. Webb, A.L. Energy retrofits in historic and traditional buildings: A review of problems and methods. *Renew. Sust. Energ. Rev.* **2017**, *77*, 748–759. [CrossRef]
15. de Santoli, L.; Lo Basso, G.; Nastasi, B. Innovative Hybrid CHP systems for high temperature heating plant in existing buildings. *Energy Procedia* **2017**, *133*, 207–218. [CrossRef]
16. Kneiske, T.M.; Braun, M.; Hidalgo-Rodriguez, D.I. A new combined control algorithm for PV-CHP hybrid systems. *Appl. Energy* **2018**, *210*, 964–973. [CrossRef]
17. Kneiske, T.M.; Niedermeyer, F.; Boelling, C. Testing a model predictive control algorithm for a PV-CHP hybrid system on a laboratory test-bench. *Appl. Energy* **2019**, *242*, 121–137. [CrossRef]
18. Kneiske, T.M.; Braun, M. Flexibility potentials of a combined use of heat storages and batteries in PV-CHP hybrid systems. *Energy Procedia* **2017**, *135*, 482–495. [CrossRef]
19. Lee, Z.; Lee, K.; Choi, S.; Park, S. Combustion and Emission Characteristics of an LNG Engine for Heat Pumps. *Energies* **2015**, *8*, 13864–13878. [CrossRef]
20. Hepbasli, A.; Erbay, Z.; Icier, F.; Colak, N.; Hancioglu, E. A review of gas engine driven heat pumps (GEHPs) for residential and industrial applications. *Renew. Sust. Energ. Rev.* **2009**, *13*, 85–99. [CrossRef]
21. Elgendy, E. Analysis of Energy Efficiency of Gas Driven Heat Pumps. Ph.D. Thesis, Otto-von-Guericke-Universität Magdeburg, Magdeburg, Germany, 2011.
22. Elgendy, E.; Schmidt, J. Optimum utilization of recovered heat of a gas engine heat pump used for water heating at low air temperature. *Energy Build.* **2014**, *80*, 375–383. [CrossRef]
23. Protopapadaki, C.; Saelens, D. Heat pump and PV impact on residential low-voltage distribution grids as a function of building and district properties. *Appl. Energy* **2017**, *192*, 268–281. [CrossRef]

24. Protopapadaki, C.; Saelens, D. Corrigendum to "Heat pump and PV impact on residential low-voltage distribution grids as a function of building and district properties" [Appl. Energy 192 (2017) 268–281]. *Appl. Energy* **2017**, *205*, 1605–1608. [CrossRef]

25. Sichilalu, S.; Xia, X.; Zhang, J. Optimal Scheduling Strategy for a Grid-connected Photovoltaic System for Heat Pump Water Heaters. *Energy Procedia* **2014**, *61*, 1511–1514. [CrossRef]

26. Carvalho, A.D.; Moura, P.; Vaz, G.C.; de Almeida, A.T. Ground source heat pumps as high efficient solutions for building space conditioning and for integration in smart grids. *Energy Convers. Manag.* **2015**, *103*, 991–1007. [CrossRef]

27. Roselli, C.; Diglio, G.; Sasso, M.; Tariello, F. A novel energy index to assess the impact of a solar PV-based ground source heat pump on the power grid. *Renew. Energy* **2019**, *143*, 488–500. [CrossRef]

28. Fischer, D.; Madani, H. On heat pumps in smart grids: A review. *Renew. Sust. Energ. Rev.* **2017**, *70*, 342–357. [CrossRef]

29. Longfei, M.A.; Long, G.; Li, X.; Chen, Y.; Gong, C.; Wang, W.; Xu, H. Research on influence of large-scale air-source heat pump start-up characteristics to power grid. In Proceedings of the 2017 IEEE Conference on Energy Internet and Energy System Integration (EI2), Beijing, China, 26–28 November 2017; pp. 1–4. [CrossRef]

30. Bernath, C.; Deac, G.; Sensfuß, F. Influence of heat pumps on renewable electricity integration: Germany in a European context. *Energy Strategy Rev.* **2019**, *26*, 100389. [CrossRef]

31. Klyapovskiy, S.; You, S.; Cai, H.; Bindner, H.W. Incorporate flexibility in distribution grid planning through a framework solution. *Int. J. Electr. Power Energy Syst.* **2019**, *111*, 66–78. [CrossRef]

32. Thurner, L.; Scheidler, A.; Schäfer, F.; Menke, J.H.; Dollichon, J.; Meier, F.; Meinecke, S.; Braun, M. Pandapower—An Open Source Python Tool for Convenient Modeling, Analysis and Optimization of Electric Power Systems. *IEEE Trans. Power Syst.* **2018**, *33*, 6510–6521. [CrossRef]

33. Scheidler, A.; Thurner, L.; Braun, M. Heuristic optimisation for automated distribution system planning in network integration studies. *IET Renew. Power Gener.* **2018**, *12*, 530–538. [CrossRef]

34. Schaefer, F.; Menke, J.H.; Braun, M. Comparison of Meta-Heuristics for the Planning of Meshed Power Systems. *arXiv* **2020**, arXiv:2002.03619.

35. Sieberichs, M.; Ashrafuzzaman, R.; Moser, A. Implications of optimization strategies on expansion planning in medium- and low-voltage networks. In Proceedings of the 2017 6th International Conference on Clean Electrical Power (ICCEP), Santa Margherita Ligure, Italy, 27–29 June 2017; pp. 236–241. [CrossRef]

36. Bolgaryn, R.; Scheidler, A.; Braun, M. Combined Planning of Medium and Low Voltage Grids. In Proceedings of the 2019 IEEE Milan PowerTech, Milan, Italy, 23–27 June 2019; pp. 1–6. [CrossRef]

37. Büchner, D.; Thurner, L.; Kneiske, T.M.; Braun, M. Automated Network Planning including an Asset Management Strategy: Conference Center, Bonn. In Proceedings of the International ETG Congress 2017, Bonn, Germany, 28–29 November 2017; VDE Verlag: Berlin, Germany, 2017; Volume 155.

38. Yan, J.; Zhou, K.; Deng, C.; Huang, J. A GIS Based Service-Oriented Power Grid Intelligent Planning System. In Proceedings of the 2011 Asia-Pacific Power and Energy Engineering Conference, Wuhan, China, 25–28 March 2011; pp. 1–4. [CrossRef]

39. Mueller, F.; Zimmerlin, M.; de Jongh, S.; Suriyah, M.R.; Leibfried, T. Comparison of multi-timestep Optimization Methods for Gas Distribution Grids. In Proceedings of the 2019 54th International Universities Power Engineering Conference (UPEC), Bucharest, Romania, 3–6 September 2019; pp. 1–6. [CrossRef]

40. Then, D.; Spalthoff, C.; Bauer, J.; Kneiske, T.M.; Braun, M. Impact of Natural Gas Distribution Network Structure and Operator Strategies on Grid Economy in Face of Decreasing Demand. *Energies* **2020**, *13*, 664. [CrossRef]

41. Then, D.; Hein, P.; Kneiske, T.M.; Braun, M. Analysis of Dependencies between Gas and Electricity Distribution Grid Planning and Building Energy Retrofit Decisions. *Sustainability* **2020**, *12*, 5315. [CrossRef]

42. Khani, H.; El-Taweel, N.A.; Farag, H.E.Z. Power Loss Alleviation in Integrated Power and Natural Gas Distribution Grids. *IEEE Trans. Ind. Informat.* **2019**, *15*, 6220–6230. [CrossRef]

43. Shao, C.; Shahidehpour, M.; Wang, X.; Wang, X.; Wang, B. Integrated Planning of Electricity and Natural Gas Transportation Systems for Enhancing the Power Grid Resilience. *IEEE Trans. Power Syst.* **2017**, *32*, 4418–4429. [CrossRef]

44. Drauz, S.R.; Spalthoff, C.; Würtenberg, M.; Kneikse, T.M.; Braun, M. A modular approach for co-simulations of integrated multi-energy systems: Coupling multi-energy grids in existing environments of grid planning operation tools. In Proceedings of the 2018 Workshop on Modeling and Simulation of Cyber-Physical Energy Systems (MSCPES), Porto, Portugal, 10 April 2018; pp. 1–6. [CrossRef]

45. Pambour, K.; Cakir Erdener, B.; Bolado-Lavin, R.; Dijkema, G. Development of a Simulation Framework for Analyzing Security of Supply in Integrated Gas and Electric Power Systems. *Appl. Sci.* **2017**, *7*, 47. [CrossRef]

46. Farrokhifar, M.; Nie, Y.; Pozo, D. Energy systems planning: A survey on models for integrated power and natural gas networks coordination. *Appl. Energy* **2020**, *262*, 114567. [CrossRef]

47. Barati, F.; Seifi, H.; Nateghi, A.; Sepasian, M.S.; Shafie-khah, M.; Catalão, J.P.S. An integrated generation, transmission and Natural Gas Grid Expansion Planning approach for large scale systems. In Proceedings of the 2015 IEEE Power Energy Society General Meeting, Denver, CO, USA, 26–30 July 2015; pp. 1–5. [CrossRef]

48. Talebi, A.; Sadeghi-Yazdankhah, A.; Mirzaei, M.A.; Mohammadi-Ivatloo, B. Co-optimization of Electricity and Natural Gas Networks Considering AC Constraints and Natural Gas Storage. In Proceedings of the 2018 Smart Grid Conference (SGC), Sanandaj, Iran, 28–29 November 2018; pp. 1–6. [CrossRef]

49. Jooshaki, M.; Abbaspour, A.; Fotuhi-Firuzabad, M.; Moeini-Aghtaie, M.; Lehtonen, M. Multistage Expansion Co-Planning of Integrated Natural Gas and Electricity Distribution Systems. *Energies* **2019**, *12*, 1020. [CrossRef]

50. Alhamwi, A.; Medjroubi, W.; Vogt, T.; Agert, C. FlexiGIS: an open source GIS-based platform for the optimisation of flexibility options in urban energy systems. *Energy Procedia* **2018**, *152*, 941–946. [CrossRef]

51. Alhamwi, A.; Medjroubi, W.; Vogt, T.; Agert, C. GIS-based urban energy systems models and tools: Introducing a model for the optimisation of flexibilisation technologies in urban areas. *Appl. Energy* **2017**, *191*, 1–9. [CrossRef]

52. Alhamwi, A.; Medjroubi, W.; Vogt, T.; Agert, C. Development of a GIS-based platform for the allocation and optimisation of distributed storage in urban energy systems. *Appl. Energy* **2019**, *251*, 113360. [CrossRef]

53. OpenStreetMap: Database Contents License (DbCL) 1.0. Available online: http://www.openstreetmap.org/ (accessed on 2 January 2019).

54. Weiß, N.; Krecher, M. *Energiepotenzialstudie Gemeinde Schutterwald*; Badenova AG & Co. KG: Freiburg, Germany, 2014.

55. Walberg, D.; Holz, A.; Gniechwitz, T.; Schulze, T. *Wohnungsbau in Deutschland-2011-Modernisierung oder Bestandsersatz: Studie zum Zustand und der Zukunftsfähigkeit des Deutschen Kleinen Wohnungsbaus: Band I Textband*; Arbeitsgemeinschaft für Zeitgemäßes Bauen: Kiel, Germany, 2011.

56. Verwaltungsgemeinschaft Offenburg. *Flächennutzungsplan Juli 2009: Blatt 2/West*; Voegele + Gerhardt Freie Stadtplaner und Architekten DWB SRL BDA: Offenburg, Germany; Karlsruhe, Germany, 2009.

57. planSERVICE: Leitungsauskunft Online. Available online: https://planservice.regiodata-service.de/ (accessed on 17 May 2019).

58. Institut Wohnen und Umwelt. TABULA—Entwicklung von Gebäudetypologien zur Energetischen Bewertung des Wohngebäudebestands in 13 Europäischen Ländern: Supplementary Data Tables Appendix C. Available online: https://www.iwu.de/fileadmin/user_upload/dateien/energie/werkzeuge/ TABULA-Analyses_DE-Typology_DataTables.zip (accessed on 15 June 2019).

59. Loga, T.; Stein, B.; Diefenbach, N.; Born, R. *Deutsche Wohngebäudetypologie: Beispielhafte Maßnahmen zur Verbesserung der Energieeffizienz von Typischen Wohngebäuden*, 2nd ed.; Institut Wohnen und Umwelt (IWU): Darmstadt, Germany, 2015.

60. Institut Wohnen und Umwelt (IWU). DE Germany—Country Page: Residential Building Typology. Available online: https://episcope.eu/building-typology/country/de/ (accessed on 17 February 2020).

61. Statistisches Landesamt Baden-Württemberg. *Bevölkerung und Haushalte: Gemeinde Schutterwald am 9. Mai 2011: Ergebnisse des Zensus 2011*; Statistisches Landesamt Baden-Württemberg: Stuttgart, Germany, 2014.

62. von Appen, J.; Haack, J.; Braun, M. Erzeugung zeitlich hochaufgelöster Stromlastprofile für verschiedene Haushaltstypen. In Proceedings of the IEEE Power and Energy Student Summit, Stuttgart, Germany, 22–24 January 2014.

63. Drauz, S.R. Synthesis of a Heat and Electrical Load Profile for Single and Multifamily Houses Used for Subsequent Performance Tests of a Multi-Component Energy System. Master's Thesis, RWTH Aachen, Aachen, Germany, 2016.

64. Kallert, A.; Egelkamp, R.; Schmidt, D. High Resolution Heating Load Profiles for Simulation and Analysis of Small Scale Energy Systems. *Energy Procedia* **2018**, *149*, 122–131. [CrossRef]

65. Liu, F.G.; Tian, Z.Y.; Dong, F.J.; Yan, C.; Zhang, R.; Yan, A.B. Experimental study on the performance of a gas engine heat pump for heating and domestic hot water. *Energy Build.* **2017**, *152*, 273–278. [CrossRef]

66. Staffell, I.; Brett, D.; Brandon, N.; Hawkes, A. A review of domestic heat pumps. *Energy Environ. Sci.* **2012**, *5*, 9291–9306. [CrossRef]

67. DWD Climate Data Center. Historische stündliche Stationsmessungen der Lufttemperatur und Luftfeuchte für Deutschland: Stundenwerte_TU_01602_20040701_: Version v006. Available online: ftp://ftp-cdc.dwd.de/pub/CDC/observations_germany/climate/hourly/air_temperature/historical/stundenwerte_TU_01602_20040701_20171231_hist.zip (accessed on 24 September 2018).

68. Petoukhov, V.; Semenov, V.A. A link between reduced Barents-Kara sea ice and cold winter extremes over northern continents. *J. Geophys. Res.* **2010**, *115*. [CrossRef]

69. Streblow, R.; Ansorge, K. *Genetischer Algorithmus zur Kombinatorischen Optimierung von Gebäudehülle und Anlagentechnik: Optimale Sanierungspakete für Ein- und Zweifamilienhäuser*; Arbeitspapier 7; Gebäude-Energiewende: Berlin, Germany, 2017.

70. Institut für Energie- und Umweltforschung Heidelberg GmbH; Fraunhofer Institut für Energiewirtschaft und Energiesystemtechnik IEE; Consentec GmbH. *Wert der Effizienz im Gebäudesektor in Zeiten der Sektorenkopplung*; Agora Energiewende: Berlin, Germany, 2018.

71. Hinz, E. *Kosten energierelevanter Bau- und Anlagenteile bei der Energetischen Modernisierung von Altbauten*, 1st ed.; Institut Wohnen und Umwelt (IWU): Darmstadt, Germany, 2015. [CrossRef]

72. Henning, H.M.; Palzer, A. *Was kostet die Energiewende?—Wege zur Transformation des Deutschen Energiesystems bis 2050*; Fraunhofer Institut für Solare Energiesysteme (ISE): Freiburg, Germany, 2015.

73. Icha, P. *Entwicklung der Spezifischen Kohlendioxid-Emissionen des Deutschen Strommix in den Jahren 1990–2017*; Number 11/2018 in Climate Change; Umweltbundesamt (UBA): Dessau-Roßlau, Germany, 2018; p. 9.

74. Umweltbundesamt. Submission under the United Nations Framework Convention on Climate Change and the Kyoto Protocol 2019: National Inventory Report for the German Greenhouse Gas Inventory 1990–2017. Available online: https://www.umweltbundesamt.de/sites/default/files/medien/1410/publikationen/2019-05-28_cc_24-2019_nir-2019_en_0.pdf (accessed on 23 July 2019).

75. Gemeindewerke Schutterwald. GWS-Wärmestrom (Grundversorgung) Wärmepumpenanlage mit mit 3×2 Stunden Sperrzeit/Tag bei getrennter Messung. Available online: https://www.gemeindewerke-schutterwald.de/fileadmin/Dateien/Dateien/Tarifblaetter_2019/2019-Waermestrom-GV-WP_3x2.pdf (accessed on 4 December 2018).

76. Badenova AG & Co. KG. Tarife & Preise Badenova Erdgas PUR/BIO 10/BIO 100. Available online: https://www.badenova.de/mediapool/pdb/media/dokumente/produkte_1/erdgas_3/erdgas_pur/Tarife_und_Preise_badenova_Erdgas_PUR_BIO_10_BIO_100_ab_01012017.pdf (accessed on 9 May 2019).

77. Cao, K.K.; Pregger, T.; Scholz, Y.; Gils, H.C.; Nienhaus, K.; Deissenroth, M.; Schimeczek, C.; Krämer, N.; Schober, B.; Lens, H.; et al. *Analyse von Strukturoptionen zur Integration Erneuerbarer Energien in Deutschland und Europa unter Berücksichtigung der Versorgungssicherheit (INTEEVER): Schlussbericht: BMWi–FKZ 03ET4020 A-C*; Deutsches Zentrum für Luft- und Raumfahrt: Stuttgart, Germany; Universität Stuttgart, Institut für Feuerungs- und Kraftwerkstechnik: Stuttgart, Germany; Fraunhofer Institut für Energiewirtschaft und Energiesystemtechnik IEE: Kassel, Germany, 2019.

78. Jurich, K. *CO2-Emissionsfaktoren für Fossile Brennstoffe*; Umweltbundesamt (UBA): Dessau-Roßlau, Germany, 2016.

79. Meinecke, S.; Thurner, L.; Braun, M. Review and Classification of Published Electric Steady-State Power Distribution System Models: Under review. *arXiv* **2020**, arXiv:2005.06167.

80. Pandapower. Available online: https://www.pandapower.org/ (accessed on 4 November 2019).

81. Macho, W. QGIS QChainage Plugin. Available online: https://github.com/mach0/qchainage (accessed on 23 October 2019).

82. Tveite, H. QGIS NNJoin Plugin. Available online: https://github.com/havatv/qgisnnjoinplugin (accessed on 23 October 2019).

83. Nagel, H. *Systematische Netzplanung*, 2nd ed.; VDE-Verlag and VWEW Energieverlag: Berlin, Germany; Frankfurt am Main, Germany, 2008.

84. Wörthmüller, S.; Fischer-Uhrig, F. *STANET: Netzberechnung*, Version 10.0.37 64-Bit; Ingenieurbüro Fischer-Uhrig: Berlin, Germany, 2019.

85. Cronbach, D.; Lohmeier, D.; Drauz, S.R. Pandapipes. Available online: https://github.com/e2nIEE/pandapipes (accessed on 27 February 2020).

86. Statistisches Bundesamt (Destatis). *Bauen und Wohnen—Baugenehmigungen und Baufertigstellungen von Wohn- und Nichtwohngebäuden (Neubau) nach Art der Beheizung und Art der verwendeten Heizenergie: 1980–2017*; Lange Reihen, Statistisches Bundesamt (Destatis): Wiesbaden, Germany, 2018.

87. Energynautics GmbH; Öko-Institut e.V.; Bird & Bird LLP. *Verteilnetzstudie Rheinland-Pfalz*; Energynautics GmbH, Öko-Institut e.V., Bird & Bird LLP: Darmstadt, Germany, 2014.

88. Schwechater Kabelwerke GmbH. Preisliste 01.06.2017. Available online: https://www.skw.at/upload/Downloads/SKW_Brutto_Preisliste_gueltig_ab__01_06_2017.xls (accessed on 14 March 2018).

89. bnNETZE. Ergänzende Bedingungen der bn Netze GmbH zur Niederdruckanschlussverordnung (NDAV) Gültig ab 1. January 2018. Available online: https://bnnetze.de/web/Downloads/Kunden/Netzkunden/Netzanschluss/Erdgas/Preise-Netzanschluss-Erdgas/Ergänzende-Bedingungen-zur-NDAV-bnNETZE.pdf (accessed on 16 October 2018).

90. Cerbe, G. *Grundlagen der Gastechnik: Gasbeschaffung, Gasverteilung, Gasverwendung*, 6th ed.; Hanser: Munich, Germany, 2004.

91. Traber, T.; Fell, H.J. *Natural Gas Makes No Contribution to Climate Protection*; Energy Watch Group: Berlin, Germany, 2019.

92. Bundesnetzagentur für Elektrizität, Gas, Telekommunikation, Post und Eisenbahnen. Festlegung von Eigenkapitalzinssätzen Nach § 7 Abs. 6 StromNEV: BK4-16-160. Available online: https://www.bundesnetzagentur.de/DE/Service-Funktionen/Beschlusskammern/1_GZ/BK4-GZ/2016/2016_0001bis0999/2016_0100bis0199/BK4-16-0160/BK4-16-0160_Beschluss_Strom_BF_download.pdf?__blob=publicationFile&v=1 (accessed on 17 May 2019).

93. Deutsche Bundesbank. Umlaufsrenditen nach Wertpapierarten (Monats- und Tageswerte): Umlaufsrenditen inländischer Inhaberschuldverschreibungen/Insgesamt/Monatsdurchschnitte: BBK01.WU0017. Available online: https://www.bundesbank.de/dynamic/action/de/statistiken/zeitreihen-datenbanken/zeitreihen-datenbank/759778/759778?listId=www_skms_it01 (accessed on 17 May 2019).

94. Gemeinde Schutterwald. Haushaltsplan 2019. Available online: https://www.schutterwald.de/fileadmin/Dateien/Dateien/Rathaus___Service/Haushaltssatzung_und_HH-Plan_2019.pdf (accessed on 17 May 2019).

 *energies*

Article

Coordinated Flexibility Scheduling for Urban Integrated Heat and Power Systems by Considering the Temperature Dynamics of Heating Network

Wei Wei [1], Yaping Shi [1], Kai Hou [1,*], Lei Guo [2], Linyu Wang [2], Hongjie Jia [1], Jianzhong Wu [3] and Chong Tong [4]

[1] Key Laboratory of Smart Grid of Ministry of Education, Tianjin University, Tianjin 300072, China; weiw@tju.edu.cn (W.W.); shiyaping@tju.edu.cn (Y.S.); hjjia@tju.edu.cn (H.J.)
[2] State Grid (Suzhou) City & Energy Research Institute, State Grid Energy Research Institute Co., Ltd., Beijing 102209, China; guolei@sgeri.sgcc.com.cn (L.G.); wanglinyu@sgeri.sgcc.com.cn (L.W.)
[3] School of Engineering, Cardiff University, Cardiff CF24 3AA, UK; wuj5@cardiff.ac.uk
[4] State Grid Corp China, Jiangsu Elect Power Co., Ltd., Suzhou 215007, China; chong.tong@smartpdg.com
* Correspondence: khou@tju.edu.cn; Tel.: +86-158-2297-5162

Received: 23 May 2020; Accepted: 22 June 2020; Published: 24 June 2020

Abstract: The coordinated heat-electricity dispatch of the urban integrated energy system (UIES) helps to improve the system flexibility, thereby overcoming the adverse effects caused by the random fluctuations of renewable energy (RE) and promoting the penetration of RE. Among them, the dynamic characteristics of the urban heat network (UHN) are important features that need to be considered for the operating scheduling of the UIES. This paper aims to establish a flexibility scheduling model for UIES based on the dynamic characteristics of the UHN. First, the typical structure and key equipment model of the urban integrated heat and power system (UIHPS) with the dynamic characteristics of the UHN is proposed. Then, the definition and model of the UIHPS flexibility and the assessment index of the flexibility are developed. Moreover, a flexibility scheduling model for a UIHPS that considers the dynamic characteristics of a UHN is established. Finally, the validity of the proposed model is validated by case studies, and the applicability of flexibility scheduling and the effect of heat load (HL) are analyzed.

Keywords: urban integrated heat and power system; random fluctuations of renewable energy; flexibility scheduling; temperature dynamics of the urban heat network

---

1. Introduction

With problems such as energy shortages and environmental protection becoming increasingly prominent during the development of society, it is an inevitable decision to greatly develop renewable energy (RE). Electricity is the main use of RE. However, as the proportion of RE integration into the power system increases, its variability and uncertainty have brought new challenges to the security of power system operations [1]. Considerable attention has been paid toward research on power system flexibility resulting from the variability of RE generation [2]. By increasing the system flexibility, the adverse effect on the power system operation brought by the high penetration of RE can be effectively coped with, and therefore, the utilization of RE will be improved. Unfortunately, it is difficult to meet the growing demand for integrating RE by only deploying resources in electric power systems. An integrated energy system (IES) can effectively enhance the system flexibility by utilizing the complementary and synergistic relationships between various energy vectors, such as electricity, heat and gas, thus promoting a scaled development of RE [3,4]. In some cities in northern Europe and northern China, the urban heating network (UHN) and electric network are jointly constructed as a city-level integrated electricity-heat system through combined heat and power (CHP) units [5,6]. The system flexibility will be greatly improved by the coordinated scheduling of urban integrated heat and power systems (UIHPSs), therefore further facilitating the integration of RE.

Previous research on flexibility is mainly restricted to the power system itself, including the definition and evaluation of power system flexibility [7,8], as well as optimal dispatching [9,10]. Reference [7] proposes the insufficient ramping resource expectation (IRRE) as a metric to assess the power system flexibility over different time horizons and in different directions. Reference [8] proposed an index that could evaluate the maximum uncertainty a system could accept. In terms of flexibility scheduling, [9] described the operational flexibility of electric power systems from three dimensions: ramp rate, power and energy and the developed flexibility dispatch based on this. Regarding the European power market, the optimal procurement of flexible ramping products was studied in [10], and a deterministic, flexible power system dispatching model was proposed.

With the introduction of IES in recent years, there have been a number of research advancements that have improved the absorption of RE (e.g., wind and solar) by utilizing multi-energy coupling devices [11–17]. In [11], distributed electric heat pumps (HPs) were introduced into a wind-thermal power system as a heat source (HS) and spinning reserve to increase wind power utilization. A model for determining the operational flexibility of CHP with thermal energy storage was established in [12], and it was found that both a more powerful CHP and a larger buffer could increase flexibility. Reference [13] improved the flexibility of CHP units using electric boilers and heat storage tanks to better integrate wind power. The potential of HPs applied for demand side management and wind power integration in the German electricity market was studied in [14]. Reference [15] investigates the trends of district heating technologies in Europe, and indicates that the district heating development requires more flexible energy systems with building automation, RE and prosumers' participation, to improve the RE utilization and energy efficiency. Greater RE penetration was achieved by integrating heat pumps into district heating in [16,17].

Although existing research has made some achievements in improving power system flexibility through the integrated operation of electricity and heat, little attention has been paid to the influence of the dynamic characteristics of heating networks, such as transmission delay and temperature dynamics. For UHNs, the transmission time of the heat medium from the HS to users can be up to several hours. Hence, the system flexibility can be greatly increased by rationally using the transmission delay characteristic of UHNs. On the other hand, the supply and return temperatures of each node in a UHN can vary within a certain range [18]. By reasonably using this characteristic, the operational flexibility of the UHN will be further enhanced. Therefore, for urban integrated energy systems (UIESs), the utilization of the above dynamic characteristics of UHNs through coupling components (e.g., CHPs) can better cope with fluctuations of RE (e.g., wind power) and significantly improve the system flexibility. This paper aims to establish a flexibility scheduling model for a UIES based on the dynamic characteristics of a UHN. First, we build the typical structure of UIHPSs with the dynamic characteristics of a UHN, as well as the main equipment models in UIHPSs. Then, the flexibility definition and model for UIHPSs are developed, and the assessment metrics of flexibility are also provided. Subsequently, the flexibility scheduling model for a UIES that considers the dynamic characteristics of a UHN is established. Finally, the effectiveness of the proposed model is verified by case studies, and the applicability of flexibility scheduling and the influence of HL are analyzed.

2. UIHPS

UIES is a typical application form of IES [4,5], which refers to the unified design and operation of different energy vectors (e.g., electricity, heat, gas and hydrogen) within an urban area to achieve a safe, efficient and green supply for various energy demands. Figure 1 shows a typical structure of a UIHPS. At the level of UIES, the electric supply mainly comes from high-grade electric networks, local conventional thermal power plant (TPP) units, CHP units and RE generation equipment such as wind turbines, while the heat supply is largely from CHP units and coal-fired or gas-fired boilers.

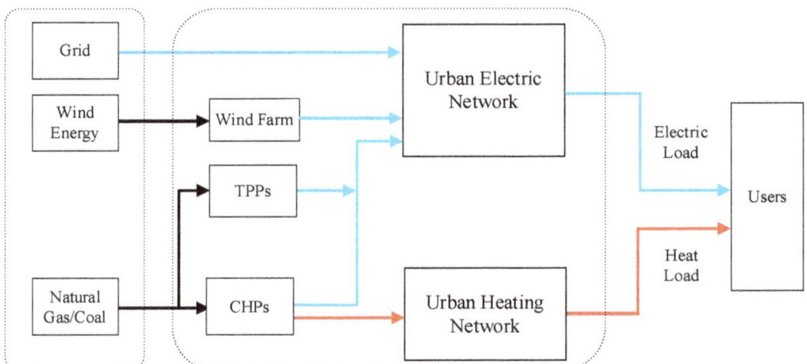

Figure 1. Schematic diagram of an urban integrated heat and power system (UIHPS).

In the traditional heat-electricity separate operation mode, CHP units, as the coupling link, separately execute the power instructions of the electric and heat dispatching departments and coordinate with them when conflicts arise. Under this mode, the regulated resources of urban thermal and power systems have not been fully utilized, and there is still room for improvement in increasing RE integration and energy efficiency.

2.1. Energy Supply Equipment

2.1.1. CHP Units

CHP units can simultaneously produce heat and electricity, with high energy conversion efficiency [19]. In particular, extraction condensing CHP units have more flexible thermoelectric operating characteristics and are widely used in UIESs. This study is based on the extraction condensing CHP unit (hereinafter referred to as the CHP unit).

The feasible operating region of a CHP unit is shown in Figure 2. When the thermal output of the CHP unit changes, the corresponding upper and lower electric power limits also change. Any operating point in this feasible region can be represented by a convex combination of corner points [20], as shown in Equations (1)–(3):

$$H_{g,t}^{chp} = \sum_{k=1}^{NK_g} \left(\alpha_{g,t}^k \cdot H_g^k \right) \tag{1}$$

$$P_{g,t}^{chp} = \sum_{k=1}^{NK_g} \left(\alpha_{g,t}^k \cdot P_g^k \right) \tag{2}$$

$$\sum_{k=1}^{NK_g} \alpha_{g,t}^k = 1,\ 0 \le \alpha_{g,t}^k \le 1 \tag{3}$$

The operating cost of a CHP unit can be expressed as follows:

$$C_{g,t}^{chp} = \sum_{k=1}^{NK_g} \left(\alpha_{g,t}^k \cdot C_g^k \right) \tag{4}$$

The relationship between the heat output of the CHPs connected to the HS j and the supply and return temperatures of the HS j is shown as Equation (5).

$$\sum_{g \in CHP_j} H_{g,t}^{chp} = c\dot{m}_j^S \left(T_{j,t}^{Ss} - T_{j,t}^{Sr} \right) \times 10^{-3} \quad \forall j \in N^{HS} \tag{5}$$

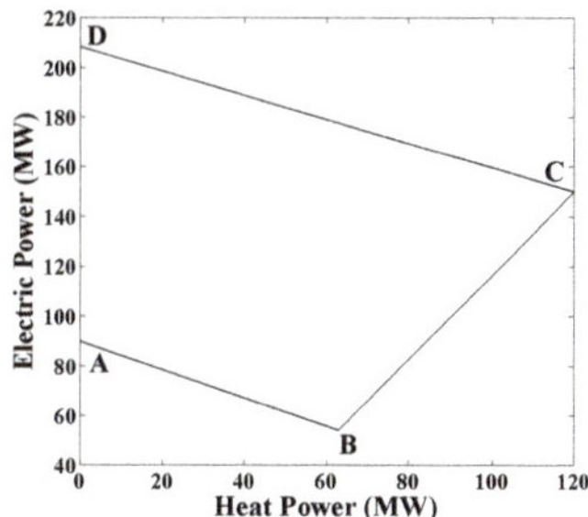

Figure 2. Feasible operating region of combined heat and power (CHP).

The supply and return temperatures at the HS should be within a certain range to guarantee the heating quality:

$$T^{Ss}_{j,\min} \leq T^{Ss}_{j,t} \leq T^{Ss}_{j,\max} \tag{6}$$

$$T^{Sr}_{j,\min} \leq T^{Sr}_{j,t} \leq T^{Sr}_{j,\max} \tag{7}$$

The ramping constraints of the CHP units are shown in Equation (8):

$$R^{chp}_{d,g} \cdot \Delta t \leq P^{chp}_{g,t} - P^{chp}_{g,t-1} \leq R^{chp}_{u,g} \cdot \Delta t \tag{8}$$

2.1.2. TPP Units

The operating cost of TPP units can be described as follows:

$$C^{tpp}_{g,t} = a_g \left(P^{tpp}_{g,t} \right)^2 + b_g P^{tpp}_{g,t} + c_g \tag{9}$$

The generated energy output constraints of TPP units are shown as follows:

$$P^{tpp}_{g,\min} \leq P^{tpp}_{g,t} \leq P^{tpp}_{g,\max} \tag{10}$$

The ramping constraints of TPP units are expressed as Equation (11).

$$R^{tpp}_{d,g} \cdot \Delta t \leq P^{tpp}_{g,t} - P^{tpp}_{g,t-1} \leq R^{tpp}_{u,g} \cdot \Delta t \tag{11}$$

2.2. UHN

The UHN is usually divided into the primary network and the secondary network, which are connected with each other through heat exchangers. The dynamic characteristics of the secondary heating network have no direct effect on the cooperative operation of the UIHPS. In this paper, the heat exchanger and the secondary heating network are equivalent to the HL of the primary heating network, and $T^{Ls}_{i,t}$ refers to the temperature of mass flowing from the HL. The structure of the UHN is shown in Figure 3.

The quality regulation mode (CF-VT) [21] is one of the frequently used control strategies for urban heating systems. Under this regulation mode, the mass flow of the UHN remains constant, while the thermal dispatching department optimizes the supply temperature of the CHP units to meet the HL in different periods [22]. This strategy decouples the control of the hydraulic condition and the thermal condition in the heating system and has achieved good results in industrial practice. In this paper, our study is based on the CF-VT strategy of a UHN.

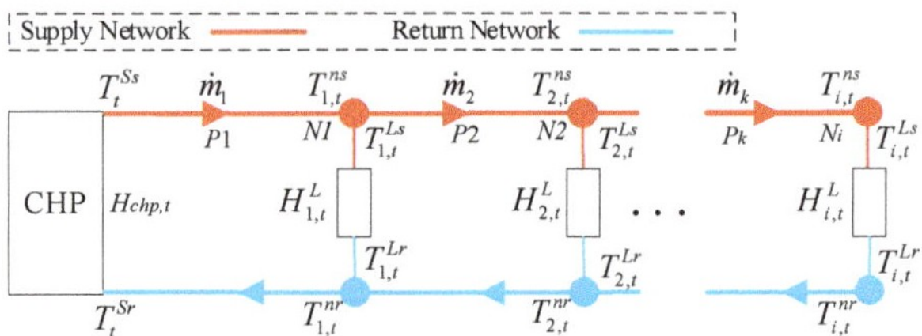

Figure 3. Schematic diagram of an urban heat network (UHN).

2.2.1. Dynamic Characteristics of the UHN

For a city-level thermoelectric IES, there is generally a delay varying from tens of minutes to several hours during the heat energy transmission due to the limitation of hot water velocity. Figure 4 shows the transmission delay of the UHN.

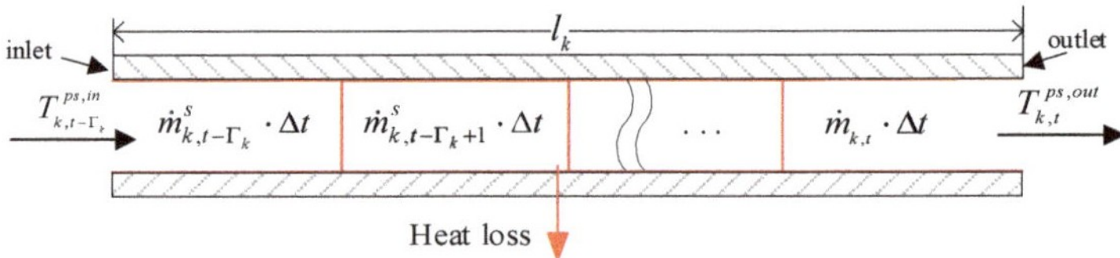

Figure 4. Schematic diagram of the transmission delay.

Considering that the return network has the same properties as the supply network and is symmetrically distributed, this paper takes the supply pipeline as an example to illustrate the dynamic characteristics of a UHN. The transmission delay from the input to the output of a pipe is proportional to the total length of the pipe and is inversely proportional to the hot water velocity. Under the CF-VT strategy, the transmission delay of each pipeline is constant (the "t" in the mass flow subscript is omitted below), shown as follows:

$$\Gamma_k^s = round\left[\frac{\pi(D_k/2)^2 \rho l_k}{\dot{m}_k^s \Delta t \times 3600}\right] \quad \forall k \in P^s \tag{12}$$

The function *round* [·] in Equation (12) represents rounding-off. Considering the heat loss of the hot water transported by the pipeline [23], the outlet temperature of a pipe at time t can be calculated as

$$T_{k,t}^{ps,out} = \psi_k\left(T_{k,t-\Gamma_k}^{ps,in} - T_a\right) + T_a \quad \forall k \in P^s, \psi_k = e^{-\frac{\lambda_k l_k}{cm_k}} \tag{13}$$

According to the first law of thermodynamics, the energy flowing into a node is identical to the energy flowing out. Thus, the mixed temperature at any node can be expressed as

$$\sum_{k \in P_i^u}\left(\dot{m}_k^s T_{k,t}^{ps,out}\right) = \left(\sum_{k \in P_i^u} \dot{m}_k^s\right) T_{i,t}^{ns} \quad \forall i \in N^s \tag{14}$$

The mass temperature flowing from a node to any downstream pipe is equal to the mixed temperature at the node:

$$T_{b,t}^{ps,in} = T_{i,t}^{ns} \quad \forall i \in N^s, b \in P_i^d \tag{15}$$

According to the model of the UHN described above, the temperature calculation formula for each node in the supply network, while considering the network topology, can be given as

$$T_{i,t}^{ns} = \sum_{j \in N^{Ss}} \sum_{v \in \{1,\dots,n_{j,i}\}} \left(F_{j,i}^{v} \Psi_{j,i}^{v} \left(T_{j,t-\Gamma_{j,i}^{v}}^{Ss} - T_a \right) \right) + T_a \quad \forall i \in N^s \tag{16}$$

where

$$F_{j,i}^{v} = \prod_{h \in N_{j,i}^{v}} \alpha_h, \quad \alpha_h = \frac{\dot{m}_g^s}{\sum_{k \in P_h^u} \dot{m}_k^s}, \quad g = P_{j,i}^{v} \cap P_h^u \tag{17}$$

$$\Psi_{j,i}^{v} = \prod_{k \in P_{j,i}^{v}} \psi_k \tag{18}$$

$$\Gamma_{j,i}^{v} = \sum_{k \in P_{j,i}^{v}} \Gamma_k \tag{19}$$

2.2.2. Equivalent Pipe Model of the HL

As the load of the urban primary heating network, the HL of the heat exchange station can be calculated as follows:

$$H_{i,t}^{L} = c \dot{m}_i^{L} \left(T_{i,t}^{Ls} - T_{i,t}^{Lr} \right) \times 10^{-3} \quad \forall i \in N^{HL} \tag{20}$$

The supply and return temperatures of HLs should be within a certain range to guarantee their normal operation. The temperature constraints of HLs are shown as follows:

$$T_{i,\min}^{Ls} \leq T_{i,t}^{Ls} \leq T_{i,\max}^{Ls} \tag{21}$$

$$T_{i,\min}^{Lr} \leq T_{i,t}^{Lr} \leq T_{i,\max}^{Lr} \tag{22}$$

To facilitate the establishment of the dynamic characteristic model of the UHN, the HL is regarded as the pipe connecting the supply network and the return network [24]. The mass flow rate of the equivalent pipe is equivalent to the mass flow rate of the HL, and the transmission delay is equal to 0. The heat loss of the equivalent pipe is equal to the HL, which is shown as follows:

$$T_{i,t}^{PL,out} = T_{i,t}^{PL,in} - \frac{H_{i,t}^{L}}{c \dot{m}_i^{L}} \times 10^3 \quad \forall i \in N^{HL} \tag{23}$$

According to the temperature loss calculation (Equation (23)) of the HL equivalent pipe and the node temperature in the supply network (Equation (16)), the outlet temperature of a HL equivalent pipe can be expressed as

$$T_{i,t}^{PL,out} = \sum_{j \in N^{Ss}} \sum_{v \in \{1,\dots,n_{j,i}\}} \left(F_{j,i}^{v} \Psi_{j,i}^{v} \left(T_{j,t-\Gamma_{j,i}^{v}}^{Ss} - T_a \right) \right) - \frac{H_{i,t}^{L}}{c \dot{m}_i^{L}} \times 10^3 + T_a \quad \forall i \in N^{HL} \tag{24}$$

Based on the dynamic characteristic equations of the UHN and the equivalent pipe model of the HL, the temperature at each node in the return network can be written as

$$T_{i',t}^{nr} = \sum_{i \in N^{Ls}} \sum_{w \in \{1,\dots,n_{i,i'}\}} \left(F_{i,i'}^{w} \Psi_{i,i'}^{w} \left(T_{i,t-\Gamma_{i,i'}^{w}}^{PL,out} - T_a \right) \right) + T_a \quad \forall i' \in N^r \tag{25}$$

2.3. Urban Electric Network

To improve the computational performance of the simulated optimal dispatching for the UIHPS, the linearization model of the urban electric network is used as follows [25]:

$$P_i = \sum_{j \in i} G_{ij} U_j - \sum_{j \in i} B'_{ij} \theta_j \tag{26}$$

$$Q_i = -\sum_{j \in i} B_{ij} U_j - \sum_{j \in i} G_{ij} \theta_j \tag{27}$$

where B'_{ij} is the imaginary part of the element in the i-th row and j-th column of the node admittance matrix without grounding branches.

The linear model of the branch power flow is shown as follows:

$$P_{ij} = g_{ij}(U_i - U_j) - b_{ij}(\theta_i - \theta_j) \tag{28}$$

The active and reactive power injected into an electric bus can be expressed as

$$P_{i,t} = P_{i,t}^{chp} + P_{i,t}^{tpp} + P_{i,t}^{wind} - P_{i,t}^{load} \tag{29}$$

$$Q_{i,t} = Q_{i,t}^{chp} + Q_{i,t}^{tpp} + Q_{i,t}^{wind} - Q_{i,t}^{load} \tag{30}$$

The constraints of the node voltage amplitude and node voltage phase angle can be described as

$$U_{\min} \leq U_i \leq U_{\max} \tag{31}$$

$$\theta_{\min} \leq \theta_i \leq \theta_{\max} \tag{32}$$

The constraints of the branch power flow can be described as follows:

$$P_{ij,\min} \leq P_{ij} \leq P_{ij,\max} \tag{33}$$

3. UIES Flexibility

Power system flexibility is defined by the International Energy Agency (IEA) as the ability of a power system to quickly respond to predictable and unpredictable changes and to cope with large fluctuations in both supply and demand (i.e., flexibility demand) while remaining within the system boundary constraints [1]. For the UIES with high renewable penetration, random fluctuations in RE (e.g., wind power) have a great impact on the safe operation of the power system, which is the focus of flexibility studies. The flexibility of a UIHPS is an extension of the power system flexibility, which focuses on how to use the inherent dynamic characteristics of the UHN to quickly deal with fluctuations of RE generation through the coordinated operation of the urban thermal and power networks. Without a loss of generality, this paper focuses on the research of wind generation fluctuations.

The flexibility of the UIES has the following features:

(1) The flexibility demand of the UIES is directional. The power system requires an instantaneous supply–demand balance. When wind generation increases or decreases unexpectedly, there is a downward and upward flexibility demand, respectively. In this case, the system is required to have corresponding downward and upward flexibility. When the actual wind power is greater than what is predicted, it will lead to wind curtailment if the system has insufficient downward adjustable resources; likewise, when the actual wind generation is less than forecasted, there will be load shedding due to insufficient upward available capacity;

(2) The flexibility of the UIES is related to the type of units. Various types of generating units are the main flexibility resources, and their flexibility is shown as the upward and downward adjustable

generation capacity, which is mainly limited by the electric output limits and the ramp rate. The upper and lower generation limits of TPP units are relatively fixed. In contrast, the electric output limits of CHP units are connected with the current heat output, and thus the upward and downward available generation capacity is related to both the electrical and thermal output of units;

(3) The flexibility of the UIES is related to the level of heat and electric load, and the flexibility in different directions should be considered for different periods. Specifically, the effect of power load on flexibility is more direct. In the electric load valley period, due to the low power demand, the power output of TPP and CHP units is closer to the low limit, and the whole system may be faced with insufficient downward flexibility in response to a sudden increase in wind power generation. Similarly, the challenge during peak power loads is the lack of upward flexibility under the condition of unpredicted decreases in wind energy generation. Thus, the flexibility of the UIES focuses on the downward flexibility during the electric load valley period and the upward flexibility during the electric load peak period;

(4) The flexibility of the UIES is affected by the dynamic characteristics of the UHN. Since the transmission delay of the hot water needs to be considered in the UHN, the heat output of CHP units does not need to maintain an instantaneous balance with the current HL. Moreover, the supply and return temperatures that directly determine the thermal output of CHPs can vary within a certain range, which will have a great impact on the flexibility of the UIES.

Therefore, the UIES flexibility can be further divided into downward flexibility during the electric load valley period and upward flexibility during the electric load peak period, which can be calculated as follows:

$$f_t^d = \sum_{g=1}^{N_g} \min\left(P_{g,t} - P_{g,min}, \Delta t \cdot R_{d,g}\right) \tag{34}$$

$$f_t^u = \sum_{g=1}^{N_g} \min\left(P_{g,max} - P_{g,t}, \Delta t \cdot R_{u,g}\right) \tag{35}$$

To assess the system flexibility for different periods, we propose the insufficient rate of flexibility as the metric. The insufficient flexibility rate in different dispatching periods is expressed as the downward flexibility deficiency rate Δf_t^d and the upward flexibility deficiency rate Δf_t^u:

$$\Delta f^d = \frac{\sum\limits_{t \in T_1}\left(\Delta P_t^u - f_t^d\right)}{\sum\limits_{t \in T_1} \Delta P_t^u} \times 100\% \tag{36}$$

$$\Delta f^u = \frac{\sum\limits_{t \in T_2}\left(\Delta P_t^d - f_t^u\right)}{\sum\limits_{t \in T_2} \Delta P_t^d} \times 100\% \tag{37}$$

The fluctuations of wind power in Equations (36) and (37) (i.e., ΔP_t^u and ΔP_t^d) can be obtained by comparing the actual wind power output with the forecasted data.

4. Flexibility Scheduling Model Based on the Temperature Dynamics of the UHN

According to the characteristics of UIES flexibility, this paper proposes a coordinated flexibility scheduling model for a UIHPS that considers the temperature characteristics of the UHN. The scheduling interval is 15 min, and the scheduling period is 24 h.

4.1. Objective Function

The objective of the coordinated flexibility dispatching of the UIHPS is to maximize the total flexibility during the electric load peak and valley periods, that is, to maximize the upward and downward flexibility for different periods, respectively.

In the electric load valley period, the objective is to maximize the system downward flexibility:

$$\max f_1 = \sum_{t \in T_1} f_t^d \tag{38}$$

In the electric load peak period, the objective is to maximize the system upward flexibility:

$$\max f_2 = \sum_{t \in T_2} f_t^u \tag{39}$$

In summary, the objective function of the proposed coordinated flexibility scheduling model is

$$\max f = f_1 + f_2 \tag{40}$$

The decision variables of the dispatch model are the power output of each scheduling interval of TPP units, the power output of each scheduling interval and the supply temperature of CHP units.

4.2. Constraints

(1) Constraints of CHP units

The constraints of CHP units are shown in Equations (1)–(8).

(2) Constraints of TPP units

The constraints of TPP units are shown in Equations (9)–(11).

(3) Constraints of the UHN

The constraints of the UHN include the node temperature calculations (Equations (16) and (25)), the HL outlet temperature calculation (Equation (24)) and the upper and lower limits of the supply and return temperatures at the HL (Equations (21) and (22)).

(4) Constraints of the electric network

The electric network constraints are expressed in Equations (26)–(33).

The proposed model is a large-scale linear programming model that can be solved by established mathematical software such as CPLEX and Gurobi. The model in this paper is implemented based on MATLAB R2013a for coding and calls Gurobi to obtain solutions.

5. Case Studies

First, this study uses a simple test to verify the effectiveness of the proposed model and analyzes various influencing factors. Then, a practical example is used to further illustrate the efficiency of the method.

5.1. Small-Scale System

This case uses the system of a six-bus electric network and six-node heat network, found in [26] and shown in Figure 5; additionally, the detailed data of transmission lines and heat pipes are given in [27]. The system includes two TPP units (TPP1, TPP2), a CHP unit and a wind farm (W) with an installed capacity of 70 MW. The parameters of the TPP units and the CHP units are given in

Appendix A. In this work, the wind abandonment penalty coefficient is USD 100/MWh [28], and the load-cutting cost is USD 1000/MWh [29].

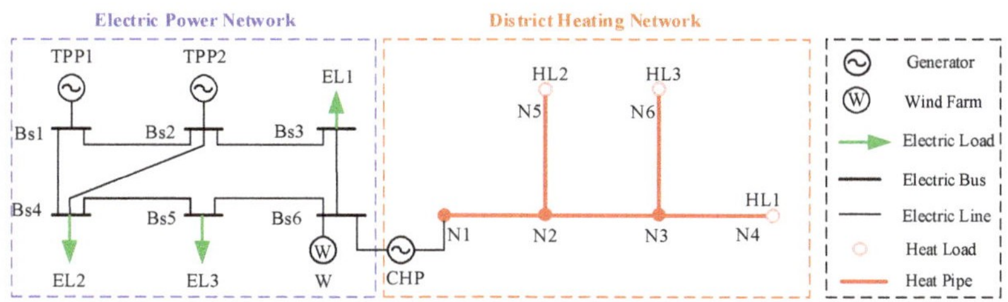

Figure 5. Modified six-bus power system with a six-node heat system.

The electric and heat loads of the system are shown in Figure 6a, and the load data can be found in [13]. The electric load valley period (T1) refers to 0:00–6:00, and the peak period (T2) is 10:00–20:00. The forecasted values of wind power are derived from the actual operating data, and the maximum value is 50 MW, which is approximately 20% of the peak electric load. To verify the effectiveness of the proposed method, a Weibull distribution is used to simulate the actual wind power based on the forecasted wind power. The predicted and actual wind power profiles are presented in Figure 6b.

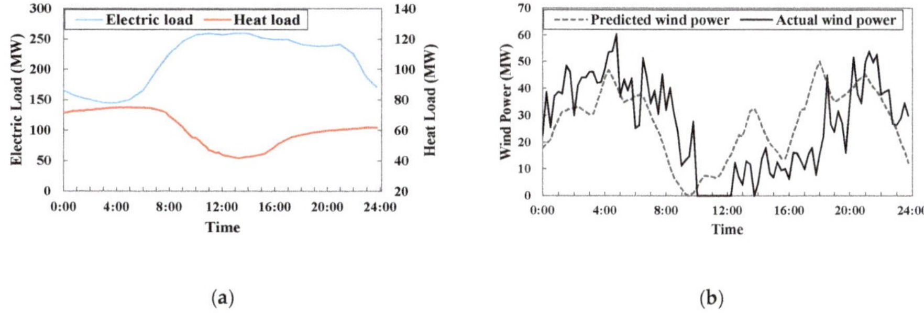

(a)   (b)

Figure 6. Power profiles: (**a**) electric and heat loads and (**b**) predicted and actual wind power.

5.1.1. Influence of the Transmission Delay on Flexibility Scheduling

To analyze the impact of the transmission delay in the UHN on flexibility scheduling, the following three scenarios are set for a comparative analysis:

Case 1: Regardless of the dynamic characteristics of the UHN, optimal scheduling is carried out to maximize system flexibility;

Case 2: Considering the dynamic characteristics of the UHN and keeping the supply temperature of the HS in Case 1 unchanged, the flexibility scheduling plan of Case 1 is analyzed;

Case 3: For the model proposed in this paper, the dynamic characteristics of the UHN are considered, and optimal scheduling is performed with the goal of maximizing system flexibility.

Figure 7 shows the supply temperature of the HS for Cases 1 and 2 and the corresponding system flexibility.

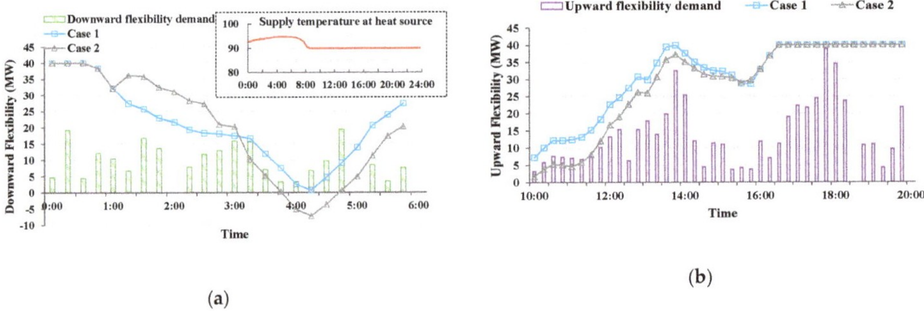

Figure 7. Comparison of system flexibility between Case 1 and Case 2: (**a**) downward flexibility in the electric load valley period and (**b**) upward flexibility in the electric load peak period.

According to Figure 7, for the scheduling results of Case 1 (without considering the transmission delay of the UHN), the system flexibility can meet the demand except for the 4:00–5:00 period, with an insufficient downward flexibility rate of 9.8% in the electric load valley period and an insufficient upward flexibility rate of 0% in the electric load peak period. However, in actual operation, due to the transmission delay of the UHN, there will be some deviation between the actual thermal output of the CHP units and the planned thermal output, and there may be a large change in the actual system flexibility, which may even affect the operational safety of the thermal system. For Case 2, the heating system is scheduled based on the optimized supply temperature in Case 1, and it can be observed from its flexibility curve that, in the electric load valley period, the system downward flexibility decreases at 3:00–6:00 and cannot meet the flexibility requirements; during the electric load peak period, the system upward flexibility is reduced and cannot meet the flexibility demand from 10:00 to 11:30. The actual insufficient downward flexibility rate during the electric load valley period is 29.3%, and the actual insufficient upward flexibility rate in the electric load peak period is 2.2%. Overall, these results show that ignoring the delay characteristics of the UHN will cause a significant difference between the actual system flexibility and planned system flexibility and accordingly goes against the penetration of wind power and the sufficient supply of electric load. In particular, the actual values of downward flexibility are negative at 4:00–4:45, indicating that the scheduling plan without considering the transmission delay of the UHN is infeasible.

Moreover, flexibility scheduling that does not consider the transmission delay may also destroy the operational security of the heating system. Figure 8 shows the supply and return node temperatures in Case 1 and Case 2.

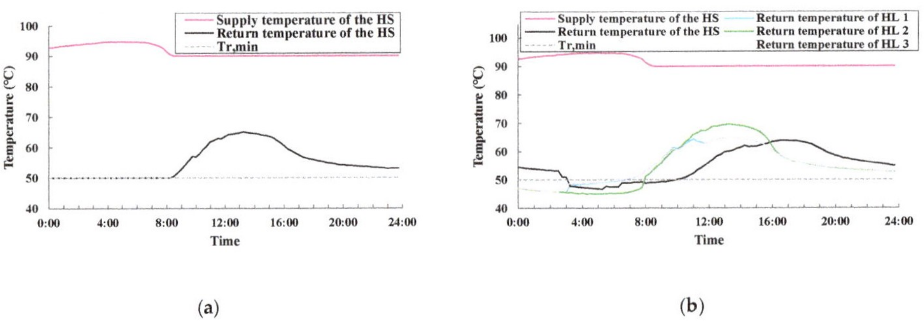

Figure 8. Node temperatures: (**a**) Case 1 and (**b**) Case 2.

According to Figure 8a, the dispatched return temperature of the HS is within the allowable range when the transmission delay of the UHN is not taken into account, and it should be noted that all the node return temperatures are the same in Case 1 because the dynamic characteristics of the UHN are not considered. However, in the actual operation, and due to the transmission delay of the UHN, the return temperatures of HL 1 and HL 3 exceed the lower limit of the return temperature at 0:00–6:30, the return temperature of HL 2 crosses the lower limit at 0:00–8:00 and the return temperature of the HS crosses the lower limit during 3:00–10:00, thus breaking the operational safety of the heat system, as shown in Figure 8b.

The thermal output of the CHP unit changes following the change in the HS return temperature. Figure 9 shows the scheduling results of the CHP unit output in Case 1 and Case 2 at 4:00 and 10:00. According to the supply temperature dispatched in Case 1, the actual thermal output of the CHP unit will exceed its feasible region, which further illustrates that a flexibility scheduling plan that does not consider the dynamic characteristics of the UHN may not be feasible in practice.

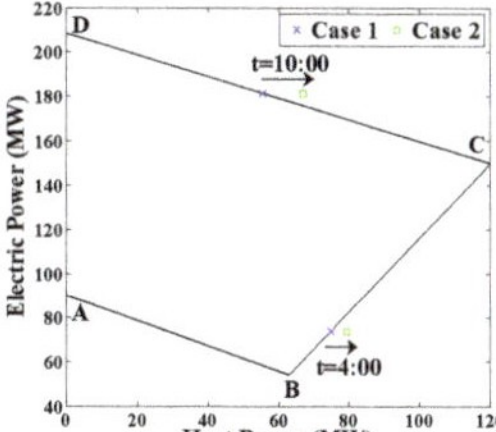

Figure 9. Comparison of the CHP unit heat output in Case 1 and Case 2 at typical times.

Next, we analyzed the results of the flexibility scheduling model proposed in this paper. Figure 10 shows the scheduling results of the proposed model (Case 3) and the scheduling results of Case 1.

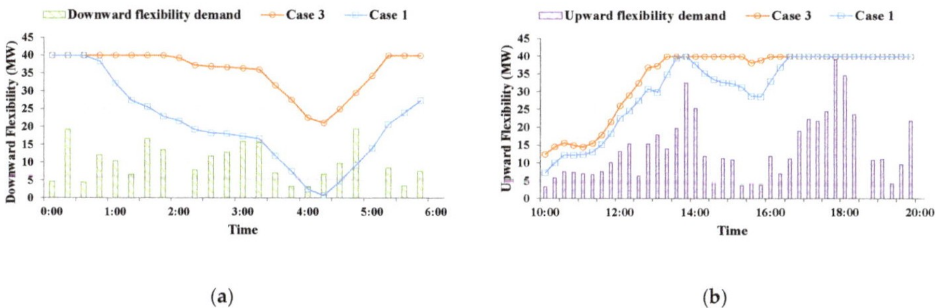

Figure 10. Comparison of system flexibility in Case 1 and Case 3: (**a**) downward flexibility in the electric load valley period and (**b**) upward flexibility in the electric load peak period.

Compared with Case 1, the model proposed in this paper reduces the insufficient rate of downward flexibility during the electric load valley period from 9.8% to 0%, and the upward flexibility during the electric load peak period is also improved. Among them, at 0:00–0:45 and 4:30–10:00, the system flexibility of Case 1 and Case 3 is kept at 40 MW, which is the sum of the maximum upward or downward climbing rate of all units within 15 min.

The next section of the research is concerned with the reasons for the improvement in the flexibility of the proposed model, which is analyzed from the temperature dynamic characteristics of the UHN and the power output changes of the CHP units. Figure 11 shows the supply and return temperatures of the HS and operational status of the CHP units.

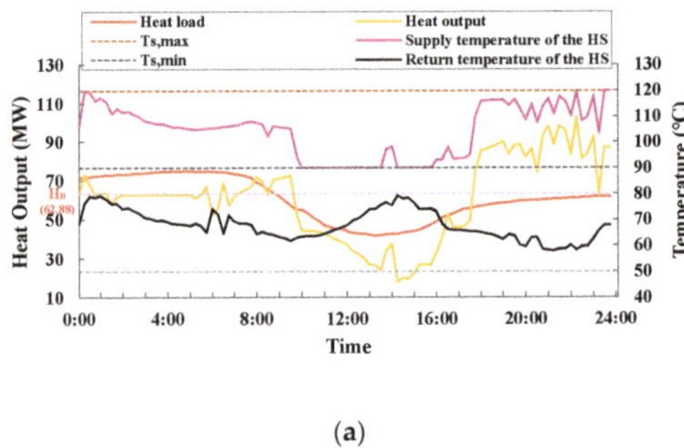
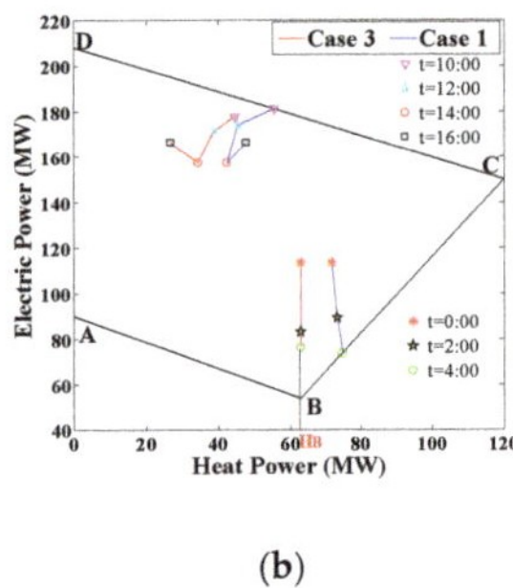

(a)

(b)

Figure 11. (**a**) Dynamic temperature characteristics of the UHN and (**b**) operational status of the CHP units.

In Figure 11a, Ts,max and Ts,min are the upper and lower limits of the HS supply temperature, respectively, and H_B (62.88 MW) is the thermal power value corresponding to point B in the feasible region of the CHP unit. After considering the dynamic characteristics of the UHN, the thermal output of the CHP units is no longer required to be consistent with the HL at all times, whereas the thermal output of the CHP units is adjusted according to the flexibility requirements at different times, thereby improving system flexibility. According to Figure 11a, in the 0:00–4:00 electric load valley period, by adjusting the supply temperature of the HS, the thermal output of the CHP unit is basically kept near 62.88 MW (H_B), where the CHP has the largest downward adjustable electric capacity; during 10:00–16:00 of the peak period, the HS supply temperature is basically maintained at the minimum value to minimize the heat output of the CHP unit and increase the upward adjustable electric capacity.

The results in this section indicate that after considering the actual dynamic characteristics of the UHN, the thermal output of the CHP unit can be adjusted within a certain range by reasonably regulating the supply temperature of the HS. Thus, the proposed model provides a greater margin for power adjustment and provides an adequate flexibility system to cope with RE fluctuations.

5.1.2. Comparative Analysis of Flexibility Scheduling and Economic Scheduling

Economic dispatch is the main dispatch mode of the UIES. This section will analyze the results of flexibility scheduling and economic scheduling and investigate the applicability of the proposed flexibility scheduling model. Among them, the economic dispatch model takes the minimum operating cost of each unit as the objective function, and the cost coefficients of each unit come from [26]. Figure 12 compares the system downward flexibility of the proposed model and the economic dispatch model during the valley load period and the system upward flexibility during the peak load period.

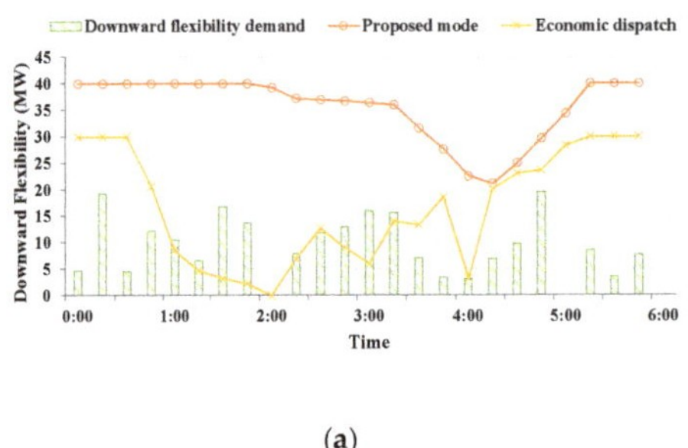
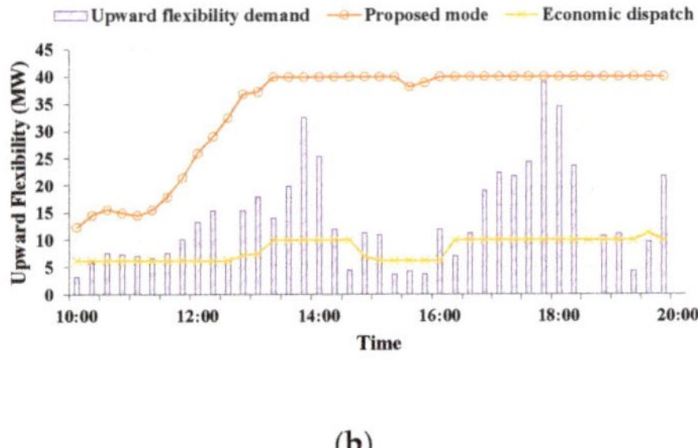

(a)

(b)

Figure 12. System flexibility in the proposed model and in the economic dispatch model: (**a**) downward flexibility in the electric load valley period and (**b**) upward flexibility in the electric load peak period.

It can be seen from Figure 12 that according to the economic dispatch model, the insufficient downward flexibility rate of the system during the valley load period is 20.5%, and wind curtailment is required for some periods. Moreover, during the peak load period, the downward flexibility rate of the system is 44.9%, and most of the time, part of the load needs to be removed. It indicates that in this case, the effect of the flexibility scheduling is better than the economic scheduling.

Next, the system flexibility differences between the two scheduling models are analyzed. Figure 13 shows the system flexibility of the proposed model and the economic dispatch model in the valley and peak load periods.

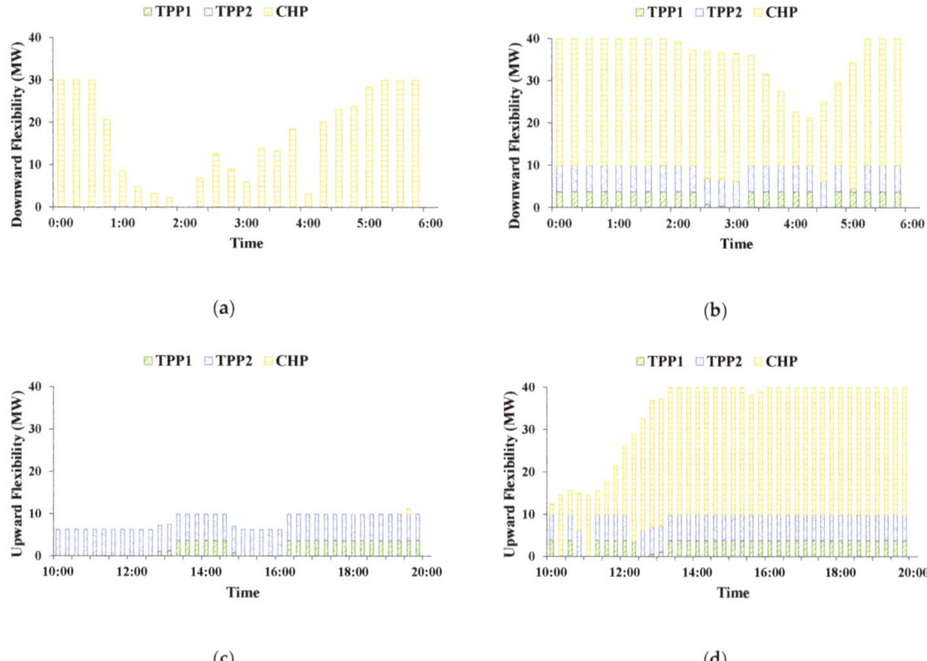

Figure 13. Composition of system flexibility in the economic dispatch and proposed models: (**a**) downward flexibility for the economic dispatch model, (**b**) downward flexibility for the proposed model, (**c**) upward flexibility for the economic dispatch model and (**d**) upward flexibility for the proposed model.

Figure 13 shows that, since CHP units supply both electricity and thermal energy, the power generation cost of CHP units is lower than that of TPP units, and thus CHP power generation will be given priority for the economic dispatch. During the electric load valley period, the electric demand is mainly supplied by the CHP units, which can provide certain downward flexibility, while the TPP units all operate at the minimum output, and there is no downward adjustable capacity. The system cannot meet the flexibility requirements at 1:00–4:00. During the peak load period, the CHP unit basically operates at the upper limit of power generation, having no upward adjustable capacity, while the upward adjustable capacity of TPP units cannot meet the flexibility requirements. In the flexibility dispatch, during the valley load period, the thermal output of the CHP unit is adjusted to H_B, which has the largest downward adjustable capacity. At the same time, the TPP units have a certain downward adjustable capacity within their climbing rate limits by reasonably dispatching the electric output of the TPP units; during the peak load period, the thermal output of the CHP unit is reduced to increase its upward adjustable capacity, while at the same time, the TPP units also have a larger upward adjustable capacity. Thus, the proposed model can meet both the flexibility demands in the valley load period and peak load period. Taking the valley load period as an example, Figure 14 shows the unit output status of the economic dispatch model and the proposed flexibility dispatch model.

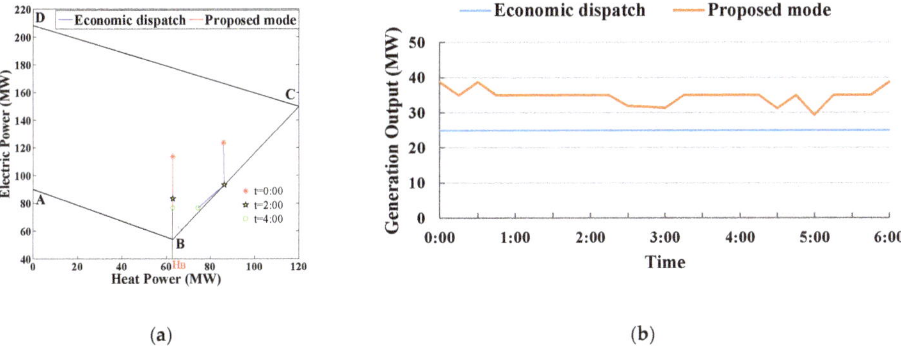

(a)   (b)

Figure 14. Unit output during the electric load valley period in the economic dispatch and proposed models: (**a**) electric and heat output of CHP units and (**b**) electric output of thermal power plant (TPP) units.

The effect of flexibility scheduling is related to the prediction error of wind power. When the prediction accuracy of wind power is approximately unchanged, the prediction error is directly related to the level of wind power. The following is the analysis of the comprehensive operating costs of the economic dispatch and flexibility dispatch models under different levels of forecasted wind power. Figure 15 shows the overall cost structure of the two scheduling models when the predicted wind power generation relative to the peak load ratio varies from 5% to 25%. Under different ratios of wind generation, the shapes of the predicted wind power curve and the actual power curve remain unchanged, as shown in Figure 6b.

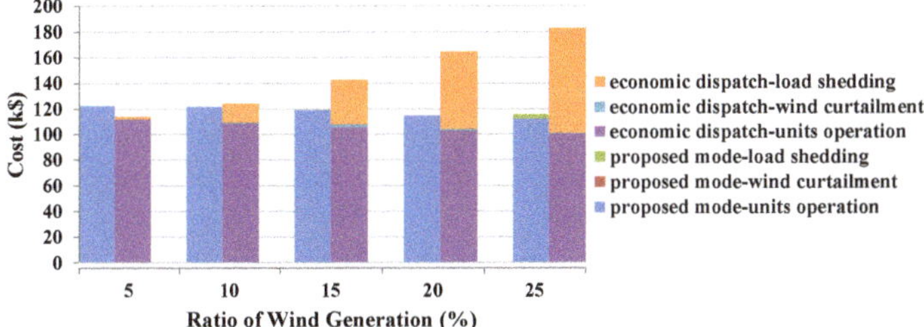

Figure 15. Composition of overall cost with different ratios of wind power generation in the proposed and economic dispatch models.

As shown in Figure 15, under any proportion of predicted wind power, the unit operating cost of the economic dispatch model is less than that of the flexibility dispatch model. However, as the ratio of wind power generation increases, the economic operating model will lead to wind curtailment and load shedding during actual operation. When wind power generation accounts for 10% or more of the maximum electric load, the superiority of the flexibility scheduling model gradually becomes more apparent. In particular, the load-shedding cost will account for 80.9% of the unit operating cost when the percentage of wind power generation reaches 25%. It shows that for the UIES with a high installed capacity of wind generation, when the forecasted wind power is large the next day, the flexibility scheduling model can be used to make the operation plan of each unit to avoid the loss caused by insufficient flexibility.

5.1.3. Impact Analysis of the HL Type

To analyze the impact of the type of HL on system flexibility, the HL type in Case 3 is changed from residential user to commercial user, which is set as Case 4 to compare with Case 3. Figure 16 shows the HLs of commercial and residential users, as well as the electric load. It can be seen that the peak and valley periods of the commercial HL are basically consistent with the electric load, which is completely opposite to that of residential users.

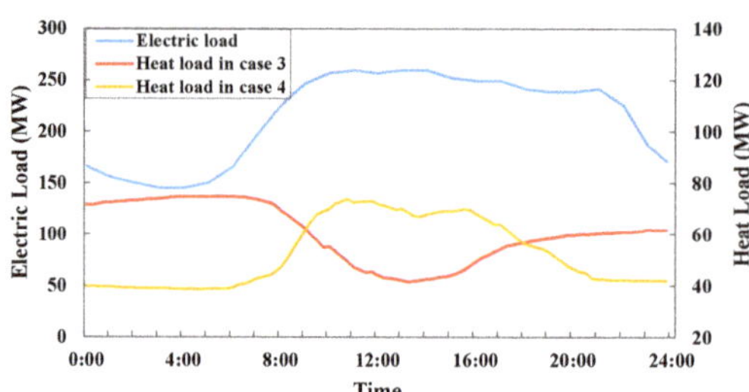

Figure 16. Heat loads in Case 3 and Case 4.

Figure 17 compares the system flexibility in Case 3 and Case 4 during the valley and peak load periods.

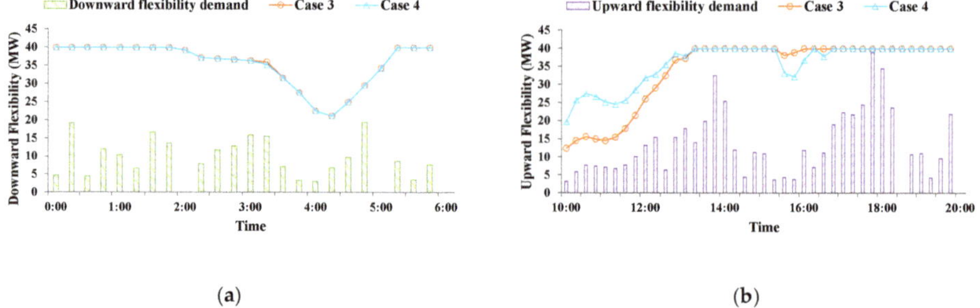

(**a**) (**b**)

Figure 17. System flexibility in Case 3 and Case 4: (**a**) downward flexibility in the electric load valley period and (**b**) upward flexibility in the electric load peak period.

As can be observed from Figure 17, there is little difference in the level of flexibility between the two scenarios, and the system flexibility demands in each period can be met.

Subsequently, the two cases were analyzed from the change in the node supply and return temperatures. Figure 18 compares the supply and return temperatures at the HS, as well as the heat output between the pre- and post-changing HLs.

Figure 18 shows that there is a certain difference in both the HS supply and return temperature curves between Case 4 and Case 3, whereas there is little change in the temperature difference between the HS supply and return temperatures, resulting in a small change in the thermal output, which makes the change in system flexibility under the two cases small.

Overall, these results indicate that the type of HL has little influence on the system flexibility. The heat output of the CHP unit at different time periods can be adjusted by reasonably optimizing the HS supply temperature for different users, thus meeting the downward and upward flexibility demands of the system.

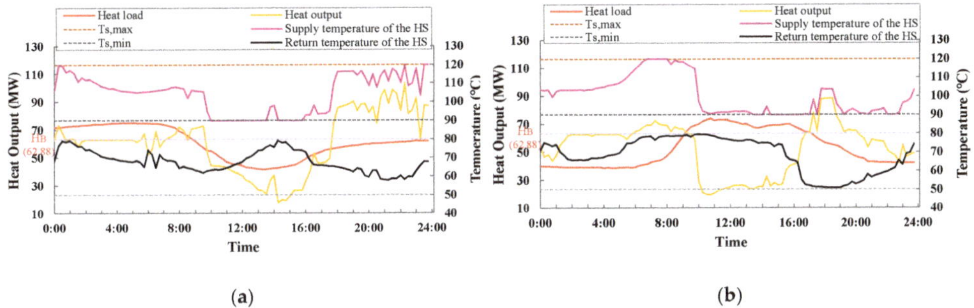

Figure 18. Dynamic temperature characteristics of the UHN and heat output: (**a**) Case 3 and (**b**) Case 4.

5.2. Practical-Scale System

To further verify the effectiveness of the proposed model, this section uses the heat and power networks of a real city for the simulation verification. The network structure is shown in Appendix B. The heating network parameters come from [30], and the detailed data of the thermoelectric system and unit parameters are provided in Appendix B. The total power capacity of this system is 1.44 GW, including 0.82 GW from CHP units, and 0.3 GW from wind turbines. The trend of electrical and thermal load as well as the wind power is consistent with the small-scale case, where the maximum thermal load is 331 MW, the maximum electrical load is 1073 MW and the maximum predicted wind power is 200 MW.

Figure 19 compares the system flexibility of the proposed model and the economic scheduling model at different periods.

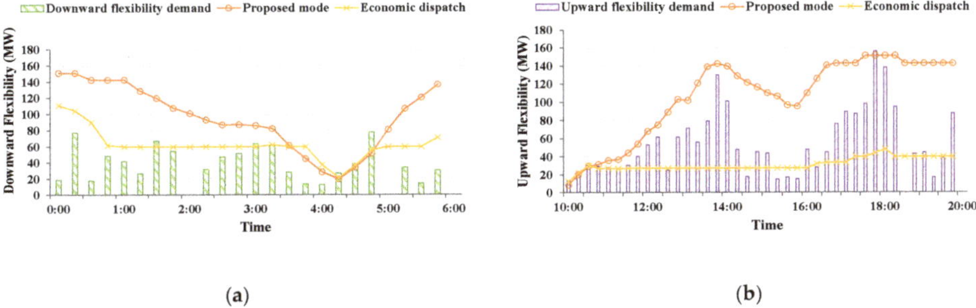

Figure 19. System flexibility in the proposed and economic dispatch models: (**a**) downward flexibility in the electric load valley period and (**b**) upward flexibility in the electric load peak period.

For the proposed model, the insufficient downward flexibility rate during the electric load valley period is 4.32%, and the insufficient upward flexibility rate in the peak load period is 0.60%. For the economic dispatch model, it has an insufficient downward flexibility rate of 4.57% in the valley load period, and the insufficient upward flexibility rate during peak load periods is 47.13%.

Table 1 shows the overall cost composition of the two scheduling models.

According to Table 1, the unit operating cost of the proposed model is USD 601,821, which is USD 40,353 higher than that of the economic dispatch model, but the load-shedding cost is USD 250,790 lower than the economic dispatch model, and the overall cost is 25.78% lower than that of the economic dispatch model.

Table 1. Overall cost for different cases.

Cost (k$)	Proposed Model	Economic Dispatch Model
Unit operation	601.821	563.749
Wind curtailment	0.956	1.321
Load shedding	3.235	256.947
Total	606.012	822.016

6. Conclusions

Aiming at a UIHPS with a high percentage of RE, this paper establishes a flexibility scheduling model for a UIHPS considering the dynamic characteristics of the UHN by combining the coupling characteristics of multi-energy system with the dynamic characteristics of a UHN. Through theoretical research and the analysis of case studies, the following conclusions are drawn:

(1) In the UIES, due to the obvious transmission delay of the UHN, it must be considered in the flexibility scheduling model, otherwise the operational safety of the system may be broken. By making full use of the transmission delay and temperature dynamics of the UHN, the overall flexibility can be improved through the cooperative scheduling of the urban heat and electricity systems, thereby dealing with the random fluctuations of renewable generation effectively;

(2) For the UIES with high penetration of RE, when the forecasted renewable generation is relatively large, the flexibility scheduling model can be used to make the operation plan of each unit. Although the unit operating cost is higher than that of the economic dispatching model, the huge cost of load shedding caused by fluctuations of renewable generation can be effectively avoided in the proposed model;

(3) The type of HL has little effect on the flexibility dispatching results for the UIHPS. For different types of HL, the heat output of the CHP units can be optimized by adjusting the supply temperature of the HS to reduce the impact of different HL peak and valley periods on the system flexibility and effectively meet the flexibility demands of the system.

In the proposed model, we mainly utilize the temperature dynamics of the primary UHN and the coupling characteristics of CHP units to improve the system flexibility. In actual operation, when the prediction accuracy of the HL is rather low, it will cause the actual heat output of CHP units to largely deviate from the planned, and thus affect the system flexibility and RE integration. Furthermore, the thermoelectric coupling characteristics of the substations and the end users also have an effect on the system flexibility. Therefore, we will investigate to improve the forecast accuracy of the HL, and cooperatively consider CHP substations user-side energy equipment to enhance the system flexibility in future research.

Author Contributions: Conceptualization, supervision, writing-review & editing and funding acquisition, W.W.; methodology, software and writing—original draft, Y.S.; writing—review and editing, K.H., H.J. and J.W.; funding acquisition, L.G. and L.W.; resources, C.T. All authors have read and agreed to the published version of the manuscript.

Funding: This research was funded by State Grid Science & Technology Project "Research on simulation analysis model and comprehensive evaluation technology of urban energy internet" (No. 5400-201957186A-0-0-00).

Nomenclature

Abbreviations		
UIES	Urban integrated energy system	
RE	Renewable energy	
UHN	Urban heat network	
UIHPS	Urban integrated heat and power system	
CHP	Combined heat and power	
IRRE	insufficient ramping resource expectation	
HP	Heat pump	
HS	Heat source	
TPP	Thermal power plant	
HL	Heat load	
CF-VT	The quality regulation mode	
IEA	International Energy Agency	

Indices and sets	
N^{Ss}/N^{Sr}	Set of HS nodes in the supply/return network
N_k^s/N_k^e	Index of starting and ending nodes of the pipeline k
$N_{j,i}^v$	Set of nodes of the v-th path from node j to node i
$P_{j,i}^v$	Set of pipelines of the v-th path from node j to node i
N^{HS}	Set of HSs
CHP_j	Set of CHPs connected to the HS j
P^s/P^r	Set of supply/return pipelines
P_i^u/P_i^d	Set of upstream/downstream pipelines of node i
N^s/N^r	Set of nodes in the supply/return network
N^{HL}	Set of HLs
T_1/T_2	Index of electric load valley/peak period

Symbol	Description
$P_{g,\min}^{tpp}/P_{g,max}^{tpp}$	Minimum/maximum electric output of the TPP unit g, MW
$R_{d,g}^{tpp}/R_{u,g}^{tpp}$	Downward/upward ramp rate of the TPP unit g, MW/h
ρ	Density of the hot water, kg/m^3
Γ_k^s	Transmission delay of the pipeline k
D_k	Diameter of the pipeline k, m
l_k	Length of the pipeline k, m
$\dot{m}_k^s$	Mass flow rate of the supply pipeline k, kg/s
ψ_k	Temperature drop coefficient of pipeline k
T_a	Ambient pipeline temperature, °C
α_h	Confluence coefficient related to the node h
$n_{j,i}$	Number of paths from node j to node i
$F_{j,i}^v$	Product of node confluence coefficients in the v-th path from node j to node i
$\Psi_{j,i}^v$	Product of pipeline temperature drop coefficients in the v-th path from node j to node i
$\Gamma_{j,i}^v$	Sum of pipeline transmission delays in the v-th path from node j to node i
$\dot{m}_i^L$	Mass flow rate of HL i, kg/s
$T_{i,\max}^{Ls}/T_{i,\min}^{Ls}$	Maximum/minimum supply temperature of HL i, °C
$T_{i,\max}^{Lr}/T_{i,\min}^{Lr}$	Maximum/minimum return temperature of HL i, °C
$P_{ij,\min}/P_{ij,\max}$	Minimum/maximum power of branch ij, MW
$U_{\min}/U_{\max}$	Minimum/maximum node voltage amplitude of bus i
$\theta_{\min}/\theta_{\max}$	Minimum/maximum node voltage phase angle of bus i
N_g	Number of adjustable generation units

Variables	
$H_{g,t}^{chp}/P_{g,t}^{chp}$	Heat/electric output of the current operating point of the CHP unit g at time t, MW
$\alpha_{g,t}^k$	Combination coefficient corresponding to the k-th corner point in the feasible region of the CHP unit g at time t
$T_{j,t}^{Ss}/T_{j,t}^{Sr}$	Supply/return temperature of the HS j at time t, °C

Parameters

NK_g — Number of corner points in the feasible region of the CHP unit g

H_g^k / P_g^k — Heat/electric output corresponding to the k-th corner point in the feasible region of the CHP unit g, MW

C_g^k — Cost corresponding to the k-th corner point of the CHP unit g, \$

c — Specific heat capacity of water, kJ/(kg·°C)

$\dot{m}_j^S$ — Mass flow rate of the HS j, kg/s

$T_{j,min}^{Ss} / T_{j,max}^{Ss}$ — Minimum/maximum supply temperature of the HS j, °C

$T_{j,min}^{Sr} / T_{j,max}^{Sr}$ — Minimum/maximum return temperature of the HS j, °C

$R_{d,g}^{chp} / R_{u,g}^{chp}$ — Downward/upward ramp rate of the CHP unit g, MW/h

Δt — Dispatch time step, h

a_g, b_g, c_g — Cost coefficients of the TPP unit g

$P_{g,t}^{tpp}$ — Electric output of the TPP unit g at time t, MW

$T_{k,t}^{ps,in} / T_{k,t}^{ps,out}$ — Inlet/outlet temperature of the pipeline k at time t, °C

$H_{i,t}^{L}$ — Heat load of the heat exchange station i at time t, MW

$T_{i,t}^{Ls} / T_{i,t}^{Lr}$ — Supply/return temperature of the HL i at time t, °C

$T_{in,i,t}^{PL} / T_{out,i,t}^{PL}$ — Inlet/outlet temperature of the equivalent pipe of the heat load i, °C

$P_{i,t}^{tpp} / Q_{i,t}^{tpp}$ — Active/reactive power output of TPP units at bus i at time t, MW/Mvar

$P_{i,t}^{wind} / Q_{i,t}^{wind}$ — Active/reactive power output of the wind farm at bus i at time t, MW/Mvar

$P_{i,t}^{load} / Q_{i,t}^{load}$ — Active/reactive power load at bus i at time t, MW/Mvar

f_t^d / f_t^u — Downward/upward flexibility of UIES at time t, MW

$\Delta P_t^u / \Delta P_t^d$ — Upward/downward fluctuation of wind power at time t, MW

$\Delta f_t^d / \Delta f_t^u$ — Downward/upward flexibility deficiency rate

Appendix A. Data of Small-Scale Case Study

Table A1. Units parameters.

Unit	Bus	Pmax (MW)	Pmin (MW)	Qmax (MW)	Qmin (MW)	RU (MW/h)	RD (MW/h)	a ($/MW^2)	b ($/MW)	c ($)	Startup ($)
1	1	30	10	70	−40	15	15	0.0005	16.83	220.58	125
2	2	50	15	200	−80	25	25	0.0013	40.62	161.87	374
3	6	208.2	54	150	−120	40	40	0.0044	3.60	100	600

Table A2. Feasible region and corresponding cost of CHP.

Point	Heat Output (MW)	Electric Output (MW)	Cost ($)
A	0	90	2040
B	62.88	54	1770
C	120	150	3330
D	0	208.2	2910

Appendix B. Data of Practical-Scale Case Study

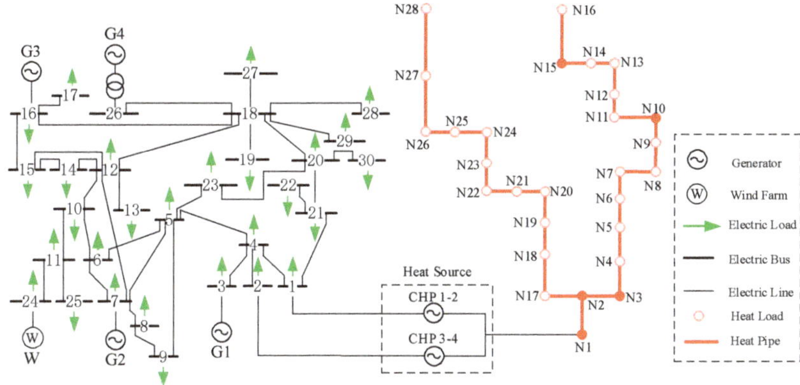

Figure A1. Schematic diagram of the electricity and heating networks in the practical system.

Table A3. Heat network parameters.

id	F_node	T_node	Length (m)	Diameter (m)	Conductivity (W/(m·°C))	Roughness (m)	Flowrate (kg/s)
1	1	2	1000	1	0.12	0.0005	1757.012
2	2	3	2264.5	1	0.12	0.0005	596.784
3	3	4	865	1	0.12	0.0005	596.784
4	4	5	1939	1	0.12	0.0005	489.276
5	5	6	2531	1	0.12	0.0005	397.893
6	6	7	315	1	0.12	0.0005	320.219
7	7	8	300	0.9	0.12	0.0005	254.195
8	8	9	990	0.9	0.12	0.0005	198.075
9	9	10	689	0.9	0.12	0.0005	150.373
10	10	11	259	0.9	0.12	0.0005	150.373
11	11	12	200	0.8	0.12	0.0005	109.826
12	12	13	300	0.8	0.12	0.0005	75.361
13	13	14	260	0.6	0.12	0.0005	46.066
14	14	15	402	0.6	0.12	0.0005	21.166
15	15	16	1600	0.35	0.12	0.0005	21.166
16	2	17	2500	1	0.12	0.0005	1160.228
17	17	18	2500	1	0.12	0.0005	957.334
18	18	19	2050	1	0.12	0.0005	784.874
19	19	20	1050	1	0.12	0.0005	638.282
20	20	21	1800	0.9	0.12	0.0005	513.680
21	21	22	1750	0.9	0.12	0.0005	407.768
22	22	23	2600	0.9	0.12	0.0005	317.743
23	23	24	1900	0.9	0.12	0.0005	241.221
24	24	25	2400	0.8	0.12	0.0005	176.178
25	25	26	1900	0.8	0.12	0.0005	120.891
26	26	27	2800	0.6	0.12	0.0005	73.898
27	27	28	3600	0.6	0.12	0.0005	33.953

Table A4. Electric network parameters

F_bus	T_bus	r (p.u)	x (p.u)	b (p.u)	F_bus	T_bus	r (p.u)	x (p.u)	b (p.u)
1	4	0.00031	0.00180	−0.08495	14	15	0.00267	0.02106	−0.01189
1	4	0.00031	0.00180	−0.02242	15	16	0.00051	0.00630	−0.01189
2	4	0.00031	0.00180	−0.03879	15	16	0.00051	0.00630	−0.00915
3	4	0.00031	0.00180	−0.02000	16	17	0.00024	0.00299	−0.90000
4	5	0.00062	0.00367	−0.02130	16	17	0.00024	0.00299	−0.01200
4	5	0.00062	0.00367	−0.01000	16	18	0.00710	0.03739	−0.07777
5	12	0.00327	0.01706	−0.01000	16	18	0.00710	0.03739	−0.05500
5	6	0.00060	0.00310	−0.09215	19	18	0.00139	0.01094	−0.01204
5	7	0.00026	0.00204	−0.04100	19	18	0.00139	0.01094	−0.00862
5	9	0.00793	0.04177	−0.10824	20	18	0.00560	0.02948	−0.03318
5	23	0.00499	0.02630	−0.03660	20	18	0.00560	0.02948	−0.04575
8	9	0.00292	0.02310	−0.12654	20	23	0.00728	0.03833	−0.00165
7	8	0.00131	0.01031	−0.12654	20	21	0.00197	0.01559	−0.20000
6	7	0.00011	0.00064	−0.15090	20	21	0.00197	0.01559	−0.20000
6	7	0.00011	0.00064	−0.16410	22	21	0.00165	0.00544	−0.15833
6	10	0.00104	0.00543	−0.06128	1	21	0.00148	0.01171	−0.01372
10	11	0.00299	0.01573	−0.00044	1	21	0.00148	0.01171	−0.01372
10	11	0.00299	0.01573	−0.13150	24	11	0.00207	0.00207	−0.43500
10	12	0.00250	0.01315	−0.03596	25	11	0.00207	0.00207	−0.39500
13	12	0.00466	0.01535	−0.57500	26	18	0.00207	0.00207	−0.75000
12	15	0.00400	0.03163	−0.82500	27	18	0.00207	0.00207	−0.75000
12	14	0.00057	0.00452	−0.01696	28	18	0.00207	0.00207	−0.00522
12	18	0.00470	0.02475	−0.03431	29	18	0.00207	0.00207	−0.00695
12	18	0.00470	0.02475	−0.03250	30	20	0.00207	0.00207	−0.01958

Table A5. Units parameters.

Unit	Bus	Pmax (MW)	Pmin (MW)	Qmax (MW)	Qmin (MW)	RU (MW/h)	RD (MW/h)	a ($/MW^2)	b ($/MW)	c ($)	Startup ($)
1	1	240	98	300	−300	120	120	0.0044	3.60	100	1700
2	1	240	98	300	−300	120	120	0.0044	3.60	100	1700
3	2	170	60	300	−300	102	102	0.0044	3.60	100	1300
4	2	170	60	300	−300	102	102	0.0044	3.60	100	1300
5	3	0.01	0	200	−250	0	0	0.0141	16.08	212.31	1200
6	7	50	20	200	−25	25	25	0.0141	16.08	212.31	1200
7	16	50	20	700	−700	25	25	0.0527	43.66	781.52	1500
8	26	220	60	999	−999	110	110	0.0527	43.66	781.52	2100

Table A6. Feasible region and corresponding cost of CHPs.

Point	CHP 1−2			CHP 3−4		
	Heat Output (MW)	Electric Output (MW)	Cost ($)	Heat Output (MW)	Electric Output (MW)	Cost ($)
A	0	100	2753	0	70	1927
B	102	98	3662	50	60	2124
C	135	190	4875	70	154	3483
D	0	240	4130	0	170	2926

References

1. Alizadeh, M.I.; Moghaddam, M.P.; Amjady, N.; Siano, P.; Sheikh-El-Eslami, M.K. Flexibility in future power systems with high renewable penetration: A review. *Renew. Sustain. Energy Rev.* **2016**, *57*, 1186–1193. [CrossRef]
2. Agency, I.E. *Empowering Variable Renewables—Options for Flexible Electricity Systems*; OECD Publishing: Paris, France, 2009; pp. 13–14.

3. Wu, S.Y.; Wang, P.; Yang, J.; Li, Z.N.; Ouyang, M. Review on interdependency modeling of integrated energy system. In Proceedings of the IEEE Conference on Energy Internet and Energy System Integration, Beijing, China, 26–28 November 2017.

4. Wu, J.; Yan, J.; Jia, H.; Hatziargyriou, N.; Djilali, N.; Sun, H. Integrated energy systems. *Appl. Energy* **2016**, *167*, 155–157. [CrossRef]

5. Mancarella, P. MES (multi-energy systems): An overview of concepts and evaluation models. *Energy* **2014**, *65*, 1–17. [CrossRef]

6. Niemi, R.; Mikkola, J.; Lund, P.D. Urban energy systems with smart multi-carrier energy networks and renewable energy generation. *Renew. Energy* **2012**, *48*, 524–536. [CrossRef]

7. Lannoye, E.; Flynn, D.; O'Malley, M. Evaluation of Power System Flexibility. *IEEE Trans. Power Syst.* **2012**, *27*, 922–931. [CrossRef]

8. Zhao, J.Y.; Zheng, T.X.; Litvinov, E. A Unified Framework for Defining and Measuring Flexibility in Power System. *IEEE Trans. Power Syst.* **2016**, *31*, 339–347. [CrossRef]

9. Ulbig, A.; Andersson, G. Analyzing operational flexibility of electric power systems. *Int. J. Electr. Power Energy Syst.* **2015**, *72*, 155–164. [CrossRef]

10. Marneris, I.G.; Biskas, P.N.; Bakirtzis, E.A. An Integrated Scheduling Approach to Underpin Flexibility in European Power Systems. *IEEE Trans. Sustain. Energy* **2016**, *7*, 647–657. [CrossRef]

11. Yang, Y.L.; Wu, K.; Long, H.Y.; Gao, J.C.; Yan, X.; Kato, T.; Suzuoki, Y. Integrated electricity and heating demand-side management for wind power integration in China. *Energy* **2014**, *78*, 235–246. [CrossRef]

12. Nuytten, T.; Claessens, B.; Paredis, K.; Van Bael, J.; Six, D. Flexibility of a combined heat and power system with thermal energy storage for district heating. *Appl. Energy* **2013**, *104*, 583–591. [CrossRef]

13. Chen, X.Y.; Kang, C.Q.; O'Malley, M.; Xia, Q.; Bai, J.H.; Liu, C.; Sun, R.F.; Wang, W.Z.; Li, H. Increasing the Flexibility of Combined Heat and Power for Wind Power Integration in China: Modeling and Implications. *IEEE Trans. Power Syst.* **2015**, *30*, 1848–1857. [CrossRef]

14. Papaefthymiou, G.; Hasche, B.; Nabe, C. Potential of Heat Pumps for Demand Side Management and Wind Power Integration in the German Electricity Market. *IEEE Trans. Sustain. Energy* **2012**, *3*, 636–642. [CrossRef]

15. Sayegh, M.A.; Danielewicz, J.; Nannou, T.; Miniewicz, M.; Jadwiszczak, P.; Piekarska, K.; Jouhara, H. Trends of European research and development in district heating technologies. *Renew. Sustain. Energy Rev.* **2017**, *68*, 1183–1192. [CrossRef]

16. Sayegh, M.A.; Jadwiszczak, P.; Axcell, B.P.; Niemierka, E.; Brys, K.; Jouhara, H. Heat pump placement, connection and operational modes in european district heating. *Energy Build.* **2018**, *166*, 122–144. [CrossRef]

17. Wang, J.D.; Zhou, Z.G.; Zhao, J.N.; Zheng, J.F. Improving wind power integration by a novel short-term dispatch model based on free heat storage and exhaust heat recycling. *Energy* **2018**, *160*, 940–953. [CrossRef]

18. Gu, W.; Wang, J.; Lu, S.; Luo, Z.; Wu, C.Y. Optimal operation for integrated energy system considering thermal inertia of district heating network and buildings. *Appl. Energy* **2017**, *199*, 234–246. [CrossRef]

19. Salta, M.; Polatidis, H.; Haralambopoulos, D. Industrial combined heat and power (CHP) planning: Development of a methodology and application in Greece. *Appl. Energy* **2011**, *88*, 1519–1531. [CrossRef]

20. Lahdelma, R.; Hakonen, H. An efficient linear programming algorithm for combined heat and power production. *Eur. J. Oper. Res.* **2003**, *148*, 141–151. [CrossRef]

21. Pirouti, M.; Bagdanavicius, A.; Ekanayake, J.; Wu, J.Z.; Jenkins, N. Energy consumption and economic analyses of a district heating network. *Energy* **2013**, *57*, 149–159. [CrossRef]

22. He, P.; Sun, G.; Wang, F.; Wu, H. *District Heating Engineering*, 4th ed.; China Architecture and Building Press: Beijing, China, 2009; pp. 285–289.

23. Liu, X.Z.; Wu, J.Z.; Jenkins, N.; Bagdanavicius, A. Combined analysis of electricity and heat networks. *Appl. Energy* **2016**, *162*, 1238–1250. [CrossRef]

24. Zheng, J.F.; Zhou, Z.G.; Zhao, J.N.; Wang, J.D. Integrated heat and power dispatch truly utilizing thermal inertia of district heating network for wind power integration. *Appl. Energy* **2018**, *211*, 865–874. [CrossRef]

25. Yu, D.L.; Tu, C.W.; Wang, Z.L.; Lv, C.; Wang, H.F. Optimal Energy Flow of Combined Electrical and Heating Multi-energy System Considering the Linear Network Constraints. *Proc. CSEE* **2019**, *37*, 1933–1944.

26. Li, Z.G.; Wu, W.C.; Wang, J.H.; Zhang, B.M.; Zheng, T.Y. Transmission-Constrained Unit Commitment Considering Combined Electricity and District Heating Networks. *IEEE Trans. Power Syst.* **2016**, *7*, 480–492. [CrossRef]

27. Illinois Inst. Technol. Test Data of 6-Bus System for UC-CEHN. Available online: http://motor.ece.iit.edu/data/UCCEHN_6bus.xls (accessed on 14 August 2015).
28. Song, Z.R.; Shen, F.; Nan, Z.; Zhang, Y.B.; Zhao, L.; Deng, X.Y.; Zhang, N.; Li, H.; Zhang, Z.X.; Ye, W.; et al. Power Grid Planning Based on Differential Abandoned Wind Rate. *J. Eng.* **2017**, *13*, 1055–1059. [CrossRef]
29. Awad, A.S.A.; EL-Fouly, T.H.M.; Salama, M.M.A. Optimal ESS Allocation and Load Shedding for Improving Distribution System Reliability. *IEEE Trans. Smart Grid* **2014**, *5*, 2339–2349. [CrossRef]
30. Zhou, S.J. Operational Parameters Prediction and Optimization Research of District Heating System Based on Pipe Network Dynamic Model. Ph.D. Thesis, Shandong University, Jinan, China, 2012.

Article

An Optimal Day-Ahead Thermal Generation Scheduling Method to Enhance Total Transfer Capability for the Sending-Side System with Large-Scale Wind Power Integration

Yuwei Zhang [1],*, Wenying Liu [1], Yue Huan [1], Qiang Zhou [2] and Ningbo Wang [2]

[1]   State Key Laboratory of Alternate Electrical Power System with Renewable Energy Sources, North China Electric Power University, Beijing 102206, China; liuwyls@126.com (W.L.); a636huanyue@163.com (Y.H.)
[2]   State Grid Corporation of Gansu Province, Lanzhou 730000, China; fdzxzhouqiang@126.com (Q.Z.); wangningbo1963@163.com (N.W.)
*   Correspondence: zhangyuweincepu@126.com; Tel.: +86-188-1177-4465

Received: 11 April 2020; Accepted: 6 May 2020; Published: 9 May 2020

Abstract: The rapidly increasing penetration of wind power into sending-side systems makes the wind power curtailment problem more severe. Enhancing the total transfer capability (TTC) of the transmission channel allows more wind power to be delivered to the load center; therefore, the curtailed wind power can be reduced. In this paper, a new method is proposed to enhance TTC, which works by optimizing the day-ahead thermal generation schedules. First, the impact of thermal generation plant/unit commitment on TTC is analyzed. Based on this, the day-ahead thermal generation scheduling rules to enhance TTC are proposed herein, and the corresponding optimization models are established and solved. Then, the optimal day-ahead thermal generation scheduling method to enhance TTC is formed. The proposed method was validated on the large-scale wind power base sending-side system in Gansu Province in China; the results indicate that the proposed method can significantly enhance TTC, and therefore, reduce the curtailed wind power.

Keywords: enhance total transfer capability; day-ahead thermal generation scheduling; reduce curtailed wind power

---

1. Introduction

Large-scale wind power bases are mostly located in the areas with rich wind energy and far from the load center. Because of the small capacity of local load, a vast amount of wind power needs to be delivered through a transmission channel to the load center. Wind power is highly intermittent; during the periods of large availability of wind energy, the lack of consumption space in local areas and the restriction of total transfer capability (TTC) of the transmission channel are two main reasons for wind power curtailment.

Many relevant studies focus on expanding consumption space for wind power in the local areas of large-scale wind power bases. A pair of studies [1,2] suggest that the curtailed wind power could be consumed by the energy-intensive load, which is located close to the large-scale wind power base and has better regulation flexibility than residential load. Two papers [3,4] studied the potential of deep peak regulation of thermal generation units and proposed a wind-thermal peak regulation trading mechanism to help consume wind power. However, through economic analysis [5] revealed that blindly reducing the wind power curtailment using the supply side and demand side resources may not be cost-effective. References [6,7] show that a wind power generation system equipped with energy storage can smooth the fluctuation of wind power, but it needs a fairly large investment. Using the methods given above, during the periods of large availability of wind energy, wind power can be further consumed in the local areas by regulating energy-intensive load upward, regulating the output of thermal generation units deep downward and storing wind energy.

The above studies have provided methods that can effectively improve the wind power curtailment situation. However, the installed capacity of wind power in a large-scale wind power base is far too large; the curtailed wind power cannot be fully consumed through the above-mentioned methods. Delivering wind power through the transmission channel to the load center is still an important way to consume large-scale wind power. Reference [8] shows that the flexible demand response in the load center can shift the load to wind power peak periods, which enables the receiving-side system to spare more available consumption space for wind power. However, large-scale wind power delivered through the transmission channel makes the transmission channel vulnerable [9]. Therefore, enhancing TTC and making full use of the transmission channel is of great significance for wind power consumption.

TTC of the transmission channel is usually limited by its transient stability-constrained total transfer capability (still referred to as TTC) [10], so TTC can be enhanced by improving system transient stability. Some studies have found that flexible AC transmission system (FACTS) devices could be utilized to improve transient stability besides their main function of controlling power flow [11–14]. FACTS devices including the unified power flow controller (UPFC) and the thyristor switched series capacitor (TSSC) use silicon controlled rectifiers (SCRs) instead of traditional mechanical switches. The development of silicon-coated gold nanoparticle technology enables FACTS devices to achieve super-resolution and quicker adjustment according to the system instructions [15,16]. Reference [11] modulates the active power and reactive power using UPFC to improve the first swing stability and achieves enhancing TTC of long-distance transmission lines. The authors of [12] designed a power oscillation damping controller which enables TSSC to have continuous reactance, and by decreasing the reactance between the sending and receiving ends, TTC can be enhanced. However, the main function of FACTS devices is not to improve the system's transient stability; what is more, for the systems that do not have FACTS devices, installing them requires additional costs.

Transient stability analysis shows that some electrical parameters, such as the inertia constant and grid reactance, have great impacts on system transient stability [17]. Due to the differences in the electrical parameters of thermal generation units, system electrical parameters and TTC change with thermal generation unit commitment. For a system with large-scale wind power integration, day-ahead thermal generation scheduling is necessarily made to ensure power balance, mostly with the objective of minimizing the generation cost, sometimes considering the objectives of maximizing the reliability of the power system and minimizing the emission [18,19]. However, no previous work on enhancing TTC by optimizing thermal generation schedules has been reported.

This paper presents a new method with which to enhance TTC based on the existing grid structure, which is by optimizing the day-ahead thermal generation schedules, and it can help reduce curtailed wind power for the sending-side system with large-scale wind power integration. This method is especially suitable for the sending-side system with thermal generation plants and large-scale wind power integration and the transmission channel in its existing grid structure, which is the common structure of large-scale wind power bases. The resources that are used in this method only involve the existing thermal generation plants, and it requires no further installment of devices. The main contributions are as follows:

1. The mechanism of the impact of thermal generation plant/unit commitment on TTC is revealed.
2. TTC is enhanced by optimizing the day-ahead thermal generation schedules, requiring no investment in installing additional devices. With the enhanced TTC, more wind power is allowed to be delivered through the transmission channel to the load center; therefore, the curtailed wind power is reduced.

The rest of the paper is organized as follows. In Sections 2 and 3, the impact of enhancing TTC on reducing curtailed wind power and the mechanism of the impact of thermal generation plant/unit commitment on TTC are analyzed. In Section 4, the optimal day-ahead thermal generation scheduling method to enhance TTC is proposed. In Section 5, the case analysis is presented. Finally, the conclusions are given in Section 6.

2. The Impact of Enhancing TTC on Reducing Curtailed Wind Power

Figure 1 shows a schematic diagram of a sending-side system with large-scale wind power integration, a receiving-side system and the transmission channel between them. Assuming that during the periods of large availability of wind energy, wind power is curtailed, the impact of enhancing TTC on reducing curtailed wind power is analyzed below.

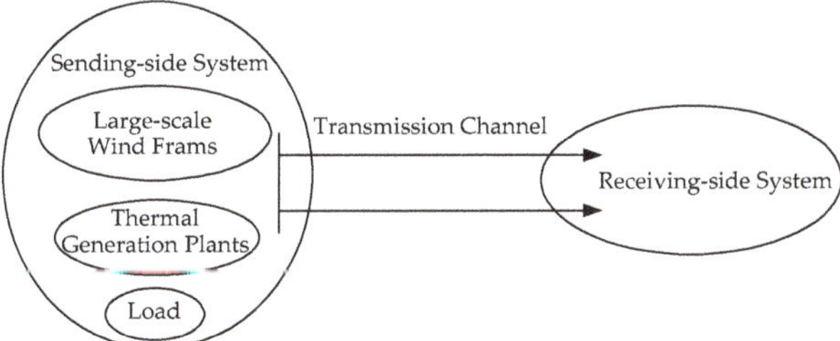

Figure 1. Schematic diagram of sending-side and receiving-side systems and a transmission channel.

For the sending-side system, the wind power output is regarded as a negative load, and the power delivered through the transmission channel is regarded as a positive load. The equivalent load in the sending-side system is as follows:

$$P_{EL} = P_{SL} + P_D - P_W \tag{1}$$

where P_{SL} is the total load in the sending-side system, P_D is the power delivered through the transmission channel and P_W is the wind power output.

The equivalent load is balanced with the thermal generation in the sending-side system as follows:

$$P_T - P_{EL} = 0 \tag{2}$$

where P_T is the thermal generation in the sending-side system.

During the periods of large availability of wind energy, the power delivered through the transmission channel reaches TTC, and if the minimum allowable thermal generation P_T^{min} is greater than the equivalent load P_{EL}, wind power needs to be curtailed, and the curtailed wind power is the sum of Area 1 and Area 2 during periods $t_1 \sim t_2$ in Figure 2.

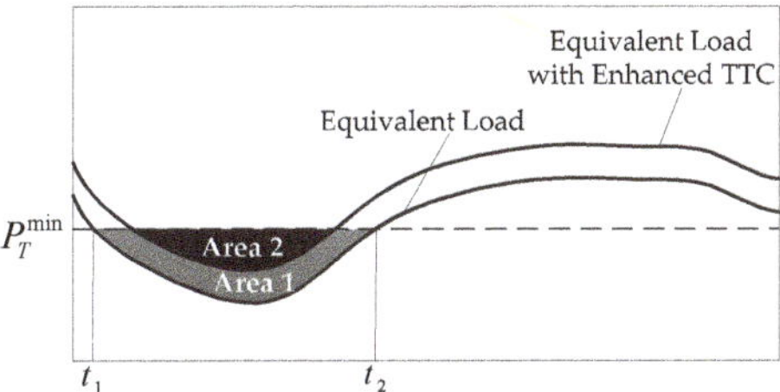

Figure 2. Schematic diagram of the impact of enhancing total transfer capability (TTC) on reducing curtailed wind power.

If TTC is enhanced, the power delivered through the transmission channel P_D can be increased; according to Equation (1), the equivalent load P_{EL} is increased accordingly. Therefore, the curtailed wind power reduces from the sum of Area 1 and Area 2 to Area 2 in Figure 2.

3. The Impact of Thermal Generation Plant/Unit Commitment on TTC

The TTC of the transmission channel is usually limited by its transient, stability-constrained total transfer capability, which is considered as a security constraint while making generation schedules. In fact, due to the differences of thermal generation plants and units in terms of their electrical parameters and their electrical distances from the transmission channel, thermal generation plant/unit commitment has a great impact on TTC.

As shown in Figure 3, for a multi-generator sending-side system, while analyzing the impact of thermal generation plant/unit commitment on TTC, the wind farms and load are omitted. The online thermal generation units in the sending-side system are equivalent to one synchronous generator, and the receiving-side system is equivalent to an infinity bus system.

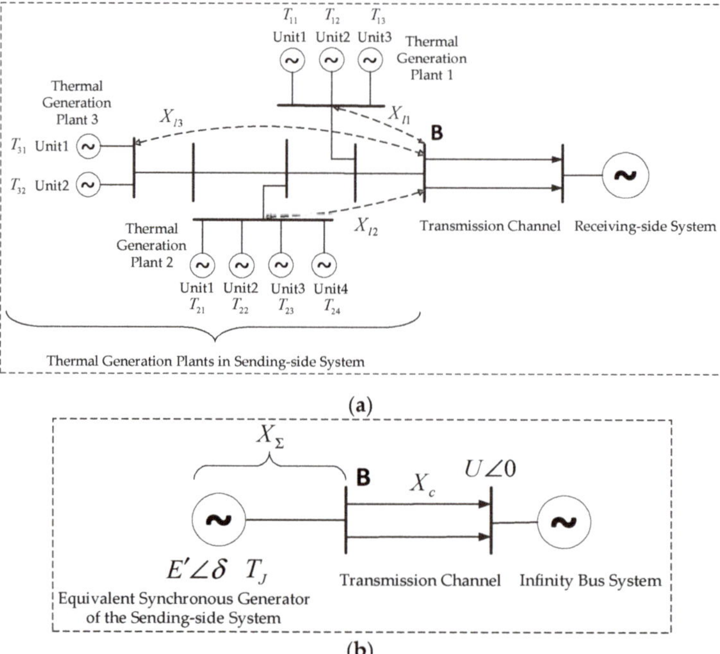

Figure 3. (a) Diagram of a multi-generator sending-side system and receiving-side system and transmission channel; (b) diagram of an equivalent sending-side system and receiving-side system and transmission channel.

Under the condition that the system is able to stay transiently stable after a fault occurs, the maximum steady state power that can be delivered through the transmission channel in Figure 3b represents TTC (it is not the actual TTC, but it can be used for qualitatively analyzing the impact of thermal generation plant/unit commitment on TTC). The system transient stability is evaluated by the stability of the synchronous generator rotor angle in the first swing after a fault occurs. The rotor motion equation of the equivalent synchronous generator of the sending-side system is shown in Equations (3) and (4):

$$\overset{''}{\delta} = \frac{d^2\delta(t)}{dt^2} = \frac{1}{T_J}(P_m - P_e) \tag{3}$$

$$P_e = \frac{E'U}{X_\Sigma + X_c}sin\delta(t) = P_e^{max}sin\delta(t) \tag{4}$$

where δ, E', T_J, P_m, P_e are the rotor angle, internal voltage, inertia constant, mechanical and electrical power of the equivalent synchronous generator of the sending-side system, respectively. $\ddot{\delta}$ is the acceleration of δ. P_e^{max} is the maximum of P_e. P_m equals the steady state value of P_e, and it represents the steady state power delivered through the transmission channel. U is the magnitude of the infinity bus voltage, and the phase angle of the infinity bus voltage remains 0. Reactance X_Σ is the equivalent electrical distance between the transmission channel and the equivalent synchronous generator of the sending-side system. X_c is the reactance of the transmission channel.

According to equal area criterion [20], if the rotor's acceleration area A during the fault is equal to its maximum possible deceleration area D after the fault is removed, the system is in the transient stability critical state, and P_m reaches its maximum value P_m^{max} which represents TTC, as shown in Figure 4.

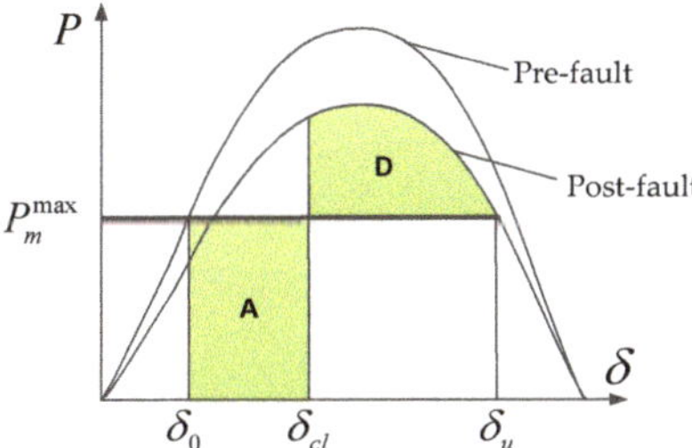

Figure 4. Diagram of power-angle curves when the system is in a transient stability critical state.

According to Equations (3) and (4), X_Σ and T_J are both important factors in the system's transient stability evaluation, so they have an impact on TTC. Due to the differences of thermal generation plants and units in their electrical distances from the transmission channel and their inertia constants, as shown in Figure 3a, thermal generation plant/unit commitment has a great impact on X_Σ and T_J, and therefore, on TTC, and the detailed analysis is as follows.

3.1. The Impacts of Thermal Generation Plant Commitment on X_Σ and TTC

The impact of X_Σ on TTC is analyzed through Figure 5. If X_Σ decreases from $X_{\Sigma 0}$ in Figure 5a to $X_{\Sigma 1}$ in Figure 5b, according to Equation (4), P_e^{max} will increase from P_e^{max0} to P_e^{max1} (superscript "0" and "1" correspond to Figure 5a,b, respectively). TTC in Figure 5a is P_m^{max0}, and assuming the power delivered through the transmission channel in Figure 5b remains P_m^{max0}, it can be seen that $A_1 \approx A_0$ and $D_1 > D_0$, which means the system in Figure 5b has not reached the transient stability critical state, and there is still space for P_m to increase; therefore, $P_m^{max1} > P_m^{max0}$. In conclusion, TTC increases with the decrease of X_Σ.

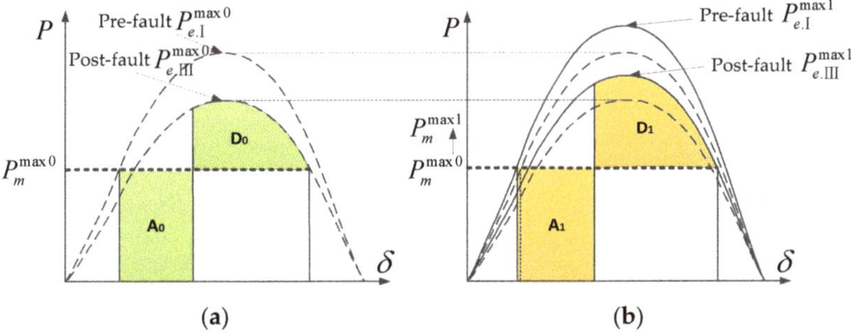

Figure 5. (a) Diagram of power-angle curves when $X_\Sigma = X_{\Sigma 0}$. (b) Diagram of power-angle curves when $X_\Sigma = X_{\Sigma 1}$.

For the three thermal generation plants in the sending-side system in Figure 3a, assuming $X_{l1} < X_{l2} < X_{l3}$ and ignoring the influences of other parameters, if only two thermal generation plants need to be online, then the plant commitment of Plant 1 and Plant 2 can obtain the minimum X_Σ and the maximum TTC.

3.2. The Impacts of Thermal Generation Unit Commitment on T_J and TTC

The impact of T_J on TTC is analyzed through Figure 6. If T_J increases from T_{J0} in Figure 6a to T_{J2} in Figure 6b, according to Equation (3), $\ddot{\delta}$ will decrease, so the fault removal rotor angle δ_{cl} will decrease from δ_{cl}^0 to δ_{cl}^2 (superscript "0" and "2" correspond to Figure 6a,b, respectively). TTC in Figure 6a is P_m^{max0}, and assuming the power delivered through the transmission channel in Figure 6b remains P_m^{max0}, it can be seen that $A_2 < A_0$ and $D_2 > D_0$, which means the system in Figure 6b has not reached the transient stability critical state, and there is still space for P_m to increase; therefore, $P_m^{max2} > P_m^{max0}$. In conclusion, TTC increases with T_J.

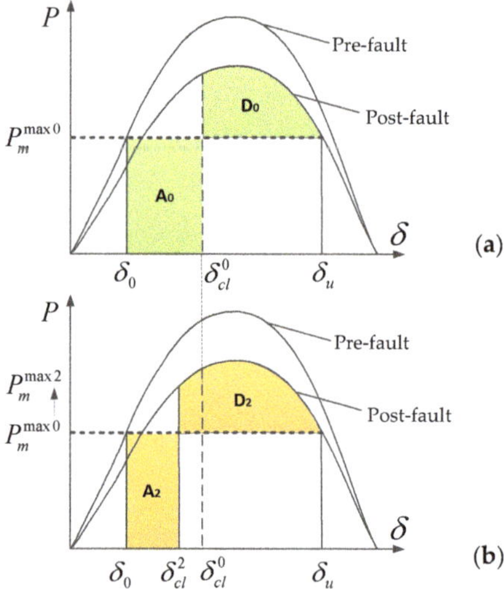

Figure 6. (**a**) Diagram of power-angle curves when $T_J = T_{J0}$. (**b**) Diagram of power-angle curves when $T_J = T_{J2}$.

For each thermal generation plant that is scheduled to be online, its equivalent inertia constant changes with the thermal generation unit commitment inside the plant. TTC increases with the equivalent inertia constant of each online thermal generation plant.

4. Optimal Day-Ahead Thermal Generation Scheduling Method to Enhance TTC

4.1. Day-Ahead Thermal Generation Scheduling Rules

Based on the impact of thermal generation plant/unit commitment on TTC, the day-ahead thermal generation scheduling rules to enhance TTC are proposed as follows:

Rule 1: For thermal generation plant commitment, the plants with shorter electrical distances from the transmission channel take priority.

Rule 2: For thermal generation unit commitment inside the thermal generation plant which is scheduled to be online, the units with bigger inertia constants and lesser generation costs take priority.

Rule 3: According to the above rules, optimize the day-ahead thermal generation schedules to enhance TTC, and then increase wind power output based on the enhanced TTC to reduce curtailed wind power.

4.2. Day-Ahead Thermal Generation Plant Commitment Optimization

4.2.1. Day-Ahead Thermal Generation Plant Commitment Optimization Model

According to Rule 1, the day-ahead thermal generation plant commitment optimization model is established as follows:

1. Objective

    The objective is to obtain the minimum equivalent electrical distance between the transmission channel and the equivalent synchronous generator of the sending-side system.

    $$\min X_{\Sigma}(U_P) \tag{5}$$

    where $U_P = (U_{P1}, \cdots, U_{PN_P})$ is the vector of the online/offline state of the thermal generation plants and N_P is the number of the thermal generation plants in the sending-side system. If the ith thermal generation plant is online, then $U_{Pi} = 1$; otherwise, $U_{Pi} = 0$; and once the state is settled, it does not change in one day. X_{Σ} is the function of U_P, and it is calculated using the Gaussian elimination equivalent method as follows:

    $$X_{\Sigma} = \mathrm{Im}\left(\frac{1}{Y_{PB}^T E - Y_{PB}^T Y_{PP}^{-1} Y_{PB}}\right) \tag{6}$$

    where $\mathrm{Im}(*)$ is the imaginary part of $*$. Y_{PP} is the thermal generation plant's node admittance matrix in the sending-side system; B is the border node in the sending-side that the transmission channel connects to, as shown in Figure 3. Y_{PB} is the column vector of mutual admittance between node B and the nodes in the sending-side system. E is a unit column vector with the same dimensions as Y_{PB}.

2. Constraint:

    The online thermal generation plants should be able to supply enough generation for the equivalent load and spare an appropriate amount of backup and reserve capacity.

    $$\sum_{i=1}^{N_P} U_{Pi} P_{pi}^{max} \geq P_{EL.forcast}^{max} + P_B \tag{7}$$

    where P_{Pi}^{max} is the maximum allowable output of the ith thermal generation plant. $P_{EL.forcast}^{max}$ is the maximum equivalent load forecast; $P_{EL.forcast}^{max} = \max\limits_{t \in T_d} P_{EL.forcast}^t$; T_d is the set of day-head scheduling periods, and the equivalent load forecast $P_{EL.forcast}^t$ can be obtained from Equation (1), with the known total load forecast and wind power forecast in the sending-side system and the initial scheduled power delivered through the transmission channel. P_B is the backup and reserve capacity, and each online thermal generation plant should at least have one spare generation unit for emergency. After the thermal generation plant commitment is settled, the generation schedule for each online thermal generation plant is in proportion to its maximum allowable output.

    $$P_{Pi}^t = \frac{U_{Pi} P_{Pi}^{max}}{\sum_{i=1}^{N_P} U_{Pi} P_{Pi}^{max}} P_{EL.forcast}^t \tag{8}$$

    where P_{Pi}^t is the generation schedule of the ith thermal generation plant in period t.

4.2.2. Solution Algorithm

The day-ahead thermal generation plant commitment optimization model is a 0–1 optimization problem, and the number of thermal generation plants in the large-scale wind power base sending-side system is finite and relatively small; therefore, the implicit enumeration algorithm is appropriate to solve this model [21]. The steps are as follows:

1. Use the priority method to give the initial feasible solution U_P^0, and calculate the objective of the initial feasible solution, which is $X_\Sigma\left(U_P^0\right)$.
2. Add the filter constraint $X_\Sigma(U_P) \leq X_\Sigma\left(U_P^0\right)$.
3. For the remaining feasible solutions, first check whether each satisfies the filter constraint. If not, it is considered infeasible. Otherwise, check whether it satisfies constraint (7). If it satisfies constraint (7), take it as the current optimal solution, and calculate its objective as the new filter constraint.
4. Repeat step 3 until all feasible solutions are traversed, and output the optimal solution, which is the optimized day-ahead thermal generation plant commitment.

4.3. Day-Ahead Thermal Generation Unit Commitment and Schedule Optimization

4.3.1. Day-Ahead Thermal Generation Unit Commitment and Schedule Optimization Model

Based on the optimized day-ahead thermal generation plant commitment and generation schedules from Section 4.2, according to Rule 2, the day-ahead thermal generation unit commitment and schedule optimization model for each online thermal generation plant is established as follows:

1. Objectives

   The objectives of thermal generation plant unit commitment and schedule optimization for the ith ($i = 1, \cdots, N_p$) online thermal generation plant are maximizing the equivalent inertia constant and minimizing the generation cost. Objective A:

$$\max T_i = \sum_{j=1}^{N_i} u_{ij} T_{ij} \frac{S_{ijN}}{S_B} \tag{9}$$

   Objective B:

$$\min C_i = \sum_{t=1}^{T_d} \sum_{j=1}^{N_i} u_{ij}\left(a_{ij} + b_{ij} P_{ij}^t + c_{ij}(P_{ij}^t)^2\right)\Delta t \tag{10}$$

   where T_i, C_i and N_i are the equivalent inertia constant, generation cost and number of generation units of the ith thermal generation plant, respectively. u_{ij}, T_{ij}, S_{ijN} and a_{ij}, b_{ij}, c_{ij} are the online/offline state, inertia constant, rated capacity and generation cost parameters of the jth generation unit in the ith thermal generation plant, respectively. $u_i = \left[u_{i1}, \cdots, u_{iN_i}\right]$ is the vector of u_{ij}; if the jth generation unit in the ith thermal generation plant is online, then $u_{ij} = 1$; otherwise, $u_{ij} = 0$; and once the state is settled, it does not change in one day. P_{ij}^t is the generation schedule of the jth generation unit in the ith thermal generation plant in period t.

2. Constraints:

$$\sum_{j=1}^{N_i} u_{ij} P_{ij}^t = P_{Pi}^t \tag{11}$$

$$u_{ij}P_{ij}^{min} \leq P_{ij}^{t} \leq u_{ij}P_{ij}^{max} \tag{12}$$

where P_{ij}^{min}, P_{ij}^{max} are the minimum and maximum allowable outputs of the jth generation unit in the ith thermal generation plant, respectively. If P_{Pi}^{t} is less than $\sum_{j=1}^{N_i} u_{ij}P_{ij}^{min}$, then $P_{Pij}^{t} = u_{ij}P_{ij}^{min}$.

4.3.2. Solution Algorithm

The day-ahead thermal generation unit commitment and schedule optimization model is a multi-objective optimization problem. In engineering applications, the multi-objective particle swarm optimization algorithm (MOPSO) can effectively solve this kind of problem [22].

For the ith ($i = 1, \cdots, N_p$) thermal generation plant, here are the steps of solving the above model using MOPSO:

1. Initialize a swarm of n particles.

    The vector of the generation schedules of the thermal generation units in the ith thermal generation plant $x_k = [P_{i1k}^1, \cdots, P_{i1k}^{T_d}, \cdots, P_{iN_ik}^1, \cdots, P_{iN_ik}^{T_d}]$ is the position of the kth particle. If $P_{ijk}^1 = \cdots = P_{ijk}^{T_d} = 0$, then $u_{ijk} = 0$; otherwise, $u_{ijk} = 1$. The velocity of x_k is $v_k = [v_k^1, \cdots, v_k^{T_d \times N_i}]$. x_k is initialized to the original day-ahead generation schedules of the thermal generation units in the ith thermal generation plant which are obtained based on the lowest generation cost rule. v_k is initialized to 0.

2. Update the personal best position.

    Take objective A and objective B as $F_1 = -T_i$ and $F_2 = C_i$, respectively. For the particle positions x and y, if $F_m(x) \leq F_m(y)$ ($m = 1, 2$) exists, and there is at least one m that satisfies $F_m(x) < F_m(y)$, then it is established that x dominates y. For the kth particle, its personal best position is $x_{k.best}$; if it is dominated by its current position, then update $x_{k.best}$ with its current position.

3. Update the set of global best positions.

    For the set of global best positions G_{best}, in each generation of particles, find the non-dominated particle position and add it to the set.

4. Update the position and velocity of the particles.

    For the k th particle, its velocity and position are updated as follows.

$$v_{k+1}^l = \omega v_k^l + c_1 r_1 \left(x_{k.best}^l - v_k^l\right) + c_2 r_2 \left(g_{s.best}^l - v_k^l\right) \tag{13}$$

$$x_{k+1}^l = x_k^l + v_{k+1}^l \tag{14}$$

where ω is inertia weight; c_1, c_2 are coefficient parameters. r_1, r_2 are random numbers in $[0,1]$, and they are generated by uniformly random sampling. $x_{k.best}^l$ is the lth index of $x_{k.best}$, and $g_{s.best}^l$ is the lth index of the randomly selected global best position.

5. Repeat steps 2–4 until the maximum number of iterations is reached, and output the set of global best positions G_{best}, which is a set of Pareto optimal solutions. The maximum number of iterations is determined according to the convergence situation of the objectives.

Use fuzzy membership function to select the satisfactory optimization solution from G_{best}. For the sth Pareto optimal solution, the satisfaction of the mth objective is quantified using the fuzzy membership function as follows:

$$\mu_{ms} = \frac{F_m^{max} - F_{ms}}{F_m^{max} - F_m^{min}} \tag{15}$$

where $F_m^{\max}$, $F_m^{\min}$ are the maximum value and minimum value of the mth objective in G_{best}. The satisfaction of the sth Pareto optimal solution is as follows:

$$\mu_s = \frac{1}{2} \sum_{m=1}^{2} \mu_{ms} \tag{16}$$

Find the Pareto optimal solution with the highest satisfaction, and take it as the day-ahead thermal generation unit commitment and the generation schedules inside the ith thermal generation plant.

Using the above method, solve the day-ahead thermal generation unit commitment and schedule optimization model for each thermal generation plant that is scheduled to be online from Section 4.2, and the thermal generation schedules in the sending-side system are obtained.

4.4. Optimal Day-Ahead Thermal Generation Scheduling Method

In summary, the optimal day-ahead thermal generation scheduling method to enhance TTC for the sending-side system with large-scale wind power integration is as follows, and the flow diagram of this method is shown in Figure 7.

1. Collect the following data: total load forecast, wind power forecast, grid topology parameters, thermal power plant and unit parameters, original day-ahead generation schedules in the sending-side system and original TTC.
2. With the objective of minimizing the equivalent electrical distance between the transmission channel and the equivalent synchronous generator of the sending-side system, establish the day-ahead thermal generation plant commitment optimization model. Solve the model and obtain the optimized day-ahead thermal generation plant commitment, and the generation schedules.
3. Based on the optimized day-ahead thermal generation plant commitment and generation schedules from step 2, for each online thermal generation plant, with the objectives of maximizing the equivalent inertia constant and minimizing the generation cost, establish the day-ahead thermal generation unit commitment and schedule optimization model. Solve the model and get the optimized day-ahead thermal generation unit commitment and generation schedules.
4. Based on the optimized day-ahead thermal generation schedules, calculate the enhanced TTC. According to Rule 3, increase wind power output based on the enhanced TTC to reduce curtailed wind power.

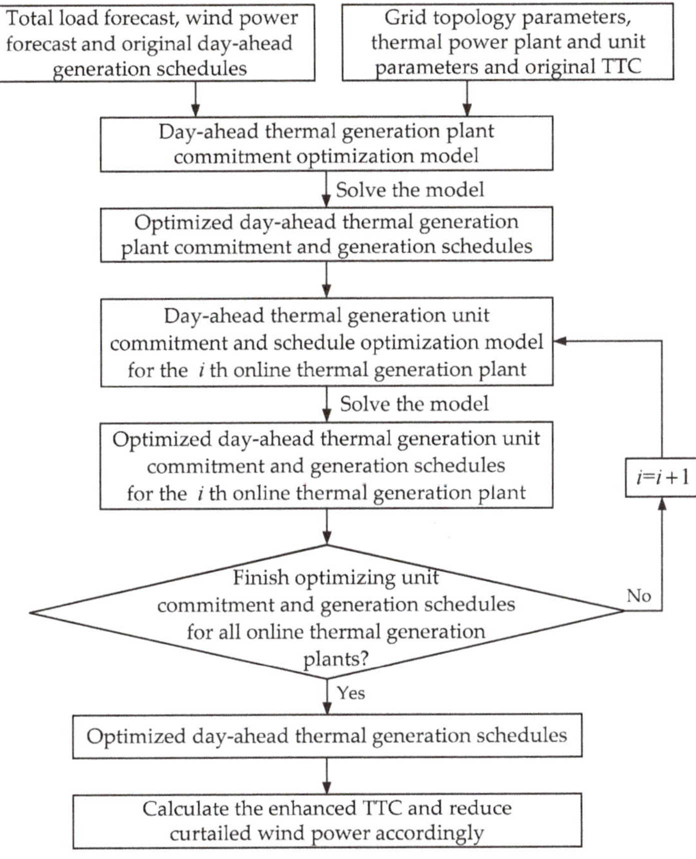

Figure 7. Flow diagram of the optimal day-ahead thermal generation scheduling method.

5. Case Analysis

The proposed optimal day-ahead thermal generation scheduling method to enhance TTC was applied to the large-scale wind power base sending-side system in Gansu Province in China. The optimization models were solved using the above algorithms in MATLAB. TTC was calculated using the transient, stability-constrained continuation power flow method [23], using the Power System Analysis Software Package (PSASP), which is widely used for power system calculations and simulations in China.

5.1. Test System Description

The simplified diagram of the large-scale wind power base sending-side system in Gansu Province is shown in Figure 8. The receiving-side system is equivalent to an infinity bus system. Hexi Substation is the border node in the sending-side that the transmission channel connects to. Four thermal generation plants are involved; the capacities of the units in each plant are shown in Table 1; and the relative parameters of each kind of unit are shown in Table 2. The electrical distance parameters that describe the grid topology are shown in Table 3 (the resistances of the lines are small and ignored). The base power is $S_B = 100\ MVA$ in the system. While calculating the transient, stability-constrained TTC of the transmission channel, the transient models of the involved electronic components are used, and the relative parameters and the TTC calculation process are shown in the Appendix A.

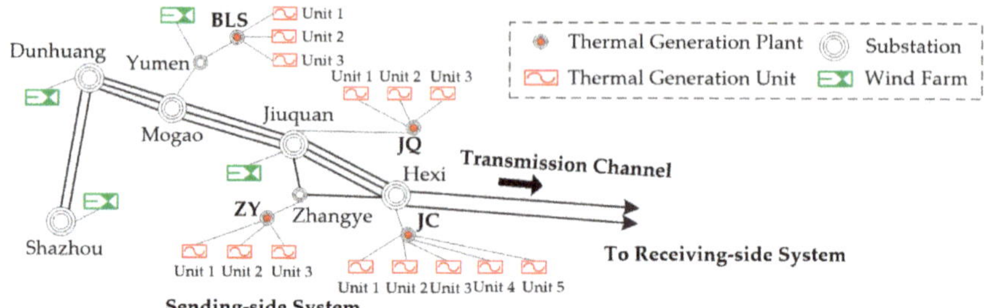

Figure 8. Simplified diagram of the large-scale wind power base sending-side system in Gansu Province.

Table 1. Capacities of units in each thermal generation plant.

Thermal Generation Plant	Unit Number	Unit Capacity/ (MW)
JC	Unit 1	500
	Unit 2	500
	Unit 3	500
	Unit 4	300
	Unit 5	200
ZY	Unit 1	200
	Unit 2	100
	Unit 3	100
JQ	Unit 1	500
	Unit 2	300
	Unit 3	200
BLS	Unit 1	200
	Unit 2	100
	Unit 3	100

Table 2. Parameters of each kind of thermal generation unit.

Unit Capacity/ (MW)	Inertia Constant/ (s/100 MW)	Generation Cost Parameters		
		a/($)	b/($/MWh)	c/($/(MW)^2 h)
500	1.50	200	38	0.040
300	2.63	230	45	0.045
200	3.12	260	46	0.051
100	3.59	300	65	0.080

Table 3. Electrical distance parameters.

Lines	Reactance/(p.u.)	Lines	Reactance/(p.u.)
Hexi-JC	0.080	Jiuquan-Mogao	0.250
Hexi-Zhangye	0.433	Mogao-Shazhou	0.071
Hexi-Jiuquan	0.426	Mogao-Yumen	0.135
Zhangye-Jiuquan	0.519	Mogao-Dunhuang	0.013
Zhangye-ZY	0.530	Yumen-BLS	0.061
Jiuquan-JQ	0.22		

The original day-head generation schedules are shown in Table 4. The four thermal generation plants are all online, and the thermal generation unit commitment and the generation schedule in each plant are based on the lowest generation cost rule. The corresponding original TTC is 4404 MW. Due to the restriction of TTC of the transmission channel, during period 1–10 (each period is one hour), the minimum allowable thermal generation is greater than the equivalent load, so the wind power is curtailed. The curtailed wind power is 5935 MWh.

Table 4. Original day-head generation schedules and curtailed wind power/(MW).

Day-Head Scheduling Period	JC		ZY	JQ	BLS	Curtailed Wind Power
	Unit 1	Unit 2	Unit 1	Unit 1	Unit 1	
1	300	300	120	300	120	201
2	300	300	120	300	120	436
3	300	300	120	300	120	700
4	300	300	120	300	120	920
5	300	300	120	300	120	920
6	300	300	120	300	120	892
7	300	300	120	300	120	920
8	300	300	120	300	120	500
9	300	300	120	300	120	350
10	300	300	120	300	120	96
11	368	368	147	368	147	0
12	366	366	146	366	146	0
13	367	367	147	367	147	0
14	411	411	164	411	164	0
15	371	371	148	371	148	0
16	374	374	149	374	149	0
17	381	381	152	381	152	0
18	447	447	179	447	179	0
19	441	441	177	441	177	0
20	400	400	160	400	160	0
21	413	413	165	413	165	0
22	379	379	152	379	152	0
23	396	396	158	396	158	0
24	409	409	164	409	164	0

5.2. Analysis of the Results of Day-Ahead Thermal Generation Schedule Optimization

Establish the day-ahead thermal generation plant commitment optimization model and solve the model using the method in Section 4.2. The optimized day-ahead thermal generation plant commitment is obtained as shown in Table 5.

Table 5. Optimized day-ahead thermal generation plant commitment.

Thermal Generation Plant	On/Off State	Equivalent Reactance/(p.u.)
JC	1	
ZY	1	0.1074
JQ	1	
BLS	0	

For the online thermal generation plants of JC, ZY and JQ, establish the day-ahead thermal generation unit commitment, schedule optimization models and solve the models using the method in Section 4.3, respectively. The optimized day-ahead thermal generation unit commitment and generation schedules are shown in Tables 6 and 7.

The comparisons of the relative electrical parameters and generation costs of the sending-side system and the corresponding TTC before and after optimizing the day-head generation schedules are shown in Table 8. The detailed process of how to get the results in Table 8 is provided in the supplementary materials.

Table 6. Optimized day-ahead thermal generation unit commitment.

Thermal Generation Plant	Unit Number	On/Off State	Equivalent Inertia Constant /(s)
JC	Unit 1	1	
	Unit 2	0	
	Unit 3	0	21.63
	Unit 4	1	
	Unit 5	1	
ZY	Unit 1	0	
	Unit 2	1	7.18
	Unit 3	1	
JQ	Unit 1	1	
	Unit 2	0	13.74
	Unit 3	1	

Table 7. Optimized day-ahead thermal generation schedules/(MW).

Day-Head Scheduling Period	JC			ZY		JQ	
	Unit 1	Unit 4	Unit 5	Unit 2	Unit 3	Unit 1	Unit 3
1	300	180	120	60	60	300	120
2	300	180	120	60	60	300	120
3	300	180	120	60	60	300	120
4	300	180	120	60	60	300	120
5	300	180	120	60	60	300	120
6	300	180	120	60	60	300	120
7	300	180	120	60	60	300	120
8	300	180	120	60	60	300	120
9	300	180	120	60	60	300	120
10	300	180	120	60	60	300	120
11	368	221	147	74	74	368	147
12	366	219	146	73	73	366	146
13	367	220	147	73	73	367	147
14	411	247	164	82	82	411	164
15	371	223	148	74	74	371	148
16	374	224	149	75	75	374	149
17	381	228	152	76	76	381	152
18	447	268	179	89	89	447	179
19	441	265	177	88	88	441	177
20	400	240	160	80	80	400	160
21	413	248	165	83	83	413	165
22	379	228	152	76	76	379	152
23	396	238	158	79	79	396	158
24	409	245	164	82	82	409	164

Table 8. Comparison of relative electrical parameters, generation cost and TTC before and after optimizing the day-head generation schedules.

	Equivalent Reactance/(p.u.)	Equivalent Inertia Constant/(s)	Thermal Generation Cost/($)	TTC/(MW)
Before	0.1074	34.98	1,732,578	4404
After	0.0591	42.55	1,918,591	5105

As shown in Table 8, by optimizing the day-ahead thermal generation schedules, the equivalent electrical distance between the transmission channel and the equivalent synchronous generator of the sending-side system decreases by 0.0483 p.u. and the equivalent inertia constant of the sending-side system increases by 7.57 s. The corresponding TTC is enhanced from 4404 to 5105 MW. The explanation of the TTC enhancement is as follows: Compared to the original day-head thermal generation schedules, the optimized generation schedules choose the thermal generation plants which are electrically closer to the transmission channel to be online, and shut down the BLS thermal generation plant which is the farthest plant from the transmission channel; therefore, the equivalent electrical distance between the transmission channel and the equivalent synchronous generator of the sending-side system is decreased; what is more, the optimized day-ahead thermal generation unit commitment inside each online plant chooses the thermal generation units with bigger inertia constants to be online; therefore, the equivalent inertia constant of the sending-side system is increased. Due to the decreased equivalent electrical distance, the electrical connection between the sending-side system and the transmission channel becomes tighter, and due to the increased equivalent inertia constant, the rotor angle accelerates slowly after a fault occurs, and this improves the transient stability of the system. Therefore, while the power delivered through the transmission channel is still 4404 MW, the system has not reached the transient stability critical state, and the power delivered through the transmission channel can increase to 5105 MW, which is the TTC after the day-head generation schedules are optimized.

Using the enhanced TTC, the curtailed wind power can be reduced. The comparison of curtailed wind power before and after optimizing the day-head generation schedules is shown in Figure 9.

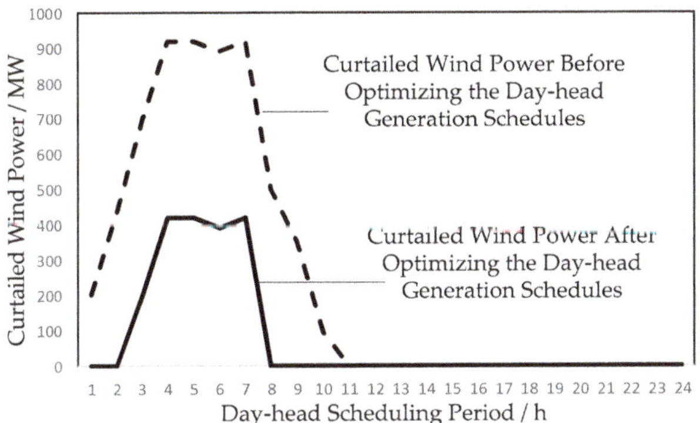

Figure 9. Comparison of curtailed wind power before and after optimizing the day-head generation schedules.

As shown in Figure 9, the curtailed wind power is reduced by 4083 MWh, and the benefit from it is $274,214 assuming the tariff of wind power is 67.16 $/MWh. As shown in Table 8, the thermal generation cost increases by $186,013. In general, the economic benefit brought by optimizing the day-ahead thermal generation schedules to enhance TTC is $88,201. It can be seen that, for the purpose of enhancing TTC, some of the units which are electrically closer to the transmission channel and with bigger inertia are constant while those with higher generation costs (like the Unit 4 and Unit 5 in JC thermal generation plant) are chosen to be online, and that increases the generation cost. However, the benefit from consuming the curtailed wind power because of the enhanced TTC is more than the additional generation cost, and the general cost is reduced. Therefore, the proposed method is economically favorable.

5.3. Suitable Parameter Selection for MOPSO

While solving the day-ahead thermal generation unit commitment and schedule optimization models in Section 4.3.1, we use the MOPSO in Section 4.3.2, and the initial population, generation of random members, termination criterion and other parameters in MOPSO have certain effects on the results. Taking the day-ahead thermal generation unit commitment and schedule optimization model solution of JC thermal generation plant as an example, the suitable parameters for solving the models are selected through comparison and sensitivity analysis as follows.

1. Initial population

There are three commonly used initial solutions in the day-ahead thermal generation unit commitment and schedule optimization problem, and they are used as initial populations in MOPSO respectively, as follows:

Initial population A: day-ahead generation schedules obtained based on the lowest generation cost rule.

Initial population B: day-ahead generation schedules obtained based on the installation capacity proportion rule.

Initial population C: random day-ahead generation schedules.

The objective results of JC thermal generation plant obtained from the above three initial populations are shown in Table 9. It can be seen in Table 9 that the equivalent inertia constant results are the same; this is because the day-ahead thermal generation unit commitments obtained from these three different initial populations are the same. Initial population A can get the lowest thermal generation cost because it is originally obtained based on the lowest generation cost rule. Therefore, we chose initial population A as the initial population in Section 5.2.

Table 9. Comparison of objective results of JC thermal generation plant based on the above three initial populations.

	Initial Population A	Initial Population B	Initial Population C
Equivalent Inertia Constant/(s)	21.63	21.63	21.63
Thermal Generation Cost/($)	931,402	931,671	932,085

2. Generation of random members and inertia weight

In the velocity update equation (Equation (13)), the generation of random members r_1, r_2 and the value of inertia weight ω effect the global search capability of MOPSO. Figure 10 shows the objective results based on different generation methods of random members and different values of inertia weight. In Figure 10a,b, the random members are generated by uniform random sampling in [0,1] and Monte Carlo sampling in [0,1], respectively, and the inertia weights range from 0.4 to 2.4.

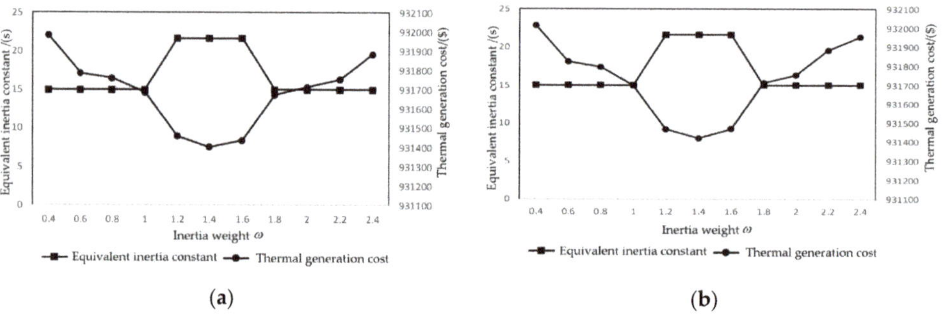

Figure 10. (a) Objective results with random members generated by uniformly random sampling; (b) objective results with random members generated by Monte Carlo sampling.

It can be seen from Figure 10 that the generation method of random members has little effect on the results. However, the value of inertia weight significantly influences the results, and when the value is near 1.4, MOPSO gets the best global search capability, and this is because if the value of inertia weight is too small, the results will fall into the local optimal solution, and if the value of inertia weight is too big, the results may miss the global optimal solution. Therefore, we used uniformly random sampling to generate random members for simplicity, and we let the inertia weight take the value of 1.4 in Section 5.2.

3. Termination criterion

The termination criterion, which is the maximum number of iterations of MOPSO should consider when converging on the objectives. Figure 11 shows the convergence situation of the objectives as the number of iterations increases. It can be seen that when the number of iterations is more than

50, both equivalent inertia constant and thermal generation cost converge to their optimal values. Therefore, the maximum number of iterations is set to 55 as the termination criterion in Section 5.2.

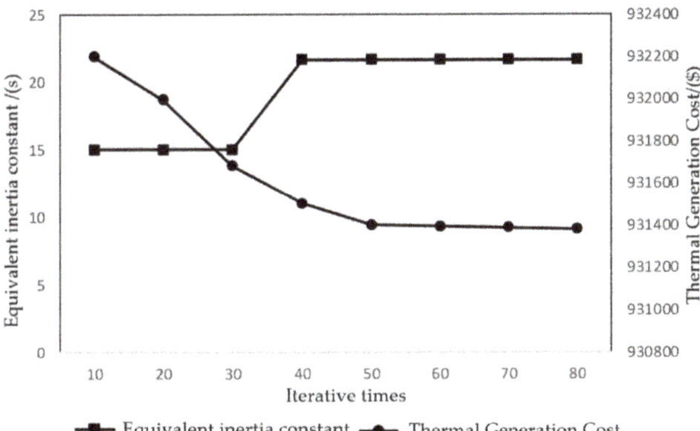

Figure 11. Convergence situation of the objectives as the number of iterations increases.

6. Conclusions

Enhancing TTC allows more wind power to be delivered through the transmission channel to the load center, and it is of great importance to help consume wind power in the large-scale wind power base sending-side system. This paper proposed a new method to enhance TTC based on the existing grid structure; namely, by optimizing the day-ahead thermal generation schedules. It can significantly enhance TTC, and therefore, reduce the curtailed wind power. The conclusions are as follows:

1. TTC increases with the decrease of the equivalent electrical distance between the transmission channel and the equivalent synchronous generator of the sending-side system. TTC increases with the inertia constant of the equivalent synchronous generator of the sending-side system.
2. Optimizing the day-ahead thermal generation plant commitment with the objective of minimizing the equivalent electrical distance between the transmission channel and the equivalent synchronous generator of the sending-side system, and then optimizing the day-ahead thermal generation unit commitment and schedule considering the objective of maximizing the equivalent inertia constant can significantly enhance TTC, and therefore, help consume the curtailed wind power.

The proposed method in this paper enhances TTC by optimizing the day-ahead thermal generation schedules; thus, it achieves a reduction of curtailed wind power, and it is of great significance to help improve wind power consumption in the sending-side system with large-scale wind power integration.

Supplementary Materials: The following are available online at http://www.mdpi.com/1996-1073/13/9/2375/s1, PDF: Some Detailed Calculation Processes of the Results.

Author Contributions: Conceptualization, Y.Z.; methodology, Y.Z. and W.L.; validation, Y.Z. and Y.H.; investigation, Y.Z. and Y.H.; writing—original draft preparation, Y.Z.; writing—review and editing, W.L. and N.W.; supervision, W.L. and Q.Z. All authors have read and agreed to the published version of the manuscript.

Funding: This research was funded by National Key Technologies R&D Program (number 2015BAA01B04) and Project of State Grid Corporation of China (number SGGSKY00FJJS1800025).

Conflicts of Interest: The authors declare no conflict of interest.

Appendix A

The generators in the sending-side system are thermal generation units and wind power generators. The thermal generation units are simulated using synchronous generators, and their transient reactance is shown in Table A1. The wind farms are simulated using doubly-fed induction generators (DFIG).

The active power control mode of the DFIG is maximum power point tracking and the reactive power control mode is constant power factor control. The transient relevant parameters are shown in Table A2.

Table A1. Transient parameters of units in each thermal generation plant.

Thermal Generation Plant	Unit Number	Transient Reactance/(p.u.)
JC	Unit 1	0.0256
	Unit 2	0.0256
	Unit 3	0.0256
	Unit 4	0.0277
	Unit 5	0.0282
ZY	Unit 1	0.0282
	Unit 2	0.0291
	Unit 3	0.0291
JQ	Unit 1	0.0256
	Unit 2	0.0277
	Unit 3	0.0282
BLS	Unit 1	0.0282
	Unit 2	0.0291
	Unit 3	0.0291

Table A2. Transient relevant parameters of DFIG.

Transient Parameters	Values
Stator resistance	0.00706 p.u.
Rotor resistance	0.005 p.u.
Stator leakage inductance	0.171 p.u.
Rotor leakage inductance	0.156 p.u.
Magnetizing inductance	2.9 p.u.

Taking the TTC calculation of the original day-head generation schedules as an example, the TTC calculation process is as follows:

1. Based on the original day-head generation schedules in Table 4, take the generation schedule of period 1 as the initial operation point.
2. Increase the outputs of the wind farms in proportion to their original outputs, and decrease the outputs of the generators in the receiving-side system equally, and perform power flow calculations and a transient stability time domain simulation to check that the system is able to stay stable when an N-1 fault occurs on the transmission channel.
3. Repeat step 2, and the maximum steady state power that can be delivered through the transmission channel is 4404 MW, provided the system is able to stay transiently stable after a fault occurs. Therefore, the TTC of the original day-head generation schedules is 4404 MW.

References

1. Zhu, D.; Liu, W.; Hu, Y.; Wang, W. A Practical Load-Source Coordinative Method for Further Reducing Curtailed Wind Power in China with Energy-Intensive Loads. *Energies* **2018**, *11*, 2925. [CrossRef]
2. Liao, S.; Xu, J.; Sun, Y.; Bao, Y.; Tang, B. Control of Energy-Intensive Load for Power Smoothing in Wind Power Plants. *IEEE Trans. Power Syst.* **2018**, *33*, 6142–6154. [CrossRef]
3. Ma, H.; Yan, Z.; Li, M.; Han, D.; Han, X.; Song, Y.; Wei, P.; Li, G.; Liu, Y. Benefit Evaluation of the Deep Peak-regulation Market in the Northeast China Grid. *IEEE Comput. Appl. Power CSEE J. Power Energy Syst.* **2019**, *5*, 533–544.
4. Peng, F.; Gao, Z.; Hu, S.; Zhou, W.; Sun, H.; Wang, Z. Bilateral Coordinated Dispatch of Multiple Stakeholders in Deep Peak Regulation. *IEEE Access* **2020**, *8*, 33151–33162. [CrossRef]

5. Yao, X.; Yi, B.; Yu, Y.; Fan, Y.; Zhu, L. Economic analysis of grid integration of variable solar and wind power with conventional power system. *Appl. Energy* **2020**, *264*, 114706. [CrossRef]

6. Lu, M.S.; Chang, C.L.; Lee, W.J.; Wang, L. Combining the Wind Power Generation System with Energy Storage Equipment. *IEEE Trans. Ind. Appl.* **2009**, *45*, 2109–2115.

7. Bitaraf, H.; Rahman, S. Reducing Curtailed Wind Energy Through Energy Storage and Demand Response. *IEEE Trans. Sustain. Energy* **2018**, *9*, 228–236. [CrossRef]

8. Hou, L.; Li, W.; Zhou, K.; Jiang, Q. Integrating flexible demand response toward available transfer capability enhancement. *Appl. Energy* **2019**, *251*, 113370. [CrossRef]

9. Fang, R.; Shang, R.; Wang, Y.; Guo, X. Identification of vulnerable lines in power grids with wind power integration based on a weighted entropy analysis method. *Int. J. Hydrogen Energy* **2017**, *42*, 20269–20276. [CrossRef]

10. Hou, G.; Vittal, V. Determination of Transient Stability Constrained Interface Real Power Flow Limit Using Trajectory Sensitivity Approach. *IEEE Trans. Power Syst.* **2013**, *28*, 2156–2163. [CrossRef]

11. Gholipour, E.; Saadate, S. Improving of Transient Stability of Power Systems Using UPFC. *IEEE Trans. Power Deliv.* **2005**, *20*, 1677–1682. [CrossRef]

12. Johansson, N.; Ängquist, L.; Nee, H.P. An Adaptive Controller for Power System Stability Improvement and Power Flow Control by Means of a Thyristor Switched Series Capacitor (TSSC). *IEEE Trans. Power Syst.* **2010**, *25*, 381–391. [CrossRef]

13. Paital, S.R.; Ray, P.K.; Mohanty, A. Comprehensive review on enhancement of stability in multimachine power system with conventional and distributed generations. *IET Renew. Power Gener.* **2018**, *12*, 1854–1863. [CrossRef]

14. Varma, R.K.; Rahman, S.A.; Vanderheide, T. New Control of PV Solar Farm as STATCOM (PV-STATCOM) for Increasing Grid Power Transmission Limits During Night and Day. *IEEE Trans. Power Deliv.* **2015**, *30*, 755–763. [CrossRef]

15. Yossef, D.; Tali, I.; Yehonatan, R.; Dror, M.; Danping, L.; Zeev, Z. Silicon-coated gold nanoparticles nanoscopy. *J. Nanophotonics* **2016**, *10*, 036015.

16. Vegerhof, A.; Barnoy, E.A.; Motiei, M.; Malka, D.; Danan, Y.; Zalevsky, Z.; Popovtzer, R. Targeted Magnetic Nanoparticles for Mechanical Lysis of Tumor Cells by Low-Amplitude Alternating Magnetic Field. *Mater. Spec. Issue Nanoprobes Imaging* **2016**, *9*, 943. [CrossRef]

17. Ariff, M.A.M.; Pal, B.C.; Singh, A.K. Estimating Dynamic Model Parameters for Adaptive Protection and Control in Power System. *IEEE Trans. Power Syst.* **2015**, *30*, 829–839. [CrossRef]

18. Jiang, Y.; Yu, S.; Wen, B. Monthly electricity purchase and decomposition optimization considering wind power accommodation and day-ahead schedule. *Int. J. Electr. Power Energy Syst.* **2019**, *107*, 231–238. [CrossRef]

19. Salkuti, S.R. Day-ahead thermal and renewable power generation scheduling considering uncertainty. *Renew. Energy* **2019**, *131*, 956–965. [CrossRef]

20. Paudyal, S.; Ramakrishna, G.; Sachdev, M.S. Application of Equal Area Criterion Conditions in the Time Domain for Out-of-Step Protection. *IEEE Trans. Power Deliv.* **2010**, *25*, 600–609. [CrossRef]

21. Romero, R.; Monticelli, A. A Zero-one Implicit Enumeration Method for Optimizing Investments in Transmission Expansion Planning. *IEEE Trans. Power Syst.* **1994**, *9*, 1385–1391. [CrossRef]

22. Pham, M.; Zhang, D.; Koh, C.S. Multi-Guider and Cross-Searching Approach in Multi-Objective Particle Swarm Optimization for Electromagnetic Problems. *IEEE Trans. Magn.* **2012**, *48*, 539–542. [CrossRef]

23. Ejebe, G.C.; Tong, J.; Waight, J.G.; Frame, J.G.; Wang, X.; Tinney, W.F. Available Transfer Capability Calculations. *IEEE Trans. Power Syst.* **1998**, *13*, 1521–1527. [CrossRef]

Article

Multi-Time Scale Optimization Scheduling Strategy for Combined Heat and Power System Based on Scenario Method

Yunhai Zhou [1,2]**, Shengkai Guo** [1,2,]*****, Fei Xu** [3,]*****, Dai Cui** [4,5]**, Weichun Ge** [4,5]**, Xiaodong Chen** [4] **and Bo Gu** [4]

[1] College of Electrical Engineering & New Energy, China Three Gorges University, Yichang 443000, China; zhouyunhai@ctgu.edu.cn

[2] Hubei Provincial Key Laboratory for Operation and Control of Cascaded Hydropower Station, China Three Gorges University, Yichang 443000, China

[3] Department of Electrical Engineering, Tsinghua University, Beijing 100000, China

[4] Jinzhou Power Supply Company, Liaoning Electric Power Co., Ltd., Shenyang 100084, China; cuidai1982@163.com (D.C.); 13804012589@126.com (W.G.); chenxd@ln.sgcc.com.cn (X.C.); g17639623818@163.com (B.G.)

[5] School of Electrical Engineering, Shenyang University of Technology, Shenyang 100084, China

***** Correspondence: shengkai_guo@163.com (S.G.), xufei@tsinghua.edu.cn (F.X.); Tel.: +86-1827-167-4293(S.G.); +86-1351-100-3897 (F.X.)

Received: 16 March 2020; Accepted: 30 March 2020; Published: 1 April 2020

Abstract: The wind–heat conflict and wind power uncertainty are the main factors leading to the phenomenon of wind curtailment during the heating period in the northern region of China. In this paper, a multi-time scale optimal scheduling strategy for combined heat and power system is proposed. Considering the temporal dependence of wind power fluctuation, the intra-day wind power scenario generation method is put forward, and both day-ahead and intra-day optimization scheduling models based on the scenario method are established to maximize the system's revenue. The case analyzes the impacts of the initial heat storage capacity of a heat storage device and different scheduling strategies on system revenue. It is verified that the scheduling strategy can better adapt to wind power uncertainty and improve the absorption capacity of wind power, while ensuring the safety and economical efficiency of system operation.

Keywords: combined heat and power system; wind power uncertainty; scenario method; temporal dependence; optimization scheduling

1. Introduction

At present, wind power (WP) is developing rapidly all over the world. By the end of June 2019, the cumulative installed capacity of wind power in China reached 193 million kW, of which 9.09 million kW was newly installed from January to June 2019. However, in the rapid development of wind power, the problem of wind power accommodation is particularly prominent. According to the National Energy Administration of China, the national average wind power utilization hours in the first half of 2019 were 1133 hours, and wind power curtailment reached 10.5 billion kW·h in the first six months [1].

The main reason for wind curtailment is the uncertainty of wind power output [2], which requires more flexible adjustment resources than conventional units. However, in the northern region of China, where wind power resources are abundant, flexible power sources such as pumped storage station and gas station account for less than 4% of the total. Peak regulation in winter is particularly difficult. In addition, due to the superposition of "three phases" during the heating period, through the period of electricity consumption and peak period of wind power generation, most of the wind energy is curtailed in the late night of the winter heating period. Furthermore, the thermal-electric coupling of the combined heat and power (CHP) unit limits the electric power regulating ability and intensifies the wind curtailment phenomenon.

In order to reduce the wind curtailment in the winter heating period, a large number of studies have been conducted to improve the regulating capacity of CHP units [3]. The authors of [4] analyzed the operating characteristics of the CHP unit configured with heat storage and the peak shaving ability, and a dispatching model for combined heat and power system (CHPS) with heat storage device (HSD) was established, which verified the effectiveness of the model for improving wind power accommodation. In [5], two aspects of scheduling model and optimization control were discussed, and the specific implementation of heat storage in CHP units to improve the system's ability to absorb wind power was studied. The authors of [6] presented a model that determined the theoretical maximum of flexibility of a combined heat and power system (CHPS) coupled to a thermal energy storage solution that can be either centralized or decentralized. The authors of [7] presented the issue of robust operation of a multicarrier energy system with electric vehicles and CHP units. However, these above studies did not consider the impact of wind power uncertainty on the scheduling results, and only the operation mode under the typical daily wind power output curve is considered. Therefore, the authors of [8] considered the randomness of wind power output, and the stochastic optimization dispatching model based on scenario method was established, and solved the uncertainty problem of wind power output well. In [9], to cope with the uncertainties of the renewable energy sources and loads, some scenarios were generated using the scenario-based analysis. Nevertheless, the prediction accuracy of wind power output is not high enough, and the forecasting error of wind power is positively correlated with time. The result in single day-ahead scheduling plan differs greatly from the day's operating output, and cannot be applied to the CHPS with large-scale wind power. In view of this, the authors of [10] established the intra-day rolling scheduling model and adopted the intra-day rolling scheduling to modify the day-ahead plan output. In [11], through day-ahead scheduling, day-in rolling, and real-time dispatch, the influence of uncertain factors on microgrid cluster's stable and economic operation could be eliminated to the most extent; whereas, in the above literatures, the impact of initial heat storage capacity of the HSD on the optimization results was not discussed, and most of them assigned a fixed value such as the total or half capacity of the HSD. Furthermore, the operation optimization of HSD is a typical multi-stage decision problem, with close relationship between different periods, but the intra-day scheduling plan cannot be globally optimized because it is a phased plan (four hours), whereas the wind power uncertainty model used in most of the current researches did not consider the temporal dependence of wind power fluctuation.

Based on problems above, a multi-time scale optimal scheduling strategy based on scenario method in this paper is proposed for the CHPS including WP unit, thermal power (TP) unit, and CHP unit with HSD. First, based on short-term wind power forecast data, the day-ahead optimization scheduling model based on the scenario method is established to maximize the benefit of CHPS. The optimal initial heat storage capacity of the HSD is determined by analyzing the impact on the overall revenue of the system. At this time, the day-ahead scheduling plan is obtained. Then, based on the day-ahead scheduling plan and the ultra-short-term prediction data of wind power, considering the temporal dependence of wind power fluctuation, the intra-day wind power scenario generation method is put forward. Moreover, the intra-day rolling scheduling model is also established to maximize the benefit of CHPS, and solved by using commercial optimization software Cplex, which is provided by IBM (International Business Machines Corporation) in Armonk, NY, USA. The expected values of the decision variables in dispatch period are used as the intra-day plan to adjust and modify the previous plan. Finally, the value of each decision variable at this moment is taken as the boundary condition of the next rolling scheduling period, and the above steps are cycled until the end of the whole day-ahead scheduling plan. The simulation is compared with the dispatch model that does not consider this, and the effectiveness to improve system revenue and promote wind power consumption is verified.

The paper is structured as follows. In Section 2, the structure of the CHPS is described. The day-ahead optimal scheduling model is established in Section 3. The intra-day rolling optimization scheduling model is proposed in Section 4. The model's solution method is described in detail in Section 5, and the case study is presented in Section 6. Finally, the conclusions are drawn in Section 7.

2. Structure of the Combined Heat and Power System

The brief structure of the CHPS including the TP units and the WP units, as well as the CHP units with HSDs, is shown in Figure 1.

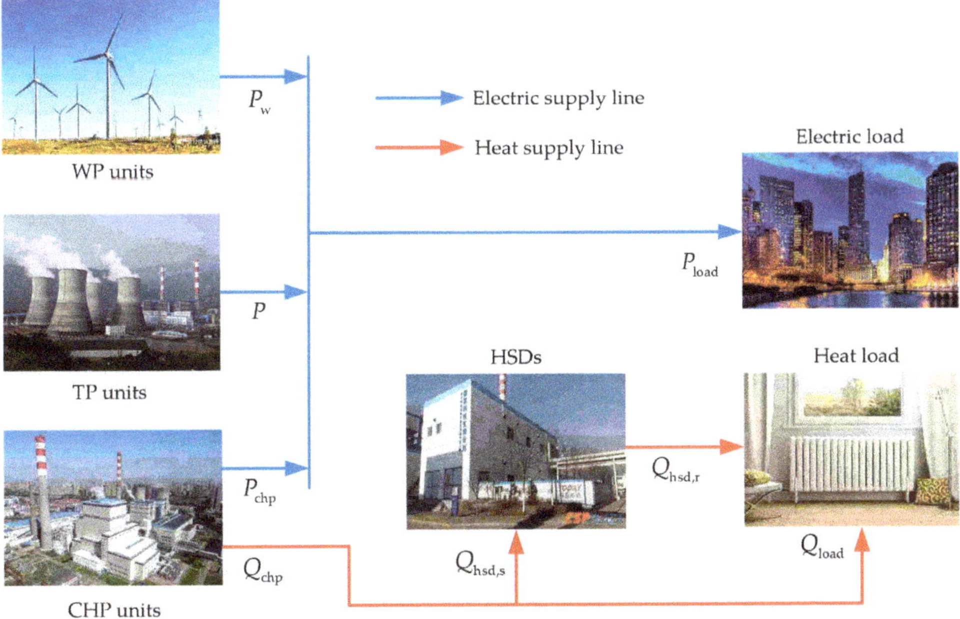

Figure 1. Structure of the combined heat and power system.

By adding the HSD, the CHPS can break the rigid constraint of the CHP unit and increase the regulating capacity. Traditional CHP units are generally constrained by "power determined by heat" during the winter heating period. With the addition of the HSD, the heat output of the CHP unit will not have to be balanced with the heat load in real time by controlling the heat storage and heat release of the HSD. In this way, the CHP unit will break the rigid constraint of "power determined by heat" and can have certain flexibility of regulation, to enhance the system's regulation capacity and the accommodation of wind power.

3. Day-ahead Optimization Scheduling Model

3.1. Objective Function

The day-ahead optimal scheduling model of CHPS takes the highest expected value of the system operating revenue under different wind power output scenarios as the objective function [12], which is composed of the power and heat supply benefits minus the operation cost of the units and penalty cost in the system. In order to encourage the system to accommodate more wind power, the cost of wind power generation is ignored.

$$\max f_1 = \sum_{k=1}^{N_k} \rho_k \sum_{t=1}^{T} \left(S_t^k - C_t^k - C_{\text{pun},t}^k \right) \tag{1}$$

where k is scenario numbering. N_k is the total number of scenarios. ρ_k is the probability of scenario k. T is the total scheduling time. S_t^k, C_t^k, and $C_{\text{pun},t}^k$ are the power and heat supply benefits, the operation cost of the units, and the penalty cost at moment t under scenario k, respectively.

In the CHPS, to encourage more wind power accommodation, the cost of wind power generation can be ignored. The system operating cost includes the cost of generating electricity from TP units, the cost of generating electricity, and heat from CHP units [13].

3.1.1. Generation Cost of TP Units

Due to the uncertainty of wind power, the peak-shaving tasks undertaken by (TP) units have been aggravated, and the operating status has also changed accordingly. The power generation cost of TP units in the dispatching model includes the start-up and shutdown cost and operating cost. As the start-up and shutdown of the unit are not considered in day-ahead dispatch, this paper only considers the operating cost of TP units.

$$F_1 = \sum_{t=1}^{T} \sum_{i=1}^{N} \left[a_i (P_{i,t}^k)^2 + b_i P_{i,t}^k + c_i \right] \tag{2}$$

where F_1 is the operating cost of TP units. a_i, b_i, and c_i are the second-order fitting coefficients of the operating cost of unit i. N is the total number of TP units. $P_{i,t}^k$ is the electrical output of unit i at moment t under scenario k.

3.1.2. Generation Cost of CHP Units

The CHP units not only bear the electric output of the system, but also undertake the heating task. Therefore, these units are in continuous operation during the heating period, so the generation cost of CHP units only includes the operating cost too. After removing the heat supply of HSD at time t, the heat output can be converted into the electric output under the pure condensing condition and then calculated by Formula (3) [14].

$$F_2 = \sum_{t=1}^{T} \sum_{i=1}^{C} \left\{ a_{ci} [P_{ci,t}^k + C_v (Q_{ci,t}^k + Q_{si,t}^k)]^2 + b_{ci} [P_{ci,t}^k + C_v (Q_{ci,t}^k + Q_{si,t}^k)] + c_{ci} \right\} \tag{3}$$

where F_2 is the operating cost of CHP units. a_{ci}, b_{ci}, and c_{ci} are the second-order fitting coefficients of the operating cost of CHP unit i. C is the total number of CHP units. C_v is the operating cost parameters of units. $P_{ci,t}^k$ and $Q_{ci,t}^k$ are the electrical output and heat output of unit i at time t under scenario k, respectively. $Q_{si,t}^k$ is the heat storage and release power (positive storage and negative discharge) of the HSD at time t under scenario k.

3.1.3. Wind Curtailment and Load Shedding Penalty

The penalty cost mainly includes the load shedding penalty when the actual wind power output being less than the planned output [15], and the wind curtailment penalty when the wind power being greater than the planned output.

$$\begin{cases} P_{wi,t}^{\text{cut}} = P_{wi,t}^{\max} - P_{wi,t} \,, \, P_{wi,t}^{\max} > P_{wi,t} \\ P_{\text{load},t}^{\text{cut}} = P_{wi,t} - P_{wi,t}^{\max} \,, \, P_{wi,t}^{\max} < P_{wi,t} \end{cases} \tag{4}$$

where $P^{\text{cut}}_{wi,t}$ and $P^{\text{cut}}_{\text{load},t}$ are the curtailed wind power and the load shedding power at moment t, respectively. $P_{wi,t}$ is the wind power output scheduled at time t. $P^{\max}_{wi,t}$ is the maximum wind power output at moment t on the day of dispatch.

The total penalty cost can be expressed as

$$C^k_{\text{pun}} = \sum_{t=1}^{T} \left(\gamma \times P^{\text{cut},k}_{wi,t} + \beta \times P^{\text{cut},k}_{\text{load},t} \right) \tag{5}$$

where γ and β are the penalties for unit curtailment of wind and load shedding, respectively. $P^{\text{cut},k}_{wi,t}$ and $P^{\text{cut},k}_{\text{load},t}$ are the curtailed wind power and load shedding power at moment t under scenario k, respectively.

3.2. Constraints

The constraints of the day-ahead scheduling model include the operating constraints of each unit, the operating constraints of HSDs, the electric load and thermal load balance constraints, and network security constraints.

3.2.1. Operation Constraints of WP Units

In this paper, a discrete scenario set is used to describe the uncertainty of wind power output. It is assumed that the maximum predicted output of WP unit i at time t under scenario k is $P^k_{wimax,t}$, and the constraints in the day-ahead scheduling model are as follows,

$$0 \le P^k_{wi,t} \le P^k_{wimax,t} \tag{6}$$

where $P^k_{wi,t}$ is the wind power output at time t under scenario k.

3.2.2. Operation Constraints of TP Units

The upper and lower limits of output constraints and the climb rate constraints of TP units are mainly considered in the day-ahead scheduling model. The specific constraints are as follows,

$$P_{imin} \le P^k_{i,t} \le P_{imax} \tag{7}$$

$$-RD_{i,max} \cdot \Delta t \le P_{i,t} - P_{i,t-1} \le RU_{i,max} \cdot \Delta t \tag{8}$$

where P_{imin} and P_{imax} are the limits of $P^k_{i,t}$. $-RD_{i,max}$ and $RU_{i,max}$ are the maximum climb rate of unit i. Δt is the interval per unit time.

3.2.3. Operation Constraints of CHP Units

This paper considers the CHP unit type of extraction, which has a coupling constraint on the heat and electric output [16]. When the heat output is constant, the unit's electric output is limited to a specific interval [8], as shown in Figure 2.

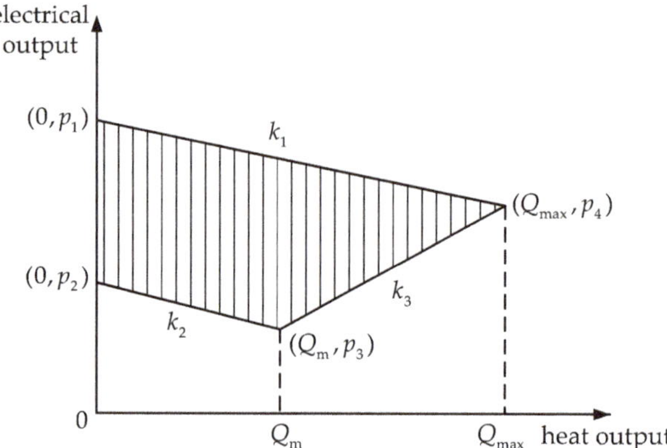

Figure 2. The output feasible range of the combined heat and power (CHP) unit.

The output feasible range of the CHP unit can be expressed as

$$\begin{cases} 0 \leq Q^k_{ci,t} \leq Q_{ci,max} \\ P^k_{ci,t} \leq p_1 + k_1 Q^k_{ci,t} \\ P^k_{ci,t} \geq p_2 + k_2 Q^k_{ci,t} \\ P^k_{ci,t} \leq p_3 + k_3(Q^k_{ci,t} - Q_m) \end{cases} \tag{9}$$

where k_1, k_2, and k_3 are the unit operation coefficients. p_1 and p_2 are the maximum and minimum values of generating power when the unit's air extraction value is zero, respectively. p_3 is the minimum electrical power of the unit. Q_m is the heating power corresponding to the minimum electrical power of the unit. $Q_{ci,max}$ is the maximum heating power of unit i.

The output change of CHP unit is determined by the amount of air extraction. Therefore, its climbing rate should be converted to the electric power constraint under the unit's pure condensing condition.

$$-RD_{ci,max} \cdot \Delta t \leq \overline{P}^k_{ci,t} - \overline{P}^k_{ci,t-1} \leq RU_{ci,max} \cdot \Delta t \tag{10}$$

where $\overline{P}^k_{ci,t}$ is the electric power of unit i converted to pure condensing condition at time t under scenario k. $RD_{ci,max}$ and $RU_{ci,max}$ are the maximum climb rate of CHP unit i.

3.2.4. Operation Constraints of HSDs.

The operation constraints of HSDs mainly include heat storage and release power constraints, the thermal storage tank capacity constraints, state constraints, and the cycle constraints of HSD [13]. The specific constraint expression is as follows,

$$\begin{cases} -Q_{fi,max} \leq Q^k_{si,t} \leq Q_{zi,max} \\ 0 \leq S^k_{eh,t} \leq S_{eh,max} \\ S^k_{eh,t} = S^k_{eh,t-1} + Q^k_{si,t} - \varphi_{eh} S^k_{eh,t-1} \\ \sum\limits_{t=1}^{T} Q^k_{si,t} = 0 \end{cases} \tag{11}$$

where $Q_{fi,max}$ and $Q_{zi,max}$ are the maximum heat release and storage power of HSDs, respectively. $S_{eh,max}$ is the maximum heat storage capacity of the thermal storage tank. φ_{eh} is the heat loss rate, which is generally taken as 0.05.

3.2.5. Electric Load and Heat Load Balance Constraints

$$\begin{cases} \sum_{i=1}^{W} P^k_{wi,t} + \sum_{i=1}^{N} P^k_{i,t} + \sum_{i=1}^{C} P^k_{ci,t} = P_{\mathrm{load},t} \\ \sum_{i=1}^{C} (Q^k_{ci,t} - Q^k_{si,t}) = Q_{\mathrm{load},t} \end{cases} \tag{12}$$

where $P_{\mathrm{load},t}$ and $Q_{\mathrm{load},t}$ are the electricity and heat load requirements at time t of the CHPS, respectively.

3.2.6. Network Security Constraints

$$P_{l,\min} \leq P^k_{l,t} \leq P_{l,\max} \tag{13}$$

where $P^k_{l,t}$ is the power of line l at time t under scenario k. $P_{l,\min}$ and $P_{l,\max}$ are the limits of $P^k_{l,t}$.

4. Intra-Day Rolling Optimization Scheduling Model

As the prediction error of wind power output is positively correlated with time [17], the intra day rolling scheduling model of the CHPS utilizes the ultra-short-term forecast output of wind power to optimize and adjust the day-ahead scheduling plan to ensure the economical efficiency of the system. However, the operation optimization of HSD is a typical multi-stage decision problem, with close relationship in different periods, and the intra-day scheduling plan cannot be globally optimized because it is a phased plan.

Based on the problem, the adapted intra-day rolling scheduling strategy is proposed in this paper, and it is shown in Figure 3. By adopting this scheduling strategy, it can ensure that the results of each rolling scheduling are the global optimal solutions.

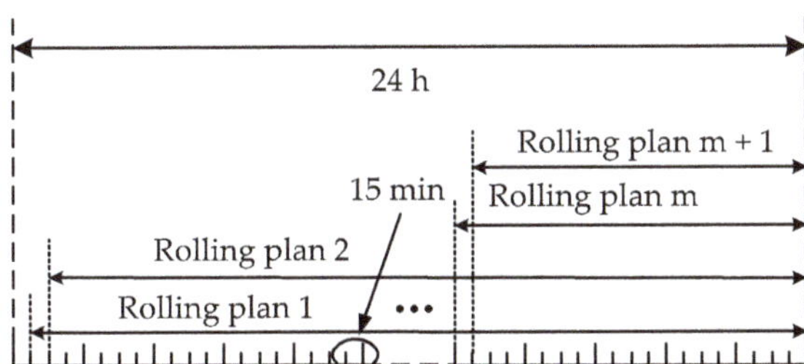

Figure 3. The adapted intra-day rolling scheduling strategy.

As the scheduling time scale is changed, the data form of wind power input is also changed. In order to obtain the wind power input applicable to the scheduling strategy, this paper is based on the ultra-short-term prediction data of wind power, and considering the temporal dependence of wind power fluctuation, the intra-day wind power scenario generation method is put forward, and it is shown in Figure 4. First, the weighted Euclidean distance method is adopted to calculate the similarity to the corresponding period of the day-ahead scenario set of wind power, and the scenarios with the high similarity degree are extracted to form a new scenario set. Then, the ultra-short-term wind power prediction data is used to update the wind power output of the corresponding period of the new scenario set. Based on ultra-short-term electric and heat load forecast data, the intra-day rolling optimization scheduling model is solved to roll and modify the day-ahead scheduling plan. At last, the optimization result obtained at this time is taken as the boundary condition of the next rolling scheduling period, and the above steps are cycled until the end of the whole day-ahead scheduling plan.

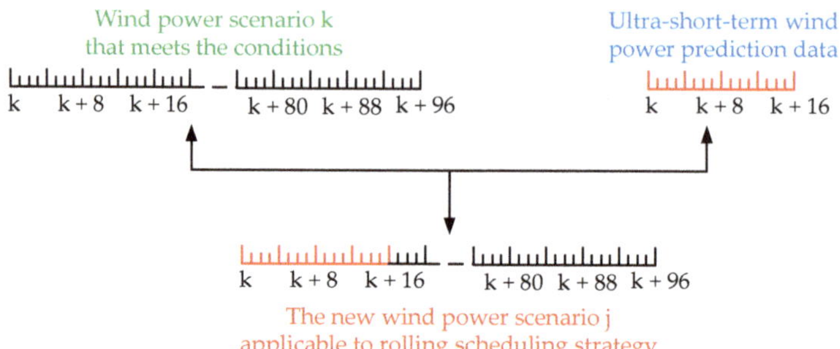

Figure 4. The intra-day wind power scenario generation method.

4.1. Intra-Day Wind Power Scenario Generation

Euclidean distance represents the meaning of distance in the general sense, and is often used to measure the degree of similarity between two variables. The closer the distance is, the higher the similarity is between the two variables. The calculation method is as follows,

$$dist_{ed}(x, y) = \left(\sum_{i=1}^{n} (x_i - y_i)^2 \right)^{\frac{1}{2}} \tag{14}$$

where x and y are the quantities. n is the dimension of the vector.

Suppose $k = [P^k_{w,1}, P^k_{w,2}, \cdots, P^k_{w,96}] \in K$ is a certain scenario data of wind power short-term forecast, and $s = [p_{t+1}, p_{t+2}, \cdots, p_{t+16}]$ is the ultra-short-term forecast data obtained four hours in advance. Considering that the wind power output at time t has time correlation characteristics with the previous time, and the longer the interval is, the weaker the correlation is; therefore, the weighted Euclidean distance is adopted to characterize this feature when calculating the similarity of two scenarios, and the wind power output change rate at the two moments is defined as

$$\Phi = \frac{\left| P_{w,t+1} - P_{w,t} \right|}{P_{w,t}} \times 100\% \tag{15}$$

calculate the change rates of $P_{w,t}$ with $P_{w,t-4}, \cdots, P_{w,t-1}$, respectively, and the sample proportion τ_i whose change rate within the range of 10% is also calculated. Then, the correlation weight coefficient ω_i at each moment is defined as

$$\omega_i = \frac{\tau_i}{\sum\limits_{i=1}^{16} \tau_i} \tag{16}$$

the Euclidean distance between $k\prime = [P^k_{w,t+1}, P^k_{w,t+2}, \cdots, P^k_{w,t+16}]$ and s is

$$dist_{ed}(k\prime, s) = \left(\sum_{i=1}^{16} \omega_i (P^k_{w,t+i} - p_{t+i})^2 \right)^{\frac{1}{2}} \tag{17}$$

then the scenario whose similarity meets certain requirements is selected:

$$\begin{cases} dist_{ed}(k\prime, s) \leq \varepsilon \\ \rho_{k\prime} = \rho_k \end{cases} \tag{18}$$

where ε is the maximum distance limit that satisfies the conditions. $\rho_{k\prime}$ is the probability of scenario k'.

Replacing the wind power output in the period of $t+1 \sim t+16$ at scenario k with s is denoted as the new scenario j. The probability of the scenario j is

$$\rho_j = \frac{\rho_{k\prime}}{\sum\limits_{k\prime=1}^{N_j} \rho_{k\prime}} \tag{19}$$

where N_j is the number of screened scenarios.

In conclusion, the intra-day scenario set J of wind power output during the day is obtained.

4.2. Objective Function

The objective function of the intra-day rolling dispatch model is still maximized by the system's revenue, taking into account the power and heat supply benefits minus the operation cost of the units and penalty cost in the system, which is the same as in Equation (1).

4.3. Constraints

The constraints of the intra-day rolling scheduling model are roughly the same as those of the day-ahead scheduling model, and only the bias constraint is added to make the scheduling result better be connected with the day-ahead dispatch plan [10].

$$\left| P_{\text{roll},t}^{j} - P_{\text{day},t}^{j} \right| \leq \zeta \tag{20}$$

where $P_{\text{roll},t}^{j}$ and $P_{\text{day},t}^{j}$ are the rolling plan and the day-ahead plan total power generation in time period t under scenario j, respectively. ζ is the limit of power deviation.

In addition, the output and climb rate constraints of each unit, the operation constraints of HSDs, the electric load and heat load balance constraints, and network security constraints are considered, as shown in Equations (6)–(13).

5. Calculating Procedures

Simultaneous to Equations (1)–(13), the day-ahead optimization scheduling model of the CHPS is obtained. The random decision variables include the power output of each unit, the heat output of CHP units, and the heat storage and release power of HSDs. First, based on short-term wind power forecast data, a large number of wind power scenes are obtained. Then, the initial heat storage of the HSD is set to 0, and the model is solved by commercial optimization software Cplex, which is provided by IBM (International Business Machines Corporation) in Armonk, NY, USA [8]. Then, the initial heat storage capacity is iteratively modified to obtain the day-ahead scheduling plan set with different values. The value of the heat storage corresponding to the subset with the highest benefits is the optimal initial heat storage capacity of the HSD, and the expected value of each decision variable in the subset is the day-ahead scheduling plan of the system.

Simultaneous to Equations (1) and (6)–(20), the intra-day rolling optimization scheduling model of the CHPS is established, in which the decision variables are the same as the day-ahead scheduling model. Based on the ultra-short-term wind power forecast data, the method introduced in section 4.1 is used to generate a scenario set of wind power. Then, based on the ultra-short-term electricity and heat load forecast data, as the input of the model, the intra-day rolling optimization scheduling model is solved, and the expected values of the decision variables in dispatch period are used as the intra-day plan to adjust and modify the previous plan. Finally, the value of each decision variable at this moment is taken as the boundary condition of the next rolling scheduling period, and the above steps are cycled until the end of the whole day-ahead scheduling plan. The solution process is shown in Figure 5.

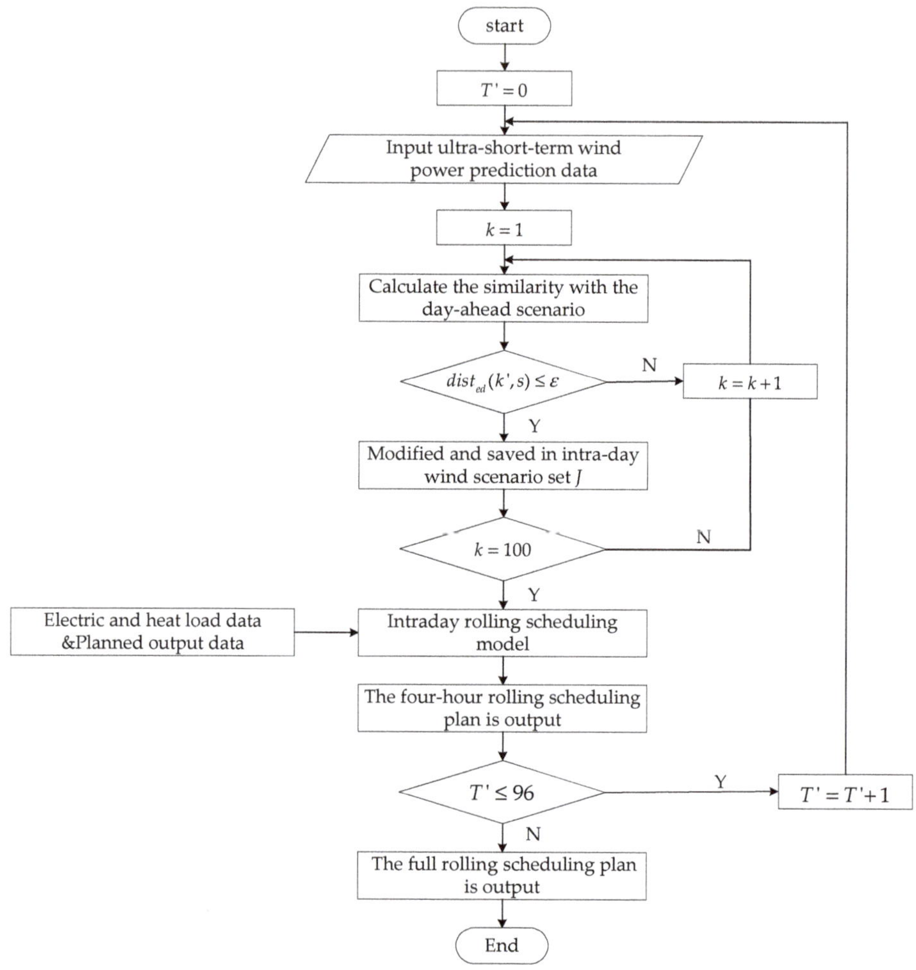

Figure 5. The calculating procedures for intra-day rolling optimization scheduling model.

6. Case Study

6.1. General Situation of Simulation

Numerical simulations are conducted on a local grid in Liaoning province, including a wind farm, a CHP unit with HSD, and two TP units. The supply price of electric and heating is shown in Table A2 of Appendix A, and the parameters of units are detailed in Table A3 of Appendix A [18,19]. The wind power scene set data is shown in Figure 6 [20,21], different colored lines represent the output of the WP unit under different scenarios. The ultra-short-term forecast data of wind power, electric load, and heat load data in the system are shown in Figure 7.

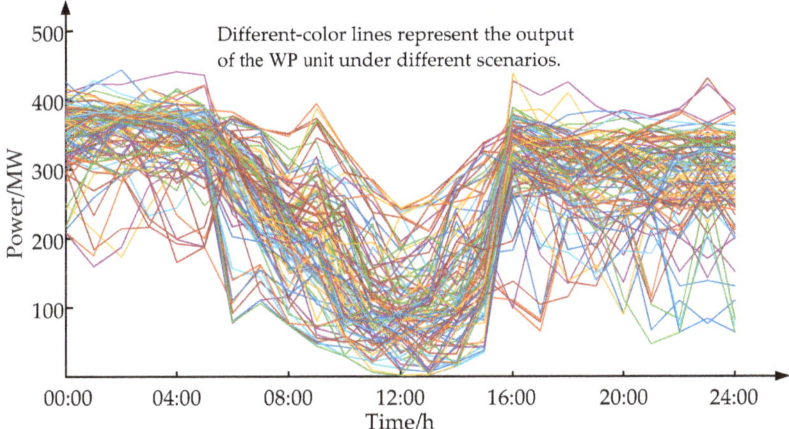

Figure 6. Day ahead wind power forecast scenario set.

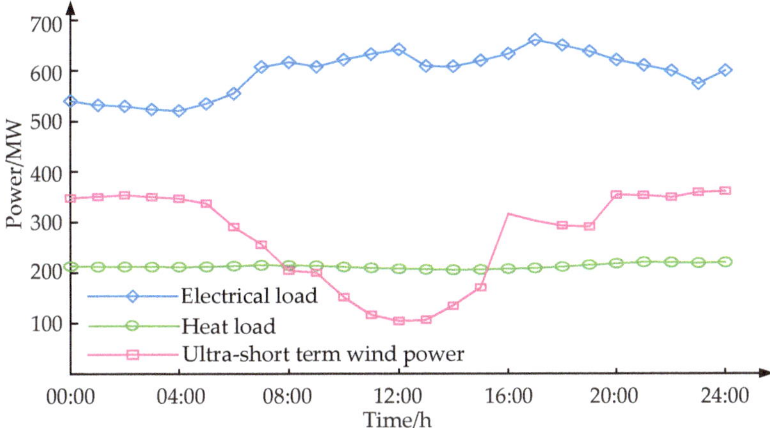

Figure 7. Intra-day wind power forecasting curve and electric/heat load curve.

6.2. Impact of Initial Heat Storage Capacity on Overall Revenue

Before the day-ahead dispatch, a certain amount of heat needs to be stored in the HSD. In order to analyze the impact of initial heat storage capacity on the overall revenue of the system, the expected value of the overall revenue under different initial heat storage capacity is calculated. It can be seen from Figure 8 that both increase with the initial heat storage capacity, until the initial value of the heat storage reaches 260 MW·h and the system's revenue is maximum. When the initial heat storage capacity is greater than 260 MW·h, the overall system revenue tends to be stable.

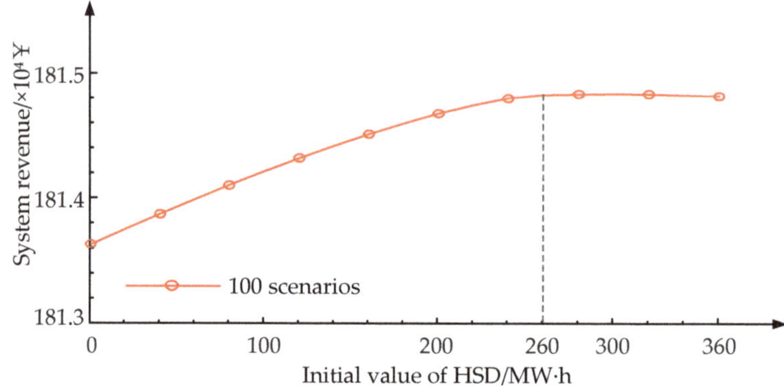

Figure 8. The system revenue under different initial heat storage capacity.

In order to analyze the effect of initial value on optimal scheduling results, two cases are considered:

Case 1: When the initial heat storage capacity is 0 MW·h.

Case 2: When the initial heat storage capacity is 260 MW·h.

Figure 9 shows the output of each unit of the system when the initial heat storage capacity is 0 MW·h and 260 MW·h. At this time, the expected revenue of the system is 1813.6 thousand yuan and 1814.7 thousand yuan, respectively. It can be seen from the figure that the wind power output of the wind farm is relatively large from 0 am to 7 am, and this is the period of low power load. If the initial heat storage capacity is low, the heat load demand can only be met by the CHP unit. However, the restriction of "determining power generation by heat" of the CHP unit will cause a large amount of wind power curtailment, and the inefficient use of renewable energy will lead to a decrease in revenue. When the initial heat storage capacity is 260 MW·h, from 0 am to 7 am, the heat load is jointly satisfied by the heat generation of the CHP unit and the HSD. At this time, the CHP unit is reduced electric output, so as to absorb more wind power to improve the total system revenue. When the initial heat storage capacity is greater than 260 MW·h, the overall system revenue tends to be stable. Therefore, considering the economics of the system, 260 MW·h is selected as the initial value of the HSD in the subsequent analysis and calculation.

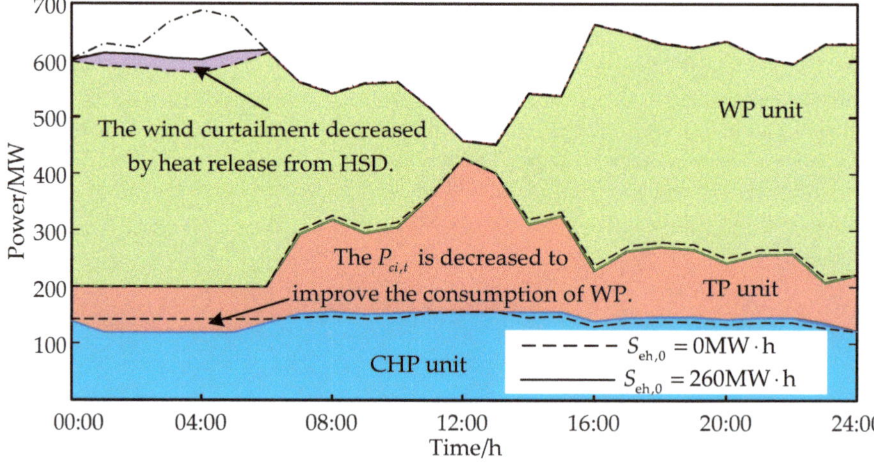

Figure 9. System operation results under different initial heat storage capacity.

6.3. Comparison of the Stochastic Model and Deterministic Model

In this part, two models are considered.

Case 3: Deterministic model.

The scheduling strategy of the deterministic model is based on the wind power output curve of the typical day to optimize and calculate the value of each decision variable. The objective function is as follows,

$$\max f_1 = \sum_{t=1}^{T} \left(S_t - C_t - C_{\text{pun},t} \right) \tag{21}$$

Case 4: Stochastic model (established in this paper).

We assume that scenario 1 is the wind power output of the typical day, and output of each unit is calculated as the day-ahead scheduling plan of the deterministic model. Moreover, the day-ahead scheduling plan of the stochastic model is calculated by the method mentioned in the paper. Then, suppose that scenario 2 to scenario 100 are the actual wind power output on the dispatching day, the output of each unit under each scenario is obtained, and compared with the scheduling plan of the stochastic model and the deterministic model. The unit penalty cost for wind curtailment and load shedding is 100 yuan/MW·h, and the penalty cost in different scenarios is calculated. The results are shown in Figure 10.

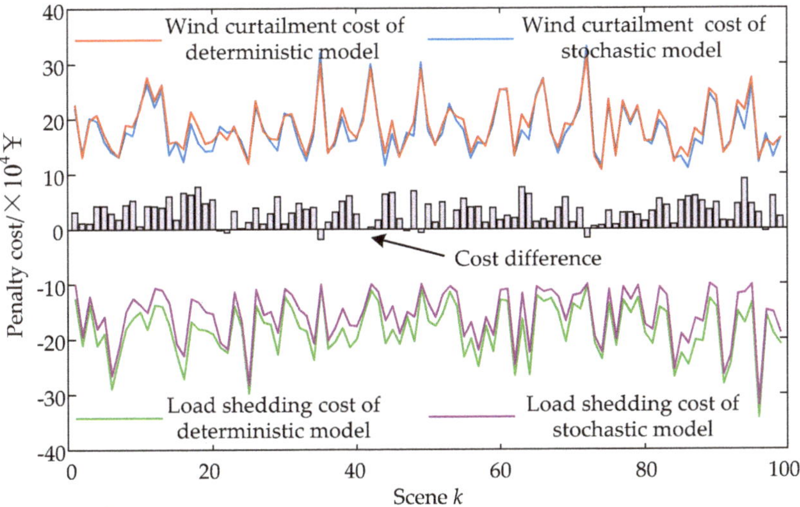

Figure 10. Comparison of penalty costs between two models in different scenarios.

It can be seen in Figure 10 that in more than 93% of the scenarios, the penalty cost of the scheduling plan calculated by the stochastic model is lower. Therefore, the scheduling plan arranged by the stochastic model can better adapt to the uncertainty of wind power.

6.4. Impact of Temporal Dependence of Wind Power Fluctuation on Overall Revenue

In this part, the two scheduling strategies are calculated separately.

Case 5: The traditional rolling scheduling strategy [22].

Case 6: The adapted rolling scheduling strategy.

The impact on the system revenue is analyzed, and the results under different scheduling strategies are shown in Table 1.

Table 1. The system's revenue before and after considering temporal dependence.

Items	Considering Temporal Dependence		Differential Values
	Before	**After**	
Electric and heat income/×10⁴¥	195.13	195.13	0
Costs of TP units/×10⁴¥	7.07	7.02	0.05
Costs of CHP units/×10⁴¥	6.01	5.94	0.07
Consumption of Wind Power/MW·h	5875.0	5941.4	66.4
Costs of Wind Curtailment penalty/×10⁴¥	3.06	2.40	0.66
Total system's revenue/×10⁴¥	178.99	179.77	0.78

When the temporal dependence is not considered, the cost of generating electricity for TP units and CHP units are 70,700 yuan and 60,100 yuan, which is an increase of 1200 yuan compared with the method of considering it. At the same time, after considering the temporal dependence, 66.4 MW·h of wind power is absorbed in the system, and the total revenue of the system increased by 7800 yuan than before.

Comparing Table 1 with Figures 11 and 12, it can be seen that when the temporal dependence is not considered, the scheduling plan in each cycle (four hours) only needs to meet the highest return during this period. Therefore, at time 8–15, the electric load is high, but the wind power output is relatively small. In order to ensure the maximum stage benefits, the thermal output of the system is so small that HSD has no excess heat storage. At times 16–17 and 22–24, the wind power output is large, and the increased heat output of the CHP unit must not only meet the needs of the heat load, but also take into account the cycle constraints of the HSD, which shrinks the space for wind power accommodation. After considering temporal dependence, the HSD stores heat at 8–15 pm. When the subsequent wind power output is large, the heat storage device cooperates with the CHP unit to meet the heat load demand, thereby reducing the power output of the CHP unit. At the same time, more wind power is accommodated, and higher system revenue is obtained.

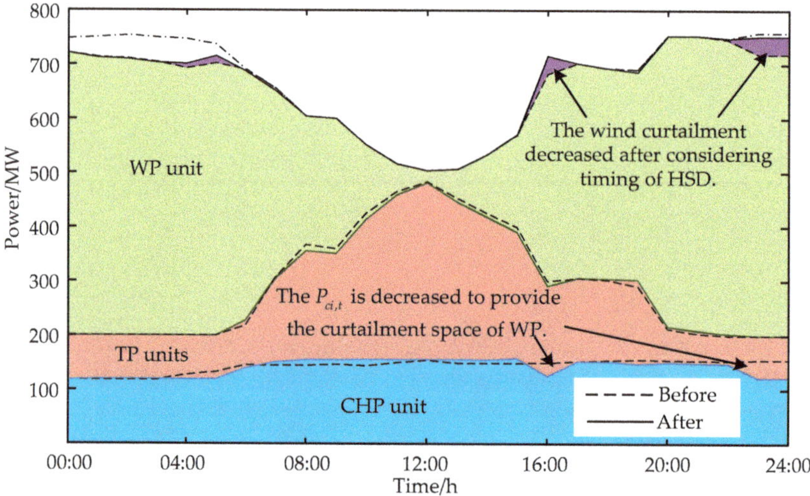

Figure 11. The output of each unit before and after considering the temporal dependence of wind power fluctuation.

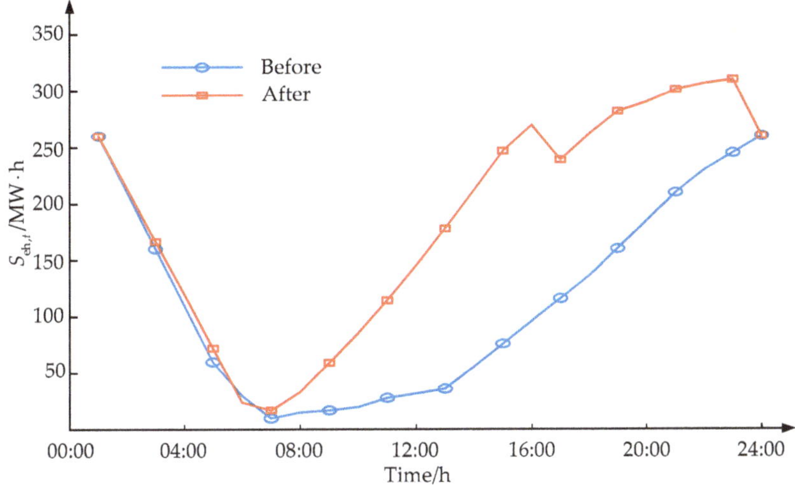

Figure 12. The changes in heat storage of HSD before and after considering temporal dependence.

Furthermore, in order to consider the impact of potential uncertainty on the system, the system revenue and the additional wind power consumption under the two strategies with different times of wind power are studied, and the results are shown in Figure 13.

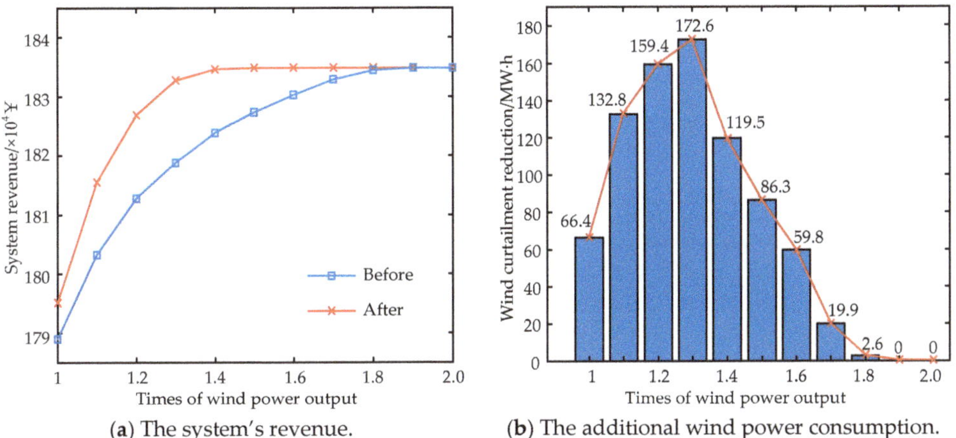

(**a**) The system's revenue.

(**b**) The additional wind power consumption.

Figure 13. The system revenue (**a**) and the additional wind power consumption (**b**) under the two strategies with different times of wind power.

As can be seen from Figure 13, with increasing wind power output, the system revenue also increases. When the wind power is 1.3 times of the original data, the system revenue difference under the two scheduling strategies reaches the maximum. When wind power is 1.4 times the original data, the wind power consumption of system under the adapted scheduling strategy approaches the peak value first, so the system revenue is almost stable with increasing of wind power. When the wind power is 1.9 times the original data, both reach the maximum wind power consumption of the system under the two scheduling strategies. It can be seen that when considering wind power fluctuation, the adapted scheduling strategy can quickly approach the maximum wind power consumption level, so it is more suitable for the system with wind power fluctuation scenarios.

6.5. Algorithm Analysis

In this paper, the primal dual interior point method optimization program is used from the Cplex optimization toolbox. The interior point method is essentially a combination of the Lagrange function method, Newton method, and logarithmic barrier function method. It starts from the initial interior point and follows the steepest descent direction, moving directly from the inside of the feasible region to the optimal solution. Its salient feature is that the number of iterations has little relation to the system size. In the paper, the decision variables need to be optimized for each wind power output scenario, and the required accuracy can be achieved within 20 iterations. In the case of the intra-day rolling optimization dispatch process, a total of 15 iterations were performed. Changing of the objective function versus the number of iterations is shown in Figure 14.

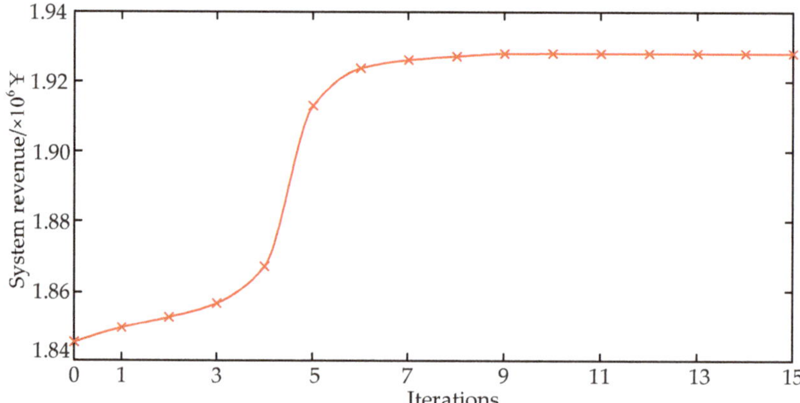

Figure 14. Changing of the objective function versus the number of iterations.

7. Conclusions

Aiming at the problem of scheduling difficulties caused by the uncertainty and fluctuation of wind power output, this paper proposes a multi-time scale optimal scheduling strategy for CHPS based on the scenario method. Both day-ahead and intra-day optimization scheduling models are established to maximize the system's revenue. The case analyzes the impacts of initial heat storage capacity of heat storage device and different scheduling strategies on system revenue. The main conclusions are as follows.

1. The initial heat storage capacity of the HSD can affect the overall revenue of the system, which increases with the initial value. When the value is 260 MW·h, the system achieves the highest yield, so 260 MW·h is used as the optimal initial heat storage capacity of the HSD for subsequent analysis.
2. In the day-ahead scheduling plan, compared with the deterministic model, it is verified that the day-ahead optimal scheduling model based on the scenario method can greatly reduce the system penalty cost, improve system benefits, and better adapt to the uncertainty of wind power.
3. Compared with the scheduling method without considering the temporal dependence of wind power fluctuation, the results show that considering temporal dependence can better reserve more wind power absorption space for the system, promote the accommodation of wind power, and further improve the overall revenue of the system.

Author Contributions: All of the authors have contributed to this research. Conceptualization, Y.Z.; methodology, Y.Z., F.X., and S.G.; software, Y.Z. and S.G.; writing—original draft preparation, S.G.; writing—review and editing, Y.Z. and S.G.; project administration, F.X.; supervision, D.C. and W.G.; data curation, X.C.; funding acquisition, B.G. All authors have read and agreed to the published version of the manuscript

Funding: This research was funded by the State Grid Liaoning Electric Power Co., Ltd. Technology Project (NO. SGTYHT / 17-JS-199).

Acknowledgments: The authors give their thanks to Pinchao Zhao, Pengxiang Huang, Yifan Zhao, and Mengyao Li for giving valuable suggestions.

Conflicts of Interest: The authors declare no conflicts of interest.

Appendix A

Table A1. Table of full names and acronyms.

Full Name	Acronym
Wind power	WP
Thermal power	TP
Combined heat and power	CHP
Heat storage device	HSD

Table A2. The supply price of electric and heating.

Supply Price ¥/MW·h	Time/h					
	0–7	8–10	11–15	16–18	19–21	22–24
electric	200	500	800	500	800	500
heating	450	450	450	450	450	450

Table A3. The parameters of units.

Power Source Type	Installed Capacity	Power Output/MW		Climb Rate/MW		Operating Parameters		
		Min	Max	Down	Up	a	b	c
WP unit	1×450 MW	0	450	0	450	-	-	-
TP unit	2×250 MW	50	250	50	50	0.0037	12.98	36
CHP unit	1×250 MW	40	210	50	50	0.0044	13.29	39
HSD	1×360 MW·h	0	360	50	60	-	-	-

References

1. China's Wind Power Grid Integration in the First Half of 2019. Available online: www.nea.gov.cn/2019-07/26/c_138259422.htm (accessed on 26 June 2019).
2. Xue, Y.S.; Lei, X.; Xue, F.; Yu, C.; Dong, C.Y.; Wen, F.S.; Jv, P. A review on impacts of wind power uncertainties on power systems. *Proc. CSEE* **2014**, *34*, 5029–5040. [CrossRef]
3. Hu, K.; Chen, Q. Overall energy efficiency and flexibility retrofit scheme analysis of heat-power integrated energy system. *Therm. Power Gener.* **2018**, *47*, 14–21. [CrossRef]
4. Lv, Q.; Chen, T.Y.; Wang, H.X.; Li, L.; Lv, Y.; Li, W.D. Combined heat and power dispatch model for power system with heat accumulator. *Electr. Power Autom. Equip.* **2014**, *34*, 79–85. [CrossRef]
5. Chen, L.; Xu, F.; Wang, X.; Min, Y.; Ding, M.S.; Huang, P. Implementation and effect of thermal storage in improving wind power accommodation. *Proc. CSEE* **2015**, *35*, 4283–4290. [CrossRef]
6. Nuytten, T.; Claessens, B.; Paredis, K.; Bael, J.V.; Six, D. Flexibility of a combined heat and power system with thermal energy storage for district heating. *Appl. Energy* **2013**, *104*, 583–591. [CrossRef]
7. Hamidreza, Z.; Seyed, A.T.; Mohammad, S. Robust operation of a multicarrier energy system considering EVs and CHP units. *Energy* **2019**, *192*, 116703. [CrossRef]
8. Yu, J.; Sun, H.B.; Shen, X.W. Optimal operating strategy of integrated power system with wind farm, CHP unit and heat storage device. *Electr. Power Autom. Equip.* **2017**, *37*, 139–145. [CrossRef]
9. Seyed, A.E.; Navid, R. A new isolated renewable based multi microgrid optimal energy management system considering uncertainty and demand response. *Int. J. Electr. Power Energy Syst.* **2020**, *118*, 105760. [CrossRef]
10. Yang, L. Research on Energy Optimization Scheduling for Combined Heat and Power System Considering Wind Consumption. Master's Thesis, Northeast Electric Power University, Jilin, China, 2018. (In Chinese).

11. Chen, L.; Zhu, X.; Xu, X.; Liu, H. Multi-time scale coordinated optimal dispatch of microgrid cluster based on MAS. *Electr. Power Syst. Res.* **2019**, *177*, 105976. [CrossRef]

12. Wu, X.; Wang, X.L.; Li, J.; Guo, J.L.; Zhang, K.; Chen, J. A joint operation model and solution for hybrid wind energy storage systems. *Proc. CSEE* **2013**, *33*, 10–17. [CrossRef]

13. Dai, Y.H.; Chen, L.; Min, Y.; Xu, F.; Hou, Y.K.; Zhou, Y. Optimal dispatch for joint operation of wind farm and combined heat and power plant with thermal energy storage. *Proc. CSEE* **2017**, *37*, 3470–3479. [CrossRef]

14. Cui, Y.; Chen, Z.; Yan, G.G.; Tang, Y.H. Coordinated wind power accommodating dispatch model based on electric boiler and CHP with thermal energy storage. *Proc. CSEE* **2016**, *36*, 4072–4080. [CrossRef]

15. Tian, L.; Xie, Y.; Hu, B. A deep peak regulation auxiliary service bidding strategy for CHP units based on a risk-averse model and district heating network energy storage. *Energies* **2019**, *12*, 3314. [CrossRef]

16. Zhang, L.; Luo, Y.; Luo, H.H.; Miao, S.H.; Ye, J.; Zhou, G.P.; Sun, L. Scheduling of integrated heat and power system considering multiple time-scale flexibility of CHP unit based on heat characteristic of DHS. *Proc. CSEE* **2018**, *38*, 985–998. [CrossRef]

17. Lu, Q.Y.; Hu, W.; Min, Y.; Wang, Z.M.; Luo, W.H.; Cheng, T. A multi-pattern coordinated optimization strategy of wind power and energy storage system considering temporal dependence. *Autom. Electr. Power Syst.* **2015**, *39*, 6–12. [CrossRef]

18. Shui, Y.; Liu, J.Y.; Gao, H.J.; Huang, S.; Jiang, Z.Z. A distributionally robust coordinated dispatch model for integrated electricity and heating systems considering uncertainty of wind power. *Proc. CSEE* **2018**, *38*, 7235–7247. [CrossRef]

19. Dinh, B.H.; Nguyen, T.T.; Quynh, N.V.; Dai, L.V. A novel method for economic dispatch of combined heat and power generation. *Energies* **2018**, *11*, 3113. [CrossRef]

20. Pinson, P. Wind energy: Forecasting challenges for its operational management. *Stat. Sci.* **2013**, *4*, 564–585. [CrossRef]

21. Bukhsh, W.A.; Zhang, C.Y.; Pinson, P. An integrated multiperiod opf model with demand response and renewable generation uncertainty. *IEEE Trans. Smart Grid* **2015**, *7*, 1495–1503. [CrossRef]

22. Wang, H.; Ai, Q.; Gan, L.; Zhou, X.Q.; Hu, F. Collaborative optimization of combined cooling heating and power system based on multi-scenario stochastic programming and model predictive control. *Autom. Electr. Power Syst.* **2018**, *42*, 51–58. [CrossRef]

 *energies*

Article

Technical Approaches and Institutional Alignment to 100% Renewable Energy System Transition of Madeira Island—Electrification, Smart Energy and the Required Flexible Market Conditions

Hannah Mareike Marczinkowski [1],* and Luísa Barros [2]

[1]   Department of Planning, Aalborg University, Rendsburggade 14, 9000 Aalborg, Denmark
[2]   ITI, LARSyS, University of Madeira, 9020-105 Funchal, Portugal; luisa.barros@iti.larsys.pt
*   Correspondence: hmm@plan.aau.dk; Tel.: +45-93-56-23-55

Received: 30 June 2020; Accepted: 25 August 2020; Published: 27 August 2020

Abstract: The integration of renewable energy (RE) in energy systems can be approached in many ways depending on local possibilities. Evaluating this in the limited context of islands, this paper presents a multi-energy system transition to a 100% RE share in a two-folded technical analysis. The case study of Madeira Island using the EnergyPLAN modeling tool is used to show strengths and weaknesses of, on the one hand, electrifying all transport and heating demands on an island, while remaining demands are supplied with biomass, and, on the other hand, additional smart charging, vehicle-to-grid, thermal collectors and storages, as well as electrofuel production and storages. Technical results indicate the potentials and advantages of the second approach with 50% less biomass and no curtailment at 1–3% higher costs, compared to the first one with 7% of production curtailed. The technical analysis is supported by the institutional analysis that highlights the balancing needs through additional flexibility and interaction in the energy system. For maximum flexibility, of both demand and grid, and successful implementation of 100% RE, investment incentives and dynamic tariffs are recommended entailing more dynamic consumer involvement and strategic energy planning.

Keywords: energy system analysis; modeling; multi-energy system; smart energy system; flexible demand; self-sufficiency; dynamic market

1. Introduction

Energy systems, both large and small, are transitioning towards higher shares of renewable energy (RE), such as from wind or photovoltaic (PV), in response to replacing fossil-fuel technologies in the fight against climate change [1]. A well-planned transition to 100% RE is the main objective in many places however especially islands present challenging systems but also potential lighthouses in the struggle to analyze and identify the best approach to transitioning [2]. Not only are small and/or developing islands sensitive or even vulnerable in terms of access to energy at a reasonable cost [3], but also the importance and difficulties of various European islands was recognized in research [4], as well as politically in the "Valletta Declaration", which proclaims remote European islands as favorable for innovation [5]. Despite the differences in energy-intensity, population and geography, the common main energy objective is higher RE shares and self-sufficiency—for islands and globally.

The potential of transitioning to 100% RE has been addressed on a global level by Ram et al. [6], as well as in a more detailed review by Hansen et al. [7]. Both present the trends and latest studies on 100% RE systems and how it has gained attention especially over the last few years. While some of the studies reviewed in [7] have included cross-border interaction in national energy systems, they are often made in island mode or are even focusing on geographical islands, which illustrates the role of island energy system perspectives. The increased shares of wind and PV often require balancing due to their fluctuations, such as through cross-border trade or storages. This is restricted in island systems since trade is often limited and storages not always available or feasible in island set-ups.

An extensive review related to this by Jurasz et al. [8] states that complementarity energetic studies are needed to evaluate the feasibility of variable RE. It shows that the design of energy management strategies and the balance between storage and cost of energy analysis can be used to increase the reliability of the system. The process of RE integration in a 100% RE system at a regional level not only has to take into account historical data but also future consideration for different interested entities [8].

Various other studies and reviews have presented high RE utilization shares specifically on islands due to their vulnerability and often heavily fossil-reliant energy systems, but also due to the high RE potential. In [9], Kuang et al. review various islands on the way to or reaching a 100% RE share by evaluating different RE technologies and strategies for improvement. Ioannidis et al. [10] present the vulnerability and argue for the urgent need to transition energy systems of islands globally, while Cross et al. [11] further present potentials on islands and their importance in tests and demonstration—both including the need and possibilities of 100% RE island systems. Finally, Meza et al. present the complex transformation of a developing island with RE and storage capacity in competition with diesel for the existing generators and buses [12].

The integration of RE from wind and PV plays an important role, as the resource is naturally occurring and does not need to be traded or imported. While RE often supplies either the electricity or the heating sector, other sectors can benefit from using the renewable electricity and heat with further cross-sector integration as a required step to enable future smart energy systems and 100% RE in an efficient way. While [6,9–11] discuss the vulnerability and potential of RE on islands, their focus lies in the electricity sector, which is, however, already expanding across other sectors and is therefore required to be acknowledged differently. This cross-sector energy system planning—which is also known under the smart energy system (SES) label [13,14]—addresses the intermittency of RE through the integration of electricity, heating and/or transport sectors to provide the best integration flexibility and maximize the impact of smart balancing technologies. This flexibility option is also referred to as multi-energy system interaction, where different energy sectors support each other, such as in [15] where the power to heat potential is presented. The otherwise often used approach of cross-border trade is not always an option for many islands, where self-sufficiency should be targeted instead. Additionally, without smart energy planning, a high use of biomass in the power plants can be the result, as this is one of the primary storable RE sources that can be stored and used directly [16]. However, using high amounts of biomass is seen critically under the energy-food nexus and should be limited to stay within locally sustainable levels [17].

Originating from the combined heat and power production experience in Northern Europe, heat and electricity have already been widely researched together, as has also been done in island perspective by comparing thermal storage and electrical batteries in two different islands [18]. Other examples of island studies using this approach have modeled isolated islands in the Azores [19], a small connected island in Croatia [20] or a large one in the Canary Islands [21]. All these articles discuss the technical and socioeconomic benefits of integrating energy sectors in the transition to 100% RE systems by making use of additional balancing and storage options.

While the research and the presented barriers in this transition are often solely technical and sometimes include socioeconomic aspects, the institutional conditions also need to be aligned to support the ever-increasing complexity of energy system planning. Similarly, concerning the integration of sectors towards a SES and required changes in the technical aspects, the market also needs to align with this transformation. The change of cost structures with increasing RE shares from, e.g., wind and PV is already widely discussed for the electricity market at the national level [22,23]. Based on the idea of SES, markets are suggested to be aligned in a 'smart way' through the concept of smart energy markets, where several sectors are technically and institutionally aligned [24]. The ongoing transition of the electricity market would benefit from considering the other sectors as their influence on each other is ever-increasing, not only technically. Existing literature, however, often includes cross-border options as well as gas infrastructure for this, which is limited in island energy systems. The design of market structures in isolated energy systems may be planned differently.

As islands present not only technically but also economically sensitive systems, this alignment of markets to support each other plays an important role in successfully transitioning to 100% RE shares. Here especially, strategic coordination of local barriers and potentials could benefit the demand as well as the supply side. Cuesta et al. [25] argue that RE analyses in small communities can be optimized by including social parameters to make the most out of the local RE resources. Similarly, local energy markets have been proposed to include multiple decentral energy sources and bidding structures to unlock local flexibility [26]. However, this previous research built upon the expansion of local district heating and neither in the newly recommended SES nor in an island context. These contextual considerations, however, should go hand in hand with the technical analyses.

Lund proposes to align the different technologies, social aspects and markets in the transition, especially if it includes radical change, which denotes situations that "involve change in existing organisations and institutions" [13]. With the limitations and potentials, aiming for 100% RE supply in a smart island energy system can be considered a radical change. An approach on how to govern such radical changes is suggested by Hvelplund and Djørup [27], who conclude that the organizational coordination of a 100% RE system requires a more active and communicative governance that coordinates the local technologies and demands. In line with this, the European Commission proposes an integrated energy market through preparedness on all levels, where the consumers are in the centre of the transition, making them an involved, stabilizing part of the market [28]. Despite acknowledging regional possibilities, islands have been mostly overlooked in the existing research and policies, even though the alignment of technologies and markets is most fragile and eminent here and may even present solutions useful in other geographical contexts.

The trends of aiming at 100% RE supply in either the separate sectors or cross-sectorally can be considered radical in any energy system, and the required alignment of technical, socioeconomic and institutional aspects has not been made for island energy systems. An example of islands evaluating different options to reach a SES with 100% RE share is through "smart technologies" in the Horizon 2020 project SMILE (smart island energy systems) [29], where technical and socioeconomic aspects have been introduced [2,30], but investigations into the institutional setup are still lacking behind. Smart technologies, such as smart charging and vehicle to grid (V2G), flexible electrofuel production or smart heating controls and storages, bring new aspects to the energy system and should be evaluated in the context of RE integration.

This article is filling the research gap for a case study, where both technical and institutional aspects are considered to integrate renewables and further reach a 100% RE share. As existing literature approaches this only to some extent, usually separate and either with a sole focus on electrification or on the whole SES, both aspects are presented in parallel in this article. Besides a comparison of technical and socioeconomic options and results through a sociotechnical analysis, the market perspective is addressed in the particular setting of an island, where local energy market barriers are identified and the potential is presented. By illustrating the required institutional setup in alliance with the technical analysis of RE and novel technologies, the paper achieves a novel angle of research.

The case study to test this novelty aims at looking at an existing energy system and its current technologies and market, and how it can be optimized to reach a 100% RE share. One of the demonstration islands from SMILE is therefore chosen, namely Madeira Island, as an exemplary isolated energy system with its own local government and in high demand of a smart and institutionally sound energy system. Hence, this research may address islands, and particularly non-interconnected ones, but also other remote and rural energy systems can benefit from it to make better use of locally produced energy.

In this paper, both technical scenarios and institutional comparisons are made to which the methodology can be found in Section 2. The corresponding results and recommendations are presented in the technical analysis in Section 3 and the market analysis in Section 4, before a discussion of both is made in Section 5. Section 6 concludes the whole paper briefly.

2. Methodology

This section introduces the methods and underlying procedures for the conducted analysis as well as the used data, which can be found in the Supplementary Material Section. This includes the use and creation of scenarios, details of the case study and the required steps to reach 100% RE systems in the technical analysis. A simplified overview of the steps and corresponding scenarios explained in Section 2.1 is illustrated in Figure 1. The modeling software EnergyPLAN that is used for these is introduced in Section 2.2. The final consequential step after the technical analysis is presented in the market analysis in Section 2.3 to support the scenarios.

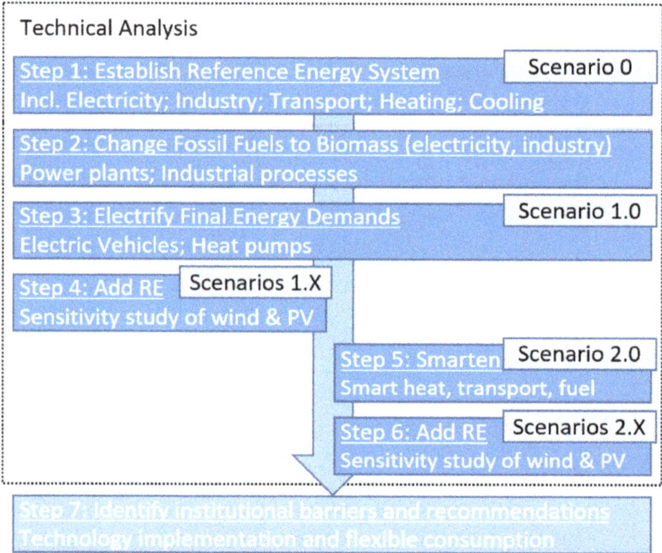

Figure 1. Step-by-step approach to scenario development and analyses.

2.1. Scenario Approach

While Section 1 presents the complexity of energy planning approaches, these are further complicated through geographical and economic boundaries. The selected choice of technical approaches would result in a variety of options and outcomes. Therefore, the following presents the introduction of the scenario development for the case study of Madeira Island, which is to test these approaches under its unique boundaries, for the reader to draw the consequences for other cases. Hence, this theory-led case study was used to test the theory that the smart approach is more beneficial than the electrification approach, and thus can be used as a valid argument based on this case [31]. The case can also be understood as an instrument to understand the suggested approaches through illustration in a certain context. Thereby, a quantitative approach was used to illustrate the explored issue of the case as well as a qualitative one where this event was portrayed.

In the following, the methodology for the creation of scenarios is shown. This approach is chosen to analyze the different impacts of the various steps that are introduced to the system. This way, alternative outcomes are created to show future development paths and corresponding implications [32]. Based on these scenarios, which present radical changes to the energy system, the market is evaluated.

2.1.1. Reference Energy System

The case study to which the above mentioned was applied was the energy system of Madeira Island and its model was based on the SMILE project. With approximately 260,000 inhabitants and an annual electricity demand of 830 GWh, it is an energy intensive island but also a completely autonomous energy system with a large service sector, complex infrastructure as well as some industry. Without a significant heat or gas network, it presents a special but real case of an isolated island to learn from [2,33].

The Madeira model was based on 2014 data (cf. Supplementary Materials) from local energy accounts and supported with data from local statistics and the electricity company of Madeira (EEM), so the production and consumption data was aligned and verified for all included sectors: electricity, heating, industry and transport [2]. In order to use this case study to evaluate the two approaches of 100% RE systems, this 2014 model was adapted to data from 2018 [34]. This resulted in the Reference Scenario 0 and concluded Step 1 of the methodology presented in Figure 1. With a high share of power supply from oil and gas based power stations, heavily oil-based transport and industry sectors, the energy system of Madeira Island was a good case study to evaluate the upcoming required transition to 100% RE in the two suggested approaches.

2.1.2. Steps to a 100% RE System

Based on Lund, the possibility of a 100% RE system is designed in three steps: energy savings, efficiency improvements and fossil fuel replacement with renewable sources [13]. The savings and efficiency were outside the scope of this paper, but the possibilities of replacing fossil fuels were evaluated within the technical analysis. This limitation confined the results to some extent but on the contrary shows what is technically possible only through the supply side. This was concluded in Step 2 by replacing the fossil fuels in the combustion processes of power stations and industry with biomass. The introduction of other RE sources in Steps 4 and 6 addresses the resulting high consumption of biomass.

Step 3 follows Step 2 with fossil fuel replacement in the transport and heating sectors. With Step 2 as an intermediate step away from fossil fuels, only Step 3 entails a scenario. Based on Lund, "oil for transportation is replaced by electricity" and "boilers are replaced by electric heating" [13]. After the introduction of electric vehicles (EVs) and heat pumps, respectively, this resulted in Scenario 1.0, which concluded the basic steps to 100% RE. This scenario was technically 100% renewable, yet potentially not sustainable, depending on the biomass availability and use. Step 4 is therefore the final necessary step to the 100% RE scenario of the electrification approach, where the RE was further explored through the expansion of PV and wind turbine capacity. This requires a sensitivity study to target scenarios where biomass is reduced by at least 50% while limiting critical excess electricity production (CEEP; often resulting in curtailment) to 100 GWh and avoiding a large increase in total annual costs (max. 25%). This resulted in Scenario(s) 1.X—with the "X" standing for a potentially limited number of scenarios.

The second approach to a 100% RE system brings us to Step 5 and additional sector integration—or the smart energy system (SES) approach, which takes Step 3 to a new level by increasing the technological complexity. Aligned with the discussion in Lund et al. [35], alternatives were introduced to the electricity-focused improvements in the energy system from Step 4, namely smart charging and V2G, thermal solar collector and thermal energy storages (TES) in combination with the flexible heat pumps, and electrofuel production [35]. The latter is illustrated in the example of electrolyzers, hydrogen storage and shifting some transport demand to hydrogen vehicles. This introduced smart technologies beyond the simple electrification scope, by making further use of smart transport, heat and fuels. While the electrification Scenarios 1.X could benefit only from electricity storage, this integrated energy system allowed system-wide balancing through TES, hydrogen and car batteries, resulting in Scenario 2.0. Compared to the electrification approach, also Step 6 makes use of a sensitivity study to compare to Scenarios 1.X and identify potential benefits over the electrification approach, resulting in Scenarios 2.X. The changes that result in the 'smart 100% RE system' were achieved through the modeling tool EnergyPLAN, explained in the following, traceable through the scenario data in the Supplementary Material.

Based on [32], the sensitivity study allowed for evaluating the impact of fluctuations of certain parameters on the outcome in a system, such as the energy system model. In the context of this study, the parameters were additional PV and wind capacity to minimize uncertainty of their contribution to the sustainability of the energy system. The capacities varied between 0 and 250 MW in eleven steps of additional capacity each, presented in 3D surface diagrams of 121 possible combinations. Besides the scenario targets in biomass, CEEP and costs, the following key performance indicators (KPIs) were added to evaluate the results of the sensitivity studies and complete the sociotechnical analysis: RE share of primary energy supply (PES) and electricity supply, sustainable share (not including biomass-based electricity) of electricity, CO_2 emissions, total and peak electricity demand.

2.2. Modeling Tool EnergyPLAN

The case study's energy system was modeled in its reference set-up, as well as through every step introduced above, with Aalborg University's EnergyPLAN v15.0 modeling tool [36]. Its ability to model all sectors, especially in relation to each other, as well as the hourly resolution makes it suitable for this research. The software allows for the characterization of the electricity, heating, transport and industry sectors as well as any fuel usage beyond these. Depending on the production, conversion and consuming units included in the manual setup of the energy systems, the sectors can be modeled separately or integrated, allowing for an evaluation of the steps and technologies accordingly. Hence, it allows for a comparable simulation of the two proposed approaches in its varying and—compared to the business as usual—increasing complexity. Related modeling material can be found in the Supplementary Material.

The KPIs can be retrieved in a collective way either for the whole energy system and the full year or by sectors and individual hours. They present the technical and socioeconomic perspectives of this research through the sociotechnical analysis, while also giving indication to the institutional ones. For the socioeconomic part, all costs are considered, including investment, fuel and operation costs as well as CO_2 costs but no taxes. When comparing to the reference system cost, the KPI only presents the total annual cost. The option in EnergyPLAN to run serial calculations facilitates the elaboration of sensitivity analyses. By allowing the sensitivity parameters to fluctuate in up to eleven calculation steps and twelve output definitions, the KPI targets are supplemented in a comprehensive way. Finally, the technical optimization option in the tool allows for optimal balance of selected RE capacity according to available technologies, e.g., dynamic EV charging or flexible hydrogen demand depending on the defined storages. Hence, the smart additions to Scenario 2.X are made within EnergyPLAN by adjusting the model and operation accordingly for the points presented in Step 5.

EnergyPLAN has been applied in multiple studies of national [37,38], regional [39,40] and island energy system perspectives [20,21]. With various combination options of supply, conversion and storage technologies (cf. flow diagrams, [36]), the energy system model can be set up as required, which results in Figure 2. Mathiesen et al. [41] present the energy flow diagrams for various technical simulations, presenting similar steps in the transition to 100% RE. These can be modeled in EnergyPLAN and adapted to illustrate the current technical analyses and may further be related to the institutional analyses. In Figure 2, the energy flow diagrams represent Scenario 0 as a traditional energy system with no apparent cross-sector interaction, and the two following flow diagrams illustrate the two approaches of Scenarios 1.X and 2.X. The 1.X diagram shows the addition of RE and electrification of transport and heating and 2.X shows the addition of V2G, solar thermal, hydrogen production and various storage options.

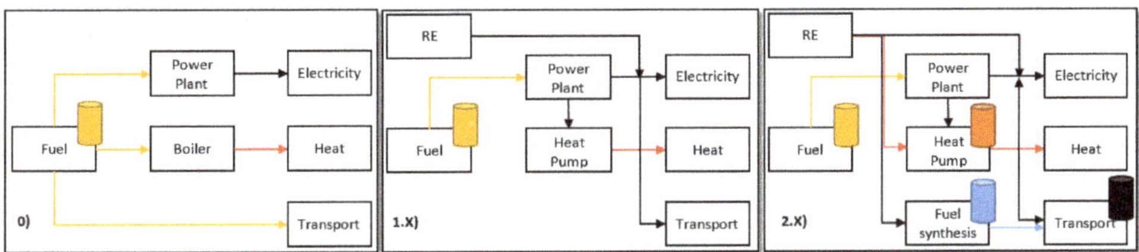

Figure 2. Energy flow diagrams for different energy system and market relation setups based on Scenarios **0**, **1.X** and **2.X**, simplified, based on [36] (barrels indicating storage options).

In relation to the technical changes between the sectors and technologies, also the market setup changes accordingly (cf. Figure 2), indicating where requirements in the market are shifting and need to be supported. Where electricity, heat and transport were mostly regulated by individual markets before, the increased complexity and interrelations of technologies in Scenarios 1 and 2 need to be reflected in the future market setup as well. These institutional perspectives lead to the methodology for the market analysis in the following.

2.3. Market Analysis Approach

Based on the technical modeling, the two approaches to 100% RE require strategical institutional support through the evaluation of the current and required future market structures. Hence, the market analysis represents Step 7, as it is based on the previous six steps but follows its own structure. The aim is to evaluate the current local barriers, potentials and requirements to support the technical scenarios with recommendations for potential changes in the energy market. As mentioned above, it is proposed to align technologies with social and market aspects [13], and to identify maximal preparedness for energy-related crises at all levels [28]. Even though islands are not specifically mentioned, they can present solutions in test scales for larger systems and markets through technical and institutional case studies.

Markets are usually defined as a variation over the theme "physical or virtual arena for the exchange of goods and services" and with particular reference to the energy sector, two types of markets are of relevance: markets for the exchange of technologies and markets for the exchange of energy or power. Thus, market analyses can address whether potential buyers in a given area have a ready access to, e.g., heat pumps or electric vehicles or vice versa whether heat pump and electric vehicle suppliers are met by willing buyers. Market analyses can also focus on how energy, e.g., electricity, biomass and electrofuels are traded with a view to ensuring a functional market that can help exploiting the flexibility options in the energy system with a view to integrating fluctuating RE sources. The main understanding applied in this paper is the latter—i.e., the trading schemes for power and/or energy, which improve the operation of the energy system. The trade of technologies is, however, a prerequisite to this and therefore touched upon as well. Furthermore, the market can be viewed from either the supplier or the consumer side, where the latter is evaluated in this paper by analyzing how the market should be designed from an end-user point of view.

Besides the system and market setup comparison in Figure 2, the following structural approach was used. Based on Hvelplund and Djørup [27], the analytical process can be divided into three phases, which relate to the sociotechnical analysis and lead to institutional considerations and alignment for this radical change. As presented above, changing the energy system's infrastructure to 100% RE supply with either approach can be considered radical. The phases are based on each other and defined as follows:

- Phase 1 presents technical scenarios with certain goals, here 100% RE, including socioeconomic data;
- Phase 2 identifies the institutional context and shows the existing barriers and benefits for the scenarios;

- Phase 3 leads to new recommendations or concrete design proposals; here applied to market structures.

The market analysis in Section 4 presents the phases as follows. After developing the scenarios in Section 2.1 (Phase 1), barriers can be identified and recommendations made. In detail, Phase 2 follows the scenario development and aims at analyzing its results in the current institutional context, e.g., if the scenarios are working in the current market, resulting in the identification of barriers and potentials. In order to do this, the context that shapes their implementation, such as current market and price structures, are taken into account. This is presented in Section 4.1. Phase 3 and Section 4.2 focus on resulting requirements for the future context by addressing the barriers from Phase 2 and supporting the technical scenarios from Phase 1, introducing recommendations that may not have been tested and/or have been blocked by institutions before; hence, new social dynamics might be the result [13]. For this, current and future regulations, considerations and proposals for short- and long-term changes in organizations and infrastructure can be named [27].

In this paper, the market analysis aimed for the 100% RE goal presented in the technical analysis, which was analyzed and discussed, and what institutional barriers need to be addressed to support its implementation was investigated. Equal to Phase 1, both Phases 2 and 3 were studied in the context of Madeira Island and specifically in the context of end-users. This was done through a quantitative review of the technical scenarios, literature review and engagement with local stakeholders [33,42,43].

3. Sociotechnical Analysis

In this section, all steps and scenarios are presented, for which the results, including the KPIs, can be found in Section 3.3. The corresponding institutional market analysis follows in Section 4.

As described in Section 2.1, the Madeira Island energy system model from 2014 formed the basis for the reference scenario and was updated in Step 1 of the analysis. This included the decommissioned hydropower station, as well as newly installed hydropower and PV capacities between 2014 and 2018. Hence, 50 kW of additional PV and 30 MW of dammed hydro power capacity with 16.5 MW pump capacity were added, while an older hydro power plant of 3.5 MW was decommissioned (based on communication with EEM in January 2020). This resulted in the Reference Scenario 0 to which later scenarios were compared. This energy system had a RE share in PES of 11% and in electricity supply of 30%.

3.1. Electrified 100% RE System

Corresponding to the scenario methodology in Section 2.1, the following Steps 2 and 3 replaced all fossil fuel with biomass or electricity. While switching to biomass in combustion processes did not entail any technical adaptions in the model and its operation, the transport and heating did. The change from petrol and diesel cars to EVs included a change of efficiency in addition to the different input. While oil-fuelled vehicles were estimated to drive 1.5 km per kWh, EVs were assumed to drive 5.9 km for the same unit of energy input (EnergyPLAN assumption). In regard to heating, all boilers still using gas, biomass and oil had efficiencies of <90%, while heat pumps had an equivalent of 300% (COP of 3.0; considered constant based on average annual value), so the overall amount of required input was reduced drastically by the end of Step 3.

With the change to biomass in power plants, the electricity sector reached a 100% renewable share, whilst the inclusion of the remaining sectors of industry, transport and heat also led to a 100% share in the PES (cf. Scenario 1.0, Table 1). The sustainable share of electricity, however, was the critical voice for when biomass was considered renewable yet not completely sustainable. As it is unknown if this amount of biomass is currently fully sustainable to harvest and combust, this was a caution factor. While the amount of biomass increased 16-fold, the CO_2 (from fossil and waste combustion) reduced by 98% from the reference scenario. Electricity demands increased by 38% and 29% for total and peak demand, respectively.

Table 1. Overview data of the main scenarios.

Analysis	Step 1	Step 2	Step 3	Step 4	Step 5	Step 6	Step 6	Step 6	
Scenario	0	-	1.0	1.1	2.0	2.1	2.2	2.3	
Additional PV capacity	0	0	0	100	0	250	50	250	(MW)
Additional wind capacity	0	0	0	200	0	75	225	225	(MW)
RE share of PES	11.0	56.1	100.0	100.0	100.0	100.0	100.0	100.0	(%)
RE share of electricity	30.2	100.0	104.8	112.7	103.9	104.0	105.3	111.8	(%)
Sustainable share of electricity	26.3	26.3	19.1	77.6	18.3	71.3	70.7	99.4	(%)
CO_2 emissions	885.4	450.6	16.4	16.4	16.4	16.4	16.4	16.4	(kt)
Biomass demand	172.9	1885.3	2762.6	1370.2	2854.2	1360.9	1419.4	716.7	(GWh)
Total electricity demand	838.9	838.9	1156.8	1220.2	1203.1	1284.7	1285.9	1315.9	(GWh)
Peak electricity demand	140.0	140.0	180.8	205.6	218.0	358.8	337.6	491.0	(MW)
Total annual costs	319.3	279.4	396.3	392.9	404.2	396.3	398.0	397.6	(B€)
CEEP	0	0	0	92.1	0	0	17.0	95.5	(GWh)

The next Step 4 is illustrated in Figure 3, where the electrified energy system was analyzed in a sensitivity study with various additional RE capacities. The results were reconciled with the target criteria, where biomass was reduced by at least 50% and CEEP was kept to <100 GWh, which resulted in only one option: Scenario 1.1.

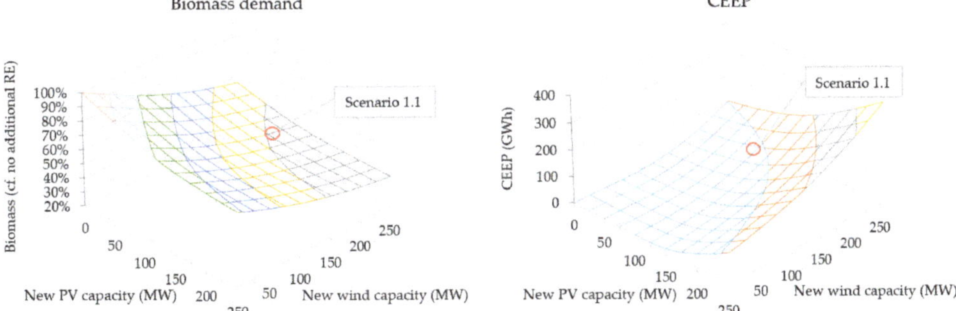

Figure 3. Sensitivity study for an electrified energy system, including optimal Scenario 1.1—resulting additional biomass demand and critical excess electricity production (CEEP).

By addressing these criteria, Step 4 resulted in the additional capacity of 100 MW PV and 200 MW wind but also increased the sustainable share of electricity to 77.6%. Hence, in technical terms, the energy system of Madeira Island was now 100% renewable and supplied by 60% RE through wind, PV, and the other previously installed RE, while the remaining 40% was supplied by biomass combustion. As can be further noted from Figure 3, the biomass consumption reduction stagnated shortly after the 50% reduction target (even with 250 MW PV and wind each, it reached 59% biomass reduction). Furthermore, only 9 out of the 121 possible scenarios resulted in 0 CEEP, but here the biomass reduction potential was below 20%. The details of Steps 1–4 and the Scenarios 1.X can be found in Table 1 next to the results from Steps 5 and 6 and its corresponding scenarios 2.X, which are explained in the following.

3.2. Smart 100% RE System

Firstly, the smarter heat entails an intensification of thermal solar collectors—double the capacity, as well as TES—to supply 100% of households and for a maximum period of 10 days, which optimizes the operation of the heat pumps. Secondly, the smartening of the transport demand including V2G requires the addition of the right grid-to-battery capacity and vice versa. The maximum share of cars charged during peak and the share of grid-connected cars was defined as 50%, and total battery storage capacity was 5 GWh. Thirdly, the smart fuel addition entailed the introduction of electrofuel through the example of an electrolyzer to produce hydrogen and 20% of the transport demand to be covered with it, as well as the option of hydrogen storage of 100 MWh. All three steps increased the electricity demand but reduced the CEEP. In EnergyPLAN, the electricity demands of EVs and the electrolyzer were flexibly modeled according to the RE production.

Step 5 resulted in Scenario 2.0, which is similar to Scenario 1.0 but slightly more expensive due to its complexity. The following sensitivity study in Step 6 therefore was required to show its full potential. Step 6 addresses the same target points as Step 4 to increase sustainability shares through reduced biomass amounts, costs and CEEP. In contrast to Step 4, where only one scenario fulfills the targets, the SES approach offered 41 possible options within the same criteria. The following proposed scenarios therefore could further address potential limitations in available land and visual impacts if required. Scenarios 2.1 and 2.2 thus demonstrated mainly PV or wind, respectively, while Scenario 2.3 did not consider any of these limitations. The results of this sensitivity study can be found in Figure 4, which shows that biomass reduction potential spanned longer than in the electrification scenario (1.X) with up to 77% biomass reduction at full additional capacities of 250 MW each. Furthermore, CEEP can be avoided in 72 out of 121 possible combinations and remained below 100 GWh in all but three options.

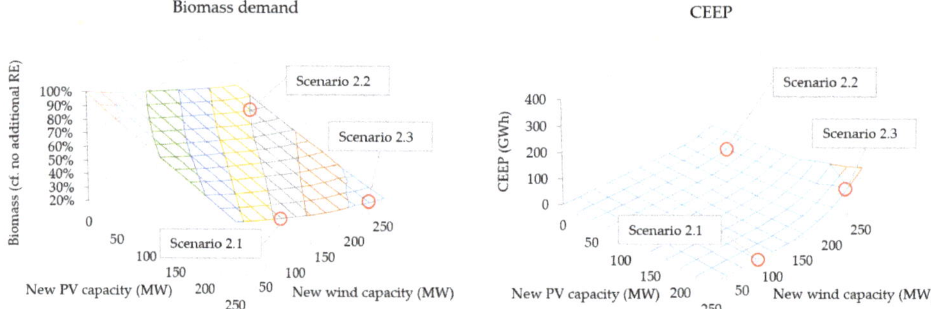

Figure 4. Sensitivity study for a smart energy system, including proposed scenarios—resulting in additional biomass demand and CEEP.

The three proposals out of the 41 options of PV–wind combinations can be pointed out in Figure 4. All proposals show a different extreme in terms of RE capacity, as well as consider potential limitations, so either one could be recommended to the island of Madeira depending on local requirements and wishes.

The first SES scenario included the maximum PV capacity within the sensitivity study and required a minimum wind capacity of 75 MW to reach the biomass reduction of >50%, which was not possible with 50 MW wind capacity (48%). It is one out of five scenarios that fulfilled the criterion and remained at 0 GWh CEEP, which was not possible with the electrified energy system scenarios. With the increasing opposition to wind turbines and potential available space on rooftops, among other challenges, this "PV Scenario 2.1" may be the preferred option to the local community. Despite only minor differences in the total annual costs, this scenario also presents the cheapest solution under the smart system approach.

The second SES scenario presents a wind-favored option, at almost maximum wind capacity and minimum PV capacity within the target criteria range. With 225 MW additional wind and 50 MW PV capacity, the CEEP was kept at 17 GWh annually, which corresponded to 2% of the electricity production. As can be seen in Table 1, the peak demand was the lowest in this scenario; however, this resulted from the peak supply delivered by the wind turbines and the smart charging and hydrogen production accordingly. This overall lowest total capacity further entailed the overall lowest number of new installations, since capacities of single wind turbines was much higher than PV systems, therefore this more centralized concentration of RE could be preferred.

The third proposed scenario, Scenario 2.3, aimed at the highest biomass reduction while staying within the CEEP limitations, and even within the cost requirements, as it is even cheaper than Scenario 2.2. With the CEEP value of 95.5 GWh, which represents 6% of the production, a biomass reduction of 75% could be achieved, resulting in a 9% share of biomass for the electricity supply. For this, 250 MW PV and 225 MW wind capacity was installed, which, however, ignores the potential space and visual limitations that might exist. It presents, however, an option if biomass is to be limited. This way, Scenario 2.3 was the most sustainable scenario with a sustainability share of 99%. For all scenarios with large PV and/or wind capacity, environmental aspects in their production could be considered.

Alternatively, if none of the above presented scenarios appears optimal after all, a compromise or combination of their aspects could be considered. For example, if suitable land for wind turbines are limited but should still be included, other scenarios could be further evaluated with Figures 4 and 5. Regarding biomass reduction, 15 combinations require <1000 GWh biomass; and regarding CEEP increase, 72 combinations are at <0.1 GWh CEEP and another 27 within 0.1–25 GWh CEEP. Finally, the target criteria of cost limitation are presented in Figure 5, where the socioeconomic cost development of the scenario is illustrated in comparison to Scenario 1.1. In contrast to the biomass consumption and CEEP, it did not develop linearly but shows the optimal scenarios at PV capacities above 175 MW and wind capacities between 50 and 175 MW. Regarding annual socioeconomic costs, Scenario 2.1 would be the most secure in this regard, but further 11 other combinations would be as well.

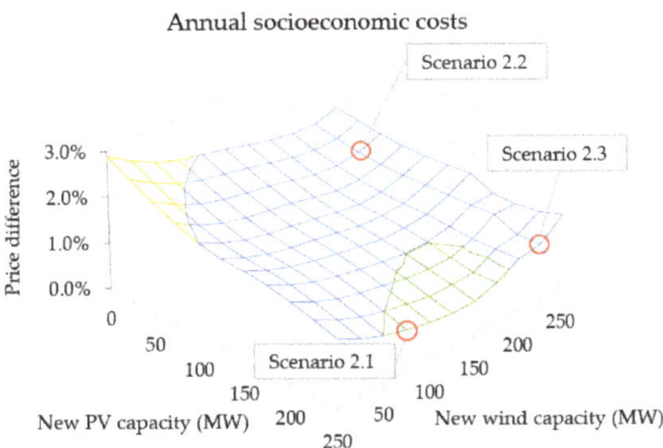

Figure 5. Sensitivity study for a smart energy system, including proposed scenarios—resulting annual socioeconomic costs, relative difference to Scenario 1.1.

3.3. Comparison

When comparing the technical analyses of transitioning to 100% RE on Madeira Island, the SES approach presented more options and thereby more benefits than the simplified electrification approach, addressing all KPIs. This was clarified with an overview in Table 1 and illustrated in Figure 6. Both alternatives achieved CO_2 reductions and higher shares of renewable and sustainable sources. Compared to the reference case, Scenario 1.1 reduced the CO_2 emission by 98% but increased the biomass consumption by 690%, the annual and peak electricity by 45% and 47%, respectively, and the annual costs by 23%. Scenarios 2.X achieved the same CO_2 reductions and similar changes compared to the reference. The remaining CO_2 emissions were related to the waste incineration.

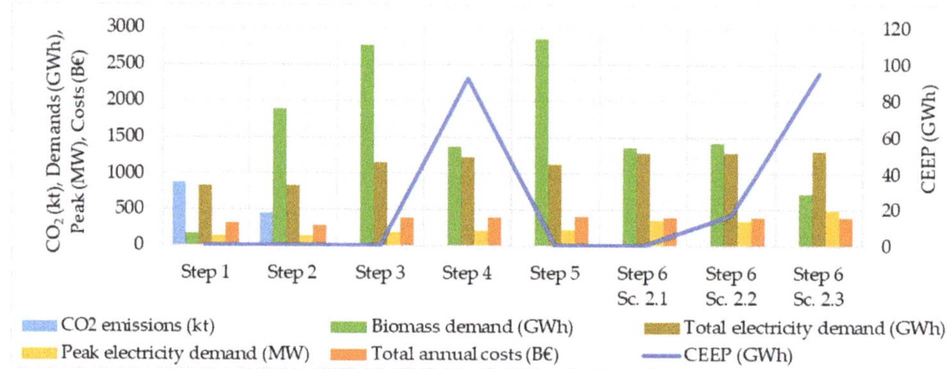

Figure 6. Overview of selected key performance indicators (KPIs) of the main scenarios.

When comparing 2.X with 1.1, however, biomass could be reduced by up to 52%, CEEP by 100% while costs increased by 0.7–2.9%. Overall, the 2.X scenarios show more possibilities for potential local restrictions. While Scenario 1.1 could achieve the biomass, CEEP (7% of total electricity production) and cost targets only in one PV-wind combination, the Scenarios 2.X present 41 combinations. If one specific target was in focus, e.g., CEEP limitations, the sensitivity study of Scenarios 2.X presents 72 combinations where CEEP can be avoided completely. This concerns only 9 out of the total 121 options for the 1.X-scenarios.

While 100% RE share of PES was reached after Step 2, the sustainable share, where biomass was used as little as possible, reached 70–99%. When taking into account the sustainability available biomass on Madeira Island, this could be elaborated. The share of RE of electricity reached values above 100% if the production exceeded the demand.

Complementary to Table 1, Figure 6 illustrates selected KPIs. While CO_2 emissions dropped drastically within the first two steps, biomass consumption increased until RE was added. Still, electricity demands were increasing with each step, similarly to the costs. While electricity demands increased by 57–350%, the costs increased by only 24.5% compared to Step 1. When including the additional CEEP values, the dependency between all KPIs is shown.

The basic electric approach is nonetheless an essential part of the transition to 100% RE share, while the additional smartening of the energy system can be understood as a potential, yet beneficial, second step. According to this, the market analysis addresses both approaches in the following section as a supplement and preliminary discussion for the implementation of proposed technical scenarios.

4. Market Analysis

Based on the above sociotechnical analysis of the two approaches to 100% RE share on Madeira Island, the institutional barriers and market requirements are addressed in the following. It becomes clear that the SES approach has more benefits in technical and sustainable terms and allows for several combinations of wind and PV capacities while being in line with the scenario targets. However, Scenario 1.1 competes with Scenarios 2.X in socioeconomic aspects and is otherwise an essential baseline for the potential uptake of smartening afterwards. The following institutional analysis therefore contributes to the comparison of the two approaches and sees Scenario 1.X as a basic first step and Scenarios 2.X as a necessary second step in the market for an optimal transition to 100% RE.

The main barriers that are not apparent in the technical scenarios in EnergyPLAN, however fundamental, are the implementation of the modeled scenarios with their corresponding technical requirements. Specifically, the island-wide change to electric heating and transport can be connected to radical changes in people's homes, public buildings and infrastructure. Additionally, the SES approach includes flexibility options that require the production of hydrogen, heat from heat pumps and the

V2G in accordance to the RE power production. While this is modeled optimally in EnergyPLAN, incentives, regulations and/or market redesign are required for implementation.

The following therefore addresses this issue on the consumer side as the key technologies can be found there and also to increase consumer participation in the energy planning. As introduced in Section 1, the European Commission proposes an integrated energy market, where the consumers are put at the centre of this transition. As this paper focuses on the opportunities of EVs and heat pumps, Section 3 shows how the consumer is required to help in this part of the transition. This may be an important aspect in order to address the implementation and flexibility issue, if planned accordingly in this institutional setup. Furthermore, the encouragement and engagement of the consumers allows not only for wider acceptance of required technologies but also results in benefits for the grid operator if their interests align through increased grid flexibility.

4.1. Barriers

As explained in Section 2.2, the technical analysis influenced the institutional one as the complexity in the technologies was also reflected in the market complexity. With more technologies on the supply, conversion and demand sides, as well as in balancing in the SES approach and improving cross-sector integration, the supporting institutional framework also needs to be addressed. Here, the decentralization in the energy sector requires a different focus to involve consumers more.

In the past—in contrary to the two proposed approaches—energy systems were dealt with separately and markets were centralized and based on fossil infrastructure, hence balancing, integration or alignment was not as complex. With the increase of fluctuating RE, this stability is affected, requiring technical and economic adjustments, as energy production and market complexity increases. When aiming at 100% RE systems, hence, the institutions and markets as we know them need to move away from the paradigm of fossil fuel being at the centre of the energy market. By coordinating the integrated market prospected by the EU, the consumers are put at the centre of the energy market and smart technologies would be allowed to become competitive.

The introduction of the smart energy markets concept [24] shows the direction these RE-based systems can have, although they must be adjusted to the limitations of islands and be more focused on the self-sufficiency of energy systems with limited resources. As transmission lines might not be available and imports can be uneconomical, the local integration through electrification of additional SES aspects is important. However, the markets of islands are usually set up in a similar manner as the national market, which does not have the same limitations. This leads to the increased necessity on islands of, on the one hand, decentralized and local utilization of RE for electricity, heating and transport demands and, on the other hand, flexible demand and balancing. Both the electrification as well as the SES approach can be evaluated accordingly.

In Madeira Island, the incentives to increase both the local utilization and the flexible demand are limited. While incentives exist for the purchase of EVs at national [44] and regional [45] levels, the number of EVs is still limited, showing that the incentive might not attract enough people. Barros et al. [42] further show that the average EV user on the island has an income above average, and that the general conception of EVs is still hesitant—either regarding battery lifetime, charging time or suitability for the island orography. Despite the current option of free public charging for EVs, their share on the island is less than 1%.

Incentives for purchasing heat pumps, TES or even electrolyzers and hydrogen vehicles are not available at all as of writing. If the consumers are to increase their use of these technologies, it is necessary to make them understand their benefits, addressing the hesitance and criticism by working on the acceptance and understanding from the end-user perspective. While this is a barrier, it can also be a potential to explore through the local consumer-involved energy market.

On the other hand, incentives for flexible consumption are in theory available for all sectors, yet with a limited reach so far. The local transmission system operator (TSO)/distribution system operator (DSO; EEM) together with the national energy services regulation entity (ERSE) introduced time-of-use tariffs to address the local issue of high electricity demand peaks during the hours 9:00–13:00 and 18:00–22:00. The result was an optional choice between three main tariffs, which only change according to season (summer and winter season) and potentially weekdays (Monday–Friday, Saturday and Sunday), see Figure 7. According to EEM, the idea is to incentivize consumers to use electricity during the night, as this is when curtailment—or in the technical analysis CEEP—most likely occurs [43].

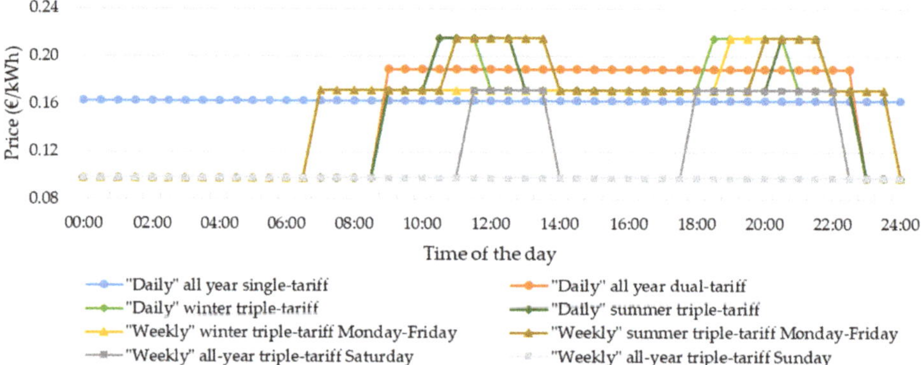

Figure 7. Time-of-use tariff options in Madeira, based on [43] ("Daily" indicating no difference between weekdays and "Weekly" indicating a difference between Monday-Friday, Saturday and Sunday).

Despite the free charging options, EV owners preferred to charge at home, either due to comfort or time constrains, but often excluding the tariffs in the considerations evaluated in [42]. On the contrary, Hashmi et al. [43] suggest that the daily tariffs are preferred without the cheaper weekend prices. While this limits the peak shaving, the local peak power contract (PPC) also has a restricting impact on the flexibility of consumption. The PPC is the maximum demand that consumers can draw from the grid and results in local power shutdowns if exceeded. Furthermore, single and dual tariffs are still the preferred option, not evidencing the benefit of EV charging at times when the electricity prices are cheaper and despite being better for the DSO/TSO.

Although incentives for more dynamic consumption are available, the current tariffs do not consider the actual sustainable share in the electricity mix. This price flexibility, yet inflexibility, has the same impact on encouraging the use of sustainable electricity in all sectors, as it does not seem to make a difference in price where the electricity is coming from. Hence, there is no difference if it is coming from a combustion plant—even though it might use biomass, it might not be sustainable due to the limited resource—or from the unlimited and sustainable sources wind or sun. Even though hydrogen is not yet much developed in Madeira Island, it plays an important role in illustrating the need for flexible production of an alternative fuel. However, examples from other islands show that there is currently no incentive to run the electrolyzer during windy/sunny times instead of during times of imported/combustion-based power, leading to the difficulty of producing green hydrogen, as is the case in the Orkney archipelago [46]. Overall, the use of best technologies at optimal times requires new approaches, which is addressed in the following.

4.2. Recommendations

The radical changes proposed and the barriers identified in the previous sections form Phase 1 and Phase 2 of the institutional analysis. In the following Phase 3, concrete changes to the market structure are suggested accordingly with a focus on end-users and energy markets though including basic aspects of the technology markets, as strategic energy planning suggests.

As evidenced above, the consumer can be economically incentivized to use electricity at certain hours—in either the electrified scenario through heat pumps and EV charging or in the smart energy scenario through additional technologies. This can be done when it is also beneficial for the grid operator, resulting in a flexible and dynamic energy market, where the fluctuating RE production is in the centre. To allow for a dynamic market, the first step is the uptake of suitable technologies on the demand side. The number of EVs, heat pumps and potential electrofuel technologies require incentives that convince consumers to invest. This can be achieved by increased support and/or clarification of the long-term benefits and by considering all income groups, addressing the social barriers mentioned. The recently introduced incentives on both the local and national level present such benefits and should be secured and potentially elaborated in the future. This represents a prerequisite to implementing the smart scenarios and goes hand in hand with suitable financial incentives for electricity consumption, aiming at the second step to market dynamics. In line with this, [47] shows the importance of incentives at the right stage as they were evaluated to be better placed at operation than at purchasing state of EVs.

Semi-flexible electricity prices have been introduced to regulate the demand and encourage consumption outside peak hours through optional tariffs, but this is not yet widely done and must be adapted to a 100% RE system and to the sensitive energy systems of islands. However, the currently used time-of-use tariffs are not suitable for high RE scenarios and more flexible tariffs are suggested to overcome the existing barriers by including equality and transparency for the consumers [48]. In comparison with Figure 7, Figure 8 indicates a potential price mechanism according to RE production, with the option to take the electricity demand (projection) into account. Similar average daily prices of 0.15 €/kWh are compared to the current tariffs, but with more extreme peaks, incentivizing dynamic consumption. This is shown in an example for 7th January, where consumption is encouraged during the windy night (and low demand), instead of the early evening, where ceasing PV and wind production (and high demand) increase the price between 15:00 and 20:00 drastically (cf. triple-tariff).

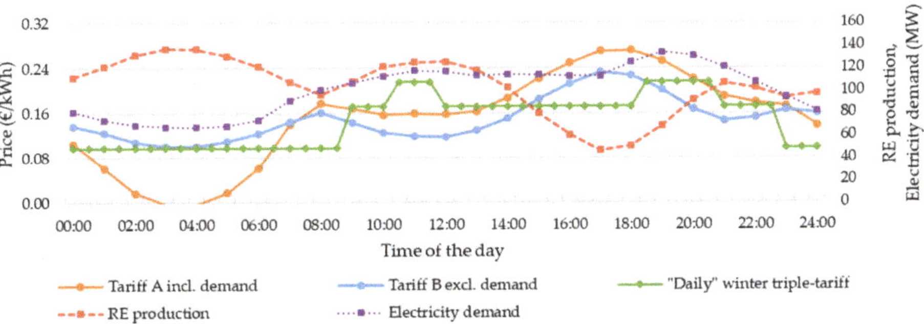

Figure 8. Proposed example of dynamic price mechanism resulting from renewable energy (RE) production including (tariff A) or excluding electricity demand (tariff B) on Madeira modeled in EnergyPLAN for 7 January.

This recommendation would make smart technologies competitive to the current use of fossil fuels. By taking into account the sustainability of the local RE production, the transmission and their costs, the price signals must be adjusted accordingly. With yet a small number of EVs but already a demand for peak shifting now, the need for scheduling is already pointed out, which will increase with additional technologies in the energy system and which will be possible with elaborating on the dynamic price mechanism proposal.

Hence, the current tariffs for end-users on Madeira Island (cf. Figure 7) can be the starting point but are recommended to become mandatory and more dynamic according to the sustainable electricity from wind and PV. Furthermore, the proximity of the consumer to the production side is to be taken into account to reflect the true transmission/distribution cost. As demonstrated in [49], proximity is a positive contribution to consumer behavior, and in [50], resulting lower prices for local consumers are argued to increase acceptance. Hence, the local TSO/DSO is to control and schedule price signals according to the technical capabilities of the overall energy system—similar to what is already done—but including further market perspectives by considering the consumer as a controllable load that requires incentives. This can be applied to EVs, heat pumps and other demands that allow flexible operation either through smart charging, TES or hydrogen storage capacity resulting from the first step. A smartly controlled or scheduled approach would also address the PPC problem and weather forecasts could also be considered in this recommendation. With increased digitalization, this and more can become an important attribute to future energy markets [51,52].

Similarly, the Madeira-specific free EV charging will not be possible for long. However, a low price for charging EVs can be negotiable according to the conditions by limiting it to hours of high sustainable energy shares. As [42] shows, some EV owners act according to price incentives when charging and they are suitable flexible consumers. This entails an expansion of charging options, not only public ones but also in business areas and smart charging support. While some solutions can be managed centrally, others may only be possible at private facilities, hence different tariff options can be considered respectively. When this is in place, a smart approach to V2G must follow to bring the full potential to the energy system by providing balance through short-term storage. This requires another economic incentive for the consumers at the household level, as their EV batteries' life cycle would be affected and the time to charge the EV battery prolonged. This is addressing the barriers of hesitance mentioned in Section 4.1 and is supported in [53], where remuneration is further elaborated. Thus, a selling price higher than the buying price is recommended, signaling the value to the market. If EV owners were not to be allowed to make a profit, this balance option would be limited to the DSO/TSO, showing an example of the importance of consumer involvement.

Though outside the scope of this paper, the limited uptake of RE capacity at the household level, where consumers become prosumers, would be further addressed with these recommendations. While there is now an injection limit from microproducers, such as PV owners, to the grid, the additional control of consumption through location-based prices could enable an easing of injection limits. This could further lead to household PV capacity increase, which may be required besides large-scale PV parks for the energy system in this paper, especially Scenario 2.1.

An overview of the two approaches, their barriers and potentials as well as recommendations can be seen in Table 2. As visible, the number of potentials is higher for the multi-energy SES approach, yet the market requirements address both approaches as the electrification still forms a first step towards the smart approach. Further discussion and recommended research can be found in the following section.

Table 2. Overview of the phases of the institutional analysis in relation to the technical analysis.

Phase	Approach	Electrification (Scenario 1.1)	Smart energy system (Scenarios 2.X)
1	Technical requirements	• RE power supply, heat pumps, EVs	• RE power supply, heat pumps, solar thermal, TES, EVs, smart charging, V2G, electrolyzer, hydrogen storage
2	Barriers	• High biomass consumption and/or CEEP, few PV-wind combinations • Limited incentives for technologies • Limited incentive for flexible demand • Limits for consumer and grid operator	• Limited incentives for technologies • Complex technological setup • Limited incentive for flexible demand
	Potentials	• Fewer technologies required • Flexibility to certain extent	• Less biomass consumption and CEEP, many PV-wind combinations • Flexibility through additional storage • Benefits for grid operator • High consumer involvement
3	Market recommendations	• Support of RE-based technologies and consumption • Dynamic end-user tariffs, according to production and location • Control and scheduling of tariffs by local entities • V2G incentive	

5. Discussion

This paper fills the previously identified research gap by showing the needs and possibilities of Madeira Island to reach a 100% RE share from both technical and institutional perspectives. To this end, two approaches are developed and compared to each other to show their strengths and weaknesses, and how they can be supported in the right way.

The importance of sector integration in a multi-energy system is presented in either approach as electricity demands can be found in the heating and transport sectors already and shows to be the way forward for the most dynamic and smart energy system as well, especially for islands. The paper shows how smart technologies and the inclusion of other energy carriers than electricity can be beneficial by providing flexibility. By using EnergyPLAN, the technical model simulates the optimal balance between the sectors and technologies. While technical analyses often do not consider the required institutional support, by adding a market analysis, the implementation potential of the technical scenarios is addressed in this work as a supplementary requirement for a full transition to high RE shares. Through consideration of the transition as being radical and analyzing market barriers accordingly, the importance of consumer involvement and demand response becomes evident. Therefore, the introduction of social parameters and local energy markets also proves to be important in islands and the presented transition.

With a three-phase institutional approach after the technical step-by-step modeling, the consumers are recommended to be more involved in the market, enabling benefits for both supply and demand sides. This entails the recommendation of institutional support, as the potential of flexible demand response can effectively be realized with dynamic pricing. This supports the implementation of the optimally modeled technical scenarios. With Madeira being a special case since its own local authority is in a position to develop this, it also shows how local power and regulation can be the key to local energy systems transitioning to 100% RE. On the contrary, if local entities are excluded, it can become a bottleneck for local energy market development.

The simplified electrification approach competes with the smart approach in socioeconomic terms by being 0.7–1.6% cheaper but at higher biomass consumption and/or CEEP. Hence, the smart technologies might be more costly, yet only marginally, while allowing for less biomass and/or CEEP. The overall increased annual costs (30% more than the reference scenario) can be attributed to large investments, such as EVs, but should also be considered in the long term, when imports are reduced and the local economy benefits from it. The benefits in reduced biomass and CEEP may require investment and exploration of electrofuels, additional storages and technologies, yet it competes with the mere electrification with various RE options and limiting its capacity. The resulting optimized utilization of fluctuating RE production allows for higher peaks, yet better acceptability, flexibility and control, both from technical and consumer perspectives.

However, for the discussion and elaboration of analysis and results, further considerations and research is recommended, for example, a discourse on the technologies to implement and the biomass resource locally available. Outside this paper's scope and not further addressed, the underlying capacity required for the scenarios entail investment, installation and maintenance. While this is reflected in the total annual cost, the implementation can be further studied. Similarly, the knowledge and resources might be locally limited, influencing the proposed scenarios. Hence, if biomass is not suitable or limited for the power production supplementing the fluctuating production from wind, PV and hydro plants, a different combination of technologies or resources is required. As hydrogen is used as an example of flexible demand with various purposes, other electrofuels or synthetic products could also be considered. Furthermore, the sustainability of biomass can be elaborated as well as its transport across the islands versus decentralized exploration close to its origin.

The technical modeling is done on an hourly basis and thereby includes seasonal and daily variations. It could, however, be improved by using an intra-hour analysis, which might be especially relevant for a sensitive island system. Additional detailed market analysis could further advance this aspect of the paper. The proposed incentives for investments in EVs or heat pumps require further examination if every consumer would require them, but some limitations might be in order. With different options for smartening the EV sector, further study is recommended of both centrally operated and private options, such as the implementation of V2G in order to increase the involvement of electricity consumers in the Madeira microgrid market. The final implementation of the tariffs would entail additional research of details not only for the electricity end-users tariffs but also for the integration of V2G. EV batteries degradation from excess of charging/discharging cycles has to be taken into account, as well as minimum EV battery range for users' daily commutes and EV users' disposal to have their vehicle available for the grid operations, specifically in hours with more RE.

Global market developments are to be taken into consideration, since the socioeconomic part of the analysis gives an impression but could be improved if prices for EVs or other investments drop. This also points out that the resulting total annual costs are very compatible with the reference since high investment costs are also included. Despite focusing on the consumers' side in this transition, energy efficiency measures are not included but could benefit it in various ways and be considered more in future research. Lastly, additional energy infrastructure, such as gas or district heating/cooling, could put further significance on a holistic energy market, making use of the various opportunities presented; and a study of other islands' consequential policies would be the final step. Overall, the complexity of energy system planning for low-carbon energy transitions and high RE shares is evident and the need for alignment of technical and institutional analyses presented through the perspective of islands but for the benefits of all.

6. Conclusions

The research gap is addressed by combining not only two different energy system planning approaches but also technical and institutional aspects of transitioning the energy system for the case of Madeira Island. The step-by-step approach to a 100% RE and highly sustainable system is modeled with EnergyPLAN and addresses market barriers and potentials in the complementary institutional analysis. The impacts of the different steps in the transition are shown and the benefits of a smart energy system pointed out. The particular settings on the island are thereby included and the limitations and sensitivity of the energy system properly addressed.

It is shown that island energy systems require a more complex energy system and smart technologies to cope with the lack of cross-border trading, where local conditions are considered and self-sufficiency improved. Without the smart energy system approach, biomass consumption and curtailment are to be expected. This, however, requires support mechanisms that take into account the fluctuations and local restrictions of RE, where the role of the consumer becomes more important. In the case of Madeira Island with its own government, better local adaption can be expected as specific tariffs and incentives already exist specifically supporting the local situation.

The general need for better alignment with market aspects is pointed out for the implementation of the technical scenarios through recommendations resulting from a market analysis. It is illustrated how important the consumer is for capacity investment and flexible demand, which can only be reached with incentives for both installation as well as use of smart technologies. Specifically, dynamic electricity prices are recommended that align with the sustainable share of electricity and the local possibilities and restrictions.

The presented research can be replicated in other energy systems and become most relevant for those with sensitive or limited infrastructure but can also be relevant for those systems that aim for higher self-sufficiency and more independence even when being well-connected. With a well-planned transition to 100% RE, as is demonstrated with this paper, the replacement of fossil fuels and the fight against climate change can be achieved.

Supplementary Materials: The following are available online at http://www.mdpi.com/1996-1073/13/17/4434/s1, EnergyPLAN data and results.

Author Contributions: Conceptualization, methodology, formal analysis, software, writing—original draft preparation, visualization, and project administration, H.M.M.; validation, investigation, resources, data curation, and writing—review and editing, H.M.M. and L.B. All authors have read and agreed to the published version of the manuscript.

Funding: This research received funding from the EU H2020 under Grant Agreement 731249.

Acknowledgments: Guidance and support from our supervisors are gratefully acknowledged.

Conflicts of Interest: The authors declare no conflict of interest.

Abbreviations

B€	Billion Euros
CEEP	Critical excess electricity production
CO_2	Carbon dioxide
COP	Coefficient of performance
DSO	Distribution system operator
EEM	Madeira electrical company
ERSE	National energy services regulation entity
EU	European Union
EV	Electric vehicle
KPI	Key performance indicator
PES	Primary energy supply
PPC	Peak power contract
PV	Photovoltaic
RE	Renewable energy
SES	Smart energy system
SMILE	Smart Island Energy system project
TES	Thermal energy storage
TSO	Transmission system operator
V2G	Vehicle to grid

References

1. Kang, J.N.; Wei, Y.M.; Liu, L.C.; Han, R.; Yu, B.Y.; Wang, J.W. Energy systems for climate change mitigation: A systematic review. *Appl. Energy* **2020**, *263*, 114602. [CrossRef]
2. Marczinkowski, H.M.; Østergaard, P.A.; Djørup, S.R. Transitioning Island Energy Systems—Local Conditions, Development Phases, and Renewable Energy Integration. *Energies* **2019**, *12*, 3484. [CrossRef]
3. Scandurra, G.; Romano, A.A.; Ronghi, M.; Carfora, A. On the vulnerability of Small Island Developing States: A dynamic analysis. *Ecol. Indic.* **2018**, *84*, 382–392. [CrossRef]
4. Skjølsvold, T.M.; Ryghaug, M.; Throndsen, W. European island imaginaries: Examining the actors, innovations, and renewable energy transitions of 8 islands. *Energy Res. Soc. Sci.* **2020**, *65*, 101491. [CrossRef]
5. Kotzebue, J.R.; Weissenbacher, M. The EU's Clean Energy strategy for islands: A policy perspective on Malta's spatial governance in energy transition. *Energy Policy* **2020**, *139*, 111361. [CrossRef]
6. Ram, M.; Bogdanov, D.; Aghahosseini, A.; Oyewo, S.; Gulagi, A.; Child, M.; Fell, H.J.; Breyer, C. *Global Energy System Based on 100% Renewable Energy-Power Sector*; Lappeenranta: Berlin, Germany, 2017.
7. Hansen, K.; Breyer, C.; Lund, H. Status and perspectives on 100% renewable energy systems. *Energy* **2019**, *175*, 471–480. [CrossRef]
8. Jurasz, J.; Canales, F.A.; Kies, A.; Guezgouz, M.; Beluco, A. A review on the complementarity of renewable energy sources: Concept, metrics, application and future research directions. *Sol. Energy* **2020**, *195*, 703–724. [CrossRef]
9. Kuang, Y.; Zhang, Y.; Zhou, B.; Li, C.; Cao, Y.; Li, L. A review of renewable energy utilization in islands. *Renew. Sustain. Energy Rev.* **2016**, *59*, 504–513. [CrossRef]
10. Ioannidis, A.; Chalvatzis, K.J.; Li, X.; Notton, G.; Stephanides, P. The case for islands' energy vulnerability: Electricity supply diversity in 44 global islands. *Renew. Energy* **2019**, *143*, 440–452. [CrossRef]

11. Cross, S.; Padfield, D.; Ant-Wuorinen, R.; King, P.; Syri, S. Benchmarking island power systems: Results, challenges, and solutions for long term sustainability. *Renew. Sustain. Energy Rev.* **2017**, *80*, 1269–1291. [CrossRef]

12. Meza, C.G.; Zuluaga Rodríguez, C.; D'Aquino, C.A.; Amado, N.B.; Rodrigues, A.; Sauer, I.L. Toward a 100% renewable island: A case study of Ometepe's energy mix. *Renew. Energy* **2019**, *132*, 628–648. [CrossRef]

13. Lund, H. *Renewable Energy Systems—A Smart Energy Systems Approach to the Choice and Modeling of 100% Renewable Solutions*, 2nd ed.; Academic Press: Massachusetts, MA, USA; Elsevier: Massachusetts, MA, USA, 2014. [CrossRef]

14. Lund, H.; Andersen, A.N.; Østergaard, P.A.; Mathiesen, B.V.; Connolly, D. From electricity smart grids to smart energy systems—A market operation based approach and understanding. *Energy* **2012**, *42*, 96–102. [CrossRef]

15. Sinha, R.; Bak-Jensen, B.; Pillai, J.R.; Zareipour, H. Flexibility from electric boiler and thermal storage for multi energy system interaction. *Energies* **2019**, *13*, 98. [CrossRef]

16. Bačeković, I.; Østergaard, P.A. Local smart energy systems and cross-system integration. *Energy* **2018**, *151*, 812–825. [CrossRef]

17. Banja, M.; Sikkema, R.; Jégard, M.; Motola, V.; Dallemand, J.-F. Biomass for energy in the EU—The support framework. *Energy Policy* **2019**, *131*, 215–228. [CrossRef]

18. Marczinkowski, H.M.; Østergaard, P.A. Evaluation of electricity storage versus thermal storage as part of two different energy planning approaches for the islands Samsø and Orkney. *Energy* **2019**, *175*, 505–514. [CrossRef]

19. Alves, M.; Segurado, R.; Costa, M. On the road to 100% renewable energy systems in isolated islands. *Energy* **2020**, *198*, 117321. [CrossRef]

20. Dorotić, H.; Doračić, B.; Dobravec, V.; Pukšec, T.; Krajačić, G.; Duić, N. Integration of transport and energy sectors in island communities with 100% intermittent renewable energy sources. *Renew. Sustain. Energy Rev.* **2019**, *99*, 109–124. [CrossRef]

21. Cabrera, P.; Lund, H.; Carta, J.A. Smart renewable energy penetration strategies on islands: The case of Gran Canaria. *Energy* **2018**, *162*, 421–443. [CrossRef]

22. Djørup, S.; Thellufsen, J.Z.; Sorknæs, P. The electricity market in a renewable energy system. *Energy* **2018**, *162*, 148–157. [CrossRef]

23. Bublitz, A.; Keles, D.; Zimmermann, F.; Fraunholz, C.; Fichtner, W. A survey on electricity market design: Insights from theory and real-world implementations of capacity remuneration mechanisms. *Energy Econ.* **2019**, *80*, 1059–1078. [CrossRef]

24. Sorknæs, P.; Lund, H.; Skov, I.R.; Djørup, S.; Skytte, K.; Morthorst, P.E.; Fausto, F. Smart Energy Markets—Future electricity, gas and heating markets. *Renew. Sustain. Energy Rev.* **2020**, *119*, 109655. [CrossRef]

25. Cuesta, M.A.; Castillo-Calzadilla, T.; Borges, C.E. A critical analysis on hybrid renewable energy modeling tools: An emerging opportunity to include social indicators to optimise systems in small communities. *Renew. Sustain. Energy Rev.* **2020**, *122*, 109691. [CrossRef]

26. Brolin, M.; Pihl, H. Design of a local energy market with multiple energy carriers. *Int. J. Electr. Power Energy Syst.* **2020**, *118*, 105739. [CrossRef]

27. Hvelplund, F.; Djørup, S. Multilevel policies for radical transition: Governance for a 100% renewable energy system. *Environ. Plan C Polit. Sp.* **2017**, *35*, 1218–1241. [CrossRef]

28. European Commission. Market Legislation: Electricity Market Design. *Eur. Comm. Energy Top.* **2019**. Available online: https://ec.europa.eu/energy/topics/markets-and-consumers/market-legislation/electricity-market-design_en (accessed on 26 August 2020).

29. Rina Consulting, S.p.A. SMart IsLand Energy Systems Project 2017. Available online: http://www.h2020smile.eu/ (accessed on 20 September 2019).

30. Marczinkowski, H.M. *Smart Island Energy Systems Deliverable D8.2*; Aalborg University: Aalborg, Denmark, 2018.

31. Leavy, P. *Oxford Handbook of Qualitative Research: Oxford Handbook of Qualitative Research*; Oxford University Press: Oxford, UK, 2014.

32. Balaman, Ş.Y.; Balaman, Ş.Y. Uncertainty Issues in Biomass-Based Production Chains. *Decis. Biomass-Based Prod. Chain.* **2019**, 113–142. [CrossRef]

33. ACIF-CCIM. *Smart Island Energy Systems Deliverable D4.1*; ACIF-CCIM: Funchal, Madeira, 2017.

34. Empresa de Eletricidade da Madeira. *EEM 2017 Annual Report*; EEM: Funchal, Madeira, 2018.

35. Lund, P.D.; Lindgren, J.; Mikkola, J.; Salpakari, J. Review of energy system flexibility measures to enable high levels of variable renewable electricity. *Renew. Sustain. Energy Rev.* **2015**, *45*, 785–807. [CrossRef]
36. Aalborg University. *EnergyPLAN Advanced Energy System Analysis Computer Model*; Aalborg University: Aalborg, Denmark, 2017; Available online: https://www.energyplan.eu/ (accessed on 26 August 2020).
37. Vanegas Cantarero, M.M. Reviewing the Nicaraguan transition to a renewable energy system: Why is "business-as-usual" no longer an option? *Energy Policy* **2018**, *120*, 580–592. [CrossRef]
38. Lund, H. Large-scale integration of optimal combinations of PV, wind and wave power into the electricity supply. *Renew. Energy* **2006**, *31*, 503–515. [CrossRef]
39. Mathiesen, B.V.; Lund, R.S.; Connolly, D.; Ridjan, I.; Nielsen, S. *Copenhagen Energy Vision 2050: A Sustainable Vision for Bringing a Capital to 100% Renewable Energy*; Aalborg University: Aalborg, Denmark, 2015.
40. Drysdale, D.; Vad Mathiesen, B.; Lund, H. From Carbon Calculators to Energy System Analysis in Cities. *Energies* **2019**, *12*, 2307. [CrossRef]
41. Mathiesen, B.V.; Lund, H.; Connolly, D.; Wenzel, H.; Østergaard, P.A.; Möller, B.; Nielsen, S.; Ridjan, I.; Karnøe, P.; Sperling, K.; et al. Smart Energy Systems for coherent 100% renewable energy and transport solutions. *Appl. Energy* **2015**, *145*, 139–154. [CrossRef]
42. Barros, L.; Barreto, M.; Pereira, L. Understanding the challenges behind Electric Vehicle usage by drivers—A case study in the Madeira Autonomous Region. In Proceedings of the ICT4S 2020: 7th International Conference on ICT for Sustainability, Bristol, UK, 21–27 June 2020.
43. Hashmi, M.U.; Pereira, L.; Bušić, A. Energy storage in Madeira, Portugal: Co-optimizing for arbitrage, self-sufficiency, peak shaving and energy backup. In Proceedings of the 2019 IEEE Milan PowerTech, Milan, Italy, 23–27 June 2019; pp. 1–15. [CrossRef]
44. República Portuguesa—Ambiente e Ação Climática—Gabinete do Ministro. Incentivo pela Introdução no Consumo de Veículos de Baixas Emissões. Diário Da República no 49/2020, Série II 2020-03-10 Despacho no 3169/2020. 2020. Available online: https://dre.pt/web/guest/pesquisa/-/search/130070443/details/normal?l=1 (accessed on 4 June 2020).
45. Região Autónoma da Madeira—Direção Regional de Economica e Transportes Terrestres. Incentivo à Mobilidade Elétrica na RAM. Funchal, Madeira 2020. Available online: https://www.madeira.gov.pt/drett/Estrutura/Mobilidade/ctl/Read/mid/4064/InformacaoId/54137/UnidadeOrganicaId/17 (accessed on 4 June 2020).
46. Zhao, G.; Nielsen, E.R.; Troncoso, E.; Hyde, K.; Romeo, J.S.; Diderich, M. Life cycle cost analysis: A case study of hydrogen energy application on the Orkney Islands. *Int. J. Hydrog. Energy* **2018**, *44*, 9517–9528. [CrossRef]
47. Kwon, Y.; Son, S.; Jang, K. Evaluation of incentive policies for electric vehicles: An experimental study on Jeju Island. *Transp. Res. Part A Policy Pract.* **2018**, *116*, 404–412. [CrossRef]
48. Bergaentzlé, C.; Jensen, I.G.; Skytte, K.; Olsen, O.J. Electricity grid tariffs as a tool for flexible energy systems: A Danish case study. *Energy Policy* **2019**, *126*, 12–21. [CrossRef]
49. Kalkbrenner, B.J.; Yonezawa, K.; Roosen, J. Consumer preferences for electricity tariffs: Does proximity matter? *Energy Policy* **2017**, *107*, 413–424. [CrossRef]
50. Funcke, S.; Ruppert-Winkel, C. Storylines of (de)centralisation: Exploring infrastructure dimensions in the German electricity system. *Renew. Sustain. Energy Rev.* **2020**, *121*, 109652. [CrossRef]
51. Specht, J.M.; Madlener, R. Energy Supplier 2.0: A conceptual business model for energy suppliers aggregating flexible distributed assets and policy issues raised. *Energy Policy* **2019**, *135*, 110911. [CrossRef]
52. Dutton, S.; Marnay, C.; Feng, W.; Robinson, M.; Mammoli, A. Moore vs. Murphy: Tradeoffs between complexity and reliability in distributed energy system scheduling using software-as-a-service. *Appl. Energy* **2019**, *238*, 1126–1137. [CrossRef]
53. Geske, J.; Schumann, D. Willing to participate in vehicle-to-grid (V2G)? Why not! *Energy Policy* **2018**, *120*, 392–401. [CrossRef]

Article

Maximizing Solar PV Dissemination under Differential Subsidy Policy across Regions

Jeongmeen Suh [1,†] and Sung-Guk Yoon [2,*,†]

[1]   Department of Global Commerce, Soongsil University, Seoul 06978, Korea; jsuh@ssu.ac.kr
[2]   Department of Electrical Engineering, Soongsil University, Seoul 06978, Korea
*   Correspondence: sgyoon@ssu.ac.kr
†   These authors contributed equally to this work.

Received: 6 May 2020; Accepted: 27 May 2020; Published: 1 June 2020

Abstract: This study investigates the effect of a renewable energy dissemination policy on investment decisions regarding solar photovoltaic (PV) installation. We present a theoretical model and conduct a simulation analysis to estimate the total capacity of solar PV generators according to a given subsidy policy. We show how the capacity maximizing subsidy policy depends on the total amount of subsidy budget, interest rate, the expected amount of solar resource and land price in each region. We particularly focus on the improvements of solar PV capacities under the same subsidy budget when the subsidy policy is changed from uniform (equal for all regions) to differential (varying over regional characteristics). This improvement is shown through a case study using Korean data.

Keywords: dissemination; renewable energy policy; renewable energy subsidies; solar PV

---

1. Introduction

To promote renewable energy, many governments have adopted various support policies such as feed-in-tariff (FIT), renewable portfolio standards (RPS), reduction in taxes, and subsidized capital costs for installments [1–6]. However, dissemination of renewable energy requires significant financial support as long as the electricity generated from renewable energy is more expensive than that generated from fossil fuels. According to the International Energy Agency's (IEA) World Energy Outlook 2013, these subsidies amounted to USD 82 billion in 2012. In many countries, it is a declared political goal to raise the market share of renewables further, and hence, it is expected that the total amount of subsidies will also rise accordingly. In the New Policies Scenario projection, which assumes that the governments will adhere to their plans and which serves as the baseline scenario, the IEA expects subsidies to reach almost USD 180 billion per year in 2035 [7].

Owing to these support policies, the generation capacity of renewables has increased rapidly. In 2018, the annual capacity addition of renewables in the world was approximately 181 GW, which is greater than that of fossil fuel-based generators [8]. Consequently, the price of electricity generated from renewables has reached that of electricity obtained from conventional generators in some major countries, i.e., grid parity, resulting in a reduction of renewable energy subsidies. For example, China, the country with the largest investment in renewable energy in the world, has abolished the subsidy for newly installed solar photovoltaic (PV) since 2018. However, a renewable energy subsidy is still required to achieve an aggressive target for the dissemination of renewable energy, e.g., 100% renewable energy.

Germany, one of the leading countries in renewables, produced approximately 40% of its electricity from renewable resources in 2018, and set its target shares of renewable energy to 65% by 2030 [9]. German environment minister said that approximately one trillion euros will be required to achieve this target by the end of the 2030s. In the United States, the state of California established a new senate bill (SB 100) that set a target of 100% carbon-free electricity by 2045 and the state has committed an investment of USD 35 million per year [10]. In South Korea, the government aimed to install new renewable generators of capacity 48.7 GW during 2018–2030 (cumulative capacity 63.8 GW) to achieve 20% share of renewable energy generation in 2030, an increase from 7.0% in 2016. The budget allocated for renewable energy dissemination projects in 2019 was KRW 267 billion (about USD 223 million), an increase of KRW 40.3 billion (about USD 33.6 million) from the previous year, and the amount of support increased by 2.67 times over the last three years.

In this article, we address how to achieve the renewable dissemination target with minimum support budget spending. We focus on a subsidy policy for solar PV generators. Solar PV, one of the most widely used renewable energies, accounted for approximately 100 GW of the total renewable capacity of 181 GW installed in 2018 [8]. Moreover, solar PV can be installed over any area because of its flexible size. For example, solar PV can be installed in a very small area such as on a rooftop or balcony in a residential home with a capacity of a couple of watts or in a large-scale generation site with a capacity of hundreds of MW. Other renewable generators such as wind, hydro, and geothermal are normally implemented on a large scale.

To achieve the aforementioned ambitious renewable targets, small-scale solar PV, as well as utility-scale solar PV, is important because it can be easily installed in urban areas [11]. According to German statistics in 2017, generators of capacity less than 100 kW account for more than half the total capacity of solar PV generators [12]. Particularly, generators of capacity less than 10 kW account for approximately 14.2% of the total solar PV capacity. Furthermore, residential PV installation capacity increased by approximately 76% in the first quarter of 2015 [13]. Unlike utility-scale solar PV, in many countries, small-scale solar PV has not become sufficiently price competitive yet. Therefore, a relevant subsidy policy is still required. Because the owner of a small-scale solar PV system mainly consumes most of the electricity generated by it, the main promotion policies are installation cost subsidies provided to the owner as tax refund or direct subsidy. Therefore, in this study, a direct subsidy policy for small-scale solar PV generators is considered. These small-scale solar PV generators are normally connected to distribution networks [14]. With a high penetration of renewable generation, distribution networks may have some problems such as reverse power flow, voltage fluctuations, power quality issues, and dynamic stability [15]. To solve them, the system operator uses several control entities and strategies including on-load tap changer (OLTP), voltage regulator, and batteries [16–19]. In this study, we assume that these technical issues raised by installing solar PV in the distribution network can be properly solved through these strategies.

We investigate how to allocate a given subsidy budget economically to achieve a renewable power capacity goal. Accordingly, we provide a theoretical model regarding the decision-making approach for each region's investment in solar PV installations at the regional-level according to the given subsidy policy. In addition, considering regional investment decisions under these subsidy policies, we theoretically describe the issue of where and how much the central government should allocate subsidy to minimize the budget. We examine the characteristics of the optimal policy through comparative static analysis, and by using real data for obtaining the results of the theoretical analysis, we estimate the additional expected capacity of solar PV installations from the change in the subsidy policy with the same budget.

The novelty of this study is as follows. First, we clearly focus on the objective of the government to maximize installed capacity through subsidy policies. Although there are several studies on the promotion of renewable energy, only a few studies explicitly consider an optimal subsidy policy to maximize renewable power capacity [20–23]. For example, Bläsi and Requate [20], and Reichenbach and Requate [21] considered studying the spillover effects of capacity production and examined how to optimize the output subsidies for renewable energy producers. However, in these models, "capacity" is simply described as the number of firms, and hence, the installation capacity of each subject is not considered. Andora and Voss comprehensively studied the properties of an optimal renewable energy subsidy policy by using the framework of a competitive peak-load pricing model [22]. According to these authors, the objective of a policy is not to maximize installed capacity, but to maximize the social welfare, which consists of consumer surplus and producer surplus. Second, to the best of our knowledge, this is the first study to allow heterogeneity in regions and examine how an optimal subsidy policy should be region-specific (referred to as differential subsidy) in the context of solar PV dissemination.

As a main result, we show how the optimal subsidy policy should reflect the heterogeneous characteristics of each region in terms of different installation costs and power generation efficiencies. Based on our simple theoretical model, we also present a case study using Korean data to show the additional PV installation capacity expected with the same subsidy budget (or the amount of budget that can be saved to achieve the same installation target) by simply switching from the uniform to differential subsidy policy. This case study confirms that the differential subsidy policy raises the total installed capacity of solar PV generators. In addition, we discuss the difference between capacity and energy maximization policies using the same Korean data.

The rest of this paper is organized as follows. Section 2 describes the theoretical model and then examines the characteristics of regionally differentiated subsidies to economize renewable energy budgets. In Section 3, numerical and case study results are presented. For the case study, we introduce the relevant data of Korea's metropolitan cities and provinces used; explain the relevant assumptions; and estimate the additional supply capacity expected under the same budget if the previous theoretical analysis results are applied. Section 4 discusses the difference between capacity and energy maximization policies. Finally, in Section 5, the paper is summarized and the implications drawn are presented.

2. Analytical Model

In this section, we present a theoretical model regarding the effect of government subsidies on the investment decisions on solar PV installation in each region. It is assumed that solar PV potentials in each region are properly estimated [24–26]. We adopt a two-stage decision-making approach. At the first stage, the government chooses its subsidy policy to maximize the nationwide capacity of solar PV. At the second stage, given the policy, each region makes its regional investment decisions on the amount of solar PV generators. Because backward induction will be used to obtain a solution, we first consider how to model the regional-level decision-making at the second stage.

2.1. Regional-Level Optimization: Investment Decisions

Under a given subsidy policy, the decisions at each regional-level can be described in another two sub-stages: in sub-stage 1, decisions are made regarding long-term investments, such as whether to invest and how much to invest in solar PV generators. Given the power capacity determined in sub-stage 1, in sub-stage 2, a decision is made regarding short-term operation to determine how much electricity is required from the main grid. The objective of each region is to minimize the sum of costs from these two sub-stages. In this study, we assume an operational period of 30 years according to the average lifespan of a solar panel, and I is the set of candidate regions for the installation of solar PV.

In sub-stage 1, i.e., the long-term investment cost of the region i, C_i^I, can be determined as

$$C_i^I(z_i, G_i; s_i) = z_i \cdot (F_i + a_i G_i^2 + b_i G_i - s_i G_i), \quad \forall i \in I \tag{1}$$

where z_i is a binary variable representing the investment decision of the region i (1 and 0 indicate investment and no investment, respectively.), F_i is the fixed cost of investment (USD), G_i is the capacity of the solar PV generators to be installed in region i (kW), and s_i denotes the subsidy in region i per capacity ($/kW). The investment cost coefficients of region i are represented by a_i and b_i. For simplicity, we set the constant F_i to be zero. Moreover, we assume that the long-term investment cost function follows a convex quadratic function. This reflects the land price component of the solar PV installation cost. With greater solar PV capacity, land having higher price per unit tends to be used. The intuitive meaning of each coefficient is as follows. The coefficient b_i can be understood as the average land price in the region i whereas a_i is its variance within the region. That is, a region with high a_i indicates the one where the investment cost is rising steeply as more capacity is installed. The subsidy s_i is paid for each installed capacity G_i, and hence, it is expressed linearly in (1).

In the sub-stage 2, the short-term operation cost of region i, C_i^O, is determined. To this end, we need to define the daily operational power supply and demand. On the supply side, the supply of electricity at each time t consists of the amount of electricity purchased from the main grid q_i^t and the amount of electricity generated by the region's own solar PV g_i^t. The amount of electricity generated by the region i's own generators is limited by the installed capacity at stage 1 G_i and the generation efficiency η_i^t. That is,

$$g_i^t \le G_i \eta_i^t, \quad \forall t, \; \forall i \in I. \tag{2}$$

On the demand side, we assume that the regional power demand in each period t, d_i^t, is exogenously given. Therefore, to satisfy the local electricity demand in region i at time t, the amount of electricity to be purchased can be obtained by

$$q_i^t = d_i^t - g_i^t. \quad \forall t, \; \forall i \in I \tag{3}$$

Accordingly, we can determine the daily operation cost C_i^O. As the amount of locally generated electricity g_i^t is determined by the installed capacity G_i, the cost at the sub-stage 2 consists only of the buying cost of electricity from the main grid. It can be expressed as the sum of the product of the purchased amount of electricity $\{q_i^t\}_t$ and its price $\{p_i^t\}_t$ over the entire period. Therefore, the daily cost of electricity at the sub-stage 2 is defined as

$$C_i^O(\{q_i^t\}_t) = \sum_t p^t q_i^t. \tag{4}$$

The total cost of region i, C_i^T, consists of the sum of the long-term investment costs and the short-term operating costs. In this study, we assume that the entire period of sub-stage 2 is a day and the operating cost at the investment decision stage can be evaluated by the time discount δ. We may consider $\delta \equiv \sum_{k=1}^{K} \frac{1}{(1+r_k)^k}$ where K is the lifespan of 30 years and r_k is the interest rate in period k. The total cost of region i can be expressed as

$$C_i^T(z_i, G_i, \{q_i^t\}_t) = C_i^I(z_i, G_i) + \delta \cdot C_i^O(\{q_i^t\}_t). \tag{5}$$

Finally, we can define the cost minimization problem of a region i as follows:

$$\min_{G_i, \{q_i^t\}_t} \quad C_i^T \tag{6}$$
$$s.t. \quad (2) \text{ and } (3)$$

For the decision whether to invest or not, we may consider that the investment is made only if the total cost of installing solar PV is lower than the total cost of not installing it. Note that the total cost for a region under no investment is equal to the buying cost from the main grid to satisfy its demand without local power supply. Thus, the investment decision can be expressed as

$$z_i = \begin{cases} 1, & \text{if } C_i^T \le \delta \sum_t p^t d_i^t \\ 0, & \text{otherwise.} \end{cases} \tag{7}$$

To solve the above regional-level decision-making, the cost minimization problem under investment can be expressed as a Lagrange function for the cost minimization problem (6) as

$$L = a_i G_i^2 + (b_i - s_i) G_i + \delta \sum_t \left[p^t q_i^t + \beta_i^t \sum_i \left\{ (d_i^t - q_i^t) - G_i \eta_i^t \right\} \right], \tag{8}$$

where β_i^t is the Lagrangian multiplier for constraint (2). The first order conditions (FOC) of an optimal investment size G_i^* and an optimal amount of electricity purchased from the main grid q_i^{t*} are as follows:

$$FOC_{G_i} \quad : \quad 2a_i G_i + (b_i - s_i) = \delta \sum_t \beta_i^t \eta_i^t$$

$$FOC_{q_i^t} \quad : \quad p^t = \beta_i^t, \quad \forall t$$

FOC_{G_i} indicates that the optimal installation capacity G_i^* should be determined at the level where the marginal cost of the investment coincides with the marginal benefit from electricity generation realized by this investment. $FOC_{q_i^t}$ indicates that q_i^{t*} should be determined when the electricity price of the main grid p_i coincides with the marginal benefit of this purchase, that is, the shadow price of investment cost saving owing to this electricity buying β_i^t. Summarizing these FOCs and Equation (3), we can obtain G_i^* and q_i^{t*} as

$$G_i^* = \frac{1}{2a_i} \left[\delta \sum_t p_i \eta_i^t - (b_i - s_i) \right] \tag{9}$$

$$q_i^{t*} = d_i^t - \eta_i^t G_i^*, \quad \forall t \tag{10}$$

Note that, in reality, there are more factors that affects the investment of solar PV generators such as law and policies. In this work, however, we focus on the economic part to obtain the most economical solution.

Based on Equations (9) and (10), a condition for investment, i.e., Equation (7), can be summarized as follows:

Proposition 1. $z_i^* = 1 \quad if \quad \delta \sum_t p_i \eta_i^t > b_i - s_i.$

This means that region i decides to install solar PV generators when the savings from the investment of the generators $\delta \sum_t p_i \eta_i^t$ is greater than the net investment cost minus subsidies $b_i - s_i$. In addition, it shows how the investment choice is related to the subsidy s_i, which is the main variable in this study. Thus, we are ready to answer what subsidy policy can maximize solar PV dissemination with a given subsidy budget? in the following sections. Conversely, what subsidy policy can minimize the subsidy budget to achieve a given dissemination target?

2.2. Social-Level Optimization: Maximizing the Dissemination of Solar PV Capacity with a Budget Constraint

Now, let us consider the government's choice in the subsidy allocation problem at the first stage. Suppose that the objective of the government is to maximize the solar PV capacity with a given subsidy budget $\bar{S}$.

$$\max_{\{s_i\}_i} \sum_i G_i \tag{11}$$

$$\text{s.t.} \quad \sum_i s_i G_i \leq \bar{S} \tag{12}$$

$$s_i \geq 0, \quad \forall i \in I \tag{13}$$

As the optimal installation capacity of each region is determined under a given subsidy policy, it can be understood that the subsidy policies are decided by policy makers in anticipation of these regional investment decisions. Therefore, we can substitute Equation (9) into Equation (11) and express G_i^* as a function of s_i. The Lagrangian equation for Equation (11) to solve the optimal subsidy problem is

$$L = \sum_i \frac{1}{2a_i}\left(\delta\sum_t p^t\eta_i^t + s_i - b_i\right) + \lambda\left[\overline{S} - \sum_i \frac{s_i}{2a_i}\left(\delta\sum_t p^t\eta_i^t + s_i - b_i\right)\right] + \sum_i \mu_i s_i, \qquad (14)$$

where λ and μ_i are Lagrangian multipliers for (12) and (13). Hereafter, for convenience, the market value per unit of solar PV capacity in region i $\delta\sum_t p^t\eta_i^t$ will be denoted by e_i, and is referred to as the marginal benefit of investment. The solution can be summarized as follow. Note that a solution to the above problem is also a solution for minimizing the subsidy expenditure to achieve the dissemination target according to the duality.

Proposition 2. *The optimal differential subsidy policy for each region i s_i^* will be*

$$s_i^*\left(\overline{S}, \{a_j\}_j, \{b_j\}_j, \{e_j\}_j\right) = \max\left[0, \frac{1}{\sqrt{2}}\left(\frac{\overline{S} + \sum_{j\in j^+}\frac{(e_j-b_j)^2}{8a_j}}{\sum_{j\in j^+}\frac{1}{4a_j}}\right)^{\frac{1}{2}} - \frac{1}{2}(e_i - b_i)\right], \qquad (15)$$

where j^+ is a set of subsidized regions which can be formally defined as

$$j^+ = \left\{i' \in I | (e_{i'} - b_{i'})^2 < 2\left(\frac{\overline{S} + \sum_{j\in j^+}\frac{(e_j-b_j)^2}{8a_j}}{\sum_{j\in j^+}\frac{1}{4a_j}}\right)\right\}. \qquad (16)$$

For detailed proof, see Appendix A.

Proposition 2 means that once a subsidy is provided, its optimal size equals the non-zero term in Equation (15). The criteria which region will receive a non-zero subsidy can be determined from Equation (16), and it is referred to as call it the eligibility condition. To get an intuition of the optimal subsidy and the eligibility condition, let us consider a special case of $(a_i, b_i, e_i) = (a, b, e), \forall i \in I$ where all the regions are under the same condition. Then, the eligibility condition becomes $\overline{S} > 0$ which is always satisfied. Furthermore, the optimal subsidy for each region becomes

$$s^* = \frac{1}{\sqrt{2}}\left[\left(4a\frac{\overline{S}}{N} + \frac{(e-b)^2}{2}\right)^{\frac{1}{2}} - \frac{1}{\sqrt{2}}(e-b)\right]. \qquad (17)$$

From Equation (17), we can obtain some useful comparative static results. (1) The subsidy per unit increases as the subsidy budget increases ($\frac{\partial s_i^*}{\partial \overline{S}} > 0$), and (2) the subsidy per unit decrease as the number of subsidized regions N^+ increases ($\frac{\partial s_i^*}{\partial N^+} < 0$). Subsequently, to further explore the intuition of the eligibility condition from another aspect, consider the situation where $n - 1$ subsidized regions have the same variable values but one region i has different value. Then, the eligibility condition is simplified as follows.

$$\frac{(e_i - b_i)^2 - (e-b)^2}{4a} < \frac{\overline{S}}{n-1}$$

This indicates that a region whose marginal net benefit ($e_i - b_i$) is nearly smaller than that of subsidized regions ($e - b$) by as much as the amount of average subsidy ($\frac{\overline{S}}{n-1}$) is also eligible to be subsidized. Finally, we can summarize how optimal subsidies per unit solar PV capacity should be designed according to the variables related to the marginal investment cost ($\{a_i\}_i, \{b_i\}_i$) and marginal investment benefit $\{e_i\}_i$ as Corollary 1.

Corollary 1. *For a region* $i \in j^+$, $\frac{\partial s_i^*}{\partial a_i} > 0$, $\frac{\partial s_i^*}{\partial b_i} > 0$, *and* $\frac{\partial s_i^*}{\partial e_i} < 0$.

Among the subsidized regions, the larger the marginal investment cost (a_i, b_i) or the lower the marginal investment benefit e_i, the greater should be the subsidy s_i to be provided. At first glance, the opposite may be desirable, but this is a well-known public goods provision result in economics (e.g., emission trading studies), which is the optimality achieved when the marginal costs across regions are equal. In our context, a region where the net benefit of 'per-unit' investment of solar PV generator is low will install only small or no solar PV generators, and hence, the marginal cost of investment of the region remains low. This allows greater room for the region to invest until its marginal cost reaches the average marginal costs in the subsidized regions.

In addition, we can observe how the optimal subsidy needs to be adjusted according to the changes in the other variables of the market value per unit of solar PV capacity in region i e_i: they are the market interest rate δ, power purchase price p^t, and the amount of solar resources by region η_i^t; $\frac{\partial s_i^*}{\partial \delta} < 0$, $\frac{\partial s_i^*}{\partial p^t} < 0$, and $\frac{\partial s_i^*}{\partial \eta_i^t} < 0$.

From these results, we can design how subsidy provision should be customized for each region to maximize the solar PV capacity with limited budgets. In the following section, we will consider how to estimate the amount of solar PV capacity that can be additionally expanded when our differential subsidy policy is applied by using real data.

3. Evaluations

In this section, we first evaluate the proposed differential subsidy policy numerically, and then demonstrate its superior performance compared with that of the the uniform policy through two case studies. One case study uses artificial parameters to show the gain of the proposed policy, while the other case study uses real data from Korea.

3.1. Numerical Result

Figure 1 plots the optimal capacity of solar PV G^* for the cost minimization problem (6). The capacity of solar PV linearly decreases as $b - s$ increases, and it decreases with an increase in a in a $1/x$ form. Furthermore, according to Proposition 1, when $b - s$ is lower than $\delta \sum_t p^t \eta^t$, this region does not install solar PV generators. Note that, in this numerical result, we use $\delta \sum_t p^t \eta^t = 900$.

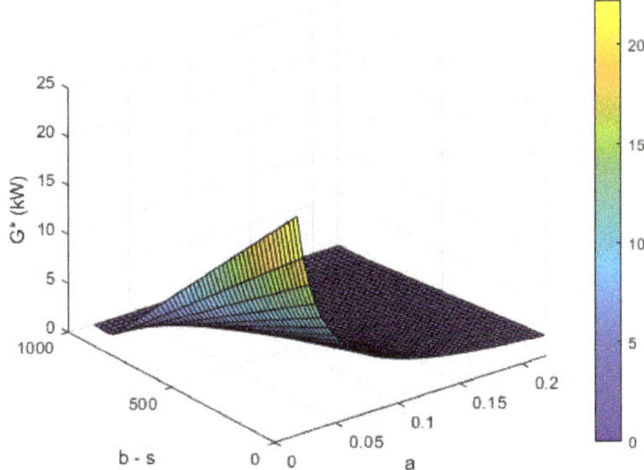

Figure 1. Installing capacity of solar PV for regional-level optimization.

The numerical results under the differential subsidy policy are shown in Figure 2. In this case, it is assumed that all the regions are under the same condition, that is, Figure 2 is a graph of (17). Intuitively, the amount of subsidy increases with an increase in $\bar{S}$ and decreases with an increase in

N as shown in Figure 2a. Figure 2b shows s^* versus a and $e - b$ with $\overline{S} = 10,000,000$ and $N = 10$. The amount of subsidy increases with an increase in a and b because the initial capacity of solar PV is low when the investment cost is high. Owing to the same reason, s^* decreases with an increase in e.

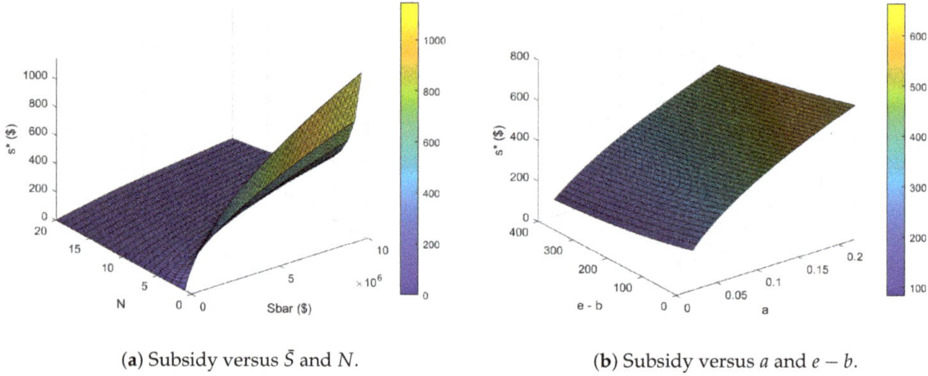

(**a**) Subsidy versus $\overline{S}$ and N.   (**b**) Subsidy versus a and $e - b$.

Figure 2. Subsidy per unit capacity for social-level optimization.

3.2. Case Study 1

We investigate two case studies for the differential subsidy policy. The first case is a simple and artificial case to observe the effect of the parameters clearly. There are three regions and three subsidy policies: no subsidy, uniform subsidy, and differential subsidy policies. The total amount of subsidy $\overline{S}$ is USD 100,000. Table 1 shows the basic parameters and results of this case study. The three regions have different characteristics: Because a and b are parameters of the investment cost C^I in Equation (1), great values of them means high investment cost in the region, i.e., regions A and B. Futhermore, because e stands for the market value per unit of solar PV capacity, a region that has high value of e means a good solar resource region, i.e., region C.

Table 1. Solar PV capacity G_s under three subsidy policies: no subsidy (w/os), uniform subsidy (eq), differential subsidy (di) policies.

	a	b	e	G_s w/o s	s_{un}	G_s w/s_{un}	s_{di}	G_s w/s_{di}
A	2	2500	2500	0	76	19	632	158
B	1	5000	2500	0	76	0	2500	0
C	1	2500	5000	1250	76	1288	0	1250

In regions A and B, $e_i \leq b_i - s_i$, and hence, they do not install solar PV with $s_i = 0$ according to Proposition 1. This indicates that only region C is suitable to invest solar PV because of an abundance of solar resource and low investment cost. Under the uniform subsidy policy, region A installs some solar PV generators as it now satisfies Proposition 1 while region B still does not invest in solar PV. Futhermore, the capacity of solar PV in region C slightly increases due to the subsidy. The total increase of the capacity of solar PV is 4.56% compared with that under no subsidy. On the other hand, the capacity of solar PV in region C remains the same under the differential subsidy policy. In this case, only region A receives a meaningful subsidy, and hence, the capacity in the region increases significantly, and the other regions experience no additional installation. The capacity increase gain is 7.74% in comparison with that of the uniform subsidy policy.

This result may seem counter-intuitive at first glance because providing higher subsidies to an economically viable region may appear to be the optimal policy to maximize solar PV capacity. However, the optimal subsidy is allocating a subsidy of USD 0 to a region with abundant solar resources in this example. The reason for this is as follows. First, economically viable regions invest in solar PV regardless of subsidies, whereas subsidies for non-viable regions convert non-installed regions into installed ones. That is, an extensive margin effect is created. Second, economically viable regions have a greater incentive to invest in more installations, and thus require greater amounts of subsidy than non-viable ones. Notice that the viable regions reach higher marginal costs because the investment cost is strictly convex. Note that, to achieve greater installation capacity with a limited subsidy budget, it is optimal to equalize the marginal costs (which include subsidies) across all the regions. That is, the optimal subsidy policy is to provide lower subsidy to more economically viable regions to dampen their marginal costs while providing higher subsidies to less viable regions to boost their marginal costs. The intuition behind this optimality condition which is to equalize marginal costs across regions, is similar to the one in the Emission Trading Scheme. For the detail, see [27], p. 215.

3.3. Case Study 2

In the second case study, we investigate the proposed differential subsidy policy using real data from Korea. Korea consists of 17 first-tier administrative divisions in terms of metropolitan cities and provinces. We assume that the total budget to promote solar PV generators is ten million dollars, i.e., $\overline{S} = 10,000,000$.

3.3.1. Parameter Settings

Generation Efficiency for Solar PV Generators η: We use solar resource data of a 20-years average value for Korea (1988–2007) based on a report published by the Korea Meteorological Administration (KMA) [28]. The average annual solar radiation was the highest at Mokpo at 5110.39 MJ/m^2 and the lowest at Seoul at 4143.82 MJ/m^2. The average yearly duration of sunshine is 2122.5 h, that is, the average daily duration of sunshine is 5.8 h. We assume that the power output of the solar PV generator is directly proportional to the solar radiation. In addition, the annual utilization rate of solar PV generators was 15.3% in 2018 according to the Ministry of Trade, Industry and Energy. The definition of the utilization rate in this work is the ratio between the actual generated amount of electricity from solar PV generators and the maximum amount of electricity from them. For example, a solar PV whose capacity is 10 kW generated 40 kWh in a day. Then, the utilization rate is 40 kWh/(10 kW × 24 h) = 16.7%. The utilization rate of 15.3% is an average value of all solar PV generators in Korea in 2018.

Using these data, we set the hourly generation efficiency η_i^t.

Installation Cost for Solar PV Generators a **and** b: Solar generator installation costs can be classified into fixed and variable costs. Examples of fixed costs are costs of solar panels, junction boxes, and inverters. On the other hand, variable costs heavily depend on the regions. If the land price is high [low], variable costs are also high [low]. To estimate the variable costs in each region, we use the official land price announced by the Ministry of Land, Infrastructure and Transport. (Public Data Portal, Standard Bulletin, https://www.data.go.kr/dataset/15004246/fileData.do) From the average value of the official land price in each region, the coefficient values of investment cost a_i and b_i are obtained.

Economic Benefits from Solar PV Generation p^t **and** δ: In this case study, it is assumed that there is no reverse power flow beyond substations. That is, all the electricity generated by solar PV is consumed in the substation area. Therefore, the economic benefit from the solar PV originates from reducing electricity bills by self-consumption, i.e., the buying price of electricity. We use a time of use (TOU) price offered by Korea Electric Power Corporation (KEPCO). The electricity price is set as USD 0.051/kWh, USD 0.07/kWh, and USD 0.096/kWh during off-peak, shoulder, and peak period, respectively. More detailed information on electricity price is given in KEPCO web site at https://home.kepco.co.kr/kepco/EN/F/htmlView/ENFBHP00102.do?menuCd=EN060201. In addition, we set the time discount δ to be 5.5% per year.

Using the above assumptions and data, we set a_i, b_i, and $e_i = \delta \sum_t p^t \eta_i^t$. These parameter settings for case study 2 are shown in Figure 3. In this graph, a_i follows the left y-axis, and b_i and e_i follow the right y-axis. Regions SU to SJ are metropolitan cities, and hence, the investment costs are higher than those in other regions resulting in high a_i and b_i. Regions SU and JN have the least $e_{SU} = 955$ and most $e_{JN} = 1149$ solar resources, respectively.

Note that the simulation analysis results in this case study suggest the basic direction for the practical application of the theoretical model. For practical applications, more realistic delimitation of relevant coefficients and the estimation of key coefficients need to be further studied.

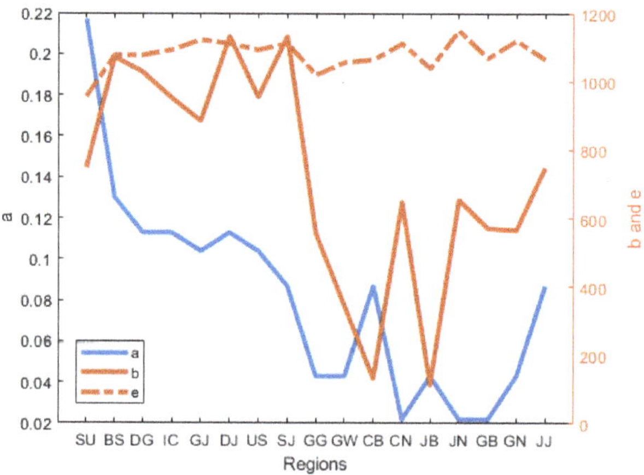

Figure 3. Parameters for case study 2.

3.3.2. Simulation Results

The result of case study 2 is shown in Figure 4. The proposed differential subsidy policies are compared with the no subsidy and uniform subsidy policies. Figure 4a shows the capacities of installed solar PV for different subsidy polices. Because of the high land price, the metropolitan cities (regions from SU to SJ) do not install a considerable number of solar PV generators. In contrast, the other regions, i.e., provinces, install a considerable number of solar PV generators. Without subsidy, regions DJ and SJ do not install solar PV at all according to Proposition 1, and the total capacity of solar PV generators is 74,580 kW. With the total subsidy of USD 10,000,000, the capacity increases by approximately 23%. Under the uniform subsidy policy, all the regions receive the same subsidy of USD 108 per kW, resulting in a total capacity of 91,851 kW of solar PV generators. In addition, the differential subsidy policy increases the total capacity further to 95,271 kW, i.e., a 3.72% increase compared with the uniform subsidy policy.

To maximize the total capacity of solar PV generators, the government needs to provide a more aggressive subsidy to regions with relatively scarce solar resources such as regions SU to SJ as shown in Figure 4b. Particularly, the two regions (DJ and SJ) that installed zero capacity under the uniform subsidy policy receive the greatest subsidies under the differential scheme. According to Proposition 2, the amount of subsidy for each region i heavily depends on $e_i - b_i$ (the marginal net benefit). This is confirmed by the simulation result of this case study. As the regions DJ and SJ have the least $e_i - b_i$, they receive the greatest subsidies. In contrast, the three regions GW, CB, and JB that have the greatest $e_i - b_i$ receive zero subsidies.

Note that the lifespan of solar panels can be changed according to maturity of the PV panel manufacture. When the lifespan increases, the time discount δ also increases resulting in an increase of e_i. Because e_i stands for the market value per unit of solar PV capacity in region i, i.e., $e_i = \delta \sum_t p^t \eta_i^t$, an increase of lifespan promotes more installation of solar PV generators G_i. Therefore, the subsidy per capacity s_i decreases because the installation capacity of solar PV generators increases.

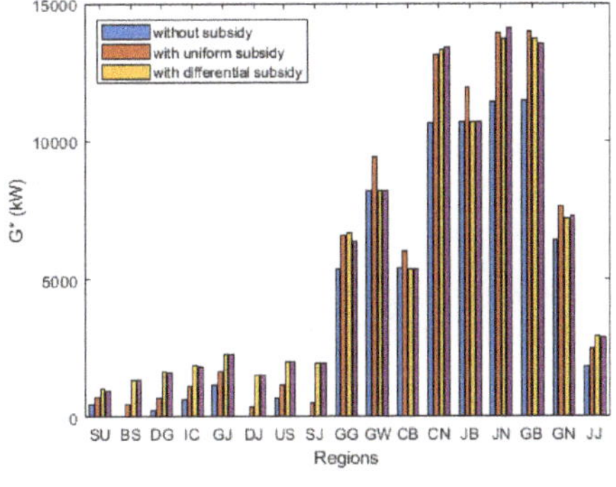

(**a**) Optimal solar PV capacity G^*

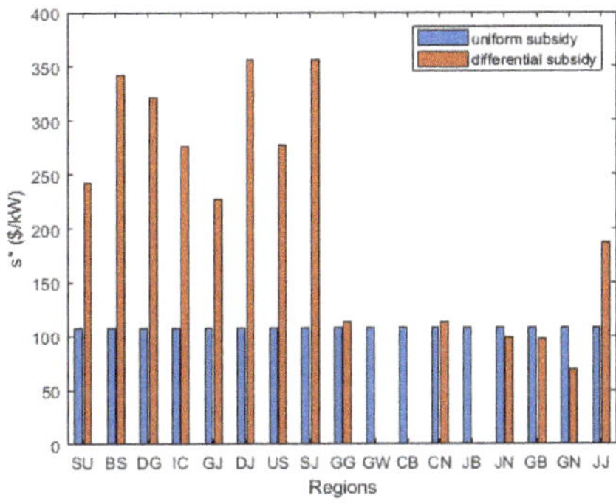

(**b**) Amount of subsidy s^*

Figure 4. Solar PV capacity and subsidy results under different subsidy policies.

4. Discussion

So far, we have considered that the objective of a dissemination policy is to maximize the total capacity of solar PV generators. Although such an assumption is a typical observation in many countries, there may be several other possible objectives of a dissemination policy such as energy generation maximization and CO_2 emission minimization. To consider how different objectives may affect our result, in this section, we focus on a case when the dissemination objective is to maximize the energy generation itself rather than its capacity. For an energy generation maximization problem, we need to convert Equation (11) into

$$\max_{\{s_i\}_i} \sum_i \sum_t G_i \eta_i^t \qquad (18)$$
$$\text{s.t.} \qquad \sum_i s_i G_i \leq \overline{S}$$
$$s_i \geq 0 \quad \forall i \in I$$

Although we do not provide a closed-form solution here, it is expected that the results are similar to the capacity maximization problem in that the energy maximization problem is a variation of it. The only difference is that the energy maximization policy places more weight on e_i, that is, it focuses more on the solar resources in the region. Thus, we can expect that it will lead to a more favorable subsidy provision to energy-efficient regions than capacity maximization does.

Figure 5 shows the difference G_i and s_i between the capacity maximization and energy maximization policies with the Korean data in Section 3.3. The differences G_i and s_i follow the left y-axis, and e_i follows the right y-axis. The difference value is defined as G_i or s_i under the energy maximization policy minus that under the capacity maximization policy. That is, a positive value indicates that the energy maximization policy installs more generators. As shown in Figure 5, e_i and the difference are highly correlated. The most abundant and scarce solar areas in case study 2 are regions SU and JN, respectively. Therefore, this policy provides higher and lower subsidies to different regions compared with the capacity maximization policy. In regions GW, CB, and JB, there is no difference between the two policies because they do not receive any subsidy under the both subsidy policies. Except the three regions, when a region i has an e_i value less [greater] than 1100, the region will install less [more] solar PV generators. The total installed capacities of solar PV generators are 95,238 kW and 95,271 kW for the energy and capacity maximization policies, respectively. However, their yearly energy outputs from the solar PV generators are 120,471 MWh and 120,430 MWh, respectively. Note that the energy output under the uniform subsidy policy is 115,972 MWh. The gain of the energy maximization policy is 3.88%.

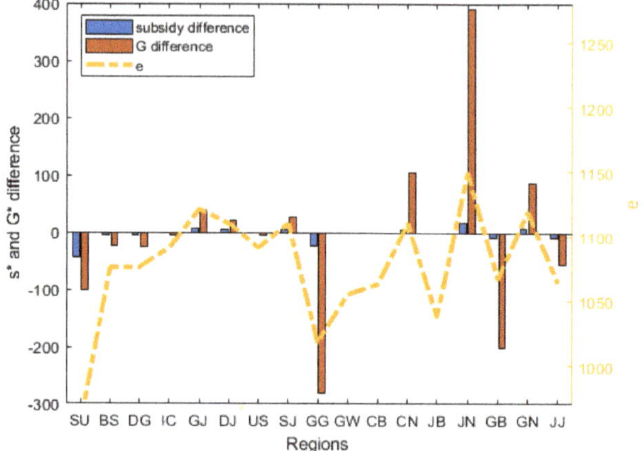

Figure 5. G_i and s_i comparison between capacity maximization and energy maximization policies.

The simulation result confirms that the energy efficiency e_i is more effective to determine the amount of subsidy. However, the difference between capacity max or energy max may not be critical in terms of the total results (i.e., total energy and total capacity) although it may depend on the solar resource distribution in each country.

5. Conclusions

We considered how a regionally differentiated subsidy policy can improve the total capacity of solar PV generators in a society under a limited subsidy budget. We showed that a simple rule of an optimal subsidy policy is to allocate subsidies to equalize regional marginal net benefits of all the eligible regions. According to this rule, it is desirable to provide higher subsidy per unit capacity to regions where investments are not made without subsidy or only small investments are expected. We also used this model to estimate the maximum possible capacity of solar PV generators under the given budget. By using the actual data, we showed the improvement of dissemination that can be achieved by a policy change from uniform to differential subsidization under the same budget. The specific subsidy rate for each region could be calculated using the total subsidy budget size, interest rate, solar resource, and land price in each region.

The limitation of the study is that our results are based on somewhat strong assumptions such as convex investment cost function and monotonically increasing marginal net benefits. As mentioned above, to apply our method to an actual policy, it would be necessary to conduct a full-scale estimation of the cost and benefit function. Nevertheless, our findings are not necessarily confined to the national level. They may also be applicable at the local municipal or even international level, such as renewable energy dissemination projects in developing countries by international organization funds.

Author Contributions: Conceptualization, J.S. and S.-G.Y.; Methodology, J.S.; Software, S.-G.Y.; Validation, J.S. and S.-G.Y.; Formal Analysis, J.S.; Investigation, J.S.; Resources, S.-G.Y.; Data Curation, S.-G.Y.; Writing—Original Draft Preparation, J.S. and S.-G.Y. All authors have read and agreed to the published version of the manuscript.

Funding: This research was supported in part by Soongsil University Research Fund (Convergence Research) of 2017, and in part by the Korea Electric Power Corporation (Grant number: R18XA04).

Conflicts of Interest: The authors declare no conflict of interest.

Appendix A. Analytic Solution for Social-Level Optimization

The Lagrangian equation for maximizing solar PV capacity with a budget constraint is

$$L = \sum_i \frac{1}{2a_i} (e_i + s_i - b_i) + \lambda \left[\overline{S} - \sum_i \frac{s_i}{2a_i} (e_i + s_i - b_i) \right] + \sum_i \mu_i s_i. \tag{A1}$$

The Karush–Kuhn–Tucker (KKT) conditions for this optimization problem are

$$\mu_i s_i = 0, \quad \forall i \in I \tag{A2}$$

$$\mu_i \geq 0, \quad \forall i \in I \tag{A3}$$

$$s_i \geq 0, \quad \forall i \in I \tag{A4}$$

$$\lambda \geq 0 \tag{A5}$$

$$\frac{\partial L}{\partial s_i} = \frac{1}{2a_i} - \lambda \frac{1}{2a_i} (e_i + 2s_i - b_i) + \mu_i, \quad \forall i \in I \tag{A6}$$

$$\lambda \left[\overline{S} - \sum_i \frac{s_i}{2a_i} (e_i + s_i - b_i) \right] = 0 \tag{A7}$$

To find a solution, let us use (A5) first. If $\lambda = 0$, $\mu_i = -\frac{1}{2a_i} < 0$ which is not possible. Therefore, we observe that $\lambda > 0$, subsequently the budget constraint is binding by (A7). That is,

$$\sum_i \frac{s_i}{2a_i} (e_i + s_i - b_i) = \overline{S}. \tag{A8}$$

For region i with $\mu_i > 0$, the optimal subsidy $s_i^* = 0$, thus $\mu_i = \frac{1}{2a_i} [\lambda (e_i - b_i) - 1]$ by (A6). For region i with $\mu_i = 0$, $s_i^* > 0$. Thus, the following condition should be satisfied by (A6):

$$\lambda = \frac{1}{e_i + 2s_i^* - b_i}. \tag{A9}$$

By using (A8) and (A9), we can determine λ. We have

$$\lambda = \frac{1}{\sqrt{2}} \left[\frac{\sum_{j \in j^+} \frac{1}{4a_j}}{\overline{S} + \sum_{j \in j^+} \frac{(e_j - b_j)^2}{8a_j}} \right]^{\frac{1}{2}} \tag{A10}$$

As $s_i^* = \frac{1}{2} \left[\frac{1}{\lambda} - (e_i - b_i) \right]$ for $i \in j^+$, the optimal subsidy of region i can be obtained as

$$s_i^* = \frac{1}{\sqrt{2}} \left[\left\{ \frac{\overline{S} + \sum_{j \in j^+} \frac{(e_j - b_j)^2}{8a_j}}{\sum_{j \in j^+} \frac{1}{4a_j}} \right\}^{\frac{1}{2}} - (e_i - b_i) \right] \quad \text{for } i \in j^+ \tag{A11}$$

$$= 0, \quad \text{otherwise.} \tag{A12}$$

References

1. Renewable Energy Target—An Australian Government Scheme. Available online: http://www.cleanenergyregulator.gov.au/RET/ (accessed on 6 May 2020).
2. RES LEGAL 2019. EEG Feed-in Tariff in Germany. European Commission. 2019. Available online: http://www.res-legal.eu/search-by-country/germany/single/s/res-e/t/promotion/aid/feed-in-tariff-eeg-feed-in-tariff/lastp/135/ (accessed on 6 May 2020).

3. Ministry of New and Renewable Energy Government of India. National Solar Mission—Gentral Financial Assistance (CFA). Available online: https://mnre.gov.in/img/documents/uploads/file_f-1585710569965.pdf (accessed on 6 May 2020).

4. International Institute for Sustainable Development. *India's Energy Transition: Mapping Subsidies to Fossil Fuels and Clean Energy in India GSI Report*; International Institute for Sustainable Development: Winnipeg, MB, Canada, 2017. Available online: https://www.iisd.org/sites/default/files/publications/india-energy-transition.pdf (accessed on 6 May 2020).

5. *Tanzania's SE4ALL Action Agenda*; Ministry of Energy & Minerals: Dar es Salaam, Tanzania, 2015. Available online: https://www.seforall.org/sites/default/files/TANZANIA_AA-Final.pdf (accessed on 6 May 2020).

6. Kwon, T.H. Policy Synergy or Conflict for Renewable Energy Support: Case of RPS and Auction in South Korea. *Energy Policy* **2018**, 123, 443–449. [CrossRef]

7. International Energy Agency. *World Energy Outlook 2013*; International Energy Agency: Paris, France, 2013. Available online: https://webstore.iea.org/world-energy-outlook-2013 (accessed on 6 May 2020).

8. REN21. *Renewables 2019 Global Status Report 2019*; REN21 Secretariat: Paris, France. Available online: https://www.ren21.net/gsr-2019/ (accessed on 6 May 2020).

9. IEA. *Germany 2020*; IEA: Paris, France. 2020. Available online: https://www.iea.org/reports/germany-2020 (accessed on 6 May 2020).

10. Senate Bill (SB) 100. 10 September 2018. Available online: https://www.energy.ca.gov/sb100 (accessed on 6 May 2020).

11. Rodrìguez, L.R.; Duminil, E.; Ramos, J.S.; Eicker, U. Assessment of the photovoltaic potential at urban level based on 3D city models: A case study and new methodological approach. *Sol. Energy* **2017**, *146*, 264–275. [CrossRef]

12. Lettner, G; Auer, H.; Fleischhacker, A.; Schwabeneder, D.; Dallinger, B.; Moisl, F.; Roman, E.; Velte, D.; Ana, H. Existing and future PV prosumer concepts. Technische Universitaet Wien, Fundacion Tecnalia Research & Innovation, with the Collaboration of PVP4Grid Consortium. Available online: https://www.pvp4grid.eu/wp-content/uploads/2018/08/D2.1_Existing-future-prosumer-concepts_PVP4Grid_FV.pdf (accessed on 6 May 2020).

13. Palmintier, B.; Broderick, R.; Mather, B.; Coddington, M.; Baker, K.; Ding, F.; Reno, M.; Lave, M.; Bharatkumar, A. *On the Path to SunShot: Emerging Issues and Challenges in Integrating Solar with the Distribution System*; Nrel/Tp-5D00-6533, Sand2016-2524 R; U.S. Department of Energy: Washington, DC, USA, 2016.

14. Appen, J.V.; Braun, M.; Stetz, T.; Diwold, K.; Geibel, D. Time in the Sun: The Challenge of High PV Penetration in the German Electric Grid. *IEEE Power Energy Mag.* **2013**, *11*, 55–64. [CrossRef]

15. Olowu, T. O.; Sundararajan, A.; Moghaddami, M. Sarwat, A.I. Future Challenges and Mitigation Methods for High Photovoltaic Penetration: A Survey. *Energies* **2018**, *11*, 1782. [CrossRef]

16. Aleem, A.A; Hussain, S.M.S.; Ustun, T. S. A Review of Strategies to Increase PV penetration Level in Smart Grids. *Energies* **2020**, *13*, 636. [CrossRef]

17. Martins, V. F.; Borges, C. L. T. Active distribution network integrated planning incorporating distributed generation and load response uncertainties. *IEEE Trans. Power Syst.* **2011**, *26*, 2164–2172. [CrossRef]

18. Mehmood, K.K.; Khan, S.U.; Lee, S.J.; Haider, Z.M.; Rafique, M.K.; Kim, C.H. A real-time optimal coordination scheme for the voltage regulation of a distribution network including an OLTC, capacitor banks, and multiple distributed energy resources. *Electr. Power Energy Syst.* **2018**, *94*, 1–14. [CrossRef]

19. Capitanescu, F.; Ochoa, L.F.; Margossian, H.; Hatziargyriou, N.D. Assessing the potential of network reconfiguration to improve distributed generation hosting capacity in active distribution systems. *IEEE Trans. Power Syst.* **2014**, *30*, 346–356. [CrossRef]

20. Bläsi, A.; Requate, T. Feed-in-tariffs for electricity from renewable energy resources to move down the learning curve? *Public Finan. Manag.* **2010**, *10*, 213–250.

21. Reichenbach, J.; Requate, T. Subsidies for renewable energies in the presence of learning effects and market power. *Resour. Energy Econ.* **2012**, *34*, 236–254. [CrossRef]

22. Andora, M.; Voss, A. Optimal renewable-energy promotion: Capacity subsidies vs generation subsidies, *Resour. Energy Econ.* **2016**, *45*, 144–158. [CrossRef]

23. Kalkuhl, M.; Edenhofer, O.; Lessmann, K. Renewable energy subsidies: Second-best policy or fatal aberration for mitigation? *Resour. Energy Econ.* **2013**, *35*, 217–234. [CrossRef]

24. Mavsar, P.; Sredenšek K.; Štumberger B.; Hadžiselimović, M.; Seme, S. Simplified Method for Analyzing the Availability of Rooftop Photovoltaic Potential. *Energies* **2019**, *12*, 4233. [CrossRef]
25. Melius, J.; Margolis, R.; Ong, S. *Estimating Rooftop Suitability for PV: A Review of Methods, Patents, and Validation Techniques*; Technical Report NREL/TP-6A20-60593; National Renewable Energy Lab.: Golden, CO, USA, 2013.
26. Lee, M.; Hong, T.; Jeong, K.; Kim, J. A Bottom-up Approach for Estimating the Economic Potential of the Rooftop Solar Photovoltaic System Considering the Spatial and Temporal Diversity. *Appl. Energy* **2018**, *232*, 640–656. [CrossRef]
27. Endres, A. *Environmental Economics: Theory and Policy*; Cambridge University Press: Cambridge, UK, 2010; ISBN 9780521173926.
28. Korea Meteorological Administration. Analysis Report of Weather Resource to Fully Utilize Solar Energy in Korea, 2008 (Wrriten in Korean). Available online: http://www.kma.go.kr/download_01/climate_energy.pdf (accessed on 6 May 2020).

Article

Cost-Effectiveness of Carbon Emission Abatement Strategies for a Local Multi-Energy System—A Case Study of Chalmers University of Technology Campus

Nima Mirzaei Alavijeh [1,*], David Steen [1,*], Zack Norwood [2], Le Anh Tuan [1] and Christos Agathokleous [1]

[1]   Department of Electrical Engineering, Chalmers University of Technology, 412 96 Gothenburg, Sweden; tuan.le@chalmers.se (L.A.T.); xristos.agathokleous@gmail.com (C.A.)

[2]   Department of Earth Sciences, Gothenburg University, 405 30 Gothenburg, Sweden; donkey@berkeley.edu

*   Correspondence: nima.mirzaei@chalmers.se (N.M.A.); david.steen@chalmers.se (D.S.)

Received: 1 March 2020; Accepted: 23 March 2020; Published: 2 April 2020

Abstract: This paper investigates the cost-effectiveness of operation strategies which can be used to abate CO_2 emissions in a local multi-energy system. A case study is carried out using data from a real energy system that integrates district heating, district cooling, and electricity networks at Chalmers University of Technology. Operation strategies are developed using a mixed integer linear programming multi-objective optimization model with a short foresight rolling horizon and a year of data. The cost-effectiveness of different strategies is evaluated across different carbon prices. The results provide insights into developing abatement strategies for local multi-energy systems that could be used by utilities, building owners, and authorities. The optimized abatement strategies include: increased usage of biomass boilers, substitution of district heating and absorption chillers with heat pumps, and higher utilization of storage units. The results show that, by utilizing all the strategies, a 20.8% emission reduction can be achieved with a 2.2% cost increase for the campus area. The emission abatement cost of all strategies is 36.6–100.2 ($€/tCO_2$), which is aligned with estimated carbon prices if the Paris agreement target is to be achieved. It is higher, however, than average European Emission Trading System prices and Sweden's carbon tax in 2019.

Keywords: multi-energy systems; local energy management systems; multi-objective optimization; rolling time-horizon; emission abatement strategies; distributed energy systems

---

1. Introduction

With growing amounts of distributed energy resources (e.g., distributed generation and energy storage) and introduction of new demands (e.g., electrical vehicles, heat pumps, etc.) in the distribution systems, local energy management systems have received attention because they provide, amongst others, the following benefits for energy systems:

- Facilitating activation of small flexibilities in the system [1]
- Reducing transmission and distribution losses [2]
- Managing congestion on the distribution level [3]
- Increasing awareness and engagement of end-users [4]

Furthermore, multi-energy systems (MESs) have been suggested to increase the synergies in the energy system as a whole by integrating and managing different energy carriers such as electricity, district heating, district cooling, and natural gas simultaneously [5]. Subsequently, analyzing the combination of local energy management systems and MESs is an interesting emerging field of research.

Within the two research areas of local energy management systems and MESs, previous studies [3,5,6] have reviewed definitions, trends, challenges, and categorization of literature providing valuable insight into the topic. For example, Grosspeithsch et al. [6] categorized the literature into four categories: general overview, model and optimization, energy management and system analysis, and case study.

One feature of the model and optimization category is that energy systems have traditionally been modeled based solely on cost minimization objectives. However, multi-criterion optimization can help broaden decision making to consider cost, the environment, reliability, social impact, utilization of renewable energy, etc. [7]. As global concerns about greenhouse gas emissions increase, carbon emissions become an increasingly important criterion to be considered in optimizing the operation of local MESs. For instance, Majidi et al. [8] proposed a cost and emission framework to assess demand response programs, and Bracco et al. [9] developed a multi-objective model to evaluate the operation of a multi-energy system considering four different building types and three energy carriers (heat, gas, and electricity). Wang et al. [10] demonstrated that a multi-objective optimization will not give one single solution but rather a set of Pareto optimal solutions. Often, the objectives are conflicting and different approaches to solve the minimization problem exist, e.g., mixed integer linear programming with weighted sums [11], evolutionary algorithms [12], game theory [13], particle swarm optimization [14], genetic algorithms [15], etc.

Furthermore, an optimization model can have a short foresight (close to real-time) or a long foresight depending on the purpose and the characteristics of the energy technologies included in the system. Optimizations with long foresight result in a more optimal management of resources especially in energy systems with seasonal storage, conventionally dispatchable units, and perfect foresight. However, such long-term optimizations require long-term forecasts and can be computationally expensive as the size and complexity of the model increases [16].

On the other hand, optimization with a short foresight lowers the computational time (which would be of value when simulating complex systems) [17] and have lesser challenges with quality of forecasts. This is especially important for systems with a large share of renewable energy sources (RES) since as the share of intermittent RES increases in the system, their stochastic nature starts to affect forecasts, availability, and prices of energy carriers. Therefore, if the model represents a system including a large share of RES, or reacts in response to the energy prices from a system with a large share of RES, a close to real-time modeling approach with short foresight can represent agents' (energy technologies') behavior closer to reality [18].

As an example from the research area of modeling and optimization of MESs, Wu et al. [11] investigated the simultaneous optimization of annual cost and CO_2 emissions in the design and operation of a distributed energy network where distributed energy sources (DESs) can exchange heat with each other through pipelines. A mixed integer linear programming (MILP) model with a weighted sum approach is used in this multi-objective optimization. Di Somma et al. [19] also used a weighted sum approach in developing a multi-objective linear programming model considering both cost and emissions. The contribution various energy technologies have on achieving the objective function was evaluated by sensitivity analyses. A limitation of this study is that it was carried out only on one customer and not a community of customers. Falke et al. [20] developed a multi-objective model for design and operation of distributed energy systems using a heuristic optimization approach. The model decomposes into three sub-model of heating network planning, buildings' renovation planning, and operation simulation. However, cooling loads and district cooling is not considered in this study. Yan et al. [21] studied the operation optimization of multiple distributed energy systems where the emissions are considered in the form of monetary costs through a carbon tax. The DESs, in this model, can exchange electricity and thermal energy with each other and electricity can be sold back to the grid. Although emissions cost is considered through carbon tax in this study, the trade-off between the emissions and the monetary costs are not discussed. In [12], an evolutionary algorithm is used to solve a multi-objective isolated MES model with high share of renewables, including investment in RES as decision variables. The paper shows that different operational approaches may be beneficial for different seasons. In [13], a scheduling method for MES is proposed using game theory. The model was found to be robust and useful for solving problems with uncertainties in, for example, wind production. In [22], a MES model is developed which includes possible constraints in the energy flow within the MES. For the electricity network, this is accomplished using a DC load flow model, and a pipeline load flow model is used for the natural gas network.

In summary, the literature contains multi-objective optimization considering both cost and emissions with different energy carriers. However, to the best of the authors' knowledge, a study which specifically extracts emission abatement strategies from multi-objective optimization models and evaluates the abatement cost for these strategies is lacking. Such a study could provide insights regarding carbon pricing and investigate the possibilities of operating local MESs in a more environmentally responsible manner. In addition, no previous study has been found which multi-objectively optimizes the three energy carriers (electricity, district heating, and district cooling) using a short foresight rolling horizon over a whole year. The benefit of considering these three energy carriers is that we can capture the synergies for emission abatement through technologies such as heat pumps.

This study was carried out within an European Union project called Fossil-free Energy District [23] whose purpose was to build a testbed using Chalmers University's campus to implement and evaluate MESs alongside a local energy market for electricity, heating, and cooling. As local energy management systems, MESs and environmental aspects are gaining momentum in the research field of energy systems, this study aims mainly to investigate:

- what operation strategies can be utilized to abate CO_2 emissions in a local multi-energy system; and
- how cost-effective these strategies can be.

To the best of our knowledge, the combination of the following aspects is what distinguishes this study from other studies within the field:

- Defining emission abatement strategies in MESs' operation and evaluating cost-effectiveness of these strategies across different carbon prices
- Optimizing a local multi-energy system over a year with a short foresight rolling time horizon
- Multi-objective optimization of three energy carriers: district heating (DH), district cooling and electricity.

In this study, the emission abatement strategies are introduced by a short foresight rolling horizon, multi-objective optimization model which is formulated with the weighted sum approach, considering costs and emissions as the objectives. The problem is solved with a rolling time horizon of ten hours to assess the real-time behavior of different technologies. Afterwards, the cost of these strategies is evaluated against the carbon prices to provide insight on the cost effectiveness of these strategies. The study includes a coupled system of district heating, district cooling, and electricity networks. Moreover, a wide range of distributed energy technologies such as biomass combined heat and power (CHP), biomass boilers, electrical heat pumps (HP), absorption chillers, battery energy storage (BES), building inertia thermal energy storage (BITES), cold water basins (CWB), and solar photovoltaic panels (PV) are considered.

The Paper's Structure

The rest of this paper is organized as follows. Section 2 presents the Chalmers University campus energy system model, its general formulation, and its input data. Section 3 presents the emission abatement strategies and their corresponding costs. In Section 4, the incentives behind each strategy and their emission abatement costs are discussed. Finally, the study is concluded and summarized in Section 5.

2. Multi-Energy System Model of Chalmers University of Technology Campus

The model developed and used in this study is modeled after the energy system of Chalmers University of Technology's Johannesburg campus. The model is open-source and available on Github [24]. The model is implemented using General Algebraic Modeling System (GAMS) [25] and Matlab [26]. Only the objective function and the structure of the model are presented in detail in the main body of this article; however, the formulations for the energy technologies are presented in Appendix A.

Figure 1 shows the flow chart of the dispatch model structure. The flow chart consists of three main modules, which are the measurement module, the MatLab module, and the GAMS module.

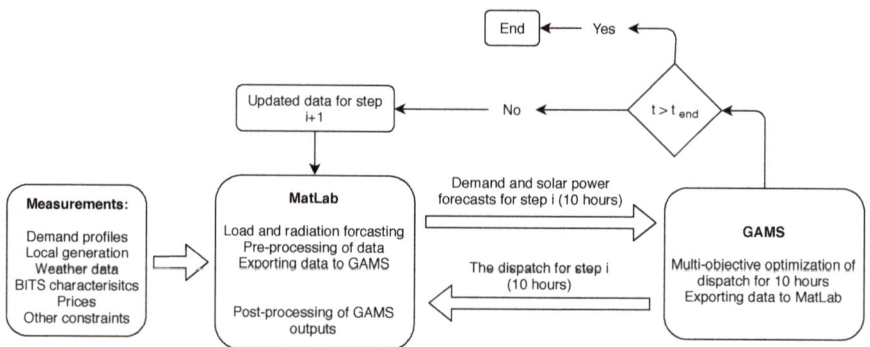

Figure 1. Flow chart of the dispatch model structure.

The first module or the measurement module reads input data to the dispatch model, including demand data, generation data, BITES data, price data, solar irradiance, outdoor temperature data, and energy storage data.

The second module of the model consists of two parts. In the first part, forecasts of the input data to the dispatch model over a given time horizon are made and the input data are then pre-processed to be exported to the GAMS optimization model. The second part is post-processing of the GAMS outputs.

The third module is the GAMS optimization model. This model takes forecast data, measurement data, and network data from the Matlab module and simulates the optimal dispatch of the local generating units, the storage units, and demand side flexibility for a given objective. The optimization and dispatch of units is performed for the desired forecast horizon, e.g., 10 h. The model is designed to perform the optimizations with a rolling time horizon, i.e., after the first optimization, the forecast is updated with new data and a new optimization is performed in order to redispatch the demand and generation units. Between each optimization, the results are stored to be used for evaluation and comparison with the actual dispatch of the system.

To better understand the formulations and results, a simplified overview of the structure of the campus's energy system is provided in Figure 2. In this figure, HP_R, HP_S, and $HP_{W1,2}$ are the four heat pumps in the system. HP_R is a refrigeration heat pump, HP_S is a heat pump used in the summer time, and $HP_{W1,2}$ are the heat pumps used during the heating season. Note that the absorption chiller can only receive heating from the external district heating network. Further note that the CHP unit has the flexibility to be used purely for heating or partly heating with electricity production. B+FGC represents campus's biomass boiler (B) with a flue gas condenser (FGC). There are three storage types in the system. Thermal energy is stored in buildings' thermal inertia (BITES), a cold water basin (CWB) stores chilled water, and battery energy storage (BES) is used to store electricity.

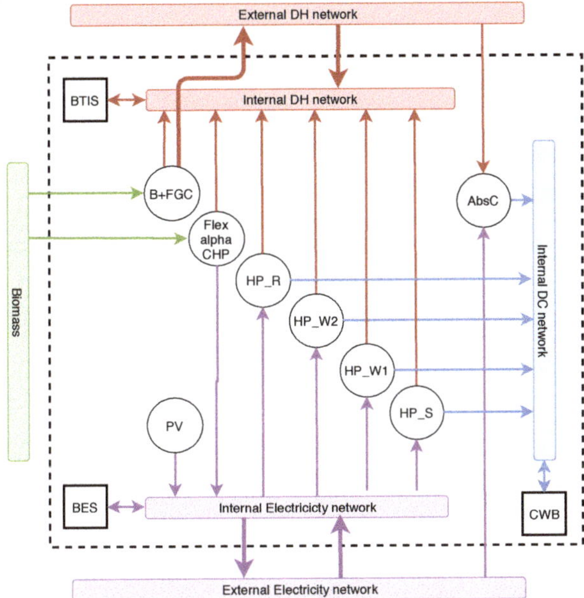

Figure 2. Campus' energy system structure. HP_R, HP_S, and $HP_{W1,2}$ are the four heat pumps. AbsC is the absorption chiller, and B+FGC represents the biomass boiler (B) with a flue gas condenser (FGC). BITES is the thermal energy storage in the buildings' inertia, CWB is cold water basin storage, and BES is the battery energy storage.

In Chalmers' campus energy system model, district heating, district cooling, and electricity networks are coupled by means of heat pumps, absorption chillers, and a combined heat and power (CHP) unit:

- Heat pumps: This technology can be placed between the three energy carriers (district heating, district cooling, and electricity) by assigning the hot source to district heating network and cold sink to district cooling network (see Figure 3).
- Absorption chiller: This technology can provide flexibility in the system by using heat from district heating network to produce cooling for the district cooling.
- Combined Heat and Power: The biomass boiler and turbine can provide electricity and heating simultaneously to the electricity and district heating networks. Flexibility is introduced by the power-to-heat ratio (alpha) that varies the amount of fuel to be converted to electricity or heating.

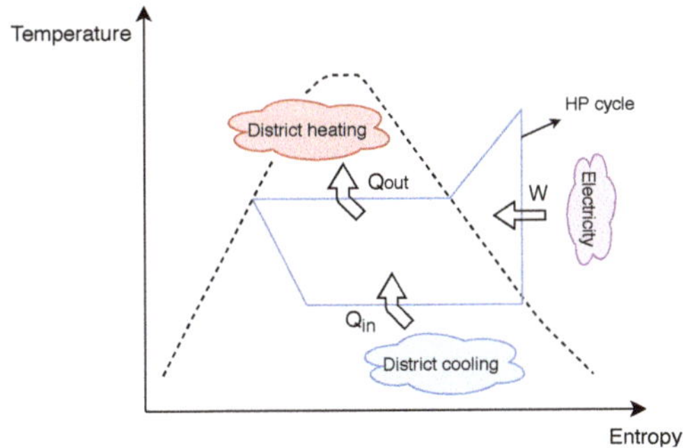

Figure 3. Heat pumps' thermodynamic cycle in district heating and district cooling coupled systems.

Although coupling of the heat pumps to the district heating and cooling systems does provide flexible and efficient production of either cooling or heating, it also imposes limitations on the load factor of heat pumps. This limitation is necessary to maintain balance between energy production and consumption in the district heating and district cooling networks. For example, if the heating demand is high while cooling demand is low, full capacity usage of heat pumps would not be possible to prevent the formation of frost in the district cooling network. This trade-off between energy efficiency and load factor is discussed more in Section 4.

In this study, a close to real-time simulation (one hour ahead) with a rolling time horizon of 10 h was chosen. This forecast horizon and optimization window were chosen so as to be technically and realistically achievable in actual energy systems. Figure 4 shows how the optimizations are carried out under the rolling-time horizon.

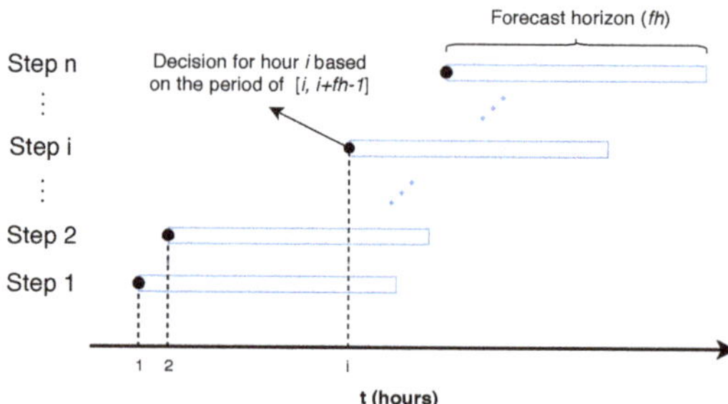

Figure 4. Modeling close to real-time with a short foresight rolling horizon.

In each step, the dispatch of units is decided based on the assumed forecast horizon (fh). In this way, the results are more closely based on real-time decision making and the solver will not be able to see the hours with less accurate forecasts. The steps are continued until the whole simulation period (one full year in this study) is covered.

2.1. Model formulation of Chalmers Campus Energy System

In this section, the main formulations of the model such as objective function, energy balances, and general constraints of the energy technologies are presented. More detailed formulations of the energy technologies can be found in Appendix A.

2.1.1. Objective Function

To consider both economic and environmental objectives, the weighted sum method has been used. The objective function is calculated based on Equation (1).

$$\min \sum_{t=i}^{i+fh-1} (\alpha \cdot C_t + \beta \cdot E_t) \qquad \forall i \in hours \tag{1}$$

In Equation (1), C_t and E_t represent the cost and the emissions of the system at hour t. The total cost and emissions are summed and minimized for each time interval of i with the span of fh hours. α and β are the weighting factors for each objective. The cost weighting factor α is fixed at 1 while the environmental weighting factor β is changed from zero to 1. Thus, in other words, β can be seen as an additional carbon tax on the emissions. This approach for changing α and β is chosen to make the carbon emission reductions comparable with carbon prices. The units for C_t and E_t are SEK and gCO_2.

In Equation (2), C_t includes fixed costs of the assets (C_{fix}), variable costs (C_{var}), revenues from exporting electricity and heat to the external networks (R_{exp}), and costs for importing from external networks. Subscript j represents the energy technologies shown in Figure 2. The cost for electricity or district heating input for technologies is counted in C_{imp} and not in C_{var} nor C_{fuel}.

$$C_t = \sum_j (C_{fix,j,t} + C_{var,j,t} + C_{fuel,j,t}) - R_{exp,t} + C_{imp,t} \tag{2}$$

E_t, which is calculated according to Equation (3), accounts for the emission for the imports from the outer grids and the assets inside the MES.

$$E_t = (P_{imp,t} - P_{exp,t}) \times EF_{exg,p,t} + (Q_{imp,t} - Q_{exp,t}) \times EF_{exg,q,t} + \sum_j E_{j,t} \tag{3}$$

P and Q represent the electric power and the heating power. EF_{exg} is the external grid's emission factor. E_j accounts for emissions from each energy technology.

2.1.2. Energy Balances

Demand–supply balances are presented in Equations (4)–(6) for electricity, heating, and cooling. In the heating system (Equation (5)), over-production is possible with help of a cooling tower which is available in the campus' energy system. The *dem* subscript represents the demand of the campus.

$$\sum_j P_{j,t} + P_{imp,t} - P_{exp,t} = P_{dem,t} \tag{4}$$

$$\sum_j Q_{j,t} + Q_{imp,t} - Q_{exp,t} \geq Q_{dem,t} \tag{5}$$

$$\sum_j K_{j,t} + K_{imp,t} - K_{exp,t} = K_{dem,t} \tag{6}$$

2.1.3. Energy Technologies

As shown in Figure 2, there are different generation and storage technologies in the system. The formulation of energy technologies are provided in Appendix A and their characteristics are presented in Appendix B. In this section, only the general explanation for the technologies is provided.

All the generation units are limited by their maximum capacity. For the case study at the Chalmers campus, some site-specific limitations were set to better represent the actual operating condition at the campus. For example, during weekday working hours, the biomass boiler and CHP units are not dispatchable as they are assumed to be used for research purposes at those times. Moreover, exporting district heating to the external grid is limited to the capacity of biomass boiler because of the network structure. Electricity production from the CHP unit is flexible, so that the heat production of its boiler can feed the turbine or go directly to the internal district heating network. The absorption chiller is only operational during the cooling season and can only receive heating from the external district heating network due to the network's design. The production of heat pumps is calculated based on Equations (7) and (8), where the cooling and heating is produced at the same time. COP_k and COP_q are the coefficient of performance for cooling and heating, respectively. HP_S is used only in cooling season while HP_{W1-2} are available in heating seasons. HP_R is a refrigeration machine and only 20% of its heat production can be used in the system due to the network configuration.

$$Q_{hp,t} = COP_q \times P_{hp,t} \tag{7}$$

$$K_{hp,t} = COP_k \times P_{hp,t} \tag{8}$$

For both storage units, the cold water basin and battery energy storage, charge/discharge efficiency has been considered. Moreover, these technologies are constrained by their charging/

discharging power and energy storage capacity. BITES has been modeled based on a methodology using two interlinked storages, a shallow and a deep storage [27].

2.2. Input Data

Input data for energy prices, marginal emissions, demand profiles, weather data ,and energy technologies are presented in this section.

Both the district heating and the electricity prices considered in this study are hourly dynamic prices. Electricity prices are obtained from Nordpool day ahead market for region SE3 year 2016–2017. To present a more realistic price on the consumer side, the sum of the spot price, network tariffs, average profits, and (when applicable) electricity certificates are used. Currently, the pricing system in the local district heating network is based on seasonal heating prices with a peak demand charge and an efficiency rebate [28]. However, to create more dynamic interactions between energy carriers, hourly district heating prices are modeled based on a typical generation mix in a Swedish district heating system (see Appendix B).

The marginal emissions of the external electricity grid are obtained from Electricity Map [29]. This methodology uses a statistical method to predict the marginal emissions for every hour based on historical data. For district heating, the marginal emissions are calculated based on the typical marginal unit in the generation mix which is in turn determined by the hourly district heating prices (see Appendix B).

Demand profiles for electricity, heating, and cooling are provided by the primary real estate owners on the campus, Akademiska Hus and Chalmers Fastigheter. The demand profiles are presented in Appendix B.

Weather data for outdoor temperature are from a local weather station and solar radiations are based on [30].

The capacity, performance, and similar energy technology parameters are obtained from the MES operator at Chalmers campus which are presented in Appendix B.

3. Key Results

In this section, the results are presented showing the trade-off between cost and emissions and the corresponding abatement strategies. Discussion of the results can be found in Section 4.

The trade-off between cost and emissions are presented in Figure 5. It can be seen that, at lower β values, higher emissions reductions can be achieved with smaller changes in the cost. The relative changes in cost and emissions in different beta values are presented in Table 1.

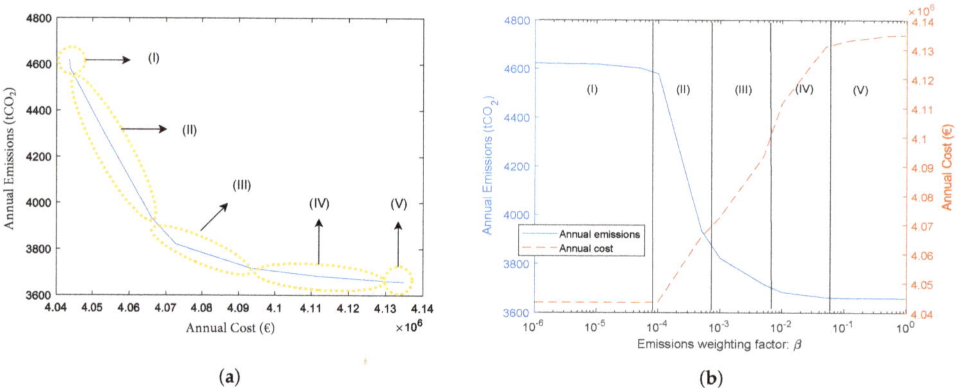

Figure 5. Trade-off between emissions and the cost: (**a**) the annual cost versus the total annual emissions; and (**b**) the annual cost and the annual emissions at different β. The roman numerals are the different abatement phases explained in Section 3.1.

Table 1. The average changes in cost and emissions in different β periods. The change is compared to pure economic optimization (i.e., $\beta = 0$).

Phase	(I)	(II)	(III)	(IV)	(V)
β	$[0, 10^{-4})$	$[10^{-4}, 10^{-3})$	$[10^{-3}, 10^{-2})$	$[10^{-2}, 10^{-1})$	$[10^{-1}, 1)$
ΔCost (%)	$4e^{-4}$	0.3	1.0	1.9	2.2
ΔEmissions (%)	-0.2	-7.9	-18.5	-20.5	-20.8

3.1. Emission Abatement Strategies

To show the emission abatement strategies, the aggregated annual change in usage of each technology is presented in Figure 6 for heating, cooling, electricity, and storage technologies. The values represent the difference in usage (kWh/year) at each β value, compared to the cost optimum case (where $\beta = 0$). The lines for coupling technologies (e.g., heat pumps and the absorption chiller) can represent consumption or production in different energy carriers. For example, the heat pump lines in the heating and cooling graphs are the production of heat pumps while in the electricity graph they show the electricity consumed by the heat pumps. Note the very different scales used in the graph for each energy carrier.

Moreover, at different β values, there is a mix of different strategies for abatement. The β spectrum is divided into five phases (Phases (I)–(V)) as in Figure 6 based on when these strategies start and end. The range of β in each phase can be seen in Table 1. The dominant strategies at each phase are summarized in Figure 7. The observed strategies in each phase are:

- Phase (I): In this phase, although the COP of HP_{w2} is higher than HP_{w1}, the MES optimizer decides to replace HP_{w2} with HP_{w1}. Moreover, the electricity and heat production from the CHP unit is increased which leads to a decrease in electricity and district heating imports.
- Phase (II): The previous actions are continued. Additionally, the heat production from the biomass boiler rapidly increases and causes an increase in the district heating exports. Moreover, the usage of the absorption chiller decreases and is replaced by HP_S.
- Phase (III): The actions in the second phase are continued in this phase as well. Additionally, the usage of HP_{w2} is further decreased and substituted by HP_R which has a lower COP. Another strategy started in this phase is the increase in usage of storage. Furthermore, it's observed that the operation of CHP starts to move towards relatively greater heat production than electricity and consequently more electricity imports.
- Phase (IV): From this phase on, no considerable change is observed in the heating and cooling systems except the increase in usage of BITES and slight increase in usage of HP_R (instead of HP_{w2}). At the beginning of this phase, further increase in the usage of the biomass boiler and further decrease in the usage of the absorption chiller no longer contributes to the objective function. The CHP's transition from electricity to heat production is continued in this phase as well.
- Phase (V): In this phase, the MES optimizer cannot do much to abate more emissions. Only a slight increase in storage usage and a small decrease in electricity production from the CHP is observed.

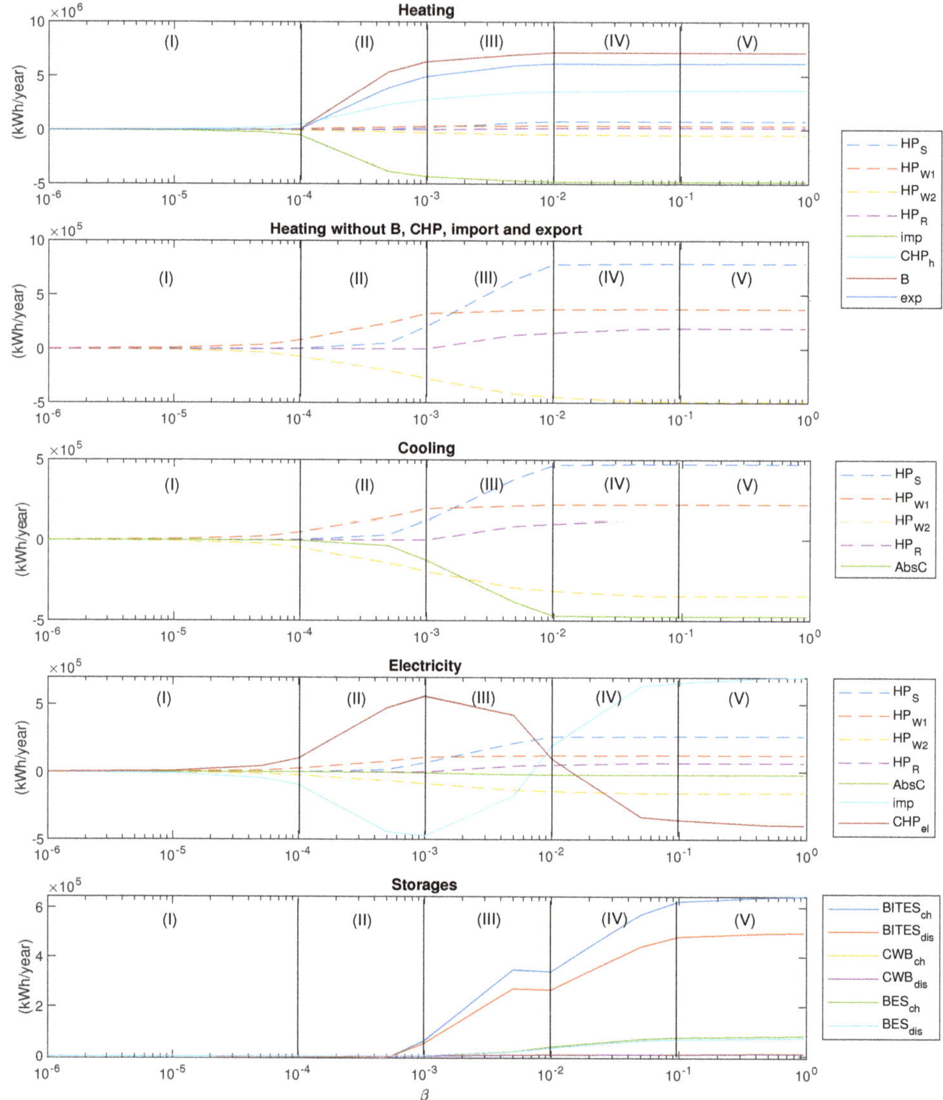

Figure 6. The change in the total energy per year (consumed, produced or stored) of each technology compared to the cost optimal case versus beta for heating, cooling, electricity, and storage.

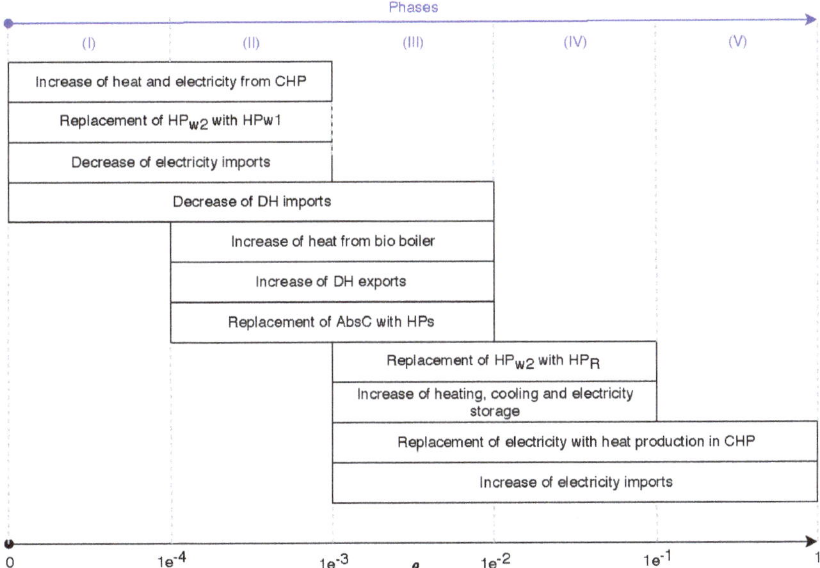

Figure 7. CO$_2$ abatement strategies at different phases.

Figure 8 is an example of the simulated dispatch of the campus energy system at three example *beta* values of 0, 0.001, and 1. A few different abatement strategies can be observed in this figure such as increase in heat production from the biomass boiler, increase and then decrease of electricity production from the CHP unit, increase in heating exports, replacement of HP_W2 with HP_W1, replacement of HP_W2 with HP_R, and increase in utilization of the cold water basin and building inertia thermal energy storage. The strategy related to increase in using battery energy storage cannot be noticed in this figure due to its small scale compared to the electricity demand. Moreover, replacement of the absorption chiller with heat pumps cannot be seen because the absorption chiller is not in operation during the winter.

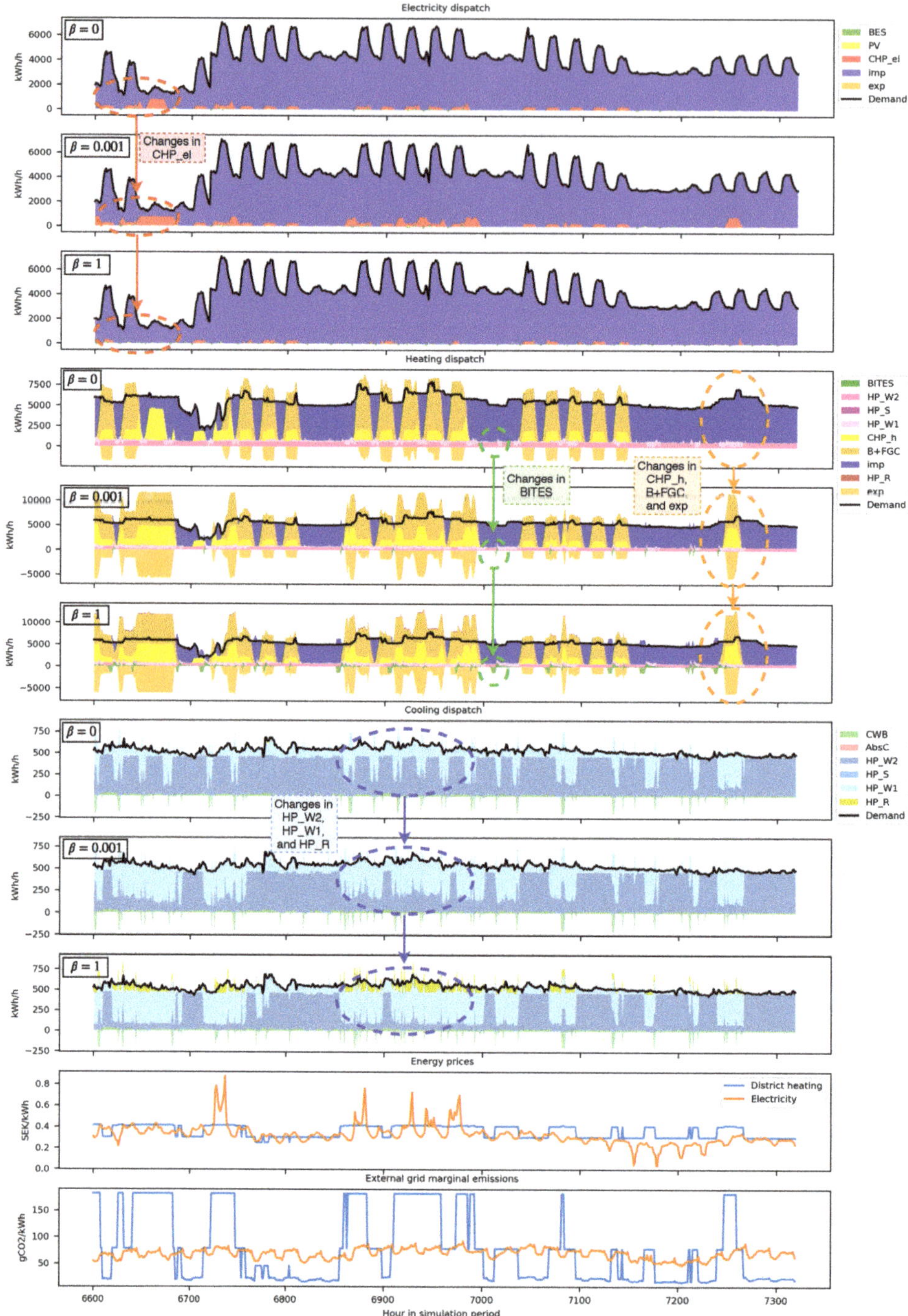

Figure 8. Dispatch time series for December 2016 for three β values of 0, 0.001, and 1 demonstrate some of the abatement strategies. Negative values in the dispatch plots represents exporting or charging of the storage units.

3.2. Cost of Strategies

The cost abatement in each phase is presented in the form of total abatement cost and marginal abatement cost in Table 2. Total abatement cost at each emission weighting factor (β) is calculated based on Equation (9). It includes the total abated CO_2 over the year divided by added costs compared to the cost minimization case. Marginal abatement cost of each β is calculated by Equation (10), which is the same as total abatement cost but, instead, compared to the previous β.

$$TAC_{\beta_i} = \frac{E_{\beta_i} - E_{\beta_0}}{C_{\beta_i} - C_{\beta_0}} \qquad \forall \beta_i \in [0, 1) \tag{9}$$

$$MAC_{\beta_i} = \frac{E_{\beta_i} - E_{\beta_{i-1}}}{C_{\beta_i} - C_{\beta_{i-1}}} \qquad \forall \beta_i \in [0, 1) \tag{10}$$

Table 2. Cost of abatement strategies in different phases. $\overline{TAC}$ is the average total abatement cost in each phase and $\overline{MAC}$ is the average marginal abatement cost in each phase.

Phase	(I)	(II)	(III)	(IV)	(V)
β	$[0, 10^{-4})$	$[10^{-4}, 10^{-3})$	$[10^{-3}, 10^{-2})$	$[10^{-2}, 10^{-1})$	$[10^{-1}, 1)$
$\overline{TAC}$ (€/tCO$_2$)	-3.3	-36.6	-67.6	-97.2	-100.2
$\overline{MAC}$ (€/tCO$_2$)	-4.4	-49.2	-398.5	-955.9	-1692.5

In Table 2, it can be seen that, as β increases, the cost effectiveness of strategies decreases as it gets more costly to avoid further emissions. Moreover, the difference between total abatement cost and marginal abatement cost shows that most of the emission reductions have been carried out in the cheaper Phases (II) and (III). This can be seen in Figure 5b.

4. Discussion

In this section, first the incentives behind each strategy are explained and then the cost of CO_2 emission abatement is discussed.

4.1. Incentives Behind The Strategies

It has been observed that one of the strategies was switching from HP_{w2} to HP_{w1} and later to HP_R. This strategy is interesting because the COP values for HP_{w2} are higher than the unit it is substituted with. The reason for this switching is connected to the strict coupling of district heating and district cooling system. As shown in Table A1, $\frac{COP_q}{COP_c}$ for HP_{w2} is lower than the other two which means less heating can be obtained for each unit of cooling produced. From analyzing the input data by Equation (11), we know that there is 39% higher potential for emission savings in the heating network than in the electricity network.

$$\begin{cases} \frac{\sum_t \Delta EF_t \quad \forall t : \Delta EF > 0}{\sum_t |\Delta EF_t| \quad \forall t : \Delta EF < 0} = 1.39 \\ where : \Delta EF_t = EF_{exg,q,t} - EF_{exg,p,t} \end{cases} \tag{11}$$

Therefore, the optimizer decides to use the heat pumps with higher $\frac{COP_q}{COP_c}$ and produce as much heat as possible with cooling production as low as possible to prevent violating energy balance of internal cooling network and thus forming frost in the district cooling network. Moreover, if the district cooling network is cooled extra, the temperature would fall, and this would cause the COPs to be reduced and also cause non-linearity in the model.

This behavior therefore shows the system to be limited in the low grade heat that is available from the district cooling system, which artificially constrains the dispatch of the available heat pumps.

This energy system would therefore benefit from bore holes or other low grade heat sources which would lead to more dispatch of the higher efficiency heat pumps.

Note, furthermore, that, although the district heating emissions in summer are lower than electricity in the summer, the absorption chiller is replaced by HP_S. This is because the COP of the heat pump is higher than that of the absorption chiller, so much so that it compensates for the lower emissions in the district heating system. In this study, the COP_c of the absorption chiller is assumed to be 0.5 compared to the assumed COP value of 1.8 for HP_S.

We also see that the emissions in the external district heating system are higher than the biomass boiler emissions. This causes increased production and even export from the boiler to the district heating network to reduce overall emissions. This strategy plays a large role in emission abatement since 93.3% of the total emission in the emission minimization case ($\beta = 1$) are achieved during the phases which this strategy is utilized (i.e., Phases (II) and (III)).

As mentioned above, the heat production from the CHP unit for all β values is higher than the cost minimization case because there is a high potential to reduce emissions in the heating system. Regarding electricity production from the CHP, however, the increase in usage is observed only up to the end of Phase (II). From Phase (III) onwards, the electricity production is replaced by heat production because β is large enough to outweigh the loss of revenues from electricity production and more emissions can be saved on heating than electricity.

The use of storage technologies, mainly begins to increase from Phase (III). The incentive behind this trend is that storage is used to gain upon the variations in emission levels between consecutive hours. This happens later in the phases because the round-trip storage losses are considerable compared to the abatement gained from mitigating the volatility in emission levels.

4.2. Cost-Effectiveness of Abatement Strategies

The cost of abatement for MES can be evaluated at different magnitudes, locations and from different viewpoints by comparing three different carbon pricing schemes. The first is the prices from carbon markets (e.g., European Emission Trading System (EU ETS)), the second is the carbon taxes and the third is the emission abatement cost of similar pilot projects.

The average carbon price on EU ETS in 2019 was 24.9 ($€/tCO_2$) [31]. According to the current EU ETS regulations [32], "combustion of fuels in installations with a total rated thermal input exceeding 20 MW" are considered under the cap. This means the installations on this campus would not be regulated under the current system. However, if the regulations were to change in the future to include smaller installations in the cap and trade system, then the prices of carbon allowances would have to increase to the total abatement cost levels in Table 2 for abatement strategies to be cost efficient. Otherwise, it would be more cost efficient to buy emission allowances. According to current EU ETS prices, the emission abatement strategies would not currently be cost-effective. However, according to Stiglitz et al. [33], the carbon prices to achieve Paris agreement's goal need to be 36–73 ($€/tCO_2$) by 2020 and 45–91 ($€/tCO_2$) by 2030. This suggests that the discussed abatement strategies up to Phase (III) can be economically competitive measures if the Paris agreement target is to be achieved.

In this study, the carbon taxes are included in the pricing of energy carriers. Therefore, the introduced abatement strategies can be economically beneficial if the tax levels are increased to or above the total abatement cost levels of each phase plus the current considered tax level. The carbon tax is applicable to the installations which are not under the EU ETS system, such as small MESs.

Another comparison can be with similar pilot projects aiming to reduce emissions. Klimatklivet is a program in Sweden, which supports such projects [34]. The focus of this program is the non-ETS sector and the funding is appointed to projects which have the highest abatement per investment. Iseberg et al. [34] evaluated Klimatklivet projects and found that the abatement cost is within a range of 10–80 ($€/tCO_2$) over the other policies such as a carbon tax. These values show that the discussed abatement strategies until Phase (III) would be cost competitive with other pilot projects.

5. Conclusions

This study set out to find operation strategies which can be utilized for carbon emission abatement in a local multi-energy system and investigate the cost-effectiveness against carbon prices in carbon markets, carbon tax scheme, and cost-effectiveness of similar pilot projects.

The results of the case study show that, by utilizing all the abatement strategies, a 20.8% emission reduction can be achieved with a 2.2% increase in cost. The abatement strategies which were extracted from the optimization results include: more usage of biomass boilers in heat production, substitution of district heating and absorption chillers with heat pumps, and higher utilization of storage units. It should be noted that the system was shown to be limited in the low grade heat that was available from the district cooling system, which artificially constrained the dispatch of the available heat pumps. This system would therefore benefit from bore holes or other low grade heat sources which would lead to more dispatch of the higher efficiency heat pumps. Furthermore, the utilization of the CHP unit has shown to be sensitive to the relative weighting of emissions vs. cost in the objective function. The relative share of electricity production from the CHP unit is also shown to decrease at higher emissions weighting factors (above 10^{-3}) due to the relatively higher emissions in the district heating system compared to the electricity system.

This analysis demonstrates that across all abatement strategies the total carbon dioxide abatement cost is 36.6–100.2 (€/tCO$_2$), which is higher than both the average carbon price in EU ETS and carbon tax prices in Sweden in 2019, but at the same level as similar pilot projects in Sweden. Furthermore, all strategies up to Phase (IV) are cost competitive if the carbon prices are adjusted to reach the Paris agreement target.

The results from this study can provide insight that carbon prices might not yet be high enough to make the multi-objective operation of similar local MESs cost-effective. However, the suggested abatement strategies for similar local MESs can be cost-competitive in achieving Paris agreement's goal if carbon prices were aligned to reach the agreement's goal. Of course general conclusions can hardly be made from only one case study, and the presented conclusions are only based on this case study.

Future studies can investigate more generalized abatement strategies and cost-effectiveness evaluation for local multi-energy systems. A similar methodology can be utilized for case studies in different countries and with different characteristics, and the results can be compared to provide a broader insight into the research questions. Moreover, studying abatement strategies considering future scenarios for emissions and prices in electricity and district heating systems can lead to further insights in the operation planning of local multi-energy systems.

Supplementary Materials: The following are available online at http://www.mdpi.com/1996-1073/13/7/1626/s1, the simulation model's code, input data files for the model, and the detailed calculations for emission factors. Further details are provided in "Readme.txt" at the address above.

Author Contributions: N.M.A., D.S., and L.A.T. contributed to the work conceptualization. D.S. and Z.N. contributed to the FED project conceptualization, and the model methodology development. N.M.A., D.S., Z.N., and C.A. contributed to the model development, model validation, and formal analysis. N.M.A. did the investigation and visualization of the results. All authors contributed in writing, reviewing, and editing of the draft. The supervision of the work was carried out by D.S. and L.A.T. L.A.T., D.S., and Z.N. contributed to acquisition of project funding that supports the study. All authors have read and agreed to the published version of the manuscript.

Funding: This research was funded by the Fossil-free Energy District project, funded by the European Urban Innovative Actions program (project No. UIA01-209), Västra Götaland Region (MN 2017-00126 and RUN 2017-00522), and m2M-GRID project funded by ERANET Smart Grids Plus program, with support from the European Union's Horizon 2020 research and innovation program under grant agreement No 646039.

Acknowledgments: The authors would like to acknowledge the project partners in Fossil-free Energy District (FED) for their collaboration through the project. Moreover, we would like to acknowledge Alexander Kärkkäinen, Kalid Yunus, and Anna Boss for their contributions to the model development.

Conflicts of Interest: The authors declare no conflict of interest.

Abbreviations

The following abbreviations are used in this manuscript:

MES	Multi-energy system
MILP	Mixed integer linear programming
RES	Renewable energy source
DES	Distributed energy sources
DH	District heating
CHP	Combined heat and power
B	Biomass boiler
FGC	Flue gas Condenser
HP	Electrical heat pump
AbsC	Absoption chiller
BES	Battery energy storage
BITES	Building inertia thermal energy storage
CWB	Cold water basin
PV	Photovoltaic panel
GAMS	General Algebtic Modeling System
COP	Coefficient of performance
TAC	Total abatement cost
MAC	Marginal abatement cost
EU ETS	European Emission Trading System
EXP	export to external grid
IMP	import from external grid

Appendix A. Technology Equations

In this section, the equations for different technologies which are not presented in the article are provided.

Appendix A.1. General Constraints

All technologies are limited in their power output with their capacity $Q_{cap,j}$ according to Equation (A1).

$$
\begin{cases}
Q_j \leq Q_{cap,j} \\
P_j \leq P_{cap,j} \\
K_j \leq K_{cap,j}
\end{cases}
\tag{A1}
$$

As mentioned in Section 2.1.3, the biomass boiler and the CHP unit are not always dispatchable. Therefore, they are regularly fixed to specific levels (Equation (A2)). Moreover, HP_S, HP_{W1}, HP_{W2}, and the absorption chiller are only in operation during specific periods of the year, as also represented by Equation (A2).

$$
Q_{j,t} = Q_{fix,j,t} \qquad \forall t \in t_{undispatchle}
\tag{A2}
$$

Appendix A.2. Biomass Boiler (B) and Flue Gas Condenser (FGC)

The heat produced in the boiler (Equation (A3)) is related to the fuel input ($Q_{fuel,B,t}$) and the efficiency of the boiler (η_B) . Besides, the heat recovered from the flue gas condenser (Equation (A4)) is related to the heat not absorbed by boiler and the efficiency of the condenser (η_{fgc}).

Since the flue gas condenser and the biomass boiler are in the same unit, their heat is aggregated and presented as one unit in this study (Equation (A5)).

$$
Q_{B,t} = Q_{fuel,B,t} \times \eta_B
\tag{A3}
$$

$$
Q_{fgc,t} = Q_{fuel,B,t} \times (1 - \eta_B) \times \eta_{fgc}
\tag{A4}
$$

$$Q_{B+fgc,t} = Q_{B,t} + Q_{fgc,t} \tag{A5}$$

Moreover, the changes in boiler's output is limited to the ramp-up (RU_B) and ramp-down (RD_B) limits (Equation (A6)).

$$RU_B \leq Q_{B,t} - Q_{B,t-1} \leq RD_B \tag{A6}$$

As shown in Figure 2, the export to external district heating can only come from the biomass boiler and therefore is limited to its production (Equation (A7)).

$$Q_{exp,t} \leq Q_{B,t} \tag{A7}$$

Appendix A.3. Combined Heat and Power

The CHP unit has a boiler and a turbine. The system can regulate how much of the heat produced in the CHP boiler (Q_{chp_b}) is to be sent directly to the local district heating network ($Q_{chp_{b2g}}$) or to the turbine ($Q_{chp_{tr}}$). Moreover, the heat which is not converted to power in the turbine can be recovered and sent to the district heating grid afterward ($Q_{chp_{htr}}$).

In this study, the heat from the CHP unit is the sum of the heat sent directly from the boiler to the grid and the heat from the turbine (Equation (A8)). The heat produced by the boiler is related to the fuel input and the efficiency of the boiler (Equation (A9)). Moreover, the power production in the turbine is calculated based on the efficiency of the turbine and how much heat is sent to it (Equation (A11)).

$$Q_{chp,t} = Q_{chp_{b2g},t} + Q_{chp_{htr},t} \tag{A8}$$

$$Q_{chp_b,t} = Q_{fuel,chp_b,t} \times \eta_{chp_b} \tag{A9}$$

$$Q_{chp_{htr},t} = Q_{chp_{tr},t} \times (1 - \eta_{tr}) \tag{A10}$$

$$P_{chp_{tr},t} = Q_{chp_{tr},t} \times \eta_{tr} \tag{A11}$$

Appendix A.4. Absorption Chiller

The cooling production level of the absorption chiller (K_{absc}) is related to the cooling coefficient of performance ($COP_{k,absc}$) and presented in Equation (A12).

$$K_{absc,t} = COP_{k,absc} \times Q_{absc,t} \tag{A12}$$

The absorption chiller has a minimum operation limit during the cooling season (Equation (A13)). The cooling season is presented in Table A2 and $K_{min,absc,t}$ is considered to be 200 kW in this case study.

$$K_{absc,t} \geq K_{min,absc,t} \tag{A13}$$

Appendix A.5. Refrigeration Heat Pump (HP$_R$)

Due to the local system's limitations, the heating produced as the byproduct of this heat pump is limited to 20% of the heating load of the building it is situated in ($Q_{dem,x,t}$). This constraint is presented in Equation (A14).

$$Q_{hp_r,t} \leq 0.2 \times Q_{dem,x,t} \tag{A14}$$

Appendix A.6. Storages

All of the storages are limited to their respective energy capacities (EN_{cap}). This constraint is presented in Equation (A15).

$$EN_{j,t} \leq EN_{cap,j} \tag{A15}$$

The energy level of the storage unit (Equation (A16)) is related to the previous energy state, the amounts of charging ($P_{ch,j,t}$) and discharging ($P_{dis,j,t}$), and the efficiencies of charging (η_{ch}) and discharging (η_{dis}). Charging and discharging efficiencies are considered to be equal in each unit. Moreover, for the cold water basin, instead of P, K is used in Equation (A16).

$$EN_{j,t} = EN_{j,t-1} + P_{ch,j,t} \times \eta_{ch} - P_{dis,j,t} \times \frac{1}{\eta_{dis}} \tag{A16}$$

Appendix A.6.1. Cold Water Basin (CWB)

Due to local system's limitations, discharging of the cold water basin is only limited to the cooling load of the building it is situated in ($K_{dem,y,t}$). This constraint is presented in Equation (A17).

$$K_{dis,cwb,t} \leq K_{dem,y,t} \tag{A17}$$

Moreover, it has been observed that only considering the charging and discharging efficiencies will not stop simultaneous charge and discharge of the cold water basin. The reason is that the cooling demand is lower than heating demand in the district as a whole and there is not enough cooling storage to handle the excess produced cooling. Therefore, the optimizer would theoretically curtail cooling by charging and discharging at the same time, causing additional losses in the cooling system.

To prevent simultaneous charging and discharging, binary variables are used (Equation (A18)) where b is the binary variable and M is a big number (10^6kW in this case study).

$$\begin{cases} b_{ch,cwb,t} + b_{dis,cwb,t} \leq 1 \\ K_{ch,cwb,t} \leq b_{ch,cwb,t} \times M \\ K_{dis,cwb,t} \leq b_{dis,cwb,t} \times M \end{cases} \tag{A18}$$

Appendix A.6.2. Battery

For the battery units, an additional constraint is considered on the minimum state of charge (SOC_{min}). This constraint (Equation (A19)) is to prevent high degradation costs.

$$EN_{bes,t} \geq EN_{cap,bes} \times SOC_{min} \tag{A19}$$

Appendix B. Input Data

In this section, the input data for energy technologies, marginal emissions, marginal prices, and energy demand of the campus are provided. The parameters for energy technologies are presented in Table A1.

The heating and cooling seasons, which are considered for the heat pumps' and absorption chiller's operation is presented in Table A2.

Results from simulation models are very dependent on their assumptions and therefore other emission factors or marginal prices can affect the simulation results. With the developed model, similar studies can be performed to assess efficient abatement strategies in other locations or with different assumptions. The marginal emissions and marginal prices for external electricity and district heating networks that are used in this study are presented in Figures A1 and A2. The emission factors used for calculating the marginal emissions are presented in Tables A3 and A4.

All electricity efficiencies and emission factors are based on [35,36], except for oil and waste incineration where the work of Bertoldi et al. [37] is the source. Pumped hydro and 'Unknown' are taken from [29]. Pumped hydro is based on average Swedish emissions during times when hydro is being charged. 'Unknown' is based on the average generation plant in the Swedish electrical grid. The emission factors for CHP units are calculated based on the Benefit Allocation Method [38].

Table A1. Technology characteristics.

Biomass boiler and flue gas condenser					
$Q_{cap,b}$	$Q_{cap,fgc}$	η_b	η_{fgc}	RU_b	RD_b
8000 kW	1000 kW	0.77	0.5	1000 kW	1000 kW
Combined heat and power					
Q_{cap,chp_b}	$Q_{cap,chp_{tr}}$	η_{chp_b}	η_{tr}	RU_b	RD_b
6000 kW	800 kW	0.77	0.17	1000 kW	1000 kW

Absorption chiller		
$Q_{cap,absc}$	$COP_{k,absc}$	$K_{min,absc,t}$
2300 kW	0.5	200 kW

Cold water basin				
$K_{cap,ch,cwb}$	$K_{cap,dis,cwb}$	$EN_{cap,cwb}$	$\eta_{ch,cwb}$	$\eta_{dis,cwb}$
204 kWh/h	35 kWh/h	814 kWh	0.95	0.95

Battery energy storage (2 units)				
$P_{cap,bes}$	$EN_{cap,bes}$	$\eta_{ch,bes}$	$\eta_{dis,bes}$	SOC_{min}
100 and 50 kW	200 and 100 kWh	0.95	0.95	0.2

Thermal energy storage tank					
$Q_{cap,ch,tes}$	$Q_{cap,dis,tes}$	$EN_{cap,tes}$	$\eta_{ch,tes}$	$\eta_{dis,tes}$	$loss_{tes}$
11 MWh/h	23 MWh/h	39 MWh	0.95	0.95	0.01

Heat pumps				
	$P_{cap,hp}$	COP_q	COP_k	$\dfrac{COP_q}{COP_c}$
HP_S	216 kW	3	1.8	1.67
HP_{W1}	216 kW	3	1.8	1.67
HP_{W2}	203 kW	3.1	2.19	1.42
HP_R	263 kW	2.86	1.9	1.51

Table A2. Cooling and heating season periods.

	Starting Date	**Ending Date**
Heating Season	1st November	30th April
Cooling Season	1st May	31st October

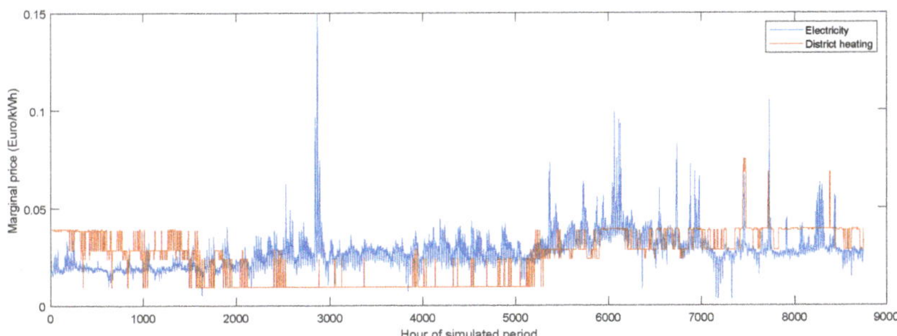

Figure A1. Marginal prices from external electricity and district heating networks over the simulated period.

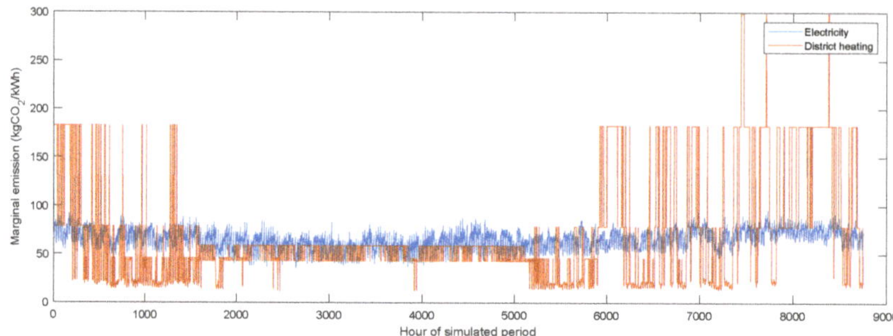

Figure A2. Marginal emissions from external electricity and district heating networks over the simulated period.

In the district heating system, the emission factor (Table A3) for the marginal unit is considered to be the marginal emission factor for every hour. The emission factors are primarily based on the energy allocation method which may differ from current practice among district heating operators in Sweden. More details on the calculation of the emissions factors can be found in Supplementary Materials based on the following references [35–39]. The hourly marginal unit data are taken from a district heating operator in Sweden. Based on the information from the district heating provider, for this dataset, the most expensive unit at each hour has been assumed to be the marginal unit. However, the common practice for marginal unit allocation in the external district heating system is not necessarily the most expensive unit because of other constraints such as the ramp-up and ramp-down limitations of the units. However, due to limited data access on the external district heating network, this assumption was considered reasonable.

Table A3. Emission factors for district heating units. HOB is the heat only boiler, RH is the waste heat from refineries, and WI is waste incineration.

Unit	HOB Biomass	CHP Biomass	HOB Gas	CHP Gas	HOB Oil	RH	CHP WI
gCO_2/kWh	79	46	299	183	339	43	59

For calculating the marginal emissions for the electricity grid, a statistical method is used to predict the marginal emissions at every hour based on historical data [40]. Firstly, the probability that each technology is on the margin based on a consumed unit of electricity in Sweden is taken from Electricity Map [29]. Secondly, the probabilities are multiplied by the relevant emission factor of each technology and summed. The emission factors for each technology can be seen in Table A4.

Table A4. Emission factors of different technologies for electricity grid.

Technology	biomass	coal	gas	hydro	nuclear	hydro-discharge
Emission factor (gCO_2/kWh)	230	820	490	24	12	46
Technology	geothermal	unknown	oil	solar	wind	hydro-charge
Emission factor (gCO_2/kWh)	38	362	782	45	11	0

The differing approaches for calculating marginal emissions in the electricity and district heating networks are due to the respective size of these systems and therefore the ability to know precisely which unit is on the margin. The district heating network of a city is usually smaller than the electricity network and limited to the boundaries of the city. Identification of the exact marginal unit can therefore be easily estimated. However, the electricity network is much larger and tangled with production units in other countries through import and export across regional and country borders. Since the exact marginal unit cannot be identified, the electricity system's marginal emissions are determined by statistical methods, as described above.

Energy demand of the campus for electricity, heating, and cooling for the simulation period is presented in Figure A3. The simulation period is selected based on availability and quality of real data from the campus energy system. The simulations are run for the period from 1 March 2016 to 28 February 2017.

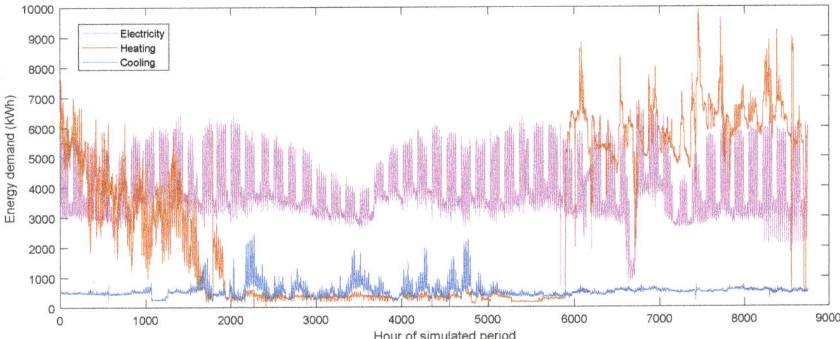

Figure A3. Energy demand of the Chalmers campus over the simulated period. The simulated period is from 1 March 2016 to 28 February 2017.

References

1. Ruester, S.; Schwenen, S.; Batlle, C.; Pérez-Arriaga, I. From distribution networks to smart distribution systems: Rethinking the regulation of European electricity DSOs. *Util. Policy* **2014**, *31*, 229–237. [CrossRef]
2. Guo, Y.; Pan, M.; Fang, Y.; Khargonekar, P.P. Decentralized Coordination of Energy Utilization for Residential Households in the Smart Grid. *IEEE Trans. Smart Grid* **2013**, *4*, 1341–1350. [CrossRef]
3. Koirala, B.P.; Koliou, E.; Friege, J.; Hakvoort, R.A.; Herder, P.M. Energetic communities for community energy: A review of key issues and trends shaping integrated community energy systems. *Renew. Sustain. Energy Rev.* **2016**, *56*, 722–744. [CrossRef]
4. Giotitsas, C.; Pazaitis, A.; Kostakis, V. A peer-to-peer approach to energy production. *Technol. Soc.* **2015**, *42*, 28–38. [CrossRef]
5. Mancarella, P. MES (multi-energy systems): An overview of concepts and evaluation models. *Energy* **2014**, *65*, 1–17. [CrossRef]
6. Grosspietsch, D.; Saenger, M.; Girod, B. Matching decentralized energy production and local consumption: A review of renewable energy systems with conversion and storage technologies. *Wiley Interdiscip. Rev. Energy Environ.* **2019**, *8*, e336. [CrossRef]
7. Perera, A.T.D.; Nik, V.M.; Mauree, D.; Scartezzini, J.L. An integrated approach to design site specific distributed electrical hubs combining optimization, multi-criterion assessment and decision making. *Energy* **2017**, *134*, 103–120. [CrossRef]
8. Majidi, M.; Nojavan, S.; Zare, K. A cost-emission framework for hub energy system under demand response program. *Energy* **2017**, *134*, 157–166. [CrossRef]
9. Bracco, S.; Dentici, G.; Siri, S. Economic and environmental optimization model for the design and the operation of a combined heat and power distributed generation system in an urban area. *Energy* **2013**, *55*, 1014–1024. [CrossRef]
10. Wang, R.; Zhou, Z.; Ishibuchi, H.; Liao, T.; Zhang, T. Localized Weighted Sum Method for Many-Objective Optimization. *IEEE Trans. Evol. Comput.* **2018**, *22*, 3–18. [CrossRef]
11. Wu, Q.; Ren, H.; Gao, W.; Ren, J. Multi-objective optimization of a distributed energy network integrated with heating interchange. *Energy* **2016**, *109*, 353–364. [CrossRef]
12. Li, G.; Wang, R.; Zhang, T.; Ming, M. Multi-Objective Optimal Design of Renewable Energy Integrated CCHP System Using PICEA-g. *Energies* **2018**, *11*, 743. [CrossRef]

13. Huang, Y.; Zhang, W.; Yang, K.; Hou, W.; Huang, Y. An Optimal Scheduling Method for Multi-Energy Hub Systems Using Game Theory. *Energies* **2019**, *12*, 2270. [CrossRef]

14. Fu, X.; Huang, S.; Li, R.; Sun, H. Electric Power Output Optimization for CCHP Using PSO Theory. *Energy Procedia* **2016**, *103*, 9–14. [CrossRef]

15. Reynolds, J.; Ahmad, M.W.; Rezgui, Y.; Hippolyte, J.L. Operational supply and demand optimisation of a multi-vector district energy system using artificial neural networks and a genetic algorithm. *Appl. Energy* **2019**, *235*, 699–713. [CrossRef]

16. Domínguez-Muñoz, F.; Cejudo-López, J.M.; Carrillo-Andrés, A.; Gallardo-Salazar, M. Selection of typical demand days for CHP optimization. *Energy Build.* **2011**, *43*, 3036–3043. [CrossRef]

17. Marquant, J.F.; Evins, R.; Carmeliet, J. Reducing Computation Time with a Rolling Horizon Approach Applied to a MILP Formulation of Multiple Urban Energy Hub System. *Procedia Comput. Sci.* **2015**, *51*, 2137–2146. [CrossRef]

18. Erichsen, G.; Zimmermann, T.; Kather, A. Effect of Different Interval Lengths in a Rolling Horizon MILP Unit Commitment with Non-Linear Control Model for a Small Energy System. *Energies* **2019**, *12*, 1003. [CrossRef]

19. Di Somma, M.; Yan, B.; Bianco, N.; Luh, P.B.; Graditi, G.; Mongibello, L.; Naso, V. Multi-objective operation optimization of a Distributed Energy System for a large-scale utility customer. *Appl. Therm. Eng.* **2016**, *101*, 752–761. [CrossRef]

20. Falke, T.; Krengel, S.; Meinerzhagen, A.K.; Schnettler, A. Multi-objective optimization and simulation model for the design of distributed energy systems. *Appl. Energy* **2016**, *184*, 1508–1516. [CrossRef]

21. Yan, B.; Somma, M.D.; Luh, P.B.; Graditi, G. Operation Optimization of Multiple Distributed Energy Systems in an Energy Community. In Proceedings of the 2018 IEEE International Conference on Environment and Electrical Engineering and 2018 IEEE Industrial and Commercial Power Systems Europe (EEEIC/I CPS Europe), Palermo, Italy, 12–15 June 2018; pp. 1–6.

22. Böckl, B.; Greiml, M.; Leitner, L.; Pichler, P.; Kriechbaum, L.; Kienberger, T. HyFlow—A Hybrid Load Flow-Modelling Framework to Evaluate the Effects of Energy Storage and Sector Coupling on the Electrical Load Flows. *Energies* **2019**, *12*, 956. [CrossRef]

23. FED—Fossil Free Energy Districts. Available online: https://www.uia-initiative.eu/en/uia-cities/gothenburg (accessed on 10 October 2019).

24. Norwood, Z.; Steen, D.; Kärkkäinen, A.; Yunus, K.; Agathokleous, C.; Mirzaei Alavijeh, N. FED Model WP4.1—Model-for-Energies. Available online: https://github.com/zacknorwood/FEDmodel/tree/model-for-energies (accessed on 31 January 2020).

25. GAMS Development Corporation. *General Algebraic Modeling System (GAMS) Release 24.7.4*; GAMS Development Corporation: Fairfax, VA, USA, 2016.

26. The MathWorks Inc.. *MATLAB Version 9.5.0.944444 (R2018b)*; The MathWorks Inc.: Natick, MA, USA, 2019.

27. Romanchenko, D.; Kensby, J.; Odenberger, M.; Johnsson, F. Thermal energy storage in district heating: Centralised storage vs. storage in thermal inertia of buildings. *Energy Convers. Manag.* **2018**, *162*, 26–38. [CrossRef]

28. Fjärrvärmepriser. Available online: https://www.goteborgenergi.se/foretag/fjarrvarme-kyla/fjarrvarmepriser (accessed on 12 November 2019).

29. Live CO_2 Emissions of Electricity Consumption. Available online: http://electricitymap.tmrow.co (accessed on 24 October 2019).

30. Strång—A Mesoscale Model for Solar Radiation. Available online: http://strang.smhi.se/ (accessed on 20 December 2019).

31. Carbon Price Viewer. Available online: https://sandbag.org.uk/carbon-price-viewer/ (accessed on 30 January 2020).

32. Smith, R. Directive 2008/94/EC of the European Parliament and of the Council of 22 October 2008. In *Core EU Legislation*; Macmillan Education: London, UK, 2015; pp. 423–426.

33. Stiglitz, J.E.; Stern, N.; Duan, M.; Edenhofer, O.; Giraud, G.; Heal, G.M.; la Rovere, E.L.; Morris, A.; Moyer, E.; Pangestu, M.; et al. *Report of the High-Level Commission on Carbon Prices*; World Bank, Washington, DC, USA, 2017.

34. Isberg, U.; Jonsson, L.; Pädam, S.; Hallberg, A.; Nilsson, M.; Malmström, C. *Utvärdering av Klimatklivet*; Naturvårdsverket: Stockholm, Sweden, 2019.

35. Bruckner, T.; Bashmakov, I.A.; Mulugetta, Y.; Chum, H.; De la Vega Navarro, A.; Edmonds, J.; Faaij, A.; Fungtammasan, B.; Garg, A.; Hertwich, E.; et al. Energy systems Climate Change 2014: Mitigation of Climate Change. In *Contribution of Working Group III to the Fifth Assessment Report of the Intergovernmental Panel on Climate Change ed OR Edenhofer et al*; Cambridge University Press: Cambridge, UK; New York, NY, USA, 2014.
36. Schlömer, S.; Bruckner, T.; Fulton, L.; Hertwich, E.; McKinnon, A.; Perczyk, D.; Roy, J.; Schaeffer, R.; Sims, R.; Smith, P.; et al. Annex III: Technology-specific cost and performance parameters. In *Climate Change*; Cambridge University Press: Cambridge, UK; New York, NY, USA, 2014; pp. 1329–1356.
37. Bertoldi, P.; Cayuela, D.B.; Monni, S.; De Raveschoot, R.P. *How to Develop a Sustainable Energy Action Plan (SEAP)*; Publications Office of the European Union: Luxembourg, 2010.
38. Siitonen, S.; Holmberg, H. Estimating the value of energy saving in industry by different cost allocation methods. *Int. J. Energy Res.* **2012**, *36*, 324–334. [CrossRef]
39. Johansson, D.; Rootzén, J.; Berntsson, T.; Johnsson, F. Assessment of strategies for CO_2 abatement in the European petroleum refining industry. *Energy* **2012**, *42*, 375–386. [CrossRef]
40. Tranberg, B.; Corradi, O.; Lajoie, B.; Gibon, T.; Staffell, I.; Andresen, G.B. Real-time carbon accounting method for the European electricity markets. *Energy Strategy Rev.* **2019**, *26*, 100367. [CrossRef]

Article

Research on Double-Layer Optimal Scheduling Model of Integrated Energy Park Based on Non-Cooperative Game

Feifan Chen [1,*], Haifeng Liang [1], Yajing Gao [2,*], Yongchun Yang [1] and Yuxuan Chen [1]

[1]   Department of Electrical Engineering, North China Electric Power University, Baoding 071003, China
[2]   Technical and Economic Consultation Center for Electric Power Construction in China Electricity Council, Beijing 100032, China
*   Correspondence: xzllcff@163.com (F.C.); 51351706@ncepu.edu.cn (Y.G.)

Received: 8 July 2019; Accepted: 14 August 2019; Published: 16 August 2019

Abstract: As the realization form of integrated energy system, integrated energy park is the key research object in the field of energy. Actual integrated energy parks are often partitioned internally. In order to take into account the interests of each zone in the optimal scheduling of integrated energy parks, a double-layer optimal scheduling model of integrated energy parks based on non-cooperative game theory is proposed. First, according to the operation of the integrated energy park, the output and cost model of the park is established. Second, with the minimum daily cost as the upper layer objective and the highest energy efficiency of the cogeneration system as the lower layer objective, a double-layer optimal scheduling model is established. Then based on non-cooperative game, the optimal operation strategy of each zone is obtained through the game among all the zones. An integrated energy park is taken as an example, the results show that the proposed model can make zones adjust their operation strategies more reasonably, thus helping to reduce the cost of the park and improve energy efficiency.

Keywords: integrated energy park; park partition; double-layer optimal scheduling; non-cooperative game; Nash equilibrium

---

1. Introduction

With the increasing demand for energy and the accelerating development of urbanization, the contradiction between the energy demand of human society and the endurance of the environment has become increasingly prominent. Seeking an energy operation mode which is characterized by energy interconnection, high efficiency, and low carbon has become the direction of future energy development [1–3]. In such a global environment, integrated energy system emerges as the times require, as the physical carrier of energy internet; it has been well known in academia and business circles [4,5]. In the *Implementation Opinions on Promoting the Construction of Multi-energy Complementary Integration and Optimization Demonstration Projects*, issued by the China National Development and Reform Commission in 2016, it is proposed that in the face of end-user demand for electricity, heat, cold, gas and other energy, the development of traditional and new energy resources should be tailored to local conditions and complementary utilization [6].

As a place where urban energy consumption gathers, the integrated energy park gathers several individuals with similar energy consumption characteristics. It has the characteristics of large energy demand, concentrated energy consumption, and gathering multiple loads of cold, heat, and electricity [7]. The day-ahead optimal scheduling of integrated energy system is an important link to ensure the optimal operation of integrated energy park. It is of great significance for energy saving, emission reduction, reducing the operation cost of integrated energy system and improving energy efficiency.

Nowadays integrated energy parks are often partitioned internally according to certain rules. As independent decision-makers, these zones exist in the integrated energy parks, which hope to maximize their own interests as far as possible. There are many researches on optimal scheduling of integrated energy system in the literatures. In literature [8], the economic emission scheduling of integrated energy system (IES) has been obtained using multi-objective optimization framework. Literature [9] presents the capabilities of the TBATS model which has no seasonality constraints, making it possible to create detailed, long-term forecasts. Literature [10] considers the characteristics of power to gas (P2G) facilities, power systems, natural gas systems, and heating systems. This paper presents an optimal dispatching model for a multi-energy system with P2G facilities. In literature [11], a two-stage stochastic mixed integer linear programming model for day-ahead energy dispatch in multi-carrier energy systems is established considering the time-varying and volatility of the renewable energy price signals. In literature [12], a new day-ahead optimal dispatching model for P2G storage and gas load management in power and natural gas markets is proposed to minimize the gas consumption cost of gas load. Most of these literatures have focused on the lowest operating cost, the best environmental protection, or other objectives, and have made in-depth studies on the day-ahead optimal scheduling of an integrated energy system. However, in the process of optimal scheduling, the interests of each zone in the integrated energy park are not considered in detail. Meanwhile, under the background of energy shortage, energy saving and emission reduction, the unit price of electricity and natural gas often changes with the increase of purchases. The unit price of electricity is affected by the relationship between supply and demand in the electricity market. The larger the quantity of electricity purchased, the higher the clearing price [13,14]. In order to reduce inefficient and unfair phenomena such as cross subsidies and inversion caused by low natural gas prices, natural gas also adopts the system that the higher the purchase volume, the higher the gas unit price [15,16]. Each zone in the integrated energy park hopes to pursue greater interests as far as possible. Because of the relationship between electricity price, gas price, and purchase volume, the interests of each zone are affected by the decision-making of other zones, so all zones constitute a non-cooperative game.

In summary, in order to take into account the interests of each zone in the optimal scheduling of integrated energy park, the dynamic model of electricity price and gas price is introduced, a double-layer optimal scheduling model of integrated energy parks based on non-cooperative game theory is proposed and used to obtain the optimal operation strategy of each zone in the park, so as to achieve Nash equilibrium solution beneficial to each zone. An integrated energy park is taken as an example, and the results show that the proposed model can make zones adjust their operation strategies more reasonably, thus helping to reduce the cost of the park and improve energy efficiency.

2. Operation Analysis of Integrated Energy Park

2.1. Typical Integrated Energy Park

Typical integrated energy parks are equipped with combined cool, heat, and power (CCHP) equipment, including combined heat and power (CHP) system, gas boiler, electric refrigeration unit, heat pump unit, absorption refrigeration unit, etc. CHP system includes gas turbine and waste heat boiler with supplementary combustion device. In addition, there are photovoltaic power generation and energy storage equipment in the park. The specific energy utilization framework is shown in Figure 1.

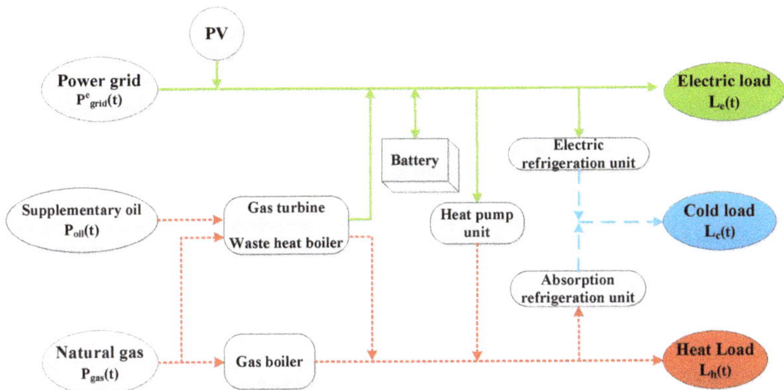

Figure 1. Energy utilization framework of a typical integrated energy park.

As can be seen from the figure above, the electric load sources of the park include power purchase from the power grid, gas turbine power generation, and photovoltaic power generation. The cold load of the park is provided by electric refrigeration units and absorption refrigeration units; as for the heat load, part of it comes from the waste heat generated by the gas turbine power generation, and the output of the waste heat can be adjusted by adding oil to the waste heat boiler. When the waste heat is insufficient, natural gas can be added to the gas boiler to supplement the heat. In addition, heat pump is also one of the sources of heat load. When the electricity price is low in the valley, the park reduces the gas purchase volume and purchases a large amount of electricity from the grid to meet the load demand and energy storage in the park; when the electricity price is high in the peak, the park purchases less electricity from the grid, and provides energy by increasing the output of gas turbines and discharging of energy storage equipment.

2.2. Equipment Output Model of Integrated Energy Park

One day is divided into 24 time periods, represented by t, $t = 1, 2, \ldots, 24$. Each time period lasts one hour, time interval is represented by T.

1. CHP system

(1) Gas turbines. Natural gas is added to the gas turbine, expands after combustion to provide mechanical energy, and then converts it into electricity. The output power of a gas turbine is

$$P^e_{CHP}(t) = \eta^e_{CHP} \cdot P^{gas}_{CHP}(t) \tag{1}$$

$P^e_{CHP}(t)$—output electric power of CHP system during time period t, W.
η^e_{CHP}—generation efficiency of CHP system.
$P^{gas}_{CHP}(t)$—energy of natural gas input into the CHP system during time period t, W.
The symbol P in the paper represents the power of the equipment, the energy provided by gas or oil per unit time, and the amount of electricity purchased per unit time.

(2) Waste heat boiler. The waste heat boiler can output waste heat generated by gas turbine, and the heat power output by the waste heat boiler is as follows:

$$P^h_{CHP}(t) = P^e_{CHP}(t) \cdot V_{CHP}(t) \tag{2}$$

$P^h_{CHP}(t)$—output electric power of CHP system during time period t, W.
$V_{CHP}(t)$—thermoelectric ratio of CHP system during time period t.

Thermoelectric ratio is the ratio of the output heat power to the output electric power of the CHP system during time period t. The thermoelectric ratio of the CHP system can be changed by supplying oil to the waste heat boiler with a supplementary combustion device to adjust the supplementary combustion rate. The supplementary combustion device can be divided into external supplementary combustion and internal supplementary combustion. Because of the high investment cost and the inconvenience of operation and maintenance of the external supplementary combustion device [17], it assumes that the internal supplementary combustion device is used in the waste heat boiler in this paper. Literature [18] provides experimental data on supplementary combustion of waste heat boilers, which are fitted by Statistical Product and Service Solutions (SPSS), and the following linear relationships are obtained.

$$V_{CHP}(t) = Y_a \cdot A(t) + Y_b \tag{3}$$

$$A(t) = Y_c \cdot \frac{P_{oil}(t)}{P_{CHP}^e(t)} + Y_d \tag{4}$$

$A(t)$—supplementary combustion rate during time period t.
$P_{oil}(t)$—energy of oil supplied during time period t, W.

The following parameters are obtained by fitting: $Y_a = 2.589$, $Y_b = 2.064$, $Y_c = 0.381$, $Y_d = 0.001$, and the results show that the R^2 of the two fitting results are 0.9883 and 0.9727, respectively (more than 0.8 means that the fitting results are relatively good).

2. Gas boiler. When the waste heat from CHP system cannot meet the demand of the heat load, natural gas can be added to the gas boiler to supplement the heat. The output heat power of the gas boiler is as follows:

$$P_{GB}^h(t) = \eta_{GB} \cdot P_{GB}^{gas}(t) \tag{5}$$

$P_{GB}^h(t)$—output heat power of gas boiler during time period t, W.
η_{GB}—heat production efficiency of gas boiler.
$P_{GB}^{gas}(t)$—energy of the natural gas input into the gas boiler during time period t, W.

3. Electric refrigeration unit. Electric refrigeration units can convert electricity to provide cold load for the park. The operation model is as follows.

$$P_e^c(t) = \eta_c^e \cdot P_c^e(t) \tag{6}$$

$P_e^c(t)$—output cold power of electric refrigeration unit during time period t, W.
η_c^e—efficiency of electric refrigeration unit.
$P_c^e(t)$—input electric power of electric refrigeration unit during time period t, W.

4. Heat pump unit. Heat pump units can convert the electric energy to provide the demand for the heat load of the park. The operation model is as follows.

$$P_e^h(t) = \eta_h^e \cdot P_h^e(t) \tag{7}$$

$P_e^h(t)$—output heat power of heat pump unit during time period t, W.
η_h^e—efficiency of heat pump unit
$P_h^e(t)$—input electric power of heat pump unit during time period t, W.

5. Absorption refrigeration unit. Absorption refrigeration units can convert heat energy to provide cold load for the park. The operation model is as follows.

$$P_h^c(t) = \eta_c^h \cdot P_c^h(t) \tag{8}$$

$P_h^c(t)$—output cold power of absorption refrigeration unit during time period t, W.
η_c^h—efficiency of absorption refrigeration unit.

$P_c^h(t)$—input heat power of absorption refrigeration unit during time period t, W.

2.3. Daily Cost Model of Park

The daily cost of the park considered in this paper includes the daily operating cost of the park and the daily environmental value cost of the park. The daily operating cost of the park does not involve the initial investment cost of each equipment; it mainly includes the daily gas purchase cost, electricity purchase cost, oil cost and the operation and maintenance cost of each equipment. The unit price of gas and electricity is related to the quantity of gas purchased and the electricity purchased in the park, which means the more the purchased quantity, the higher the unit price. Oil is regarded as fixed price because it only plays a regulatory role and its purchases are relatively small. The daily environmental value cost of the park includes the treatment cost of pollutants caused by the daily use of gas turbines, waste heat boilers, and gas boilers in the park.

2.3.1. Daily Operating Cost of the Park

1. Gas purchase cost. The park purchases gas from natural gas suppliers for gas turbines and gas boilers. The cost of purchasing gas is as follows:

$$C_{gas}(t) = c_{gas}(t) \cdot (P_{CHP}^{gas}(t) + P_{GB}^{gas}(t)) \cdot T / (\lambda_{gas} \cdot \rho_{gas}) \tag{9}$$

$C_{gas}(t)$—gas purchase cost of park during time period t, USD.
$c_{gas}(t)$—gas unit price during time period t, USD/m^3.
T—time interval, h.
λ_{gas}—calorific value of natural gas, J/kg.
ρ_{gas}—density of natural gas, kg/m^3.
The unit price of natural gas is positively correlated with the purchase amount of natural gas of the park

$$c_{gas}(t) = A_{gas} \cdot (P_{CHP}^{gas}(t) + P_{GB}^{gas}(t)) \cdot T / (\lambda_{gas} \cdot \rho_{gas}) + B_{gas} \tag{10}$$

A_{gas}—gas price parameters, USD/(m^3)^2.
B_{gas}—gas price parameters, USD/m^3.

2. Electricity purchase cost. The park purchases electricity from the power grid to meet the demand of electric load in the park and realizes the storage and conversion of power through energy storage equipment, electric refrigeration unit, and heat pump unit. The cost of power purchase is as follows.

$$C_{grid}(t) = c_{grid}(t) \cdot P_{grid}^e(t) \cdot T \tag{11}$$

$C_{grid}(t)$—electricity purchase cost of park during time period t, USD.
$c_{grid}(t)$—electricity unit price during time period t, USD/J.
$P_{grid}^e(t)$—electricity purchase of the park during time period t, W.
The unit price of electricity is positively correlated with the purchase amount of electricity of the park and is affected by the peak and valley period of electricity consumption.

$$c_{grid}(t) = a(t) \cdot P_{grid}^e(t) \cdot T + b(t) \tag{12}$$

$a(t)$—electricity price parameters during time period t, USD/(J)^2.
$b(t)$—electricity price parameters during time period t, USD/J.

3. Oil purchase cost. Oil is used for supplementary combustion, and the thermoelectric ratio of CHP system can be adjusted by adding oil to the waste heat boiler. The cost is as follows:

$$C_{oil}(t) = c_{oil} \cdot P_{oil}(t) \cdot T / \lambda_{oil} \tag{13}$$

$C_{oil}(t)$—Oil purchase cost during time period t, USD.
c_{oil}—oil unit price, USD/kg.
λ_{oil}—calorific value of oil, J/kg.

4. CCHP equipment cost. It includes the operation and maintenance costs of CHP system, gas boiler, electric refrigeration unit, heat pump unit, and the absorption refrigeration unit, as follows

$$C_{CCHP}(t) = \sum_{i=1}^{5} c_i \cdot P_i(t) \cdot T \tag{14}$$

$C_{CCHP}(t)$—total cost of operation and maintenance of CCHP equipment during time period t, USD.
c_i—operational and maintenance cost per unit output power of the above five equipment, USD/J.
$P_i(t)$—output power of five equipment during time period t, W.

5. Energy storage cost. Operation and maintenance cost of electric energy storage equipment

$$C_{sto}(t) = c_{sto} \cdot \eta_{sto} \cdot (P^e_{char}(t) + P^e_{dis}(t)) \cdot T \tag{15}$$

$C_{sto}(t)$—operation and maintenance cost of electric energy storage equipment during time period t, USD.
c_{sto}—operation and maintenance cost per unit charge and discharge volume of electric energy storage equipment, USD/J.
η_{sto}—charging and discharging efficiency of electric energy storage equipment
$P^e_{char}(t)$—charging power of electric energy storage equipment during time period t, W.
$P^e_{dis}(t)$—discharging power of electric energy storage equipment during time period t, W.

2.3.2. Daily Environmental Value Cost of the Park

1. Environmental value cost of gas turbine

$$C_{GT}(t) = P^e_{CHP}(t) \cdot T \cdot \sum_{i=1}^{5} J^i_{GT} \cdot K^i \tag{16}$$

$C_{GT}(t)$—environmental value cost of gas turbine during time period t, USD.
i—pollutants, including SO_2, NO_x, CO, CO_2, TSP.
J^i_{GT}—emissions of various pollutants per unit output electric power from gas turbine, kg/J.
K^i—environmental costs of various pollutants, USD/kg.

2. Environmental value cost of waste heat boiler

$$C_{WHB}(t) = P^h_{CHP}(t) \cdot T \cdot \sum_{i=1}^{5} J^i_{WHB} \cdot K^i \tag{17}$$

$C_{WHB}(t)$—environmental value cost of waste heat boiler during time period t, USD.
J^i_{WHB}—emissions of various pollutants per unit output heat power from waste heat boiler, kg/J.

3. Environmental value cost of gas boiler.

$$C_{GB}(t) = P^h_{GB}(t) \cdot T \cdot \sum_{i=1}^{5} J^i_{GB} \cdot K^i \tag{18}$$

$C_{GB}(t)$—environmental value cost of gas boiler during time period t, USD.
J^i_{GB}—emissions of various pollutants per unit output heat power from gas boiler, kg/J.

3. Double-Layer Optimal Scheduling Model Based on Non-Cooperative Game

Double-layer optimization is a system optimization problem with a double-layer hierarchical structure. The upper and lower layers have their own objective functions and constraints. The upper layer decision will generally affect the lower layer objective and constraints, while the lower layer will provide feedback to the decision of the upper layer, thus realizing the interaction between the decisions of upper and lower layers [19].

This paper considers a double-layer optimal scheduling model as shown in Figure 2. The upper layer aims at minimizing the daily cost of the park. The output strategy b(t) such as gas purchasing, electricity purchasing, and the output power of the equipment of the park is obtained. The lower layer aims at maximizing the energy efficiency of the CHP system in the park, and the thermoelectric ratio regulation strategy of the CHP system is obtained. The upper layer inputs the park output strategy to the lower layer, and the lower layer adjusts the thermoelectric ratio of the CHP system to optimize the energy efficiency of the CHP system, the upper layer re-optimizes according to the thermoelectric ratio adjustment strategy of the lower layer, and the upper and lower layers iterates repeatedly until the result is optimal.

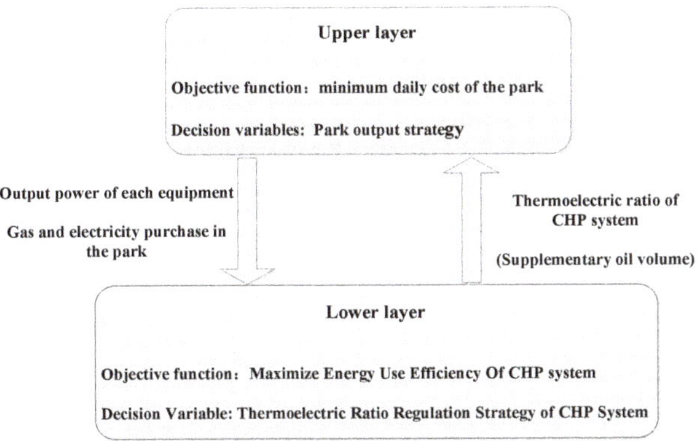

Figure 2. Double-layer optimal scheduling model.

3.1. Upper Layer Model

3.1.1. Objective Function

The upper-layer model takes the minimum daily cost of the park as the optimization objective, and the decision variables are as follows.

$$s_n(t) = \left[P_{\text{GB}}^{\text{gas}}(t), P_{\text{CHP}}^{\text{gas}}(t), P_{\text{c}}^{\text{e}}(t), P_{\text{h}}^{\text{e}}(t), P_{\text{c}}^{\text{h}}(t), P_{\text{grid}}^{\text{e}}(t), P_{\text{char}}^{\text{e}}(t), P_{\text{dis}}^{\text{e}}(t), V_{CHP}(t) \right] \tag{19}$$

The objective function of the upper layer model is

$$C_{\text{sum}} = \min \sum_{t=1}^{24} \left(C_{op}(t) + C_{en}(t) \right) \tag{20}$$

$$C_{op}(t) = C_{\text{gas}}(t) + C_{\text{grid}}(t) + C_{\text{oil}}(t) + C_{\text{CCHP}}(t) + C_{\text{sto}}(t) \tag{21}$$

$$C_{en}(t) = C_{\text{GT}}(t) + C_{\text{WHB}}(t) + C_{\text{GB}}(t) \tag{22}$$

C_{sum}—daily cost of park, USD.

$C_{op}(t)$—operation cost of park during time period t, USD.

$C_{en}(t)$—environmental value cost of park during time period t, USD.

3.1.2. Constraint Condition

1. Energy Conservation Constraints

(1) Electric load

$$P_{\text{grid}}^{\text{e}}(t) + P_{\text{CHP}}^{\text{e}}(t) + \eta_{\text{sto}} \cdot P_{\text{dis}}^{\text{e}}(t) + P_{\text{PV}}^{\text{e}}(t) = L_{\text{e}}(t) + \eta_{\text{sto}} \cdot P_{\text{char}}^{\text{e}}(t) + P_{\text{c}}^{\text{e}}(t) + P_{\text{h}}^{\text{e}}(t) \tag{23}$$

$P_{\text{PV}}^{\text{e}}(t)$—output power of photovoltaic power generation in the park during time period t, W.
$L_{\text{e}}(t)$—electric load of the park during time period t, W.

(2) Heat load

$$P_{\text{CHP}}^{\text{h}}(t) + P_{\text{GB}}^{\text{h}}(t) + P_{\text{e}}^{\text{h}}(t) = L_{\text{h}}(t) + P_{\text{c}}^{\text{h}}(t) \tag{24}$$

$L_{\text{h}}(t)$—heat load of the park during time period t, W.

(3) Cold load

$$P_{\text{h}}^{\text{c}}(t) + P_{\text{e}}^{\text{c}}(t) = L_{\text{c}}(t) \tag{25}$$

$L_{\text{c}}(t)$—Cold load of the park during time period t, W.

2. Equipment output constraints

$$0 < P_k < P_k^{\text{max}} \tag{26}$$

K—1, 2, ... , 5, represents CHP system, gas boiler, electric refrigeration unit, heat pump unit, absorption refrigeration unit. P_k—output power of various kinds of equipment, W.
P_k^{max}—maximum output power of various kinds of equipment, W.

Some equipment such as gas turbine have the problem of low efficiency at low power levels, but the efficiency can be improved by preheating gas or combustion-supporting air when the equipment is at a low power level. So the efficiency can be approximately regarded as a fixed value when power constraints are valid from 0.

3. Energy storage constraints

$$\begin{cases} 0 < P_{\text{char}}^{\text{e}}(t) < P_{\text{char}}^{\text{max}} \\ 0 < P_{\text{dis}}^{\text{e}}(t) < P_{\text{dis}}^{\text{max}} \\ P_{\text{char}}^{\text{e}}(t) \cdot P_{\text{dis}}^{\text{e}}(t) = 0 \\ \sum\limits_{t=1}^{24} \left(P_{\text{char}}^{\text{e}}(t) - P_{\text{dis}}^{\text{e}}(t) \right) \geq 0 \end{cases} \tag{27}$$

$P_{\text{char}}^{\text{max}}$—maximum charging power of electric energy storage equipment, W.
$P_{\text{dis}}^{\text{max}}$—maximum discharging power of electric energy storage equipment, W.

3.2. Lower Layer Model

3.2.1. Objective Function

The lower layer model takes the highest energy efficiency of CHP system (the ratio of output energy to input energy of CHP system) as the optimization objective, and the decision variable is V_{CHP} of CHP system, and the objective function of the lower layer model is

$$\eta_{\text{p}} = \max \frac{\sum\limits_{t=1}^{24} \left(P_{\text{CHP}}^{\text{e}}(t) + P_{\text{CHP}}^{\text{h}}(t) + P_{\text{GB}}^{\text{h}}(t) \right)}{\sum\limits_{t=1}^{24} \left(P_{\text{oil}}(t) + P_{\text{gas}}(t) \right)} \tag{28}$$

η_{p}—energy efficiency of CHP system.

$P_{gas}(t)$—the total energy of natural gas used in the park during time period t, W.

3.2.2. Constraint Condition

The CHP system in the park adopts the waste heat boiler with internal supplementary combustion device; its supplementary combustion quantity is less, thus the supplementary combustion rate constraint should be satisfied when oil is added to adjust the thermoelectric ratio [20]. The constraint is as follows.

$$0 < A(t) < A_{max} \tag{29}$$

A_{max}—maximum supplementary combustion rate of internal supplementary combustion device, 1(%).

3.3. Non-Cooperative Game

Nowadays integrated energy parks are often partitioned internally according to certain rules, such as regional function administrative level, population, economic scale, and load density. When there are multiple zones in the integrated energy park, these zones act as decision-makers, and each of them formulates the most suitable operation strategy for itself. Because the unit price of gas and electricity is related to the total purchases of gas and electricity of the park, when the operation strategy of one zone changes, the other zones will also change accordingly. Therefore, the daily cost of each zone is not only related to its own operation strategy, but also affected by the operation strategy of the other zones, thus all the zones constitutes a non-cooperative game.

It is assumed that there are N zones in the integrated energy park, and the operation strategy of each zone is as follows:

$$s_n(t) = \left[P_{GB}^{gas}(t), P_{CHP}^{gas}(t), P_c^e(t), P_h^e(t), P_c^h(t), P_{grid}^e(t), P_{char}^e(t), P_{dis}^e(t), V_{CHP}(t) \right] \tag{30}$$

n represents the nth zone and $s_n(t)$ is operation strategy of the nth zone. The daily cost of each zone is $C_n(s_n(t), S^{-n}(t))$, $S^{-n}(t)$ represents the operation strategies of $N - 1$ zones except zone n.

All zones participating in the game expect to reach the minimum of their own costs. In the process of the game, each zone will reach an equilibrium point, now for each zone, its own operation strategy is the best, and the cost will increase if the strategy is changed. So, each zone has no motive to change its own operation strategy currently. This equilibrium state is Nash equilibrium. This moment $s_n^*(t)$ of each zone is Nash equilibrium solution, and for each zone, there are operation strategy [21].

$$C_n(s_n^*(t), S^{-n*}(t)) \le C_n(s_n(t), S^{-n*}(t)) \tag{31}$$

4. Solution of the Model

The solving process of double-layer optimal scheduling model based on non-cooperative game is as follows:

1. Set the initial value of the operation strategies of N zones ($s_n(t)$) and set the precision (ε).
2. For the nth zone, the operation strategies of the remaining $N - 1$ zones ($S^{-n}(t)$) are regarded as fixed values, the optimal solution of output strategy ($b_n(t)$) and daily cost of the zone (c_n) are obtained by calculating formula (20).
3. The optimal solution of the output strategy is input to the lower layer, the optimal solution of the thermoelectric ratio regulation strategy ($V_{CHP}^{n*}(t)$) is obtained by calculating formula (28).
4. The upper objective function is recalculated according to $V_{CHP}^{n*}(t)$, and the new optimal solution of the output strategy ($b_n*(t)$) and the daily cost of the zone (c_n*) are obtained.
5. When $|c_n^* - c_n| < \varepsilon$, the currently operation strategy $s_n^* = [b_n^*(t), V_{CHP}^{n*}(t)]$ is the optimal strategy for the zone, otherwise repeat step 3.
6. Repeat steps 2–5 to find the optimal operation strategies of the remaining $N - 1$ zones.

7. Repeat steps 2–6 until the optimal operation strategies of N zones do not change, and then the operation strategy of each zone under Nash equilibrium solution is obtained.

The solution flow chart is shown in Figure 3.

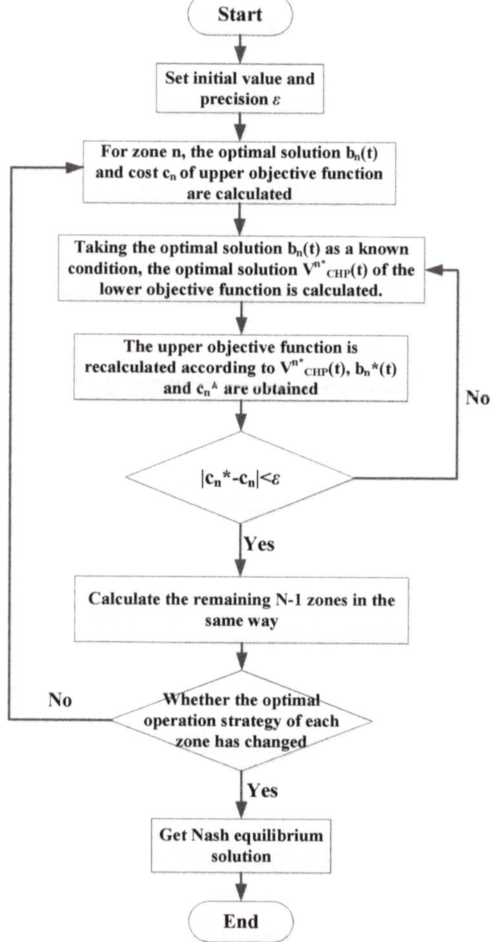

Figure 3. Solution flow chart.

5. Analysis of Examples

5.1. Basic Data and Assumptions

An integrated energy park is taken as an example, the park can be divided into four zones according to the load density. The park schematic diagram is shown in Figure 4.

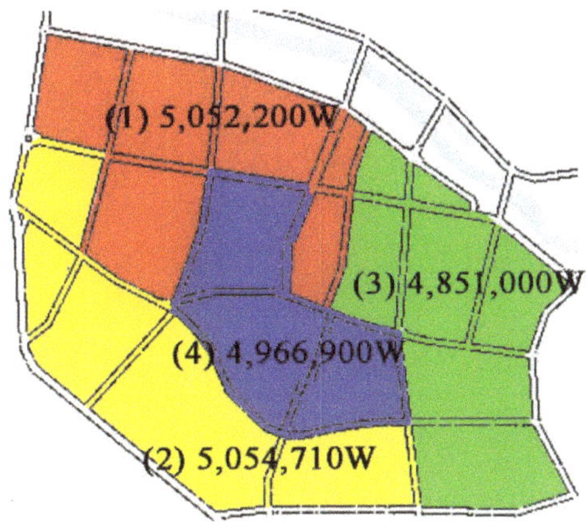

Figure 4. Schematic diagram of an integrated energy park.

The basic data of the example are as follows: $A_{gas} = 0.377$ USD/(m^3)^2, $B_{gas} = 4.35 \times 10^{-5}$ USD/m^3, $\lambda_{gas} = 4.96 \times 10^7$ J/kg, $\rho_{gas} = 0.7174$ kg/m^3. Peak-valley mechanism is adopted for electricity price parameters: valley (1 a.m.~6 a.m., 11 p.m.~12 p.m.), $a(t) = 1.57 \times 10^{-19}$ USD/(J)^2, $b(t) = 1.5 \times 10^{-8}$ USD/J; normal (7 a.m.~11 a.m., 2 p.m.~7 p.m.), $a(t) = 2.35 \times 10^{-19}$ USD/(J)^2, $b(t) = 2.94 \times 10^{-8}$ USD/J; peak (12 a.m.~1 p.m., 8 p.m.~10 p.m.), $a(t) = 3.13 \times 10^{-19}$ USD/(J)^2, $b(t) = 4.72 \times 10^{-8}$ USD/J, $c_{oil} = 0.355$ USD/kg, $\lambda_{oil} = 4.4 \times 10^7$ J/kg. The parameters of equipment efficiency and operation cost are shown in Table 1, the emission intensity and the environmental cost of various pollutants from gas turbines, waste heat boilers, and gas boilers are shown in Table 2 [22–26].

Table 1. Efficiency and operation cost of equipment in the park.

Efficiency	Numerical Value(%)	Operation Cost	Numerical Value (USD/J)
η^e_{CHP}	35	CHP system c_1	1.37×10^{-9}
η_{GB}	90	Gas boiler c_2	8.7×10^{-11}
η^e_c	400	Electric refrigeration unit c_3	3.91×10^{-10}
η^e_h	450	Heat pump unit c_4	1.05×10^{-9}
η^h_c	70	Absorption refrigeration unit c_5	3.22×10^{-10}
η_{sto}	95	Energy storage c_{sto}	7.25×10^{-11}

Note: In order to facilitate the calculation, the average efficiency of common types of equipment are chosen as the calculation data by consulting the relevant literatures.

Table 2. Pollutant emission intensity and environmental cost of pollutants in the park.

Pollutant Species	SO$_2$	NO$_x$	CO	CO$_2$	TSP
Gas turbine (kg/J)	6.4×10^{-13}	3.4×10^{-10}	0	1.1×10^{-7}	1.3×10^{-11}
Waste heat boiler (kg/J)	2.4×10^{-9}	1.1×10^{-9}	3.4×10^{-11}	2.3×10^{-7}	5.3×10^{-11}
Gas boiler (kg/J)	2.6×10^{-13}	1.6×10^{-10}	0	4.6×10^{-8}	6.0×10^{-12}
Environmental cost (USD/kg)	0.87	1.16	0.145	0.0033	0.319

The electric, heat, cold load, and photovoltaic power generation of the park in a day are shown in Figure 5.

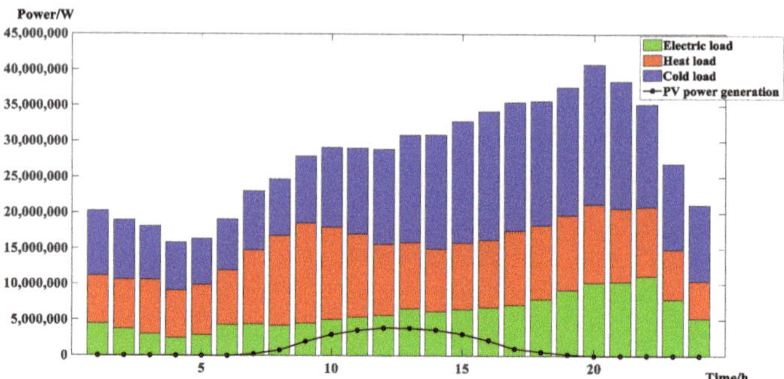

Figure 5. Load and photovoltaic power generation of the park.

5.2. Results and Analysis of Examples

According to the proposed model and solution method, the Nash equilibrium solutions of four zones are solved. An example of a zone is presented; Figure 6 is the output optimization results of electric load, heat load, and cold load of the zone. The positive value in the figure represents the input energy, such as electricity purchasing, power generation, electric energy storage discharging, boiler heat generation, etc. The negative value represents the output energy, such as electricity, heat, cold load of the zone, electric energy storage charging, etc.

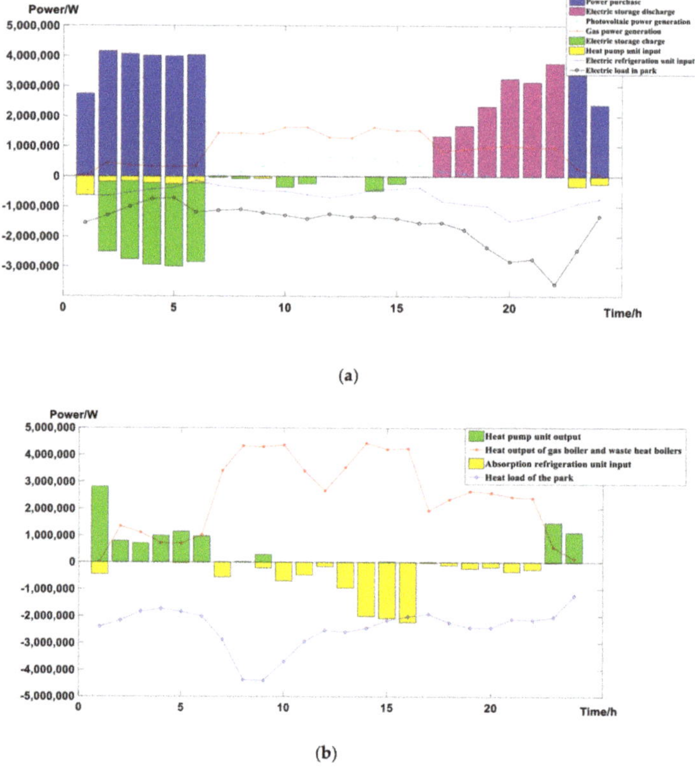

(a)

(b)

Figure 6. *Cont.*

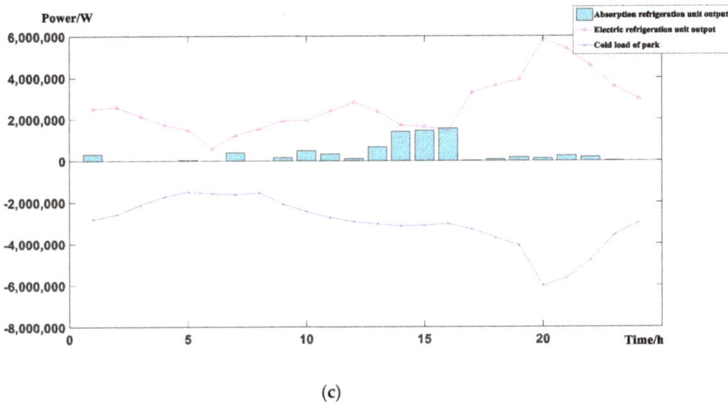

(c)

Figure 6. Optimal scheduling result of a zone. (**a**) Optimal scheduling result of electric load. (**b**) Optimal scheduling result of heat load. (**c**) Optimal scheduling result of cold load.

As can be seen from Figure 6, during late night and before dawn, the zone reduces the use of natural gas, and purchases a large amount of electricity from the grid. In addition to meeting the electric load of the zone, it also provides heat and cold loads for the zone through heat pump unit and electric refrigeration unit. In addition, excess electricity is stored by storage devices and then provides power to the zone during electricity peak. During the period from sunrise to sunset, the zone seldom purchases electricity from the grid. The energy needed is mainly from gas turbine power generation and photovoltaic power generation. The heat load of the zone almost entirely comes from the gas boiler and the waste heat boiler. As a result of the relatively large amount of heat generated at this time, the output of absorption refrigerant unit is also increased. During the period from sunset to night, the demand for electricity load increases sharply, and photovoltaic power generation almost no longer provides electricity; electricity generated by the gas turbines and stored by the electric energy storage equipment during electricity valley can provide the electric load demand of the zone.

The comparison of gas purchase and electricity purchase before and after the game is shown in Figure 7. It can be seen that the sum of gas purchase of each zone after the game is less than the gas purchase of the whole park before the game. With respect to the electricity purchase, the electricity purchase of each zone after the game are transferred from peak time and normal time to valley time, which can play the role of peak shaving and valley filling.

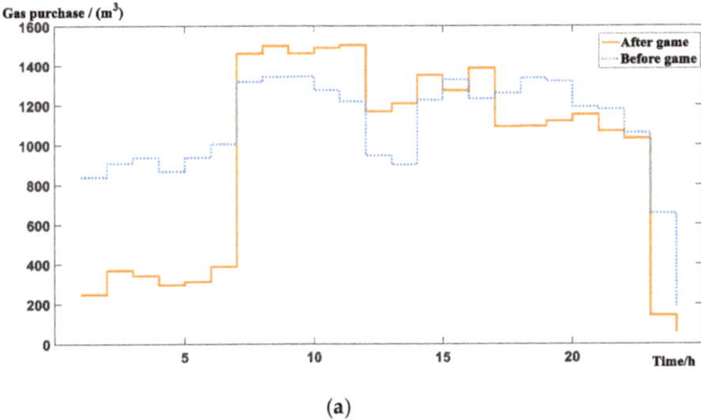

(a)

Figure 7. *Cont.*

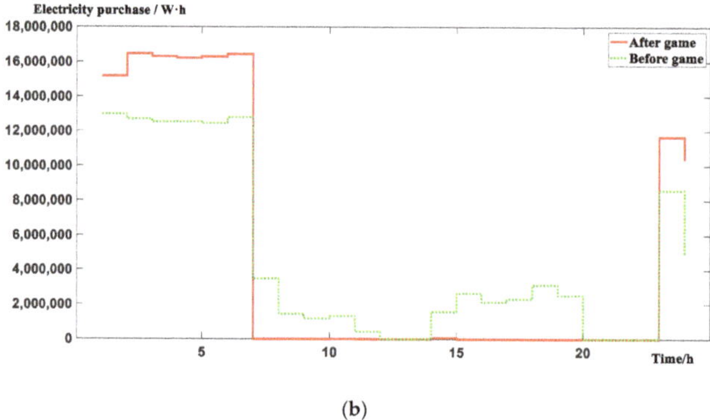

(b)

Figure 7. Comparison of gas purchase and electricity purchase of the park before and after the game. (**a**) Comparison of gas purchase. (**b**) Comparison of electricity purchase.

According to the proposed model and solution method, the daily total cost and the energy efficiency of CHP system of each zone are calculated respectively. The results are compared with those of the whole integrated energy park before the game, as shown in Table 3. The summary cost of the zones is the total cost of the four zones costs, the summary efficiency of the zones is the average efficiency of the four zones.

Table 3. Comparison of cost and energy efficiency of combined heat and power (CHP) system in the park before and after the game.

Object \ Parameter	Total Cost Csum (USD)	Efficiency (%)
Zone 1	4795.4	74.14
Zone 2	4281.8	67.49
Zone 3	4210.8	70.79
Zone 4	4648.8	64.94
Summary of zones	17,936.8	69.34
The whole park before game	19,678.82	67.33

Table 3 shows that after the non-cooperative game, the total cost of each zone is approximately 1742 USD less than the cost of the whole integrated energy park before the game, and the average energy efficiency of CHP system in various zones is about 2% higher than that in the whole park before the game.

6. Conclusions

To take into account the interests of various zones in the optimal scheduling of integrated energy parks, a double-layer optimal scheduling model of integrated energy parks based on non-cooperative game is proposed, which enables each zone to reach a balance through mutual game and get Nash equilibrium solution, thus making the most appropriate operation strategy. The following conclusions are drawn from the analysis of examples:

(1) The model proposed in this paper can make the zones adjust their operation strategies more reasonably. After the game, the total cost of all the zones is less than the cost of the whole energy park before the game, and the average energy efficiency of CHP system is improved.

(2) After the game, the amount of natural gas used by each zone has been reduced, which meets the needs of energy saving and emission reduction in the current age. At the same time, the purchase of electricity from the power grid is concentrated in the electricity valley, which has a certain role in peak shaving and valley filling.

Author Contributions: The author F.C. carried out the main research tasks and wrote the full manuscript. Y.G. proposed the original idea, provided financial support, analyzed and double-checked the results and the whole manuscript. Y.Y. contributed to data processing and summarizing the proposed ideas. H.L. provided technical and financial support throughout, while Y.C. modified the English language.

Funding: The National Nature Science Foundation of China (51607068); The Fundamental Research Funds for the Central Universities (2018MS082); The Fundamental Research Funds for the Central Universities (2017MS090).

Acknowledgments: This work is supported in part by the National Nature Science Foundation of China (51607068), the Fundamental Research Funds for the Central Universities (2018MS082) and the Fundamental Research Funds for the Central Universities (2017MS090).

Conflicts of Interest: The authors declare no conflict of interest.

Abbreviations

The following abbreviations are used in this manuscript

IES	integrated energy system
P2G	power to gas
CCHP	combined cool, heat and power
CHP	combined heat and power

References

1. Qureshi, M.I.; Rasli, A.M.; Zaman, K. Energy crisis, greenhouse gas emissions and sectoral growth reforms: Repairing the fabricated mosaic. *J. Clean. Prod.* **2016**, *112*, 3657–3666. [CrossRef]
2. Xiang, L. Energy network dispatch optimization under emergency of local energy shortage with web tool for automatic large group decision-making. *Energy* **2017**, *120*, 740–750. [CrossRef]
3. Yang, Z.; Peng, S.; Liao, Q. Non-cooperative Trading Method for Three Market Entities in Integrated Community Energy System. *Autom. Electr. Power Syst.* **2018**, *42*, 32–39.
4. Voropai, N.; Stennikov, V.; Senderov, S.; Barakhtenko, E.; Voitov, O.; Ustinov, A. Modeling of Integrated Energy Supply Systems: Main Principles, Model, and Applications. *J. Energy Eng.* **2017**, *143*, 04017011. [CrossRef]
5. Tang, B.; Gao, G.; Xia, X.; Yang, X.; Bo, T.; Gangfeng, G.; Xiangwu, X.; Xiu, Y. Integrated Energy System Configuration Optimization for Multi-Zone Heat-Supply Network Interaction. *Energies* **2018**, *11*, 3052.
6. Zeng, M. Key Support Technologies for the Future Development of Distributed Energy under the Background of Energy Internet. *Electr. Age* **2018**, *1*, 36–37.
7. Chen, F.; Gao, Y.; Liang, H.; Duan, J. Energy Consumption Characteristic Evaluation Model and Energy Supply Partition Method of Integrated Energy Park Based on Geographical Partition. *Electr. Power Constr.* **2019**, *6*, 23–32.
8. Patwal, R.S.; Narang, N. Optimal Economic Emission Scheduling of Integrated Energy Systems Using Heuristic Optimization Technique. In Proceedings of the 2018 2nd International Conference on Power, Energy and Environment: Towards Smart Technology (ICEPE), Shillong, India, 1–2 June 2018.
9. Brożyna, J.; Mentel, G.; Szetela, B.; Strielkowski, W. Multi-Seasonality in the TBATS Model Using Demand for Electric Energy as a Case Study. *Econ. Comput. Econ. Cybern. Stud. Res.* **2018**, *52*, 229–246. [CrossRef]
10. Yang, D.; Xi, Y.; Cai, G. Day-Ahead Dispatch Model of Electro-Thermal Integrated Energy System with Power to Gas Function. *Appl. Sci.* **2017**, *7*, 1326. [CrossRef]
11. Ghasemi, A.; Banejad, M.; Rahimiyan, M. Integrated energy scheduling under uncertainty in a micro energy grid. *IET Gener. Transm. Distrib.* **2018**, *12*, 2887–2896. [CrossRef]
12. Khani, H.; Farag, H.E.Z.; Farag, H.E.Z. Optimal Day-Ahead Scheduling of Power-to-Gas Energy Storage and Gas Load Management in Wholesale Electricity and Gas Markets. *IEEE Trans. Sustain. Energy* **2018**, *9*, 940–951. [CrossRef]
13. Zhang, L. *Analysis and Forecast of Market Clearing Price Law*; North China Electric Power University: Beijing, China, 2007.
14. Cao, J. *Current Electricity Market and Generator Bidding*; Xiamen University: Xiamen, China, 2001.
15. Liu, Y. Improving the natural gas pricing mechanism for residential consumers in China. *Nat. Gas Ind.* **2018**, *38*, 130–141.

16. Zhao, S.; Ying, G. On Natural Gas price mechanism given present context of China's natural gas industry. In Proceedings of the 2008 International Conference on Management Science and Engineering 15th Annual Conference Proceedings, Long Beach, CA, USA, 10–12 September 2008; pp. 1873–1879.
17. Yu, C.; Wang, J.; He, N. A Study of the Supplementary-firing Burner Unit for a Heat Recovery Steam Generator. *J. Eng. Therm. Energy Power* **2004**, *19*, 534–536.
18. Li, P.; Zang, X.; Liu, Y.; Liu, L. Rational determination to heat and power ratio for distributed CHP Plant. *Gas Turbine Technol.* **2005**, *4*, 43–46.
19. Colson, B.; Marcotte, P.; Savard, G. An overview of bilevel optimization. *Ann. Oper. Res.* **2007**, *153*, 235–256. [CrossRef]
20. Shi, J.; Xu, J.; Zeng, B.; Zhang, J. A Bi-Level Optimal Operation for Energy Hub Based on Regulating Heat-to-Electric Ratio Mode. *Power Syst. Technol.* **2016**, *40*, 2959–2966.
21. Fu, S.; Su, Z. Dynamic Power Strategy Space for Non-Cooperative Power Game with Pricing. In Proceedings of the 2017 IEEE 86th Vehicular Technology Conference (VTC-Fall), Toronto, ON, Canada, 24–27 September 2017; pp. 1–6.
22. Wu, F.; Liu, X.; Sun, Y.; Chen, N.; Yuan, T.; Gao, B. Multi-Park Game Optimizing Strategy Based on CCHP. *Autom. Electr. Power Syst.* **2018**, *42*, 68–75.
23. Yang, Z.; Zhang, F.; Liang, X.; Han, X.; Xu, Z. Economic Operation of Microgrid with Heat Pump and Energy Storage. *Power Syst. Technol.* **2018**, *42*, 1735–1743.
24. Wang, H. *Study on Optimum Configuration of Natural Gas Cogeneration System Based on Gas Turbine*; Hunan University: Changsha, China, 2015.
25. Mao, M.; Jin, P.; Zhang, L.; Ding, Y.; Xu, H. Optimization of Operation Strategies and Economic Analysis of PV Microgrid for Industries. *Trans. China Electrotech. Soc.* **2014**, *29*, 35–45.
26. Bao, N.; Zhu, R.; Ni, W. Cost of electricity analysis of the hybrid power system combining wind farm with gas turbine. *Gas Turbine Technol.* **2006**, *4*, 1–5.

Article

Optimal Low-Carbon Economic Environmental Dispatch of Hybrid Electricity-Natural Gas Energy Systems Considering P2G

Jing Liu [1,*], Wei Sun [2] and Gareth P. Harrison [2]

[1] School of Mechanical Electronic and Information Engineering, China University of Mining & Technology (Beijing), Beijing 100083, China
[2] School of Engineering, University of Edinburgh, Edinburgh EH9 3DW, UK; W.Sun@ed.ac.uk (W.S.); Gareth.Harrison@ed.ac.uk (G.P.H.)
* Correspondence: jingqisandra@163.com; Tel.: +86-010-6233-1180

Received: 8 March 2019; Accepted: 5 April 2019; Published: 9 April 2019

Abstract: Power to gas facilities (P2G) could absorb excess renewable energy that would otherwise be curtailed due to electricity network constraints by converting it to methane (synthetic natural gas). The produced synthetic natural gas can power gas turbines and realize bidirectional energy flow between power and natural-gas systems. P2G, therefore, has significant potential for unlocking inherent flexibility in the integrated system, but also poses new challenges of increased system complexity. A coordinated operation strategy that manages power and natural-gas network constraints together is essential to address such challenges. In this paper, a novel low-carbon economic environmental dispatch strategy is presented considering all the constraints in both systems. The multi-objective black-hole particle swarm optimization algorithm (MOBHPSO) is adopted. In addition to P2G, a gas demand management strategy is proposed to support gas flow balance. A new solving approach that combines the effective redundancy method, trust region method, and Levenberg-Marquardt method is proposed to address the complex coupled constraints. Case studies that use an integrated IEEE 39-bus power and Belgian high-calorific 20-node gas system demonstrate the effectiveness and scalability of the proposed model and optimization method. The analysis of dispatch results illustrates the benefit of P2G for the wind power accommodation, and low-carbon, economic, and environmental improvement of integrated system operation.

Keywords: hybrid electricity-natural gas energy systems; power to gas (P2G); low-carbon; economic environmental dispatch; trust region method; Levenberg-Marquardt method

1. Introduction

With further acceleration of the low-carbon energy process, as well as the energy crisis, environmental pollution, and other issues, the capacity of renewable energy sources has increased continuously. Due to the intermittency and uncertainty of wind power as well as the lack of peak load regulation of power system, it is likely that more and more wind power generation will have to be curtailed in order to maintain the power system reliability [1]. To solve this problem, much research is carried out to explore practical means to reduce the curtailment of wind power generation. The growing interdependence of the power system and natural-gas system and the development of power to gas technologies [2–8] creates operational interactions between the power system and natural-gas system, which could obtain additional benefits for both systems, including reducing the curtailed wind power generation. On the one hand, the power system tends to require more flexible power energy from the natural-gas system to shift peak load compared with the gas-fired units [3], which is conductive to the accommodation of wind power. On the other hand, the natural-gas system absorbs methane or hydrogen produced by P2G to guarantee the continuity of gas supply and the wind power energy will be stored and transported in the existing natural-gas system for generating low-carbon electricity or heat later [9–11], which uses the curtailed wind power directly. Therefore, the integrated electricity-natural gas energy systems with P2G have become one of the effective forms to reduce the curtailment of wind power generation.

The diagram of integrated electricity-natural gas energy systems with P2G is shown in Figure 1. It can be seen that the power system and the natural-gas system exchange the energy between P2G and gas-fired units. When the curtailed wind power is converted to hydrogen or methane through power to hydrogen facilities (P2H) or power to methane facilities (P2M), P2G which includes P2H and P2M is the load of power system and the gas source of natural-gas system. Meanwhile, the gas-fired units are the load of natural-gas system and the generators of power system. Obviously, operation parameters of P2G, power system and natural-gas system are interrelated and interactive which can affect the operation cost, CO_2 emissions, reliability, and stability of both systems. Therefore, how to deal with the interactive relationship between power system and natural-gas system and how to achieve coordinated optimal operation with economic environmental benefits are the key issues for the integrated power system and natural-gas system.

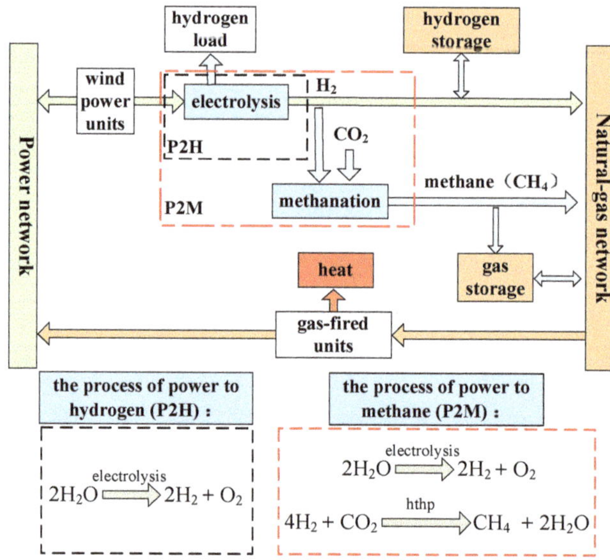

Figure 1. Diagram of integrated electricity and natural-gas energy systems with power to gas (P2G).

For the integrated electricity-natural gas energy systems, the initial research is focused on optimal power flow [12–15], unit commitment [16], optimal dispatch [17–19], steady-state analysis [20], and system planning [21]. For the calculation of optimal power flow, the total operation cost is usually considered as optimal objective and the dual interior point method [12], the Monte Carlo method [13] and the point estimation method [14] are adopted frequently. Some studies introduce an energy hub to deal with the translation of different energies in the hybrid electricity-natural gas energy systems [13,17]. For the optimization of system operation, the operation of power system and the operation of natural-gas system are mostly optimized separately using the deterministic optimization methods or stochastic optimization methods [18]. For the steady-state analysis of the hybrid electricity-natural gas energy systems, based on the steady-state analysis of power system, the analysis model of natural-gas system is realized by analogy analysis between power system and natural-gas system, and then the comprehensive steady-state analysis model of hybrid electricity-natural gas energy systems is given [20]. For the optimal system planning, a chance constrained programing approach is presented to minimize the investment cost of the integrated energy systems [21]. In these studies, P2G is not considered. As the coupling operation link of the power system and natural-gas system, P2G plays a more and more important role in wind power accommodation with broad prospects and potential for energy development [22–24]. Therefore, it is necessary to carry out the research on optimal operation of integrated electricity-natural gas energy systems considering P2G. The early studies on P2G are mainly focused on technology implementation and security application [6,25–28]. Recently, although some achievements about optimal operation of integrated electricity-natural gas energy systems considering P2G have been achieved [6–8,24,29–38], it still seems to be in the exploratory stage from the following aspects.

(1) Optimal objectives: The minimum total operation cost is mostly adopted [6,24,29–32,37]. In only a few studies, the maximum wind power accommodation [33], the minimum energy purchase cost [34], or net power demand smoothness [38] is also considered as the objective. However, environmental benefit is rarely considered. As we know, the low-carbon and emission reduction requirements become more and more important. Therefore, it is necessary to take environmental benefit into consideration.

(2) Optimal models: The operation model of power system and operation model of natural-gas system are mainly established separately based on the two-level optimal power flow structure [6,30–32]. It seems that rare consideration is given to coordinated optimization between the two energy systems.

(3) Optimal algorithms: Generally, the traditional algorithms are adopted in most studies, such as the mix-integer linear programming method [3,24], mixed-integer quadratic programming method [37], and interior point method [35]. However, the intelligent optimization algorithms with high global search ability and fast convergence speed are rarely used.

(4) Constraints handling methods: The constraints handling methods affect the operation results directly. Few articles give full details about the constraints handling methods, especially for the complicated dynamic nodal balance constraint and volume limits of gas storage in the natural-gas system.

On the above premises, this paper establishes the optimal operation model of the hybrid electricity-natural gas energy systems considering operation cost, natural-gas cost reduction due to P2G, CO_2 emissions, and SO_x emissions to achieve low-carbon, economic, and environmental benefits. The multi-objective black-hole particle swarm optimization algorithm (MOBHPSO) [39–42] is adopted. The power flow is calculated using the Newton-Raphson method. The non-linear gas flow equations are solved by the trust region method [43,44] and Levenberg-Marquardt (L-M) method [45,46], respectively. The gas demand management strategy is proposed to balance the gas flow. Moreover, the detailed handling methods of inequality constraints in natural-gas system are also given in this paper. Several case studies are carried out on a hybrid IEEE 39-bus power system and Belgian high-calorific 20-node gas system in a period of 24 h to investigate the low-carbon, economic and environmental benefits of P2G in terms of cost reduction (6.165×10^5), rate decline of wind curtailment (from 24.85% to 4.04%), CO_2 emissions reduction (3630 tons), and SO_x emissions reduction (0.254 ton).

2. Problem Formulation

The optimal low-carbon economic environmental dispatch problem of hybrid electricity-natural gas energy systems with P2G is a complicated non-convex, coupled, non-linear, multi-objective, and multi-constraint optimization problem. It contains three parts: The first one is the optimization of power system; the second one is the optimization of natural-gas system; and the last one is the coordination of the hybrid electricity-natural gas energy systems. The flow chart of this optimization problem is shown in Figure 2. Each part of the flow chart will be described in detail.

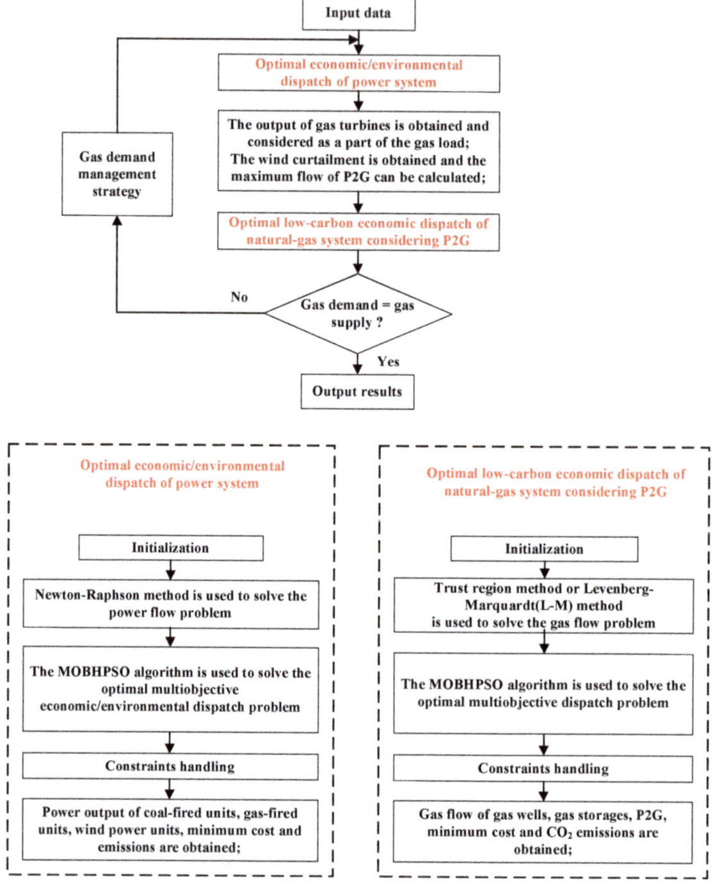

Figure 2. Flow chart.

2.1. Optimal Economic Environmental Dispatch of Power System

2.1.1. Objectives

(1)   Minimum Fuel Cost of the Power System

$$\text{Min} \quad F_p = \sum_{i=1}^{N_G} \sum_{t=1}^{T} a_i P_{Gi}(t)^2 + b_i P_{Gi}(t) + c_i \tag{1}$$

(2)   Minimum SO_x Pollutant Emissions of the Power System

$$\text{Min} \quad E_{SO_x} = \sum_{i=1}^{N_G} \sum_{t=1}^{T} \left(\alpha_i + \beta_i P_{Gi}(t) + \gamma_i P_{Gi}(t)^2 + \delta_i e^{\lambda_i P_{Gi}(t)} \right) \tag{2}$$

(3)   Minimum Load Loss Rate of the Power System

$$\text{Min} \quad L_p = \frac{\sum\limits_{t=1}^{T} \left[P_L(t) + \sum\limits_{k=1}^{N_{P2G}} P_{P2G,k}(t) - \sum\limits_{i=1}^{N_G} P_{Gi}(t) \right]}{\sum\limits_{t=1}^{T} P_L(t)} \tag{3}$$

where F_p is the fuel cost of power system; N_G is the number of power generations; T is the number of time periods; $P_{Gi}(t)$ is the power generation output at time t; a_i, b_i, c_i are coefficient of the fuel cost;

E_{SOx} is the pollutant emission of SO_x; α_i, β_i, γ_i, δ_i, λ_i are coefficient of the pollutant emission; L_p is the load loss rate presenting the reliability of power supply; N_{P2G} is the number of P2G; $P_L(t)$ is the power load at time t; $P_{P2G}(t)$ is the power supplied to the P2G facilities at time t.

The power output of gas-fired units is calculated by the product of the gas flow injected to the gas-fired units $Q_{GT}(t)$, higher heating value of natural gas HHV_g and the energy conversion efficiency $\eta_{GT}(t)$. In this paper, the last objective is converted into a constraint by being less than a given value ε.

2.1.2. Constraints

(1) Power Output Limits

$$P_{Gi}^{\min} \leq P_{Gi}(t) \leq P_{Gi}^{\max} \tag{4}$$

where $P_{Gi}^{\min}$ and $P_{Gi}^{\max}$ represent the minimum power output and maximum power output of unit i, respectively.

(2) Ramp Rate Limits

$$\begin{cases} P_{Gi}(t) \geq \max\left\{ P_{Gi}^{\min}, P_{Gi}(t-1) - \Delta P_{Gi}^{down} \right\}, & P_{Gi}(t) \leq P_{Gi}(t-1) \\ P_{Gi}(t) \leq \min\left\{ P_{Gi}^{\max}, P_{Gi}(t-1) + \Delta P_{Gi}^{up} \right\}, & P_{Gi}(t) \geq P_{Gi}(t-1) \end{cases} \tag{5}$$

where ΔP_{Gi}^{up} and ΔP_{Gi}^{down} represent the ramp up rate and the ramp down rate of unit i, respectively.

(3) Line Capacity Limit

$$S_l(t) \leq S_l^{\max} \tag{6}$$

where $S_l^{\max}$ is the maximum capacity of line l.

2.2. Optimal Low-Carbon Economic Dispatch of Natural-Gas System Considering P2G

2.2.1. Objectives

(1) Minimum the Operational Cost of Natural-Gas System

$$\text{Min} \quad C_{well} + C_{gs} + C_{P2G} - S_{P2G} \tag{7}$$

$$C_{well} = \sum_{n=1}^{N_w} \sum_{t=1}^{T} Q_{wn}(t) u_{wn}(t) \tag{8}$$

$$C_{gs} = \sum_{m=1}^{N_{gs}} \sum_{t=1}^{T} Q_{gs,m}(t) u_{gs,m}(t) \tag{9}$$

$$C_{P2G} = \sum_{k=1}^{N_{P2G}} \sum_{t=1}^{T} P_{P2G,k}(t) u_{P2G,k} \tag{10}$$

$$S_{P2G} = \sum_{k=1}^{N_{P2G}} \sum_{t=1}^{T} Q_{P2G,k}(t) u_{ave}(t) \tag{11}$$

where C_{well}, C_{gs}, and C_{P2G} represent the operation cost of gas wells, the operation cost of gas storage, and the operation cost of P2G, respectively. S_{P2G} is the saved natural-gas cost due to the P2G. N_w, N_{gs} represent the number of gas wells and the number of gas storage, respectively; $Q_{wn}(t)$ is the gas flow of gas well n; $u_{wn}(t)$ is the gas price of gas well n at time t; $Q_{gs,m}(t)$ is the gas flow of gas storage m at time t (It is positive for inflow and negative for outflow); $u_{gs,m}(t)$ is the storage price for gas storage m at time t; $u_{P2G,k}$ is the operation cost of P2G k; $Q_{P2G,k}(t)$ is the gas flow of P2G k at time t; $u_{ave}(t)$ is the average gas price (In this paper, it is the average price of gas wells).

(2) Minimum CO_2 Emissions of the Natural-Gas System

$$\text{Min} \quad E_{CO_2} = \sum_{n=1}^{N_w} \sum_{t=1}^{T} E_{wn}(t) + \sum_{m=1}^{N_{gs}} \sum_{t=1}^{T} E_{gs,m}(t) - \sum_{k=1}^{N_{P2G}} \sum_{t=1}^{T} E_{P2G,k}(t) \tag{12}$$

where E_{CO2} represents CO_2 emissions of the natural-gas system; $E_{wn}(t)$, $E_{gs,m}(t)$ are the CO_2 emissions of gas well n, gas storage m at time t, respectively; $E_{P2G,k}(t)$ is the amount of CO_2 absorbed by the methanation process of P2G k at time t.

2.2.2. Constraints

(1) Gas Flow Limits of Gas Wells

$$Q_{wn}^{min} \leq Q_{wn}(t) \leq Q_{wn}^{max} \tag{13}$$

where Q_{wn}^{min}, Q_{wn}^{max} represent the minimum gas flow and the maximum gas flow of gas well n, respectively.

(2) Gas Pressure Limits of Gas Nodes

$$M_i^{min} \leq M_i(t) \leq M_i^{max} \tag{14}$$

where $M_i(t)$ represents gas pressure of gas node i at time t. M_i^{min} and M_i^{max} are the minimum and maximum gas pressure of gas node i.

(3) Gas Flow Equation of Pipelines

The natural-gas system satisfies the mass conservation law of fluid dynamics and Bernoulli equation in the operation. The relationship between gas flow of pipelines and gas pressure of gas nodes can be modeled as follows [12,35].

$$Q_{ij}(t)\left|Q_{ij}(t)\right| = C_{ij}\left(M_i(t)^2 - M_j(t)^2 \right) \tag{15}$$

$$Q_{ij}(t) = \frac{Q_{ij}^{in}(t) + Q_{ij}^{out}(t)}{2} \tag{16}$$

where $Q_{ij}(t)$ is the average gas flow of pipeline ij (Pipeline ij is the pipeline between gas node i and gas node j); $Q_{ij}^{in}(t)$ and $Q_{ij}^{out}(t)$ are the injection and withdrawal gas flow of pipeline ij, respectively; C_{ij} is a constant related to the length, diameter, temperature and compressibility factor of pipeline ij.

(4) Line Pack Equation

Due to the compressibility of natural gas, the injection gas flow and the withdrawal gas flow of the same pipeline would be different. Some excess natural gas can be stored in the pipelines, which is called line pack. The line pack of pipeline ij is related to the average pressure and its own parameters of pipelines, which can be modeled as below [12,15].

$$L_{ij}(t) = \omega_{ij}M_{ij}(t) \tag{17}$$

$$M_{ij}(t) = \frac{M_i(t) + M_j(t)}{2} \tag{18}$$

$$L_{ij}(t) = L_{ij}(t-1) + Q_{ij}^{in}(t) - Q_{ij}^{out}(t) \tag{19}$$

where $L_{ij}(t)$ is the line pack of pipeline ij at time t; ω_{ij} is a constant related to pipeline parameters, gas constant, compressibility factor, gas density, and gas temperature.

(5) Nodal Gas Flow Balance Equation

For each gas node, the gas flows into the node must equals the gas flows out of the node.

$$\sum_{n\in i} Q_{wn}(t) + \sum_{m\in i} Q_{gs,m}(t) + \sum_{k\in i} Q_{P2G,k}(t) - \sum_{j\in Set\_I(i)} Q_{ij}^{in}(t) \\ + \sum_{j\in Set\_O(i)} Q_{ij}^{out}(t) - Q_{GT,i}(t) - Q_{Li}(t) = 0 \tag{20}$$

where, the first three items are the gas flow of gas wells, gas storage, and P2G located at gas node i at time t, respectively; $Q_{GT,i}(t)$ and $Q_{Li}(t)$ indicate the gas flow injected to gas-fired units and the gas load at gas node i at time t, respectively; $Set\_I(i)$ is the set of pipeline ij which lets gas node i as the input node; $Set\_O(i)$ is the set of pipeline ij which lets gas node i as the output node.

(6) Gas Flow Limits and Capacity Limits of Gas Storage

$$Q_{gs,m}^{min} \leq Q_{gs,m}(t) \leq Q_{gs,m}^{max} \tag{21}$$

$$V_m^{min} \leq V_m(t) \leq V_m^{max} \tag{22}$$

$$V_m(t) = V_m(t-1) + Q_{gs,m}(t) \tag{23}$$

where $Q_{gs,m}^{min}$ and $Q_{gs,m}^{max}$ are the minimum and maximum gas flow of gas storage m, respectively; $V_m(t)$, V_m^{min}, V_m^{max} are the capacity of gas storage m at time t, the minimum and maximum capacity of gas storage m, respectively. When the gas is injected to the gas storage, $Q_{gs,m}(t)$ is positive, otherwise it is negative.

(7) Compressor

The compressors are used to boost the pressure of the natural-gas network, which can help the natural gas transporting to each gas load. In this paper, the energy consumed by the compressors is calculated by using natural gas flow through the compressors. The consumed gas flow of compressor r, $Q_{cr}^{consume}(t)$, is calculated as presented below [15].

$$Q_{cr}^{consume}(t) = \beta_{cr} P_{cr}(t) \tag{24}$$

$$P_{cr}(t) = \frac{Q_{cr}(t)}{\eta_{cr} \cdot \tau} \cdot \left(\left(\frac{M_{or}(t)}{M_{ir}(t)} \right)^{\tau} - 1 \right) \tag{25}$$

where β_{cr} is energy conversion coefficient of compressor r; $P_{cr}(t)$ is the consumed energy by compressor r; $Q_{cr}(t)$ is the gas flow flowing through compressor r at time t; η_{cr} is the efficiency of compressor r; $\tau = (\alpha - 1)/\alpha$ and α is variability index of compressors; $M_{or}(t)$ and $M_{ir}(t)$ are the pressure of output node and input node of compressor r, respectively.

(8) Gas Flow Limit of P2G

$$Q_{P2G,k}^{min} \leq Q_{P2G,k}(t) \leq Q_{P2G,k}^{max} \tag{26}$$

where $Q_{P2G,k}^{min}$ and $Q_{P2G,k}^{max}$ are the minimum and maximum gas flow of P2G k, respectively.

2.3. Gas Demand Management Strategy to Coordinate the Two Energy Systems

When the pressure of some gas nodes is higher than the maximum pressure or lower than the minimum pressure, which means the gas demand and the gas supply is not balanced on these gas nodes, then the gas demand management strategy is used. The main idea is to adjust the gas flow of gas turbines to achieve the gas demand balance, which means changing the power output of gas-fired units. Then, the power output of units in power system will be adjusted.

2.4. Constraints Handling Methods

The constraints of power system are handled using the methods presented in Reference [39]. In this paper, the constraints of natural-gas system are handled by the proposed method as shown below.

2.4.1. Equality Constraints Handling Method

In this paper, the set of non-linear constraints Equations (15)–(20) of the natural-gas system are solved by the trust region algorithm [43,44] and Levenberg-Marquardt algorithm (L-M) [45,46]. Trust region and L-M methods are both simple and powerful tools for solving systems of nonlinear equations and large-scale optimization problems. They have the advantages of guaranteeing a solution whenever it exists [43–46]. In this paper, the trust region method and L-M method are used to solve the gas flow non-linear equations, respectively. The optimization results are compared in the case studies.

2.4.2. Inequality Constraints Handling Method

For the inequality constraints (13)–(14), (21)–(22), (26), the gas flow is the minimum when it is lower than the minimum value and the gas flow is the maximum when it is over the maximum value. For the gas storage volume constraint, the effective redundancy method is proposed in this paper. The details of this method are as below.

a) For gas storage m at time t;

b) If $V_m(t) \leq V_m^{min}$, calculate $\Delta V = V_m^{min} - V_m(t)$;

c) For $ii = 1{:}t$, calculate the gas flow redundancy of gas storage m at time ii. $\Delta Q_{gs}(ii) = \min\{Q_{gs,m}^{max} - Q_{gs,m}(ii), V_m^{max} - V_{gs,m}(ii)\}$. If the gas node where the gas storage m is connected with P2G, $\Delta Q_{P2G}(ii) = Q_{P2G}^{max} - Q_{P2G}(ii)$, the effective redundancy $\Delta Q(ii) - \min\{\Delta Q_{gs}(ii), \Delta Q_{P2G}(ii)\}$; else, $\Delta Q(ii) = \Delta Q_{gs}(ii)$. Then, arrange ΔQ in descending order;

d) According to the descending order, $Q_{P2G}(ii)$ and $Q_{gs,m}(ii)$ are adjusted successively until $V_m(t) \geq V_m^{min}$;

e) Update $V_m(t)$;

f) If $V_m(t) \geq V_m^{max}$, calculate $\Delta V = V_m(t) - V_m^{max}$;

g) For $ii = 1{:}t$, calculate the gas flow redundancy of gas storage m at time ii. $\Delta Q_{gs}(ii) = \min\{Q_{gs,m}(ii) - Q_{gs,m}^{min}, V_{gs,m}(ii) - V_m^{min}\}$. If the gas node where the gas storage m is connected with P2G, $\Delta Q_{P2G}(ii) = Q_{P2G}(ii) - Q_{P2G}^{min}$, the effective redundancy $\Delta Q(ii) = \min\{\Delta Q_{gs}(ii), \Delta Q_{P2G}(ii)\}$; else, $\Delta Q(ii) = \Delta Q_{gs}(ii)$. Then, arrange ΔQ in descending order;

h) According to the descending order, $Q_{P2G}(ii)$ and $Q_{gs,m}(ii)$ are adjusted successively until $V_m(t) \leq V_m^{max}$;

i) Update $V_m(t)$.

3. Case Studies Application

3.1. Description of Case Studies

The hybrid electricity-natural gas energy systems shown in Figure 3 are composed by the revised IEEE 39-bus power system [35] and Belgian high-calorific 20-node gas system [3]. The IEEE 39-bus power network has 46 branches, five coal-fired units, three gas-fired units and two wind power units, where the capacity of wind power units accounts for 35% of the total installed capacity of 3903 MW. The Belgian high-calorific 20-node gas system has 24 pipelines, two gas wells, three gas storages and two compressors. The parameters of the power system are from References [35,40] and the parameters of natural gas system are from Reference [3]. The revised parameters are shown in Tables 1 and 2 (inflow of gas storage is positive and outflow of gas storage is negative). Gas pressure limits of gas nodes are given in Table 3. Power demand and gas demand are given in Table 4. In addition, the theoretical predicted wind power output is given in Figure 4. The efficiency of P2G process is taken as 64% [6]. Wind curtailment cost is set as \$100/MWh [47]. The short-term optimal dispatch for this hybrid energy system is studied to illustrate the behavior of the proposed model, the adopted algorithm and the proposed constraints handling methods in several case studies. These case studies are simulated with a low level of initial line pack (0.5 Mm^3). In addition, all the case studies are implemented using MATLAB language programming.

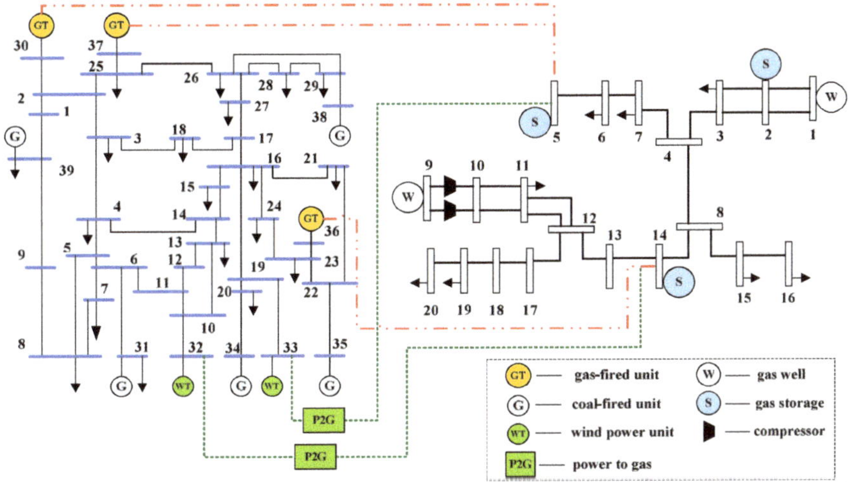

Figure 3. The hybrid electricity-natural gas energy systems.

Table 1. Parameters of power units.

Power Units	P_{max}/MW	P_{min}/MW	Ramp Up Rate/MW/h	Ramp Down Rate/MW/h
Coal-fired unit 1	470	150	80	80
Coal-fired unit 2	470	135	80	80
Coal-fired unit 3	340	73	80	80
Coal-fired unit 4	300	60	50	50
Coal-fired unit 5	243	73	50	50
Gas-fired unit 1	260	0	260	260
Gas-fired unit 2	230	0	230	230
Gas-fired unit 3	220	0	220	220
Wind power unit 1	750	0	750	750
Wind power unit 2	620	0	620	620

Table 2. Parameters of gas storage.

Gas Storage No.	Initial Capacity/Mm³	Max Capacity/Mm³	Min Capacity/Mm³	Max Gas Flow/Mm³/h	Min Gas Flow/Mm³/h
Gas Storage 1	1.5	3.5	0	0.35	-0.20
Gas Storage 2	2.0	4.5	0	0.45	-0.25
Gas Storage 3	1.5	3.5	0	0.35	-0.25

Table 3. Gas pressure limits of gas nodes.

Node No.	1	2	3	4	5	6	7	8	9	10	11	12	13	14	15	16	17	18	19	20
M_{min}/bar	30	30	30	30	10	10	30	30	50	50	30	30	30	30	15	15	25	25	15	15
M_{max}/bar	100	100	100	80	80	80	80	70	70	77	70	70	70	70	70	70	70	70	70	70

Table 4. Power demand and gas demand.

Time/h	1	2	3	4	5	6	7	8	9	10	11	12
Power demand/MW/h	1272	1188	1104	960	1080	1320	1476	1584	1740	1776	1800	1860
Gas demand/Mm³/h	1.03	0.97	0.92	0.98	0.99	1.03	1.23	1.45	1.79	1.83	1.74	1.61

Time/h	13	14	15	16	17	18	19	20	21	22	23	24
Power demand/MW/h	1680	1560	1320	1104	1416	1680	1800	2040	1860	1632	1344	1116
Gas demand/Mm³/h	1.46	1.42	1.39	1.38	1.39	1.30	1.26	1.19	1.15	1.15	1.12	0.97

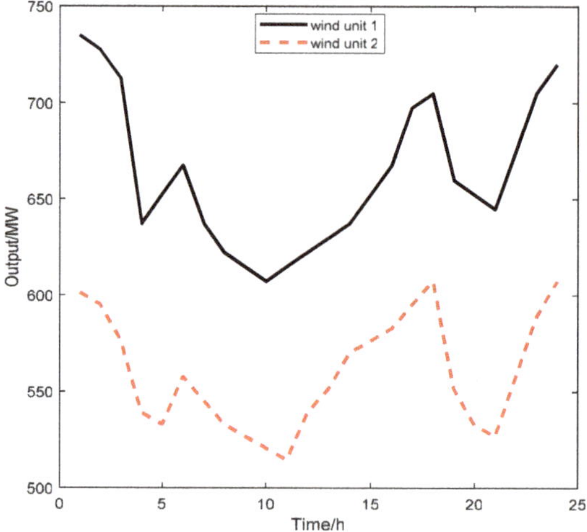

Figure 4. Predicted output of wind power units.

3.2. Analysis of Simulation Results

The Newton-Raphson method is used to obtain the power flow. Trust region method and L-M method are used to solve the non-linear equations to obtain the gas flow in natural-gas system, respectively. Furthermore, MOBHPSO [39–42] is used to optimize the multi-objective dispatch problem of hybrid electricity-natural gas energy systems based on the established models (1–3,7,12), the proposed flow chart (Figure 2), and the proposed constraints handling methods. The optimization results are shown in Tables 5 and 6. All the constraints are satisfied. The comparisons of power output and gas flow among different case studies are given in Figures 5 and 6, respectively. Moreover, it can be found the different performance of trust region method and L-M method from Figure 7 and Table 6. The wind power absorbed by P2G and the gas flow of P2G are shown in Figure 8. The volume of gas storages is given in Figure 9. The gas pressure of each gas node can be found in Appendix A.

From the obtained results, it can be seen that power output, gas flow of gas wells, gas flow of P2G, gas flow of gas storages, volume of gas storages, and gas pressure of gas nodes all satisfy their respective upper and lower bound constraints. Besides, the nodal gas flow balance equation is satisfied. Moreover, power demand and power supply are balanced which can be drawn from the calculated load loss rate $L_p = 6.37 \times 10^{-18}$. Then, the above results show that all the constraints are satisfied using the proposed constraints handling methods.

Table 5. Optimization results of the power system.

Case Studies	Fuel Cost (M\$)	SO_x Emission (ton)
Without P2G	1.080	38.193
With P2G	1.084	37.939

Table 6. Optimization results of the natural-gas system.

Case Studies	Methods	Cost of Natural-Gas/ M\$	CO_2 Emission/ 10^4 ton	Rate of Abandoned Wind Power	Operation Cost of P2G/M\$	Absorbed CO_2 by the Methanation Process/ 10^4 ton	Increased Wind Power by P2G/ MWh
Without P2G	Trust Region	0.741	5.791	24.85%	0	0	0
	L-M	0.695	5.790	24.85%	0	0	0
With P2G	Trust Region	0.732	5.727	6.71%	0.106	0.056	5321.66
	L-M	0.685	5.491	4.04%	0.122	0.064	6104.48

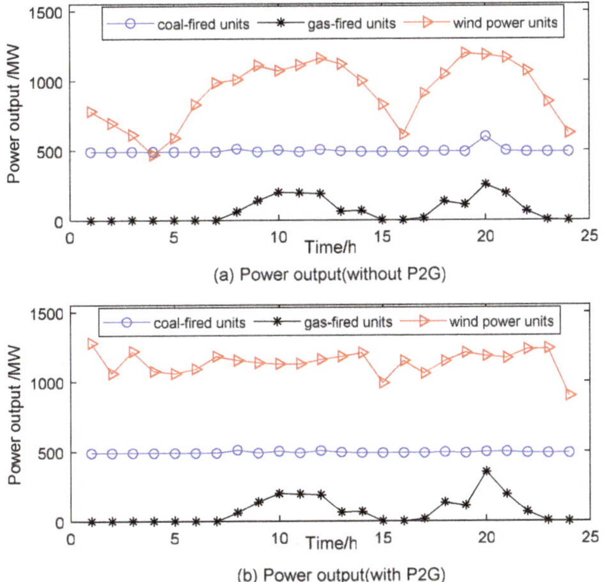

(a) Power output(without P2G)

(b) Power output(with P2G)

Figure 5. Comparison of power output without P2G and with P2G.

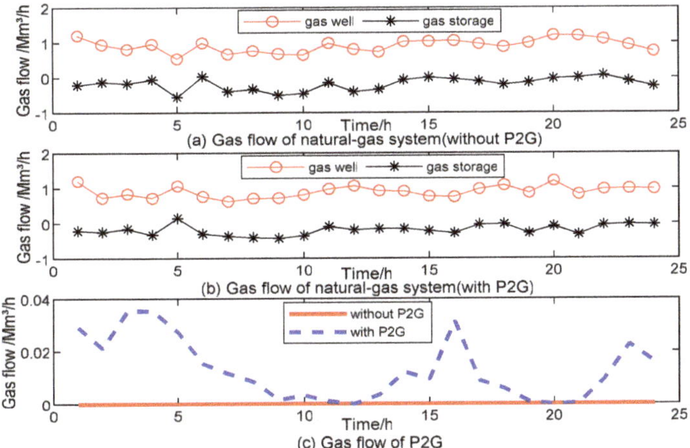

(a) Gas flow of natural-gas system(without P2G)

(b) Gas flow of natural-gas system(with P2G)

(c) Gas flow of P2G

Figure 6. Comparison of gas flow without P2G and with P2G.

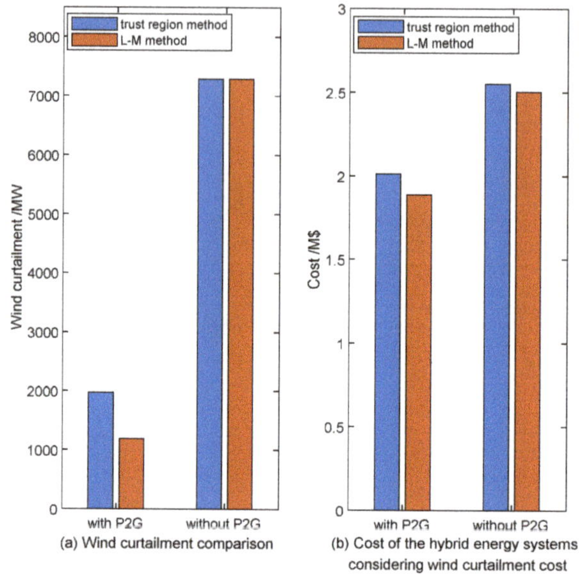

Figure 7. Results comparison of trust region method and L-M method.

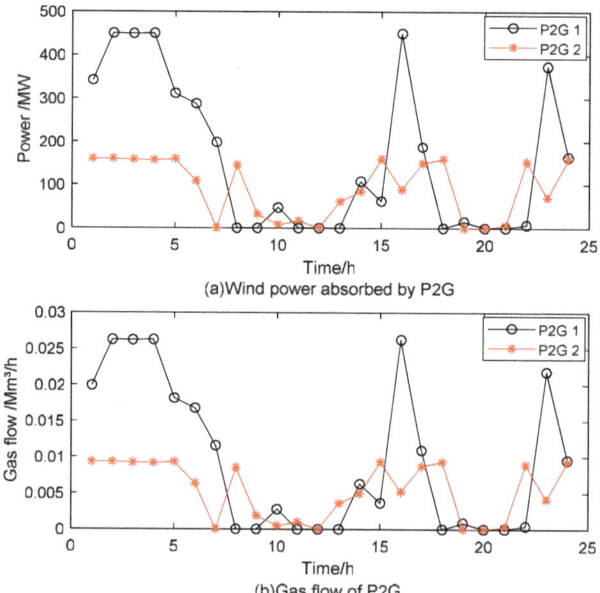

Figure 8. Wind power absorbed by P2G and the gas flow of P2G.

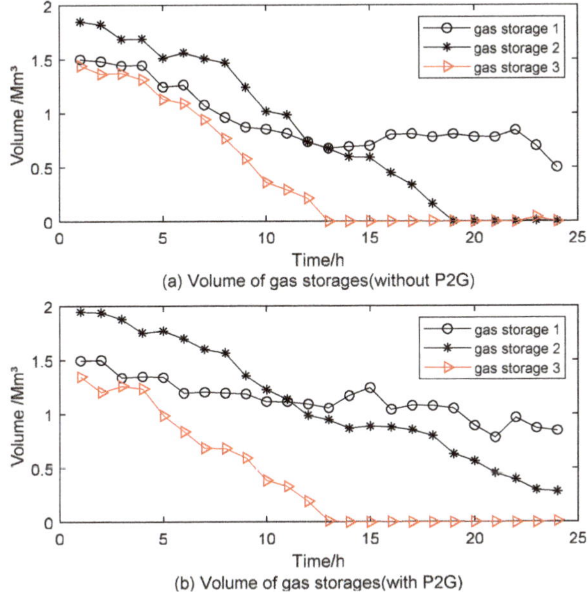

Figure 9. Volume of gas storages without P2G and with P2G.

3.2.1. Effects of P2G on the Power System

(1) From Table 5 and Figure 5, it can be seen that the fuel cost of power system with P2G is a little higher than that without P2G. At hour 20, owing to the gas injection from P2G, the pipeline pressure is higher than the maximum value, so the 'gas demand management strategy' is used and needs to increase the gas demand by increasing the output of gas-fired units connected with gas node 5 and 14. Then, to guarantee the power load balance, the output of coal-fired units would be reduced. Because the fuel cost of gas-fired units is higher than that of coal-fired units and the SO_x emissions of gas-fired units are lower than that of coal-fired units, it leads to increase of fuel cost and decline of SO_x emissions. The SO_x emissions are reduced by 0.254 ton. In addition, from Figure 8, most abandoned wind power can be absorbed by P2G. During hours 3-5, P2G works at its maximum value when the abandoned wind power is over the maximum capacity of P2G. Owing to the P2G, the wind power output is much smoother and so is the output of coal-fired units, which is propitious to the stability and reliability of the power system.

(2) From Table 6 and Figure 7a, it is obvious that the rate of abandoned wind power is declined from 24.85% to 6.71% (trust region) and from 24.85% to 4.04% (L-M), respectively; The wind power output is increased by 5321.66MWh (trust region) and 6104.48MWh (L-M), respectively.

3.2.2. Effects of P2G on the Natural-gas System

From Figures 6 and 9, it is obvious that the gas flow of gas wells and gas storages is lower when P2G is considered. In addition, the volume of gas storages with P2G is much larger than that without P2G. This is because the economic, clean, and low-carbon energy converted by P2G from wind power has the priority of use compared with that from natural gas network, which creates considerable economic and environmental benefits for the integrated energy systems. The cost benefit of P2G is evaluated in terms of the natural gas cost which it displaces. From Table 6, it can be seen the gas cost is reduced by $9000 (trust region) and $10,000 (L-M), respectively; Moreover, the environmental benefit of P2G in terms of CO_2 reduction and CO_2 absorbed in the P2G methanation process is measured. The total CO_2 emissions are declined by 1200 tons (trust region) and 3630 tons (L-M), respectively.

3.2.3. Total Cost Reduction of the Hybrid Energy Systems

The total cost of the hybrid electricity-natural gas energy systems including the wind power curtailment cost is reduced by 5.372×10^5 (trust region) and 6.165×10^5 (L-M), which can be seen from Figure 7b.

It can be concluded that the proposed model shows that the proposed constraints handling methods are effective and the feasibility of MOBHPSO algorithm for solving the multi-objective optimal dispatch problem of the hybrid electricity-natural gas energy systems is indicated. Moreover, the trust region method and L-M method are effective to solve the non-linear gas flow problem. It also can be seen that the results obtained from L-M method is much better than those obtained from trust region method.

4. Conclusions

This paper presented a multi-objective optimal dispatch model of the hybrid electricity-natural gas energy systems coupled by P2G and gas turbines in order to achieve the maximum of low-carbon economic environmental benefits. The proposed model provides not only enhanced flexibility, as it easily handles bidirectional energy flow and guarantees global optimality, but also considers the compressibility of gas, line pack of pipelines among other complicated system characteristics. The non-linear and non-convex functions of gas flow model are addressed by trust region method and L M method. The L-M method has much better performance, which can be drawn from the simulation results. Moreover, the case studies simulation results show the feasibility of MOBHPSO algorithm for solving the multi-objective optimal dispatch problem of the hybrid electricity-natural gas energy systems and the effectiveness of proposed constraints handling methods. The obtained results also illustrate that P2G can significantly benefit the operation of both power system and natural gas system in smoothing power output, cutting down gas cost, reducing CO_2 emissions and SO_x emissions as well as avoiding wind curtailment. More specifically, the gas cost is cut down up to $10,000, the total CO_2 emissions are declined up to 3630 tons and the SO_x emissions are reduced by 0.254 ton as well as the wind power curtailment is decreased up to 6104.48 MWh with the rate of abandoned wind power declined from 24.85% to 4.04%. Besides, the total cost including wind power curtailment cost is reduced up to 6.165×10^5.

Author Contributions: J.L. proposed the optimization model and algorithms, carried out case studies, completed the entire analysis, wrote and revised this paper; G.P.H. gave essential and important advice on the model of natural-gas system; W.S. gave some important suggestions on the calculation of gas flow and revised this paper.

Funding: This research was part funded by EPSRC through the Hydrogen's Value in the Energy System project, grant number EP/L018284/1 and the National Centre for Energy Systems Integration, grant number EP/P001173/1.

Acknowledgments: The authors would like to thank the financial support of the China Scholarship Council (CSC), as well as Carlos M. Correa-Posada who provided some important data used for the case studies.

Conflicts of Interest: The authors declare no conflict of interest.

Appendix A

Table A1. Gas pressure of each gas node.

Node No.	1	2	3	4	5	6	7	8	9	10
Hour 1	74.7469	73.8385	72.5356	56.7444	45.5189	41.0362	42.8709	43.8394	55.7467	61.3213
Hour 2	67.0635	66.5001	65.6495	55.1170	39.9237	38.1803	40.9818	45.8246	50.2401	55.2641
Hour 3	70.9782	70.6287	69.6258	56.8556	53.1330	46.9431	47.4426	40.3510	54.5096	59.9605
Hour 4	67.7032	67.1715	66.3961	57.2499	70.7133	54.9508	54.2037	42.9448	53.3986	58.7385
Hour 5	60.3126	59.8906	59.2212	51.7346	33.9220	33.5808	37.1390	46.9318	49.9918	54.9910
Hour 6	60.2348	60.0091	59.2823	51.2112	48.2071	40.7850	41.2236	44.1951	50.0255	55.0280
Hour 7	61.7065	61.2190	60.5111	52.7660	56.6407	46.2672	46.2665	44.8522	51.1059	56.2164
Hour 8	72.0210	71.2920	70.1714	55.5450	30.7629	30.7596	37.3143	40.5826	50.3482	55.3830

Table A1. *Cont.*

Node No.	1	2	3	4	5	6	7	8	9	10
Hour 9	59.6538	59.2193	58.5192	50.4304	63.6206	46.3675	45.9280	38.7845	52.4694	57.7163
Hour 10	63.8376	63.4229	62.5894	52.8985	40.6647	38.5312	40.3515	45.1444	53.4969	58.8466
Hour 11	70.6866	69.9749	68.8971	54.9136	27.6230	27.5911	35.2202	41.7990	50.9478	56.0425
Hour 12	68.9173	68.3366	67.3818	55.2606	41.9605	38.5358	40.9826	43.4697	52.3659	57.6024
Hour 13	64.6456	64.1633	63.3132	52.9436	32.3582	31.5162	36.1046	44.7830	50.4011	55.4412
Hour 14	68.1243	67.1946	66.2050	53.6897	45.2849	40.2787	41.5617	39.2924	51.8961	57.0857
Hour 15	77.3472	76.3396	75.1040	58.2513	28.6262	28.7076	37.7480	40.5386	50.1422	55.1564
Hour 16	74.0179	73.6698	72.5552	57.3796	36.5834	35.4363	40.2615	40.3234	50.9462	56.0408
Hour 17	72.7026	71.9010	70.8103	56.1345	35.5315	34.6661	39.3576	39.7726	50.2135	55.2349
Hour 18	75.1341	74.3651	73.1955	57.2444	37.5240	36.4759	40.8464	38.1481	50.2537	55.2791
Hour 19	70.4893	69.9243	68.9806	56.8986	63.1603	51.1618	51.1398	38.1952	53.7179	59.0897
Hour 20	90.0790	89.4097	87.8955	66.0627	30.6166	32.7061	44.5817	38.5313	51.2693	56.3962
Hour 21	72.9250	72.4852	71.4642	57.2781	44.5136	41.6715	44.0290	38.2162	56.4074	62.0482
Hour 22	69.5078	68.3668	67.3470	53.8900	37.1964	35.8036	39.0643	38.5332	54.3579	59.7937
Hour 23	70.0484	69.6090	68.6537	56.6512	59.8959	49.6890	49.7032	40.4218	53.2979	58.6277
Hour 24	56.7456	56.3199	55.5790	46.7811	33.0590	31.2034	33.3676	38.3445	64.6311	71.0942

Node No.	11	12	13	14	15	16	17	18	19	20
Hour 1	54.1081	50.5855	46.0875	44.7456	35.3047	25.7453	49.7105	35.4170	26.1208	25.9488
Hour 2	51.7236	50.3445	47.8200	46.7134	37.8536	29.1582	49.9537	40.2191	31.2779	31.1271
Hour 3	54.3675	51.3623	45.1763	40.7143	31.6600	21.1181	51.0356	43.0089	35.4623	35.3304
Hour 4	53.8731	51.4255	46.4733	43.5065	34.3087	24.3816	51.1668	44.6069	37.9960	37.8745
Hour 5	51.9467	50.9341	49.0610	48.4082	39.0532	30.5918	50.7221	45.1699	39.1410	39.0242
Hour 6	51.3066	49.8506	46.9020	45.3095	36.2103	27.2829	49.6736	44.8413	39.1515	39.0360
Hour 7	52.2445	50.6172	47.4860	45.9142	36.8016	27.8913	50.4070	45.1037	39.3228	39.2076
Hour 8	50.8360	48.7091	44.1013	41.0575	31.8666	21.3437	48.5566	44.1770	38.6343	38.5179
Hour 9	52.3018	49.4738	43.5577	39.5966	29.3700	17.2155	49.2713	44.1255	38.3480	38.2301
Hour 10	54.3000	52.1931	48.2816	46.3746	36.7013	27.3680	51.9234	45.6230	39.5052	39.3897
Hour 11	51.6829	49.6546	45.2476	42.3900	33.2849	23.2977	49.4998	44.9943	39.3993	39.2845
Hour 12	52.9842	50.8287	46.5780	44.1992	35.0247	25.4320	50.6144	45.2823	39.5316	39.4170
Hour 13	51.7051	50.2120	47.2589	45.8176	36.6459	27.6235	50.0282	45.1247	39.4647	39.3501
Hour 14	51.9730	49.3827	43.7844	39.8147	30.3185	19.0701	49.2116	44.5565	38.9702	38.8546
Hour 15	50.5211	48.3266	43.7034	40.9399	31.3684	20.1573	48.1649	43.6773	38.0939	37.9760
Hour 16	51.1041	48.7023	43.7603	40.7302	31.1775	19.9426	48.5071	43.4549	37.6395	37.5197
Hour 17	50.4220	48.1032	43.2388	40.2295	30.5039	18.9170	47.9215	43.0252	37.1790	37.0577
Hour 18	50.1500	47.5764	42.0906	38.4792	28.4809	15.5972	47.3952	42.5038	36.6134	36.4903
Hour 19	53.1562	49.8808	43.1425	38.5692	28.3725	15.1925	49.6243	43.5000	37.2428	37.1207
Hour 20	51.1983	48.4743	42.5615	38.5943	28.7701	15.8577	48.2927	43.2795	37.3198	37.1984
Hour 21	55.6171	51.8483	44.0882	38.5362	28.4171	15.2929	51.5608	44.9563	38.6616	38.5433
Hour 22	54.1611	50.9718	44.0340	39.1087	28.7767	15.8746	50.7701	45.4186	39.5082	39.3929
Hour 23	53.3906	50.6155	44.7480	40.9330	31.0949	19.5983	50.4247	45.3936	39.6974	39.5833
Hour 24	63.3091	58.1787	47.7707	39.5037	28.7623	16.1240	57.7734	49.4686	43.0386	42.9313

References

1. Mazza, A.; Bompard, E.; Chicco, G. Application of power to gas technologies in emerging electrical systems. *Renew. Sustain. Energy Rev.* **2018**, *92*, 794–806. [CrossRef]
2. Hibbard, P.J.; Schatzki, T. The interdependence of electricity and natural gas: Current factors and future prospects. *Electr. J.* **2012**, *25*, 6–17. [CrossRef]
3. Correa-Posada, C.M.; Sánchez-Martín, P. Integrated power and natural gas model for energy adequacy in short-term operation. *IEEE Trans. Power Syst.* **2015**, *30*, 3347–3355. [CrossRef]
4. Schiebahn, S.; Grube, T.; Robinius, M.; Tietze, V.; Kumar, B.; Stolten, D. Power to gas: Technological overview, systems analysis and economic assessment for a case study in Germany. *Int. J. Hydrogen Energy* **2015**, *40*, 4285–4294. [CrossRef]
5. Götz, M.; Lefebvre, J.; Mörs, F.; McDaniel Koch, A.; Graf, F.; Bajohr, S.; Reimert, R.; Kolb, T. Renewable power-to-gas: A technological and economic review. *Renew. Energy* **2016**, *85*, 1371–1390. [CrossRef]
6. Maroufmashat, A.; Fowler, M. Transition of future energy system infrastructure: Through power-to-gas pathways. *Energies* **2017**, *10*, 1089. [CrossRef]

7. Mukherjee, U.; Maroufmashat, A.; Narayan, A.; Elkamel, A.; Fowler, M. A stochastic programming approach for the planning and operation of a power to gas energy hub with multiple energy recovery pathways. *Energies* **2017**, *10*, 868. [CrossRef]

8. Eveloy, V.; Gebreegziabher, T. A review of projected power-to-gas deployment scenarios. *Energies* **2018**, *11*, 1824. [CrossRef]

9. Clegg, S.; Mancarella, P. Integrated modeling and assessment of the operational impact of power-to-gas (P2G) on electrical and gas transmission networks. *IEEE Trans. Sustain. Energy* **2015**, *6*, 1234–1244. [CrossRef]

10. Department of Energy and Climate Change. *The Future of Heating: Meeting the Challenge*; HM Government: London, UK, 2013.

11. Ball, M.B.; Wietschel, M. *The Hydrogen Economy: Opportunities and Challenges*; Cambridge University Press: Cambridge, UK, 2009.

12. An, S.; Li, Q.; Gedra, T.W. Natural gas and electricity optimal power flow. In Proceedings of the IEEE PES Transmission and Distribution Conference and Exposition, Dallas, TX, USA, 7–12 September 2003. [CrossRef]

13. Chen, S.; Wei, Z.N.; Sun, G.Q.; Wang, D.; Sun, Y.H.; Zang, H.X.; Zhu, Y. Probabilistic energy flow analysis in integrated electricity and natural-gas energy systems. *Proc. CSEE* **2015**, *35*, 6331–6340. [CrossRef]

14. Sun, G.Q.; Chen, S.; Wei, Z.N.; Chen, S.; Li, Y.C. Probabilistic optimal power flow of combined natural gas and electric system considering correlation. *Autom. Electr. Power Syst.* **2015**, *39*, 11–17. [CrossRef]

15. Osiadacz, A.J. *Simulation and Analysis of Gas Networks*; Gulf Publishing Company: Houston, TX, USA, 1987.

16. Liu, C.; Shahidehpour, M.; Fu, Y.; Li, Z.Y. Security-constrained unit commitment with natural gas transmission constraints. *IEEE Trans. Power Syst.* **2009**, *24*, 1523–1536. [CrossRef]

17. Geidl, M.; Andersson, Q. Optimal power flow of multiple energy carriers. *IEEE Trans. Power Syst.* **2007**, *22*, 145–155. [CrossRef]

18. Qadrdan, M.; Wu, J.Z.; Jenkins, N.; Ekanayake, J. Operating strategies for a GB integrated gas and electricity network considering the uncertainty in wind power forecasts. *IEEE Trans. Sustain. Energy* **2014**, *5*, 128–138. [CrossRef]

19. Chaudry, M.; Jenkins, N.; Strbac, G. Multi-time period combined gas and electricity network optimization. *Electr. Power Syst. Res.* **2008**, *78*, 1265–1279. [CrossRef]

20. Wang, W.L.; Wang, D.; Jia, H.J.; Chen, Z.Y.; Guo, B.Q.; Zhou, H.M.; Fan, M.H. Steady state analysis of electricity-gas regional integrated energy system with consideration of NGS network status. *Proc. CSEE* **2017**, *37*, 1293–1304. [CrossRef]

21. Odetayo, B.; Kazemi, M.; MacCormack, J.; Rosehart, W.D.; Zareipour, H.; Seifi, A.R. A chance constrained programming approach to the integrated planning of electric power generation, natural gas network and storage. *IEEE Trans. Power Syst.* **2018**, *33*, 6883–6893. [CrossRef]

22. Guandalini, G.; Robinius, M.; Grube, T.; Campanari, S.; Stolten, D. Long-term power-to-gas potential from wind and solar power: A country analysis for Italy. *Int. J. Hydrogen Energy* **2017**, *42*, 13389–13406. [CrossRef]

23. Liu, W.J.; Wen, F.S.; Xue, Y.S. Power-to-gas technology in energy systems: Current status and prospects of potential operation strategies. *J. Mod. Power Syst. Clean Energy* **2017**, *5*, 439–450. [CrossRef]

24. He, L.C.; Lu, Z.G.; Zhang, J.F.; Geng, L.J.; Zhao, H.; Li, X.P. Low-carbon economic dispatch for electricity and natural gas systems considering carbon capture systems and power-to-gas. *Appl. Energy* **2018**, *224*, 357–370. [CrossRef]

25. International Energy Agency. *Prospects for Hydrogen and Fuel Cell*; International Energy Agency: Paris, France, 2005.

26. De Vries, H.; Florisson, O.; Tiekstra, G. Safe operation of natural gas appliances fueled with hydrogen/natural gas mixtures (progress obtained in the naturally-project). In Proceedings of the International Conference on Hydrogen Safety, San Sebastián, Spain, 11–13 September 2007. [CrossRef]

27. Dodds, P.E.; Demoullin, S. Conversion of the UK gas system to transport hydrogen. *Int. J. Hydrogen Energy* **2013**, *38*, 7189–7200. [CrossRef]

28. Biegger, P.; Kirchbacher, F.; Medved, A.R.; Miltner, M.; Lehner, M.; Harasek, M. Development of honeycomb methanation catalyst and its application in power to gas systems. *Energies* **2018**, *11*, 1679. [CrossRef]

29. Li, Y.; Liu, W.J.; Zhao, J.H.; Wen, F.S.; Dong, C.Y.; Zheng, Y.; Zhang, R. Optimal dispatch of combined electricity-gas-heat energy systems with power-to-gas devices and benefit analysis of wind power accommodation. *Power Syst. Technol.* **2016**, *40*, 3680–3688. [CrossRef]

30. Clegg, S.; Mancarella, P. Integrated electrical gas network flexibility assessment in low-carbon multi-energy systems. *IEEE Trans. Sustain. Energy* **2016**, *7*, 718–731. [CrossRef]

31. Ye, J.; Yuan, R.X. Integrated natural gas, heat, and power dispatch considering wind power and power-to-gas. *Sustainability* **2017**, *9*, 602. [CrossRef]

32. Li, G.Q.; Zhang, R.F.; Jiang, T. security-constrained bi-level economic dispatch model for integrated natural gas and electricity systems considering wind power and power-to-gas process. *Appl. Energy* **2017**, *194*, 696–704. [CrossRef]

33. Guandalini, G.; Campanari, S.; Romano, M.C. Power-to-gas plants and gas turbines for improved wind energy dispatchability: Energy and economic assessment. *Appl. Energy* **2015**, *147*, 117–130. [CrossRef]

34. Chen, Z.Y.; Wang, D.; Jia, H.J.; Wang, W.L.; Guo, B.Q.; Qu, B.; Fan, M.H. Research on optimal day-ahead economic dispatching strategy for microgrid considering P2G and multi-source energy storage system. *Proc. CSEE* **2017**, *37*, 3067–3077. [CrossRef]

35. Wei, Z.N.; Zhang, S.D.; Sun, G.Q.; Zang, H.Y.; Chen, S.; Chen, S. Power-to-gas considered peak load shifting research for integrated electricity and natural-gas energy systems. *Proc. CSEE* **2017**, *37*, 4601–4609. [CrossRef]

36. He, C.; Liu, T.Q.; Wu, L.; Shahidepour, M. Robust coordination of interdependent electricity and natural gas systems in day-ahead scheduling for facilitating volatile renewable generations via power-to-gas technology. *J. Mod. Power Syst. Clean Energy* **2017**, *5*, 375–388. [CrossRef]

37. Shu, K.G.; Ai, X.M.; Fang, J.K.; Yao, W.; Chen, Z.; He, H.B.; Wen, J.Y. Real-time subsidy based robust scheduling of the integrated power and gas system. *Appl. Energy* **2019**, *236*, 1158–1167. [CrossRef]

38. Qu, K.P.; Zheng, B.M.; Yu, T.; Li, H.F. Convex decoupled-synergetic strategies for robust multi-objective power and gas flow considering power to gas. *Energy* **2019**, *168*, 753–771. [CrossRef]

39. Liu, J.; Luo, X.J. Environmental economic dispatching adopting multi-objective random black-hole particle swarm optimization algorithm. *Proc. CSEE* **2010**, *30*, 105–111. [CrossRef]

40. Liu, J.; Luo, X.J. Short-term optimal environmental economic hydrothermal scheduling based on handling complicated constraints of multi-chain cascaded hydropower station. *Proc. CSEE* **2012**, *32*, 27–35. [CrossRef]

41. Liu, J.; Luo, X.J. Optimal economic emission hydrothermal scheduling using a novel algorithm based on black hole theory and annual profit analysis considering fuel gas desulphurization. In Proceedings of the 1st International IET Renewable Power Generation Conference, Edinburgh, UK, 6–8 September 2011. [CrossRef]

42. Liu, J.; Lu, Q.W.; Liu, Y. Optimal capacity allocation of hybrid wind-solar-battery power system containing electric vehicles. In Proceedings of the 5th International IET Renewable Power Generation Conference, London, UK, 21–22 September 2016.

43. Byrd, R.H.; Gilbert, J.C.; Nocedal, J. A trust region method based on interior point techniques for nonlinear programming. *Math. Program.* **2000**, *89*, 149–185. [CrossRef]

44. Mohamed Abdelmageed Abdelaziz, M.; Farag, H.E.; El-Saadany, E.F.; Mohamed, Y.A.I. A novel and generalized three-phase power flow algorithm for islanded microgrids using a newton trust region method. *IEEE Trans. Power Syst.* **2013**, *28*, 190–201. [CrossRef]

45. Wilamowski, B.M.; Yu, H. Improved computation for Levenberg-Marquardt training. *IEEE Trans. Neural Netw.* **2010**, *21*, 930–937. [CrossRef]

46. Kanzowa, C.; Yamashitab, N.; Fukushimab, M. Levenberg-Marquardt methods with strong local convergence properties for solving nonlinear equations with convex constraints. *J. Comput. Appl. Math.* **2004**, *172*, 375–397. [CrossRef]

47. Zheng, J.; Wen, F.S.; Li, L.; Wang, K.; Gao, C. Transmission system expansion planning considering combined operation of wind farms and energy storage systems. *Autom. Electr. Power Syst.* **2013**, *37*, 135–142. [CrossRef]

 *energies*

Article

Reactive Power Control Method for Enhancing the Transient Stability Total Transfer Capability of Transmission Lines for a System with Large-Scale Renewable Energy Sources

Yuwei Zhang [1,*], Wenying Liu [1], Fangyu Wang [1], Yaoxiang Zhang [1] and Yalou Li [2]

[1] State Key Laboratory of Alternate Electrical Power System with Renewable Energy Sources, North China Electric Power University, Beijing 102206, China; liuwyls@126.com (W.L.); wangfangyuncepu@163.com (F.W.); zhangyaoxiangncepu@163.com (Y.Z.)

[2] China Electric Power Research Institute, Beijing 100192, China; liyaloucepri@163.com

* Correspondence: zhangyuweincepu@126.com; Tel.: +86-188-1177-4465

Received: 18 May 2020; Accepted: 15 June 2020; Published: 17 June 2020

Abstract: With the increased proportion of intermittent renewable energy sources (RES) integrated into the sending-end, the total transfer capability of transmission lines is not sufficient during the peak periods of renewable primary energy (e.g., the wind force), causing severe RES power curtailment. The total transfer capability of transmission lines is generally restricted by the transient stability total transfer capability (TSTTC). This paper presents a reactive power control method to enhance the TSTTC of transmission lines. The key is to obtain the sensitivity between TSTTC and reactive power, while the Thevenin equivalent voltage is the link connecting TSTTC and reactive power. The Thevenin theorem states that an active circuit between two load terminals can be considered as an individual voltage source. The voltage of this source would be open-circuit voltage across the terminals, and the internal impedance of the source is the equivalent impedance of the circuit across the terminals. The Thevenin voltage used in Thevenin's theorem is an ideal voltage source equal to the open-circuit voltage at the terminals. Thus, the sensitivities between TSTTC and the Thevenin equivalent voltages of the sending-end and receiving-end were firstly derived using the equal area criterion. Secondly, the sensitivity between the Thevenin equivalent voltage and reactive power was derived using the total differentiation method. By connecting the above sensitivities together with the relevant parameters calculated from Thevenin equivalent parameter identification and power flow equation, the sensitivity between TSTTC and reactive power was obtained, which was used as the control priority in the proposed reactive power control method. At last, the method was applied to the Gansu Province Power Grid in China to demonstrate its effectiveness, and the accuracy of the sensitivity between TSTTC and reactive power was verified.

Keywords: TSTTC of transmission lines; sensitivity between TSTTC and reactive power; reactive power control method

---

1. Introduction

Large-scale renewable energy source (RES) bases in China are mostly located in renewable energy-rich areas far from the load center. The reverse distribution of RES and load leads to large-scale RES power being difficult to consume locally. A large amount of intermittent RES power needs to be delivered to the load center for consumption through long-distance transmission lines. The transfer capability of transmission lines is not sufficient during the peak periods of renewable primary energy, causing a severe RES power curtailment problem [1].

The transfer capability of transmission lines is generally restricted by the transient stability total transfer capability (TSTTC) [2]. TSTTC is the ability of the system to deliver power from the sending-end to the load center through transmission lines, under the condition that the system can stay transiently stable after contingencies occur. Therefore, improving the system transient stability can help in enhancing TSTTC. Several studies proposed methods for improving the system transient stability, as well as enhancing TSTTC from different aspects. One of the effective methods is active power control strategy, which reschedules the generation plans of thermal generation units to improve the system transient stability, thereby enhancing TSTTC [3–5]. However, for the sending-end integrated with large-scale RES, the original generation plans of thermal generation units are made considering maximizing the consumption of RES power, and rescheduling generation plans may crowd out the consumption space for RES power. For this reason, studies analyzed the effect of the excitation system parameters of thermal generation units on transient stability, and they proposed methods of increasing the response ratio of excitation voltage and the force excitation threshold voltage to enhance TSTTC without changing generation plans. However, transient excitation enhancement is an open-loop control without feedback; thus, it cannot automatically adapt to the changes of system conditions, and it requires remote transmission of signals, resulting in reducing the reliability of the system [6,7]. In addition, with the application and development of flexible alternating current (AC) transmission systems (FACTS), the utilization of FACTS devices like thyristor-controlled series compensators (TCSC), static var compensators (SVC), and static synchronous compensators (STATCOM) can reduce the energy accumulation in generators during contingencies, thereby improving the system transient stability and enhancing TSTTC [8–10]. Furthermore, studies found that superconducting magnetic energy storage (SMES) can also store the excess energy in generators during contingencies with a quicker response and more flexibility [11,12]. However, for systems without FACTS devices or energy storage, installing them needs a long planning time and big investment.

In addition to the methods above, through transient stability analysis, the Thevenin equivalent voltages of the sending-end and receiving-end have a significant influence on the TSTTC, and Thevenin equivalent voltages and reactive power are highly correlated; therefore, reactive power control can also enhance the TSTTC. In a power system, reactive power control is used to control the output of reactive power sources like generators, shunt reactors, and shunt capacitors, thereby adjusting the injecting reactive power into some nodes or the reactive power flow through some lines. At present, reactive power control is mainly used for voltage regulation and voltage stability maintenance [13–15]; however, no previous work on enhancing TSTTC using reactive power control was reported.

A new reactive power control method for enhancing the TSTTC of transmission lines for a system with large-scale RES integrated into the sending-end is presented in this paper, with the following main contributions:

1. The sensitivity between TSTTC and reactive power is derived, and, with the sensitivity and TSTTC gap, the reactive power adjustment that is needed for filling the TSTTC gap can be calculated. Therefore, the TSTTC of transmission lines can be effectively enhanced by adjusting the reactive power of reactive power devices.
2. The proposed reactive power control method is especially useful for enhancing TSTTC for a system with large-scale RES integrated into the sending-end in its existing structure. Because the reactive power adjustment is available from the generators in conventional generation plants and the reactive-load compensation equipment in substations, it needs no further investment in terms of installing new devices.
3. Using reactive power control to enhance TSTTC does not involve active power changing; therefore, it will not crowd out the consumption space for RES power. With the enhanced TSTTC, the transmission lines are able to deliver more RES power to the load center, thereby reducing RES power curtailment.

This rest of the paper is organized as follows: in Section 2, the sensitivity between the TSTTC of transmission lines and the Thevenin equivalent voltage is derived using equal area criterion. In Section 3, the sensitivity between the Thevenin equivalent voltage and reactive power is derived using the total differentiation method. By connecting the above sensitivities together, the sensitivity between the TSTTC and reactive power is presented in Section 4, along with the calculation methods to get the relevant parameters needed. In Section 5, using the sensitivity between the TSTTC and reactive power as the control priority, the detailed reactive power control method for enhancing TSTTC is proposed. In Section 6, the method is applied to the Gansu Province Power Grid to demonstrate its effectiveness. Finally, some conclusions are drawn in Section 7.

2. Sensitivity between TSTTC and Thevenin Equivalent Voltages

The schematic diagram of a system composed of a sending-end and receiving-end with transmission lines connecting them is shown in Figure 1, and the sending-end is integrated with large-scale RES.

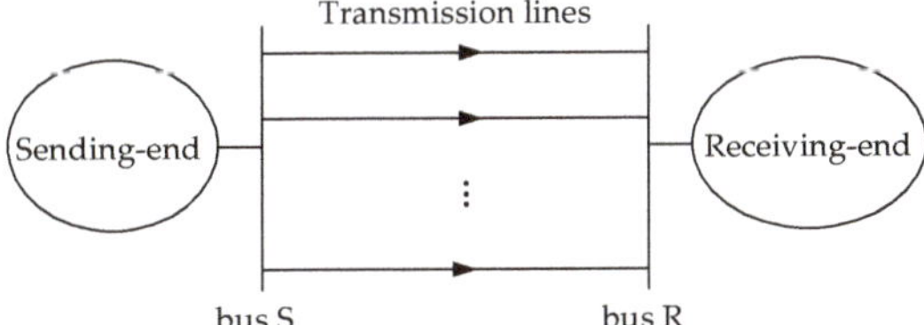

Figure 1. Schematic diagram of a system composed of the sending-end and the receiving-end with transmission lines connecting them.

For the system shown in Figure 1, according to Thevenin's theorem, at any time, looking from bus S to the sending-end, the sending-end can be equivalent to a generator (Thevenin equivalent voltage $\dot{E}_S = E_S \angle \delta_S$, Thevenin equivalent reactance X_S) [16]. Similarly, the receiving-end is also equivalent to a generator (Thevenin equivalent voltage $\dot{E}_R = E_R \angle \delta_R$, Thevenin equivalent reactance X_R). The equivalent resistance of the system is much lower than the reactance of long-distance transmission lines which is X_D; thus, so it is ignored. The equivalent model of the system is shown in Figure 2.

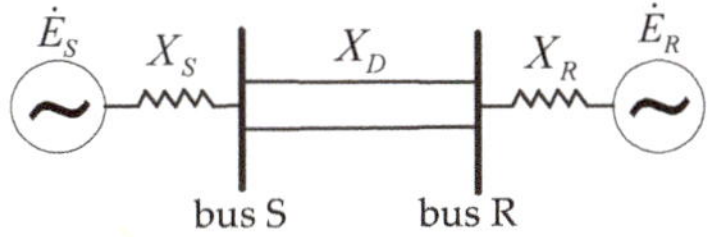

Figure 2. Equivalent model of the system.

The transmission power through the transmission lines is as follows:

$$P_{SR} = \frac{E_S E_R}{X_\Sigma} \sin(\delta_S - \delta_R), \tag{1}$$

where $X_\Sigma = X_S + X_D + X_R$.

Taking the $N - 1$ fault on the transmission lines as the expected fault, TSTTC is the maximum allowable transmission power through the transmission lines, under the condition that the system can stay transiently stable after an expected fault occurs. Let $\delta = \delta_S - \delta_R$; then, the transient stability analysis of the two-equivalent-machine system is converted into a single-equivalent-machine system whose rotor motion equation is shown in Equation (2).

$$\frac{d^2\delta(t)}{dt^2} = \frac{1}{T_J}(P_m - P_{SR}(t)),\tag{2}$$

where

$$P_m = P_{SR}(0),\tag{3}$$

$$P_{SR}(t) = \frac{E_S E_R}{X_\Sigma(t)}\sin\delta(t),\tag{4}$$

where t is the transient process moment, $X_\Sigma(t)$ is the reactance of different transient phases (pre-fault:$X_\Sigma(\mathrm{I})$, fault:$X_\Sigma(\mathrm{II})$, and post-fault:$X_\Sigma(\mathrm{III})$), and T_J is the system inertia time constant.

According to the equal area criterion, if the system is under the critical condition of transient stability, as shown in Figure 3, the equivalent rotor's acceleration area A and the biggest possible deceleration area D are equal in the first swing. In this condition, P_m gets its maximum value $P_m^{\max}$, which is the TSTTC of transmission lines [5].

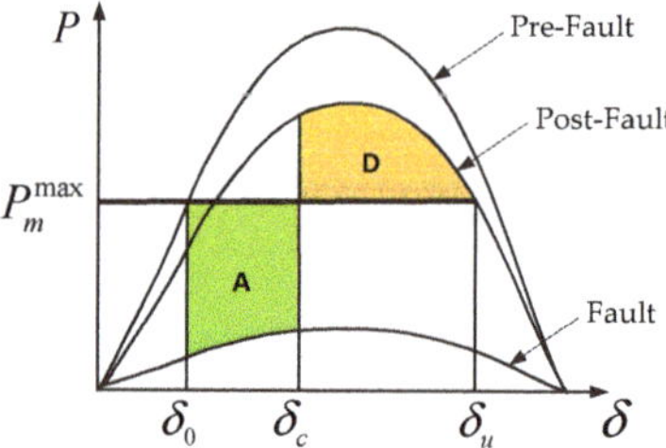

Figure 3. Transient process power–angle curve under the critical condition of transient stability.

The difference between area D and area A is zero.

$$D - A = \int_{\delta_c}^{\delta_u}[\frac{E_S E_R}{X_\Sigma(\mathrm{III})}\sin\delta(t) - P_m^{\max}]\mathrm{d}\delta(t) - \int_{\delta_0}^{\delta_c}[P_m^{\max} - \frac{E_S E_R}{X_\Sigma(\mathrm{II})}\sin\delta(t)]\mathrm{d}\delta(t) = 0,\tag{5}$$

where δ_0, δ_c, and δ_u are the initial angle, the fault clearing angle, and the critical angle, which are as follows:

$$\begin{cases} \delta_0 = \arcsin(P_m^{\max} / \frac{E_S E_R}{X_\Sigma(\mathrm{I})}) \\ \delta_c = \delta_0 + \frac{1}{2}\frac{P_m^{\max}}{T_J}t_c^2 \\ \delta_u = \arcsin(P_m^{\max} / \frac{E_S E_R}{X_\Sigma(\mathrm{III})}) \end{cases},\tag{6}$$

where t_c is the fault clearing time.

We can then take Equation (6) into Equation (5) to get the expression of the relationship between TSTTC ($P_m^{\max}$) and the Thevenin equivalent voltage (E_S, E_R) as follows:

$$\begin{aligned}
F\ &(P_m^{\max}, E_S, E_R) = 0 \\
=\ &\frac{E_S E_R}{X_\Sigma(\mathrm{III})}\{\cos[\arcsin(P_m^{\max} / \frac{E_S E_R}{X_\Sigma(\mathrm{I})}) + \frac{1}{2}\frac{P_m^{\max}}{T_J}t_c^2] - \cos\arcsin(P_m^{\max} / \frac{E_S E_R}{X_\Sigma(\mathrm{III})})\} \\
&-\frac{E_S E_R}{X_\Sigma(\mathrm{II})}\{\cos[\arcsin(P_m^{\max} / \frac{E_S E_R}{X_\Sigma(\mathrm{I})}) + \frac{1}{2}\frac{P_m^{\max}}{T_J}t_c^2] - \cos\arcsin(P_m^{\max} / \frac{E_S E_R}{X_\Sigma(\mathrm{I})})\} \\
&-P_m^{\max}[\arcsin(P_m^{\max} / \frac{E_S E_R}{X_\Sigma(\mathrm{III})}) - \arcsin(P_m^{\max} / \frac{E_S E_R}{X_\Sigma(\mathrm{I})})]
\end{aligned}\tag{7}$$

Using the implicit function derivative rule with Equation (7), the sensitivity between the TSTTC and the Thevenin equivalent voltage E_S can be calculated as shown in Equation (8). Because E_R has the same position in Equation (7) as E_S, the sensitivity between the TSTTC and E_R is similar and omitted.

$$S_{E_S}^{P_m^{max}} = \frac{\partial P_m^{max}}{\partial E_S} = -\frac{\partial F(P_m^{max}, E_S, E_R)/\partial E_S}{\partial F(P_m^{max}, E_S, E_R)/\partial P_m^{max}}. \tag{8}$$

The detailed expression of $S_{E_S}^{P_m^{max}}$ is presented in the Appendix A.

3. Sensitivity between Thevenin Equivalent Voltage and Reactive Power

The sensitivity between E_S and reactive power is derived in this section, and the sensitivity between E_R and reactive power is similar and omitted.

In Figure 2, the voltage of bus S is $\dot{U}_S = U_S \angle \theta_S$, and the complex power injected into bus S from the sending-end is $\dot{S}_S = P_S + jQ_S$. According to Kirchhoff law, $\dot{E}_S$ is as follows:

$$\dot{E}_S = \dot{U}_S + jX_S \dot{I}_S, \tag{9}$$

where $\dot{I}_S = \frac{\dot{S}_S^*}{\dot{U}_S^*}$.

We can take the module value on both sides of Equation (9) as follows:

$$E_S = \sqrt{U_S^2 + \frac{X_S}{U_S^2}(P_S^2 + Q_S^2) + 2X_S Q_S}. \tag{10}$$

While adjusting the reactive power, the active power does not change and is considered as constant. We can then use the total differentiation method on Equation (10) to obtain Equation (11) as follows:

$$\partial E_S = \left[\frac{\partial E_S}{\partial U_S}, \frac{\partial E_S}{\partial Q_S}\right]\left[\begin{array}{c}\partial U_S \\ \partial Q_S\end{array}\right], \tag{11}$$

where, according to Equation (10), $\frac{\partial E_S}{\partial U_S}$ and $\frac{\partial E_S}{\partial Q_S}$ are as follows:

$$\frac{\partial E_S}{\partial U_S} = \frac{U_S - 2X_S(P_S^2 + Q_S^2)/U_S^2}{\sqrt{U_S^2 + \frac{X_S}{U_S^2}(P_S^2 + Q_S^2) + 2X_S Q_S}}, \tag{12}$$

$$\frac{\partial E_S}{\partial Q_S} = \frac{2X_S(1 + Q_S/U_S^2)}{\sqrt{U_S^2 + \frac{X_S}{U_S^2}(P_S^2 + Q_S^2) + 2X_S Q_S}}. \tag{13}$$

Through the sensitivity between voltage and reactive power ($S_i^{U_S}$) and the reactive power transfer factor ($D_i^{Q_S}$), $\left[\begin{array}{c}\partial U_S \\ \partial Q_S\end{array}\right]$ in Equation (11) can be converted into reactive power adjustment (∂Q_i) as shown in Equation (14).

$$\left[\begin{array}{c}\partial U_S \\ \partial Q_S\end{array}\right] = \left[\begin{array}{ccc}S_{Q_1}^{U_S}, & \cdots, & S_{Q_{NG}}^{U_S} \\ D_{Q_1}^{Q_S}, & \cdots, & D_{Q_{NG}}^{Q_S}\end{array}\right]\left[\begin{array}{c}\partial Q_1 \\ \vdots \\ \partial Q_{NG}\end{array}\right]. \tag{14}$$

Taking Equation (14) into Equation (11), we get the expression of the sensitivity between E_S and reactive power as follows:

$$S_{Q_i}^{E_S} = \frac{\partial E_S}{\partial Q_i} = \frac{\partial E_S}{\partial U_S} S_{Q_i}^{U_S} + \frac{\partial E_S}{\partial Q_S} D_{Q_i}^{Q_S}. \tag{15}$$

4. Sensitivity between TSTTC and Reactive Power

In Sections 2 and 3, the sensitivity between the TSTTC and the Thevenin equivalent voltage in Equation (8) and the sensitivity between Thevenin equivalent voltage and reactive power in Equation (15) were derived. By connecting Equation (8) and Equation (15), the sensitivity between the TSTTC and reactive power is given in Equation (16).

$$S_{Q_i}^{P_m^{\max}} = \frac{\partial P_m^{\max}}{\partial E_S} \frac{\partial E_S}{\partial Q_i} + \frac{\partial P_m^{\max}}{\partial E_R} \frac{\partial E_R}{\partial Q_i} = S_{E_S}^{P_m^{\max}} S_{Q_i}^{E_S} + S_{E_R}^{P_m^{\max}} S_{Q_i}^{E_R}. \tag{16}$$

According to Equation (16), if the reactive power of node i changes by ΔQ_i, then the TSTTC changes by $S_{Q_i}^{P_m^{\max}} \Delta Q_i$ accordingly. In order to obtain $S_{Q_i}^{P_m^{\max}}$, it is known from Equations (16), (8), and (15) that two kinds of parameters are needed. The first kind can be read directly from the phasor measurement unit (PMU), the supervisory control and data acquisition (SCADA), and the system parameter database, including P_S, Q_S, $U_S \angle \theta_S$, X_D, T_J and t_c. The second kind of parameter needs to be calculated first, including the Thevenin equivalent parameters (E_S,E_R,X_S,X_R), the sensitivity between voltage and reactive power ($S_{Q_i}^{U_S}$), and the reactive power transfer factor ($D_{Q_i}^{Q_S}$). The methods to calculate them are provided below.

4.1. Thevenin Equivalent Parameter Identification

Thevenin equivalent parameters are affected by the grid topology, power generation condition, load condition, etc. Each time the reactive power is adjusted, Thevenin equivalent parameters should be updated. In this paper, the Thevenin equivalent parameters were identified using a tracking algorithm based on variation correction. The algorithm modifies the Thevenin equivalent parameters in the previous condition based on real-time conditions, thereby quickly obtaining those in current conditions [16].

Taking the sending-end shown in Figure 2 as an example, and taking the phase angle of $\dot{I}_S$ as reference, which is $\dot{I}_S = I_S \angle 0°$, by expanding Equation (9) according to the real and imaginary parts, the Thevenin equivalent parameters are expressed as follows:

$$\begin{cases} E_S \cos \delta_S = U_S \cos \theta_S \\ E_S \sin \delta_S = X_S I_S + U_S \sin \theta_S \end{cases}. \tag{17}$$

The main process is to continuously update E_S^k in the current condition (k) through deviation correction, and then use Equation (17) to obtain δ_S^k and X_S^k.

The initial E_S^k is

$$E_S^0 = \frac{E_S^{\max.0} - E_S^{\min.0}}{2}, \tag{18}$$

where

$$\begin{cases} E_S^{\max.k} = U_S^k \cos \theta_S^k / \cos \beta \\ E_S^{\min.k} = U_S^k \\ \beta = \arctan[(Z_S^0 I_S^0 + U_S^0 \sin \theta_S^0)/(U_S^0 \cos \theta_S^0)] \\ Z_S^k = |U_S^k / I_S^k| \end{cases}. \tag{19}$$

We can take E_S^0 into Equation (17) to get δ_S^0 and X_S^0. Then, with the previous values of Thevenin equivalent parameters $(E_S^{k-1}, \delta_S^{k-1}, X_S^{k-1})$ and the present values of P_S^k, Q_S^k, and $\dot{U}_S^k$, we can calculate E_S^k according to the following conditions:

If $(Z_S^k - Z_S^{k-1})(X_S^{k*} - X_S^{k-1}) > 0$, then $E_S^k = E_S^{k-1} - \varepsilon^k$,

If $(Z_S^k - Z_S^{k-1})(X_S^{k*} - X_S^{k-1}) \leq 0$, then $E_S^k = E_S^{k-1} + \varepsilon^k$,

where $\varepsilon^k = \min(E_S^{k-1} - U_S^k, E_S^{k-1} - E_S^{\max.k}, E_S^{k-1} \times \lambda)$ is the variation correction, and λ is a pre-specified parameter. X_S^{k*} is an evaluation of X_S^k considering the previous values $(E_S^{k-1}$ and $\delta_S^{k-1})$ and the present values $(P_S^k, Q_S^k$, and $\dot{U}_S^k)$.

After obtaining E_S^k, we can take it into Equation (17) to get δ_S^k and X_S^k, so that all Thevenin equivalent parameters in the sending-end are updated.

4.2. Sensitivity between Voltage and Reactive Power

The sensitivity between voltage and reactive power can be calculated from the Newton–Raphson power flow in Equation (20) [17].

$$\begin{bmatrix} \Delta P \\ \Delta Q \end{bmatrix} = \begin{bmatrix} J_H & J_N \\ J_J & J_L \end{bmatrix} \begin{bmatrix} \Delta \theta \\ \Delta U / U \end{bmatrix}, \tag{20}$$

where J_H, J_N, J_J, J_L is the Jacobian block matrix from the power flow equation, ΔP and ΔQ are the active and reactive power deviations, $\Delta \theta$ is the voltage phase angle deviation, and $\Delta U / U$ is the ratio of voltage magnitude deviation to voltage magnitude.

The active power is not changed while adjusting the reactive power; thus, $\Delta P = 0$. We can then take it to Equation (20) to get Equation (21).

$$\Delta Q = (J_L - J_J J_H^{-1} J_N) \Delta U / U. \tag{21}$$

Thus, the sensitivity between voltage and reactive power is as follows:

$$S = (J_L - J_J J_H^{-1} J_N)^{-1} / U, \tag{22}$$

where S is the set of $S_{Q_i}^{U_j}$ in Equation (15).

4.3. Reactive Power Transfer Factor

The sensitivity between the reactive power flowing through line ij (starting node i, ending node j) and the reactive power injected in node l are defined using the implicit function relationship shown in Equation (23) [18].

$$\frac{\partial Q_{ij}}{\partial Q_l} = \begin{bmatrix} \dfrac{\partial Q_{ij}}{\partial U_i}, & \dfrac{\partial Q_{ij}}{\partial U_j}, & \dfrac{\partial Q_{ij}}{\partial \theta_i}, & \dfrac{\partial Q_{ij}}{\partial \theta_j} \end{bmatrix} \begin{bmatrix} \partial U_i / \partial Q_l \\ \partial U_j / \partial Q_l \\ \partial \theta_i / \partial Q_l \\ \partial \theta_j / \partial Q_l \end{bmatrix}, \tag{23}$$

where

$$\begin{bmatrix} \dfrac{\partial Q_{ij}}{\partial U_i}, & \dfrac{\partial Q_{ij}}{\partial U_j}, & \dfrac{\partial Q_{ij}}{\partial \theta_i}, & \dfrac{\partial Q_{ij}}{\partial \theta_j} \end{bmatrix}^T = \begin{bmatrix} U_j Y_{ij} \sin(\theta_i - \theta_j) \\ U_i Y_{ij} \sin(\theta_j - \theta_i) \\ -U_i U_j Y_{ij} \cos(\theta_i - \theta_j) \\ U_i U_j Y_{ij} \cos(\theta_i - \theta_j) \end{bmatrix}, \tag{24}$$

where $\partial U_i / \partial Q_l, \partial U_j / \partial Q_l, \partial \theta_i / \partial Q_l, \partial \theta_j / \partial Q_l$ can be calculated using the Jacobian matrix in Equation (20).

The total injected reactive power (Q_S) from the sending-end into bus S is the sum of the reactive power transferring through the lines which are directly connected to bus S in the sending-end; thus, the reactive power transfer factor ($D_{Qi}^{Q_S}$) in Equation (25) is as follows:

$$D_{Qi}^{Q_S} = \frac{\partial Q_S}{\partial Q_l} = \sum_{i \in S} \frac{\partial Q_{iS}}{\partial Q_l}. \tag{25}$$

5. Reactive Power Control Method for Enhancing the TSTTC of Transmission Lines

The goal of reactive power control is to enhance the TSTTC of transmission lines effectively, under circumstances where the RES power is curtailed due to the insufficiency of TSTTC. The detailed reactive power control method is presented below. A flow diagram of the proposed method is shown in Figure 4.

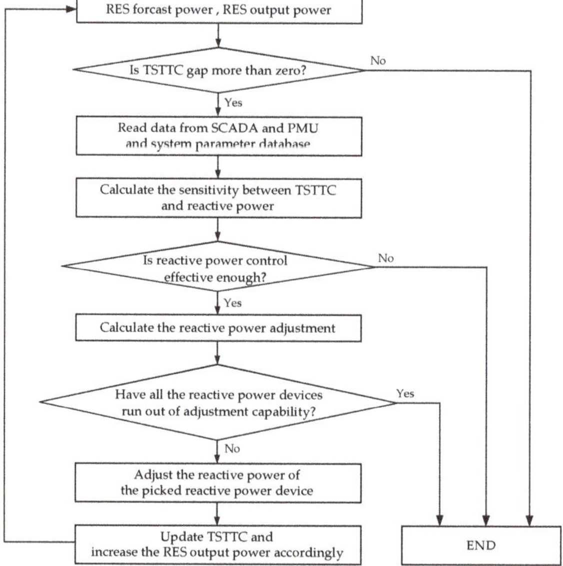

Figure 4. Flow diagram of the reactive power control method.

1. According to the RES forcast power and RES output power, the RES power curtailment is calculated, which is the TSTTC gap ($\Delta P_m^{\max}$).
2. With the data read from PMU, SCADA and the system parameter database, the method presented in Sections 2–4 is used to calculate the sensitivity between TSTTC and reactive power in Equation (16), and the reactive power devices are sorted by their sensitivities.
3. The reactive power device that is capable of reactive power adjustment and with the biggest sensitivity is selected, and its reactive power adjustment is calculated as follows:

$$\Delta Q_i = \min(\Delta P_m^{\max}/S_{Qi}^{P_m^{\max}}, Q_i^{\max} - Q_i, \Delta Q_i^U, \Delta Q_i^{US}), \tag{26}$$

where $\Delta P_m^{\max}/S_{Qi}^{P_m^{\max}}$ is the reactive power adjustment needed to fill the TSTTC gap, $Q_i^{\max} - Q_i$ is the device's maximum reactive power adjustment within its adjustment capability, ΔQ_i^U is the maximum available reactive power adjustment considering voltage level [13], and ΔQ_i^{US} is the maximum available reactive power adjustment considering voltage stability [15].

4. Each time after the reactive power is adjusted, the TSTTC is updated and the RES output power is increased accordingly. Then, all steps are repeated until there is no RES power curtailment, or the reactive power control is not effective enough, or all the reactive power devices run out of adjustment capability.

6. Case Analysis

The Gansu Province Power Grid in China was taken as an example, and the comprehensive simulation program Power System Analysis Software Package (PSASP) was used to verify the effectiveness of the proposed reactive power control method.

6.1. Test System Description

The schematic diagram of the Gansu Province Power Grid is shown in Figure 5. The Gansu Province Power Grid is composed of the sending-end which is the Hexi area grid and the receiving-end which is the Gansu main grid, while the sending-end and receiving-end are connected by the Hexi–Wusheng 750 kV transmission lines.

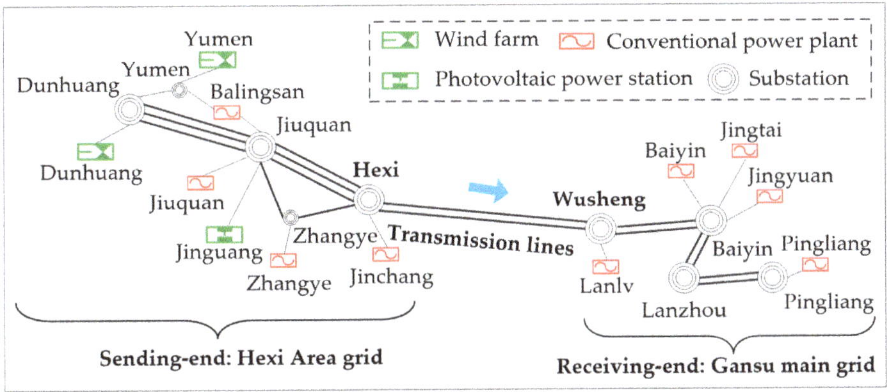

Figure 5. Schematic diagram of the Gansu Province Power Grid.

In the sending-end, at the end of 2019, the total installed capacities of wind farms, photovoltaic power stations, and conventional generation plants were 12,773 MW, 7606 MW, and 4164 MW, respectively, and the maximum load was 4940 MW. The RES power needs to be delivered to the receiving-end through the transmission lines. The TSTTC of the transmission lines is around 4400 MW in the typical winter big load operation conditions, and it is far from sufficient during the peak periods of renewable primary energy, causing severe RES power curtailment.

Taking the typical winter big load operation conditions as the original operation conditions, the RES output power, the conventional generation output power, and the total load are 6073 MW, 1900 MW, and 3569 MW, respectively. The RES forcast power is 6485 MW and the original TSTTC is 4404 MW. Therefore, there is 412 MW of RES power curtailment due to the insufficiency of the TSTTC, and the TSTTC gap is 412 MW. The expected fault for the TSTTC calculation is an $N - 1$ fault on the Hexi–Wusheng 750-kV transmission lines. The reactive power adjustment capabilities of conventional generation plants and substations are listed in Table 1.

Table 1. Reactive power adjustment capabilities of conventional generation plants and substations (sorted by the first letter of names).

Conventional Generation Plants	Reactive Power Adjustment Capabilities (MVar)	Substations	Reactive Power Adjustment Capabilities (MVar)
Baiyin	200	Baiyin	250
Balingsan	200	Dunhuang	300
Jinchang	300	Hexi	250
Jingtai	300	Jiuquan	200
Jingyuan	400	Lanzhou	250
Jiuquan	350	Pingliang	300
Lanlv	400	Wusheng	250
Pingliang	300	Yumen	250
Zhangye	350	Zhangye	300

6.2. Analysis of the Results of Reactive Power Control Method for Enhancing the TSTTC of Transmission Lines

Based on the original operation condition above, taking the first reactive power adjustment as an example, the sensitivity between the TSTTC and the reactive power of each conventional generation plant and substation is shown in Table 2.

Table 2. Sensitivity between the transient stability total transfer capability (TSTTC) and the reactive power of each conventional generation plant and substation in the first reactive power adjustment (sorted by sensitivities).

Conventional Generation Plants	Sensitivity (MW/MVar)	Substations	Sensitivity (MW/MVar)
Baiyin	0.195	Wusheng	0.248
Jingtai	0.172	Hexi	0.223
Jingyuan	0.137	Baiyin	0.199
Jinchang	0.089	Zhangye	0.078
Zhangye	0.030	Lanzhou	0.041
Lanlv	0.023	Jiuquan	0.033
Pingliang	0.021	Pingliang	0.013
Jiuquan	0.011	Yumen	0.013
Balingsan	0.010	Dunhuang	0.012

As shown in Table 2, the Wusheng substation has the biggest sensitivity, and its reactive power adjustment is 250 MW, as calculated from Equation (26). The system transient stability before and after the reactive power adjustment of Wusheng substation is shown in Figure 6, and the system transient stability is represented by the power angle difference curve of the two power plants with the largest power angle difference in the system after an $N - 1$ fault occurs. As shown in Figure 6, the system transient stability is near critical before the adjustment, and it is improved after the adjustment.

After the first reactive power adjustment, because the system transient stability is improved, the TSTTC is enhanced by 62 MW. Therefore, more RES output power can be delivered through the transmission lines, and the RES power curtailment is reduced by 62 MW to 350 MW.

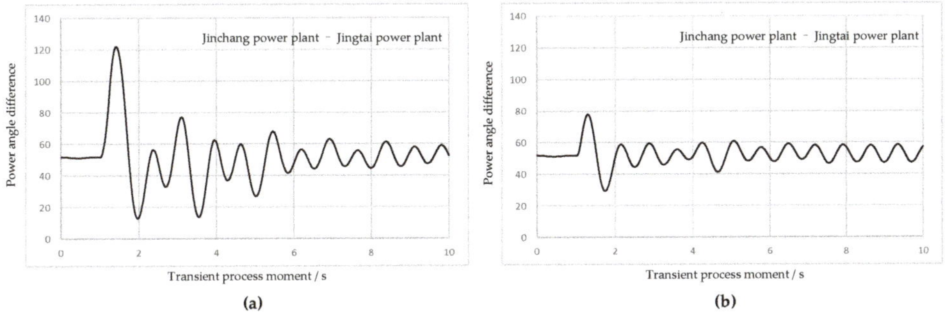

Figure 6. System transient stability of Wusheng substation: (**a**) before reactive power adjustment and (**b**) after reactive power adjustment.

The whole process of reactive power adjustment is shown in Table 3. The detailed operation condition information after each reactive power adjustment is provided in the Supplementary Materials.

Table 3. Whole process of reactive power adjustment.

Adjustment Times	Adjustment Nodes	Sensitivity (MW/MVar)	Adjustment Devices	Reactive Power Adjustment (MVar)
1	Wusheng substation	0.248	Shunt reactors	250
2	Hexi substation	0.216	Shunt capacitors	250
3	Baiyin substation	0.192	Shunt capacitors	250
4	Baiyin power plant	0.190	Generators	200
5	Jingtai power plant	0.167	Generators	300
6	Jingyuan power plant	0.135	Generators	400
7	Jinchang power plant	0.087	Generators	300
8	Zhangye substation	0.077	Shunt reactors	300
9	Lanzhou substation	0.040	Shunt reactors	250
10	Jiuquan substation	0.035	Shunt reactors	200

Figure 7 shows the changing curve of the TSTTC of the Hexi–Wusheng 750-kV transmission lines after each reactive power adjustment. Due to the reactive power adjustment, the TSTTC of Hexi–Wusheng 750-kV transmission lines is enhanced from 4404 MW to 4776 MW.

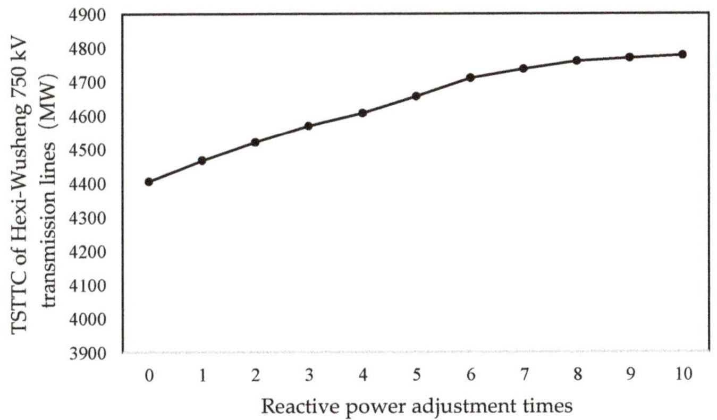

Figure 7. Changing curve of the TSTTC of Hexi–Wusheng 750-kV transmission lines.

As shown in Figure 7, after eight reactive power adjustments, the effect of reactive power adjustment is no longer obvious, and the TSTTC is 4769 MW. This is because the adjustment capacity of the devices with a high sensitivity was used up, and the remaining devices with low sensitivities have little effect on enhancing TSTTC by adjusting their reactive power; thus, the adjustment would be stopped after the eighth time.

Thereofre, in this case, if the TSTTC gap is less than 355 MW, it can be totally filled using the proposed reactive power control method. Thus, in this case, the proposed method can help to consume 355 MW of RES power curtailment by delivering it to the load center.

6.3. Accuracy of the Sensitivity between TSTTC and Reactive Power

In this paper, in order to obtain the analytical expression of the sensitivity between TSTTC and reactive power, while deriving the relationship between TSTTC and the Thevenin equivalent voltage in Section 2, the system is considered equivalent to a single-machine system. In practical operation, for a multi-machine system, the calculation of TSTTC uses the continuous power flow method with transient stability constraints [19]; thus, the accuracy of the sensitivity between the TSTTC and reactive power calculated in this paper needs to be analyzed.

Figure 8 shows the TSTTC calculated using the sensitivity proposed in this paper and the TSTTC calculated using the method in Reference [19] using 10 reactive power adjustments described above. The original TSTTC was calculated using the method in Reference [19], giving 4404 MW. The relative average error of this method compared to the continuous power flow method is 0.23%.

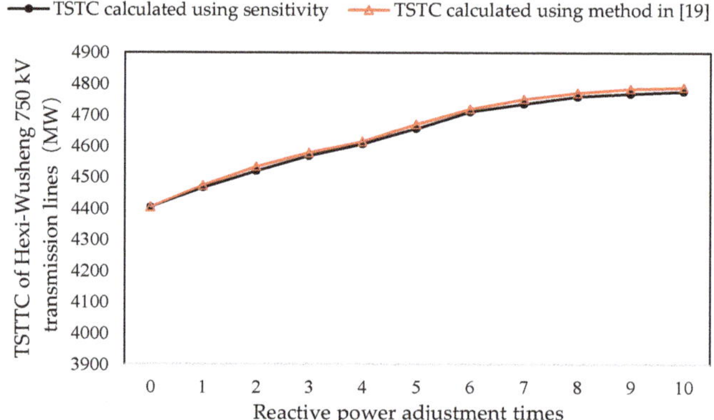

Figure 8. Comparison of TSTTC calculated using different methods.

Therefore, the sensitivity derived in this paper is accurate enough for reactive power control when enhancing the TSTTC in practical operation.

7. Conclusions

A reactive power control method for enhancing the TSTTC of transmission lines was proposed in this paper. The key is to obtain the sensitivity between TSTTC and reactive power, which is used as the control priority in the reactive power control method, so that the TSTTC can be effectively enhanced to a certain value. The detailed derivation and the calculation method to obtain the sensitivity between the TSTTC and reactive power were presented, followed by the reactive power control method. At last, the method was applied to the Gansu Province Power Grid. Results show that the reactive power control method can effectively enhance the TSTTC of transmission lines; therefore, more RES power can be delivered to the load center, and the RES power curtailment is reduced; furthermore, the derived sensitivity between the TSTTC and reactive power was verified to be accurate enough in practical operation.

The method proposed in this paper is meaningful for reducing RES power curtailment for systems with large-scale integration of intermittent RES to the sending-end.

Supplementary Materials: The following are available online at http://www.mdpi.com/1996-1073/13/12/3154/s1.

Author Contributions: Conceptualization, Y.Z. (Yuwei Zhang); methodology, Y.Z. (Yuwei Zhang) and W.L.; validation, Y.Z. (Yuwei Zhang) and F.W.; investigation, Y.Z. (Yuwei Zhang) and Y.Z. (Yaoxiang Zhang); writing—original draft preparation, Y.Z. (Yuwei Zhang); writing—review and editing, W.L and Y.L.; supervision, W.L. and F.W. All authors read and agreed to the published version of the manuscript.

Funding: This research was funded by the National Key R&D Program of China (number 2018YFE0208400) and the Fundamental Research Funds for the Central Universities (number 2019QN125).

Conflicts of Interest: The authors declare no conflicts of interest.

Appendix A

According to Equation (7), $S_{E_S}^{P_m^{\max}}$ as shown in Equation (8) in the main text can be derived as follows:

$$
\begin{aligned}
\partial F(P_m^{\max}, E_S, E_R)/\partial E_S = {} & \frac{E_R}{X_\Sigma(\mathrm{III})}(\cos\delta_c - \cos\delta_u) + \frac{E_S E_R}{X_\Sigma(\mathrm{III})}\left(\frac{\partial\delta_u}{\partial E_S}\sin\delta_u - \frac{\partial\delta_c}{\partial E_S}\sin\delta_c\right) \\
& - \frac{E_R}{X_\Sigma(\mathrm{II})}(\cos\delta_c - \cos\delta_0) - \frac{E_S E_R}{X_\Sigma(\mathrm{II})}\left(\frac{\partial\delta_0}{\partial E_S}\sin\delta_0 - \frac{\partial\delta_c}{\partial E_S}\sin\delta_c\right) - P_m^{\max}\left(\frac{\partial\delta_u}{\partial E_S} - \frac{\partial\delta_0}{\partial E_S}\right)
\end{aligned}
\tag{A1}
$$

$$
\begin{aligned}
\partial F(P_m^{\max}, E_S, E_R)/\partial P_m^{\max} = {} & \frac{E_S E_R}{X_\Sigma(\mathrm{III})}\left(\frac{\partial\delta_u}{\partial P_m^{\max}}\sin\delta_u - \frac{\partial\delta_c}{\partial P_m^{\max}}\sin\delta_c\right) \\
& - \frac{E_S E_R}{X_\Sigma(\mathrm{II})}\left(\frac{\partial\delta_0}{\partial P_m^{\max}}\sin\delta_0 - \frac{\partial\delta_c}{\partial P_m^{\max}}\sin\delta_c\right) - (\delta_u - \delta_0) - P_m^{\max}\left(\frac{\partial\delta_u}{\partial P_m^{\max}} - \frac{\partial\delta_0}{\partial P_m^{\max}}\right)
\end{aligned}
\tag{A2}
$$

where, in Equations (A1) and (A2), the expressions of δ_0, δ_c, and δ_u about E_S, E_R, and $P_m^{\max}$ are as in Equation (6) in the main text, and their derivatives of E_S and $P_m^{\max}$ are as follows:

$$
\left\{
\begin{aligned}
\frac{\partial\delta_0}{\partial E_S} &= -\frac{P_m^{\max}X_\Sigma(\mathrm{I})}{2E_S\sqrt{E_S^2 E_R^2 - (P_m^{\max}X_\Sigma(\mathrm{I}))^2}} \\
\frac{\partial\delta_c}{\partial E_S} &= -\frac{P_m^{\max}X_\Sigma(\mathrm{I})}{2E_S\sqrt{E_S^2 E_R^2 - (P_m^{\max}X_\Sigma(\mathrm{I}))^2}} \\
\frac{\partial\delta_u}{\partial E_S} &= -\frac{P_m^{\max}X_\Sigma(\mathrm{III})}{2E_S\sqrt{E_S^2 E_R^2 - (P_m^{\max}X_\Sigma(\mathrm{III}))^2}}
\end{aligned}
\right. ,
\tag{A3}
$$

$$
\left\{
\begin{aligned}
\frac{\partial\delta_0}{\partial P_m^{\max}} &= \frac{X_\Sigma(\mathrm{I})}{\sqrt{E_S^2 E_R^2 - (P_m^{\max}X_\Sigma(\mathrm{I}))^2}} \\
\frac{\partial\delta_c}{\partial P_m^{\max}} &= \frac{X_\Sigma(\mathrm{I})}{\sqrt{E_S^2 E_R^2 - (P_m^{\max}X_\Sigma(\mathrm{I}))^2}} + \frac{1}{2}\frac{t_c^2}{T_J} \\
\frac{\partial\delta_u}{\partial P_m^{\max}} &= \frac{X_\Sigma(\mathrm{III})}{\sqrt{E_S^2 E_R^2 - (P_m^{\max}X_\Sigma(\mathrm{III}))^2}}
\end{aligned}
\right. .
\tag{A4}
$$

References

1. Liu, Y.; Zhao, J.; Xu, L.; Liu, T.; Qiu, G.; Liu, J. Online TTC Estimation Using Nonparametric Analytics Considering Wind Power Integration. *IEEE Trans. Power Syst.* **2019**, *34*, 494–505. [CrossRef]
2. Gholipour, E.; Saadate, S. Improving of transient stability of power systems using UPFC. *IEEE Trans. Power Deliv.* **2005**, *20*, 1677–1682. [CrossRef]
3. Wang, Z.; Song, X.; Xin, H.; Gan, D.; Wong, K.P. Risk-Based Coordination of Generation Rescheduling and Load Shedding for Transient Stability Enhancement. *IEEE Trans. Power Syst.* **2013**, *28*, 4674–4682. [CrossRef]
4. Renedo, J.; Garcı'a-Cerrada, A.; Rouco, L. Active power control strategies for transient stability enhancement of AC/DC grids with VSC-HVDC multi-terminal systems. *IEEE Trans. Power Syst.* **2016**, *31*, 4595–4604. [CrossRef]
5. Zhang, Y.; Liu, W.; Huan, Y.; Zhou, Q.; Wang, N. An optimal day-ahead thermal generation scheduling method to enhance total transfer capability for the sending-side system with large-scale wind power integration. *Energies* **2020**, *13*, 2375. [CrossRef]
6. Kang, H.; Liu, Y.; Wu, Q.H.; Zhou, X. Switching excitation controller for enhancement of transient stability of multi-machine power systems. *CSEE J. Power Energy Syst.* **2015**, *1*, 86–93. [CrossRef]

7. Díez-Maroto, L.; Renedo, J.; Rouco, L.; Fernández-Bernal, F. Lyapunov Stability Based Wide Area Control Systems for Excitation Boosters in Synchronous Generators. *IEEE Trans. Power Syst.* **2019**, *34*, 194–204. [CrossRef]

8. Haque, M.H. Improvement of first swing stability limit by utilizing full benefit of shunt FACTS devices. *IEEE Trans. Power Syst.* **2004**, *19*, 1894–1902. [CrossRef]

9. Cvetković, M.; Ilić, M.D. Ectropy-Based Nonlinear Control of FACTS for Transient Stabilization. *IEEE Trans. Power Syst.* **2014**, *29*, 3012–3020. [CrossRef]

10. Eladany, M.M.; Eldesouky, A.A.; Sallam, A.A. Power system transient stability: An Algorithm for assessment and enhancement based on catastrophe theory and FACTS devices. *IEEE Access* **2018**, *6*, 26424–26437. [CrossRef]

11. Shi, J.; Tang, Y.; Xia, Y.; Ren, L.; Li, J.; Jiao, F. Energy function based SMES controller for transient stability enhancement. *IEEE Trans. Appl. Supercond.* **2012**, *22*, 5701304.

12. Kiaei, I.; Lotfifard, S. Tube-based model predictive control of energy storage systems for enhancing transient stability of power systems. *IEEE Trans. Smart Grid* **2018**, *9*, 6438–6447. [CrossRef]

13. Bolognani, S.; Carli, R.; Cavraro, G.; Zampieri, S. On the need for communication for voltage regulation of power distribution grids. *IEEE Trans. Control Netw. Syst.* **2019**, *6*, 1111–1123. [CrossRef]

14. Ou, R.; Xiao, X.; Zou, Z.; Zhang, Y.; Wang, Y. Cooperative control of SFCL and reactive power for improving the transient voltage stability of grid-connected wind farm with DFIGs. *IEEE Trans. Appl. Supercond.* **2016**, *26*, 1–6. [CrossRef]

15. Xinhui, C.; Zhihao, Y.; Daowei, L.; Hongying, Y.; Zeyu, L.; Debin, G.; Xuezhu, J. Silicon-coated gold nanoparticles nanoscopy a new method of sensitivity analysis for static voltage stability online prevention and control of large power grids. *Power Syst. Technol.* **2020**, *44*, 245–254.

16. Corsi, S.; Taranto, G.N. A Real-time voltage instability identification algorithm based on local phasor measurements. *IEEE Trans. Power Syst.* **2008**, *23*, 1271–1279. [CrossRef]

17. Kim, S.-B.; Song, S.-H. A hybrid reactive power control method of distributed generation to mitigate voltage rise in low-voltage grid. *Energies* **2020**, *13*, 2078. [CrossRef]

18. Hinojosa, V.H.; Gonzalez-Longatt, F. Preventive security-constrained DCOPF formulation using power transmission distribution factors and line outage distribution factors. *Energies* **2018**, *11*, 1497. [CrossRef]

19. Ejebe, G.C.; Tong, J.; Waight, J.G.; Frame, J.G.; Wang, X.; Tinney, W.F. Available transfer capability calculations. *IEEE Trans. Power Syst.* **1998**, *13*, 1521–1527. [CrossRef]

Article

Small-Signal Stability Analysis of Photovoltaic-Hydro Integrated Systems on Ultra-Low Frequency Oscillation

Sijia Wang [1], Xiangyu Wu [1,*], Gang Chen [2] and Yin Xu [1]

[1]   School of Electrical Engineering, Beijing Jiaotong University, Beijing 100044, China;
      wangsijia@bjtu.edu.cn (S.W.); xuyin@bjtu.edu.cn (Y.X.)
[2]   State Grid Sichuan Electric Power Research Institute, Chengdu 610072, China; gangchen_thu@163.com
*   Correspondence: wuxiangyu@bjtu.edu.cn

Received: 24 January 2020; Accepted: 21 February 2020; Published: 24 February 2020

Abstract: In recent years, ultralow-frequency oscillation has repeatedly occurred in asynchronously connected regional power systems and brought serious threats to the operation of power grids. This phenomenon is mainly caused by hydropower units because of the water hammer effect of turbines and the inappropriate Proportional-Integral-Derivative (PID) parameters of governors. In practice, hydropower and solar power are often combined to form an integrated photovoltaic (PV)-hydro system to realize complementary renewable power generation. This paper studies ultralow-frequency oscillations in integrated PV-hydro systems and analyzes the impacts of PV generation on ultralow-frequency oscillation modes. Firstly, the negative damping problem of hydro turbines and governors in the ultralow-frequency band was analyzed through the damping torque analysis. Subsequently, in order to analyze the impact of PV generation, a small-signal dynamic model of the integrated PV-hydro system was established, considering a detailed dynamic model of PV generation. Based on the small-signal dynamic model, a two-zone and four-machine system and an actual integrated PV-hydro system were selected to analyze the influence of PV generation on ultralow-frequency oscillation modes under different scenarios of PV output powers and locations. The analysis results showed that PV dynamics do not participate in ultralow-frequency oscillation modes and the changes of PV generation to power flows do not cause obvious changes in ultralow-frequency oscillation mode. Ultra-low frequency oscillations are mainly affected by sources participating in the frequency adjustment of systems.

Keywords: photovoltaic generation; ultralow-frequency oscillation; small-signal model; eigenvalue analysis; damping torque

---

1. Introduction

There are differences in mechanism and characteristics between ultralow-frequency oscillation and traditional low-frequency oscillation. The frequency range of low-frequency oscillation is 0.1–2.5 Hz, and frequencies of ultralow-frequency oscillation is below 0.1 Hz. At present, researchers in this field generally believe that ultralow-frequency oscillation is caused by hydropower units. In recent years, ultralow-frequency oscillation has occurred frequently. As early as 1964, a frequency oscillation with a period of about 20 s was observed in the Southwestern United States [1]. Ultralow-frequency oscillations with frequencies below 0.05 Hz have also been observed in Turkey and Bulgaria [2], but due to their small impacts, they have not attracted widespread attention from researchers. In 2016, in an asynchronous networking test conducted by Yunnan Power Grid in China, a relatively severe ultralow-frequency oscillation event occurred, which lasted about half an hour [3]. After tripping the governor of some hydropower units, the oscillation decayed. According to researchers' studies, similar possible troubles of ultralow-frequency oscillations exist in Sichuan Power Grid in China, which also contains a large number of hydropower units. This problem can be triggered after asynchronous networking [4].

Some researchers have carried out research work on ultralow-frequency oscillation. Ultralow-frequency oscillation is related to governors and turbines. The time constant of the water hammer effect and the governor parameters can change the oscillation frequency and damping [5]. When the proportion of hydropower units in a system is high, ultralow-frequency oscillation is likely to occur. Adjusting the PID parameters of governors or increasing the proportion of thermal power units can suppress this oscillation [6]. Reference [7] did the damping torque analysis and pointed out that ultralow-frequency oscillation was caused by negative damping generated by a regulating system. The improper design of governor parameters caused the negative damping torque to be very large, which affected the damping characteristics of the unit. Reference [8] used the vector margin method to analyze multimachine systems, and the results showed that thermal power units and hydropower units with small time constants of the water hammer effect can increase the vector margin of the system while hydropower units with large time constants of the water hammer effect can reduce the vector margin. Reference [9] built a small-signal model of a hydropower system and analyzed the change of the damping of the ultralow-frequency oscillation mode when the PID parameters of a governor were changed through a characteristic analysis method. Changing parameters can increase the damping ratio of the system and suppress the ultralow-frequency oscillation of the system.

With the development of distributed generation technology [10], more photovoltaic (PV) generation is connected to hydropower systems to realize an integrated system, which can make electricity complementary. As a renewable energy, solar power plays an increasingly important role in power systems. However, due to the strong correlation between the light intensity received by surface and environmental factors, the PV output power is random [11]. The output power can be maintained in a stable state, when the weather is clear and the sunlight is direct. However, when the weather is cloudy, the output power will decrease sharply in a short time. Such strong fluctuations can have a huge impact on the stability of the power system. Hydropower can quickly adjust its output power to complement the output power of solar power generation, which can realize smooth power generation for the integrated system. The access of PV changes the dynamic characteristics and power flow of the system, which may affect oscillation modes [12]. Reference [13] analyzed the impact of PV stations on a hydropower system from the perspectives of frequency characteristics, voltage characteristics, and stability. The impact of PV access on low-frequency oscillations of hydropower systems has been extensively studied.

Reference [14] pointed out that renewable energy including wind power and solar power could result in new low-frequency oscillation modes. Reference [15] focused on the damping of local-mode power system oscillations and pointed out that, through eigenvalue analysis, the impact of PV power generation on the small-signal stability of power systems can be positive or negative. Reference [16] showed that, as PV penetration increases and PV replace synchronous motors, the inertia and damping torque of a hydropower system decrease, which may reduce system damping. Reference [17] concluded that the influence factors include permeability, network topology, and disturbance patterns. Reference [18] believed that, although PV dynamics do not participate in low-frequency oscillation modes, the access of PV changes the output of the synchronous system of an original system and the power flow distribution of the system, thereby affecting the low-frequency oscillation mode. However, whether PV generations will have similar effects on ultralow-frequency oscillations has not been studied to give certain conclusions.

Motivated by the aforementioned limitations, this paper studied the impact of PV access on the ultralow-frequency oscillation mode of a hydropower system. Considering the dynamics of the PV generation, a detailed small-signal model of an integrated PV-hydro system was built. The small-signal stability analysis method was used to analyze the influence of the PV generation. Based on a two-zone and four-machine system and an actual system, the influences of different output powers and locations of the PV generation on ultralow-frequency oscillation were analyzed and explained.

The rest of this paper is organized as follows. Section 2 analyzes the damping characteristics of governors and turbines. Section 3 builds a detailed small-signal model of an integrated PV-hydro system. Section 4 analyzes the influence of PV generation on ultralow-frequency oscillation. Conclusions derived from these analyses are presented in Section 5.

2. Damping Torque Analysis

Negative damping problems of hydropower units in the ultralow-frequency band are mainly caused by the water hammer effect of a turbine and improper governor parameters. The damping torque analysis of the hydraulic turbine and the governor can obtain the damping characteristics of the ultralow-frequency band. In the following, we provide a damping torque analysis for a single hydropower unit, which reveals the basic mechanism and impact factors of ultralow-frequency oscillations [19].

The open-loop system model of a governor and a turbine is shown in Figure 1.

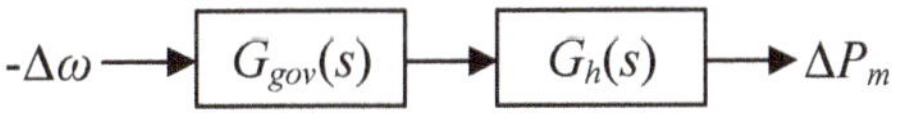

$$-\Delta\omega \longrightarrow \boxed{G_{gov}(s)} \longrightarrow \boxed{G_h(s)} \longrightarrow \Delta P_m$$

Figure 1. Open-loop system for a governor and a turbine. Symbols: ω, rotating speed; G_{gov}, governor transfer function; G_h, turbine transfer function; P_m, mechanical power.

The turbine transfer function was written as:

$$G_h(s) = \frac{1 - T_W s}{1 + 0.5 T_W s}, \tag{1}$$

where T_w is the water hammer time constant.

The governor transfer function was described as:

$$G_{gov}(s) = \frac{K_D s^2 + K_P s + K_I}{b_p K_I + s} \frac{1}{1 + T_G s}, \tag{2}$$

where K_P, K_I, and K_D are the proportional, integral, and differential parameters, respectively, b_p is the adjustment coefficient, and T_G is the time constant of the servo system.

The open-loop transfer function of the governor and turbine system was expressed as:

$$G_{OpenLoop} = G_{gov} G_h. \tag{3}$$

Decomposing Equation (3) in the $\Delta\delta - \Delta\omega$ coordinate system, Equation (4) can be obtained as:

$$-\Delta P_m = D_T \Delta\omega + S_T \Delta\delta, \tag{4}$$

where D_T is the damping torque and S_T is the synchronous torque. The torque position is shown in Figure 2. For $D_T > 0$, it provides positive damping to the system.

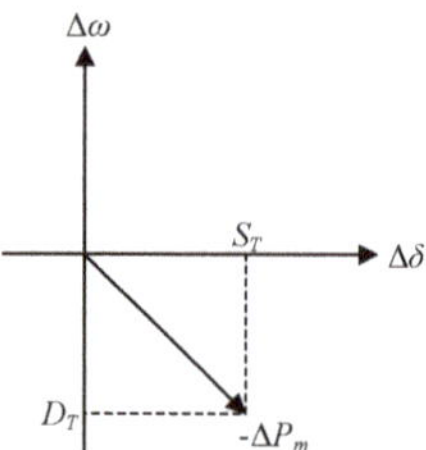

Figure 2. The position of mechanical torque.

The damping characteristics of a system composed of a governor and a turbine in a frequency range of 0–2.5 Hz are shown in Figure 3. For the water hammer time constant T_w, a larger T_w had a more negative damping torque in the ultralow-frequency band. For K_P and K_I in the PID governor, a larger value had more negative damping in the ultralow-frequency band. K_D is generally set to 0. However, the water hammer effect is an inherent characteristic of hydro turbines and cannot be changed. The primary frequency regulation ability of a governor generally requires larger K_P and K_I, which contradicts the suppression of ultralow-frequency oscillation.

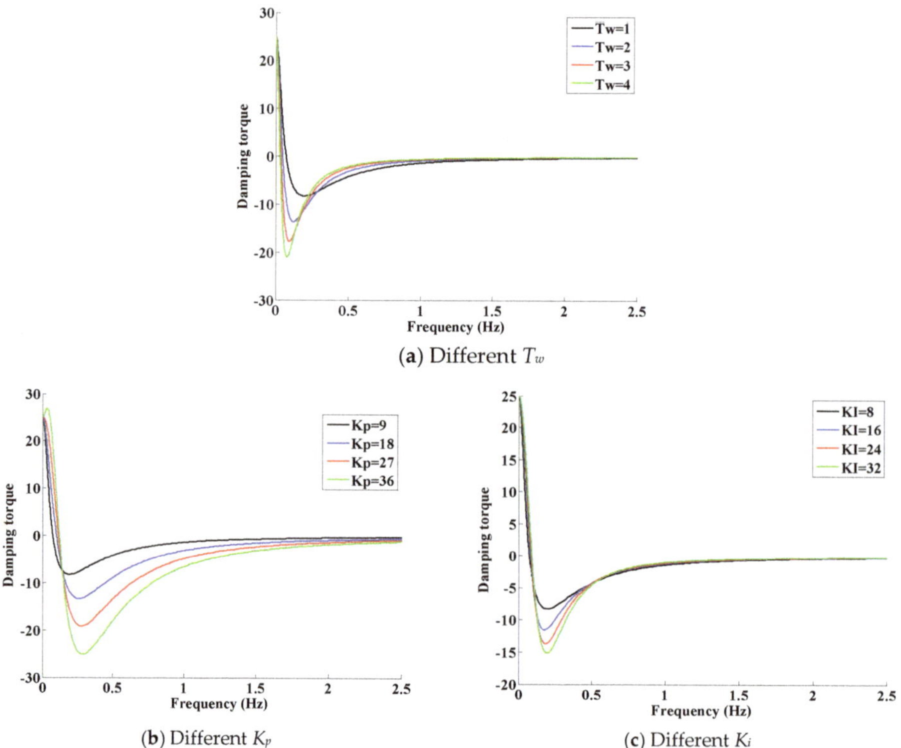

(**a**) Different T_w

(**b**) Different K_p

(**c**) Different K_i

Figure 3. The damping characteristics of the governor and the turbine in a frequency range of 0–2.5 Hz.

Although the damped torque analysis method can analyze the damping characteristics of the governor and the turbine at different frequencies, it is difficult to analyze multimachine systems and the impact of PV generation.

3. Small-Signal Dynamic Model of an Integrated PV-Hydro System

In order to analyze the impact of PV generation on the ultra-low-frequency oscillation mode of multimachine systems, a detailed small-signal model of an intergrated PV-hydro system needed to be established for small-signal stability analysis.

3.1. Modeling of PV Generation

A PV generation model mainly included a PV array, an inverter, and controllers. Figure 4 shows the structure of a PV generation model connected to a power system.

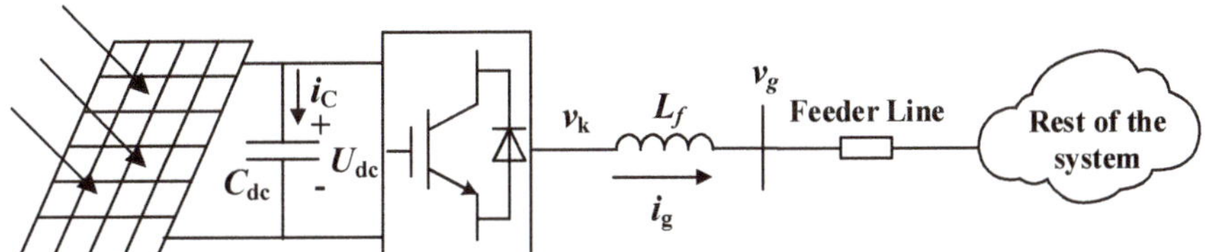

Figure 4. The structure of photovoltaic (PV) generation. Symbols: C_{dc}, direct current (DC) capacitor; U_{dc}, DC-side output voltage; i_C, the current of a DC capacitor; V_k, alternating current (AC)-side output voltage; L_f, AC inductor; i_g, AC-side output current; V_g, the voltage of the point connected with a power system.

3.1.1. PV Array

The accurate model of a PV cell is very complicated, and some parameters are difficult to measure directly [20]. Thus, it is not convenient for research and application. By simplifying calculation equations, a practical engineering model was used in this paper [21]. The standard conditions for PV cells are $S_{ref} = 1000$ W/m^2 and $T_{ref} = 25$ °C. In addition, the voltage–current equation under nonstandard conditions can be descried as:

$$I = I_{sc}[1 - C_1(e^{\frac{U}{C_2 U_{oc}}} - 1)], \tag{5}$$

$$C_2 = \frac{U_m/U_{oc} - 1}{\ln(1 - I_m/I_{sc})}, \tag{6}$$

$$C_1 = (1 - I_m/I_{sc})\exp(-U_m/C_2 U_{oc}), \tag{7}$$

where I_{sc} is the short-circuit current, U_{oc} is the open-circuit voltage, I_m and U_m are the current and the voltage at the maximum power, respectively. The parameters under nonstandard conditions can be obtained as:

$$T = T_{air} + kS, \tag{8}$$

$$I_{sc} = I_{scref}(S/S_{ref})[1 + \alpha(T - T_{ref})], \tag{9}$$

$$I_m = I_{mref}(S/S_{ref})[1 + \alpha(T - T_{ref})], \tag{10}$$

$$U_{oc} = U_{ocref}[1 - \gamma(T - T_{ref})]\ln[e + \beta(S/S_{ref} - 1)], \tag{11}$$

$$U_m = U_{mref}[1 - \gamma(T - T_{ref})]\ln[e + \beta(S/S_{ref} - 1)], \tag{12}$$

where T and T_{air} are the temperatures of the PV cell and air, S is the light intensity, U_{ocref} is the open-circuit voltage, I_{scref} is the short-circuit current, U_{mref} is the voltage of the maximum power point, I_{mref} is the current of the maximum power point in standard conditions, and k, α, β, and γ are compensation coefficients.

If the number of PV cells in series is n and the number of parallel connections is m, the voltage and the current of PV array were written as:

$$\begin{cases} U_{dc} = nU \\ I_{dc} = mI \end{cases}.$$ (13)

According to Equations (5) and (13), Equation (14) can be obtained as:

$$I_{dc} = mI_{sc}\left[1 - C_1\left(e^{\frac{U_{dc}}{nC_2U_{oc}}} - 1\right)\right].$$ (14)

3.1.2. DC Capacitor

Assume that the loss of the inverter can be ignored. Then, the output power of a PV array is equal to the sum of the power of a DC capacitor and the output power of an inverter, which can be described as:

$$U_{dc}I_{dc} = U_{dc}I_C + \frac{3}{2}v_{gd}i_{gd}.$$ (15)

The voltage of the capacitor was selected as a state variable, which can be written as:

$$C_{dc}\dot{U}_{dc} = I_C.$$ (16)

According to Equations (15) and (16), Equation (17) can be obtained as:

$$\dot{U}_{dc} = \frac{I_{dc}}{C_{dc}} - \frac{3}{2}\frac{v_{gd}i_{gd}}{C_{dc}U_{dc}}.$$ (17)

3.1.3. Inverter and Controller

The PV controller consisted of a voltage controller and a current controller, which can achieve main functions [22]. The voltage controller regulated the DC voltage to control or maximize the power extracted from the PV array. The current controller realized the control of an actual current to the current reference value. Figure 5 shows the structures of voltage and current controllers. i^*_{gq} was assigned as 0. The voltage and current control equations were given as Equations (18) and (19), respectively:

$$\begin{cases} i^*_{gd} = K_{pv}(U^*_{dc} - U_{dc}) + K_{iv}\int (U^*_{dc} - U_{dc})dt \\ i^*_{gq} = 0 \end{cases},$$ (18)

$$\begin{cases} v^*_{kd} = K_{pi}(i^*_{gd} - i_{gd}) + K_{ii}\int (i^*_{gd} - i_{gd})dt - wL_f i_{gq} + v_{gd} \\ v^*_{kq} = K_{pi}(i^*_{gq} - i_{gq}) + K_{ii}\int (i^*_{gq} - i_{gq})dt + wL_f i_{gd} + v_{gq} \end{cases}.$$ (19)

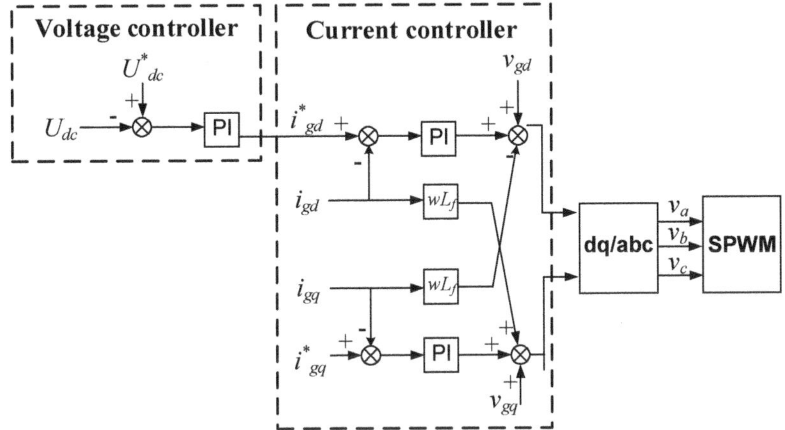

Figure 5. The structure of controllers. Symbols: U^*_{dc}, the reference value of a DC-side voltage; i^*_{gd}, the reference value of a d-axis current; i^*_{gq}, the reference value of a q-axis current; i_{gd}, the d-axis current; i_{gq}, the q-axis current; v_{gd}, the d-axis voltage; v_{gq}, the q-axis voltage; ω, the angular frequency of the system.

X_v, Y_d, and Y_q were introduced as the state variables of the controllers [23]. The dynamic equations were described as:

$$
\begin{aligned}
\dot{X}_v &= U^*_{dc} - U_{dc}, \\
\dot{Y}_d &= i^*_{gd} - i_{gd}, \\
\dot{Y}_q &= i^*_{gq} - i_{gq}
\end{aligned}
\tag{20}
$$

Considering the structure of the filter L_f, the dynamic equations of filterwere written as:

$$
\begin{cases}
L_f \dot{i}_{gd} = v_{kd} - v_{gd} + wL_f i_{gq} \\
L_f \dot{i}_{gq} = v_{kq} - v_{gq} - wL_f i_{gd}
\end{cases}
\tag{21}
$$

3.1.4. PV Generation

According to Equations (14), (17), (20), and (21), a small-signal model of a PV generation model can be obtained by linearization as following:

$$
\Delta \dot{X}_{PV} = A_{PV} \Delta X_{PV} + B_{PV} \Delta V_{gdq},
\tag{22}
$$

where $\Delta X_{PV} = [\Delta U_{dc}, \Delta X_V, \Delta Y_d, \Delta Y_q, \Delta i_{gd}, \Delta i_{gq}]^{\mathrm{T}}$, and the coefficient matrices are shown in Equations (23) and (24):

$$
A_{PV} =
\begin{bmatrix}
\dfrac{3v_{gd}i_{gd}}{2C_{dc}U_{dc}^2} - \dfrac{mI_{sc}\left[1 - C_1\left(e^{\frac{U_{dc}}{nC_2 U_{oc}}} - 1\right)\right]}{C_{dc}U_{dc}} & 0 & 0 & 0 & -\dfrac{3v_{gd}}{2C_{dc}U_{dc}} & 0 \\
1 & 0 & 0 & 0 & 0 & 0 \\
K_{pv} & K_{iv} & 0 & 0 & -1 & 0 \\
0 & 0 & 0 & 0 & 0 & -1 \\
\dfrac{K_{pi}K_{pv}}{L_f} & \dfrac{K_{iv}K_{pi}}{L_f} & \dfrac{K_{ii}}{L_f} & 0 & -\dfrac{K_{pi}}{L_f} & 0 \\
0 & 0 & 0 & \dfrac{K_{ii}}{L_f} & 0 & -\dfrac{K_{pi}}{L_f}
\end{bmatrix},
\tag{23}
$$

$$
B_{PV} =
\begin{bmatrix}
-\dfrac{3i_{gd}}{2C_{dc}U_{dc}} & 0 & 0 & 0 & 0 & 0 \\
0 & 0 & 0 & 0 & 0 & 0
\end{bmatrix}^{\mathrm{T}}.
\tag{24}
$$

3.2. Hydropower Unit

3.2.1. Synchronous Generator

All generators were synchronous generators with a fourth-order model. The model was shown as:

$$
\begin{aligned}
\dot{\delta} &= \omega_0(\omega - 1), \\
2H\dot{\omega} &= (P_m - P_e - D(1 - \omega)), \\
\dot{E}'_q &= (-E'_q - (X_d - X'_d)I_d + E_{fd})/T'_{d0}, \\
\dot{E}'_d &= (-E'_q + (X_q - X'_q)I_q)/T'_{q0}
\end{aligned}
\tag{25}
$$

where ω_0 is the base angular frequency, H is the inertia constant, P_m is the mechanical power, P_e is the electromagnetic power, D is the damping coefficient, E'_d and E'_q are the d-axis and q-axis transient voltages, respectively, X_d and X_q are the unsaturated reactances, X'_d and X'_q are the unsaturated transient reactances, I_d and I_q are the d-axis and q-axis currents, respectively, E_{fd} is the excitation voltage, and T'_{d0} and T'_{q0} are the unsaturated subtransient times. The detailed meanings of the symbols is given in [19].

3.2.2. Exciter

An excitation system is the main cause of low frequency oscillations, and it is unclear whether it has an effect on ultralow-frequency oscillations. Therefore, a detailed typical fourth-order excitation system was selected [24]. The block diagram of the excitation system is shown in Figure 6.

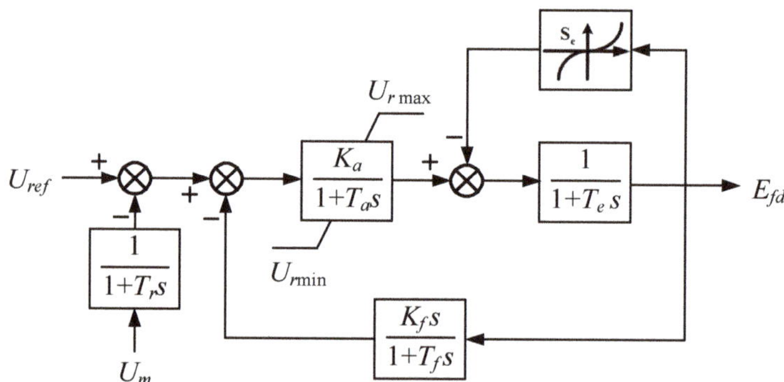

Figure 6. Block diagram of the exciter.

U_{ex1}, U_{ex2}, and U_{ex3} were selected as the state variables. The mathematical model was shown as:

$$
\begin{aligned}
\dot{U}_{ex1} &= (U_m - U_{ex1})/T_r, \\
\dot{U}_{ex2} &= (K_a(U_{ref} - U_{ex1} - U_{ex2} - K_f E_{fd}/T_f) - U_{ex2})/T_a, \\
\dot{U}_{ex3} &= -(K_f E_{fd}/T_f + U_{ex3}), \\
\dot{E}_{fd} &= -(E_{fd}(1 + S_e) - U_{ex})/T_e
\end{aligned}
\tag{26}
$$

where U_m, U_{ref}, and E_{fd} are the terminal voltage, reference input excitation voltage, and generator excitation potential, respectively, and K_a, K_f, T_a, T_f, T_r, and T_e are the amplifier gain, stabilizer gain, amplifier time constant, stabilizer time constant, measurement time constant, and excitation circuit time constant, respectively. The expressions of S_e and U_{ex} were shown as:

$$
S_e = A_e\left(e^{B_e|E_{fd}|} - 1\right),
\tag{27}
$$

$$U_{ex} = \frac{1}{2}U_{ex2}(\text{sgn}((U_{rmax} - U_{ex2})(U_{ex2} - U_{rmin})) + 1) + \frac{1}{2}U_{rmax}(\text{sgn}(U_{ex2} - U_{rmax}) + 1) + \frac{1}{2}U_{rmin}(\text{sgn}(U_{rmin} - U_{ex2}) + 1) \tag{28}$$

3.2.3. Governor and Turbine

In order to study ultralow-frequency oscillation, a detailed model of a governor and a turbine was selected [19]. It consisted of a regulating system, an electro-hydraulic servo system, and a turbine model. In an actual running system, K_D is generally set to 0. The hydraulic turbine and the PID governor are shown in Figure 7.

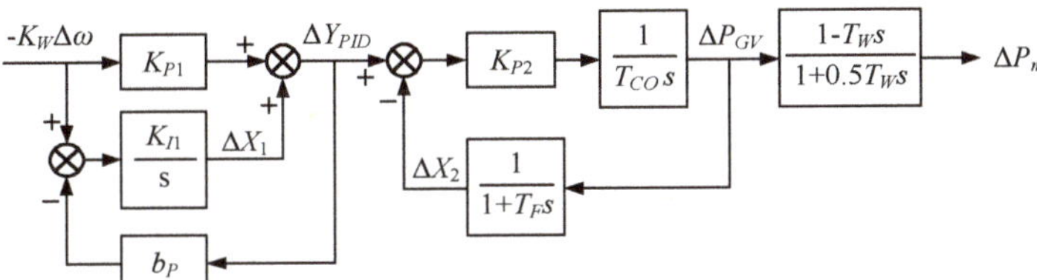

Figure 7. Block diagram of a water turbine and a PID governor. Symbols: K_W, the gain of frequency deviation; b_p, permanent difference coefficient; K_{P1}, the gain of the governor; K_{I1}, the integral gain of the governor; K_{P2}, the gain of the servo system; T_F, the time constant of stroke feedback; T_W, the time constant of the water hammer.

X_1, X_2, P_{GV}, and P_m were seleted as the state variables of the model composed of a governor and a turbine.

3.3. Small-Signal Model of the Integrated PV-Hydro System

Suppose the system has n generator nodes, one PV generation, and l connected nodes. The lines and loads of the system can be expressed by algebraic Equation (29):

$$\begin{bmatrix} \Delta I_{dq1} \\ \cdots \\ \Delta I_{dq(n+1)} \\ 0 \\ \cdots \\ 0 \end{bmatrix} = \begin{bmatrix} Y_{11} & Y_{12} \\ Y_{21} & Y_{22} \end{bmatrix} \begin{bmatrix} \Delta V_{dq1} \\ \cdots \\ \Delta V_{dq(n+1)} \\ \Delta V_{dq(n+2)} \\ \cdots \\ \Delta V_{dq(n+l+1)} \end{bmatrix}. \tag{29}$$

By eliminating the connected nodes, the nodal admittance matrix can be simplified as:

$$\begin{bmatrix} \Delta I_{dq1} \\ \cdots \\ \Delta I_{dq(n+1)} \end{bmatrix} = (Y_{11} - Y_{12}Y_{22}^{-1}Y_{21}) \begin{bmatrix} \Delta V_{dq1} \\ \cdots \\ \Delta V_{dq(n+1)} \end{bmatrix}. \tag{30}$$

By integrating the PV small-signal model into the hydropower system, a small-signal dynamic model of the integrated system can be obtained as:

$$\begin{bmatrix} \Delta \dot{x}_{w1} \\ \cdots \\ \Delta \dot{x}_{wn} \\ \Delta \dot{x}_{PV} \end{bmatrix} = A_{sys} \begin{bmatrix} \Delta x_{w1} \\ \cdots \\ \Delta x_{wn} \\ \Delta x_{PV} \end{bmatrix}, \tag{31}$$

where $\Delta X_{sys} = [\Delta X_{w1}, \ldots, \Delta X_{wn}, \Delta X_{PV}]^{T}$, A_{sys} is the complete system state matrix, and $\Delta X_{w1}, \ldots,$ ΔX_{wn} are the state variables of n hydropower units, and ΔX_{PV} is the state variables of the PV generation. By analyzing the eigenvalues and the eigenstructures of A_{sys}, the system small-signal stability can be evaluated.

4. Small-Signal Stability Analysis

According to the small-signal model above, the effect of grid-connected PV generation on ultralow-frequency oscillation was studied based on two test systems, i.e., a modified two-zone and four-machine system and an actual system.

- The two-zone and four-machine system is a typical benchmark system with standard parameters to study power system oscillations [19]. This paper selected it as a case study system and added PV generation into this system. The steam turbines of the two-zone and four-machine system were replaced by water turbines for hydropower studies.
- In order to study the effect of PV generation in an actual system, an actual integrated PV-hydro system in Sichuan Province, China was selected, so that the research has practical significance.

4.1. Modified Two-Area and Four-Machine System

Based on the two-zone and four-machine system, an integrated PV-hydro system was constructed. The structure of the integrated PV-hydro system is shown in Figure 8. The parameters of the two-zone and four-machine system can be found in Reference [19]. The characteristic matrix of the system can be obtained by Equation (31).

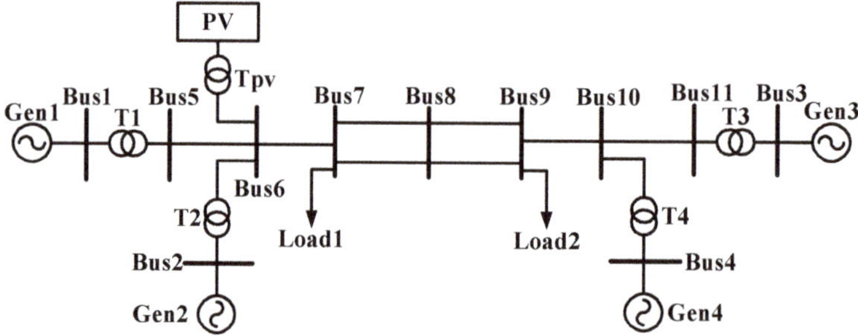

Figure 8. The structure of an integrated PV-hydro system. Gen1, Gen2, Gen3 and Gen4 are the abbreviations for the names of hydropower stations.

In order to make the damping characteristics of each hydropower unit different, different water hammer time constants were set for each hydroelectric unit. The detailed parameters of governors and turbines are shown in Table 1.

Table 1. The parameters of governors and turbines.

	Gen1	Gen2	Gen3	Gen4
T_w/s	1	1	3	3
K_P	2.6	2.6	2.6	2.6
K_I	6	6	6	6
K_D	0	0	0	0

The ultralow-frequency oscillation of the system calculated by the small-signal model is shown in Table 2. The oscillation frequency was less than 0.1 Hz, which belongs to the ultralow-frequency range.

Table 2. The ultralow-frequency oscillation.

Eigenvalues	Damping Ratio (%)	Frequency (Hz)
$-0.0127 \pm 0.1635i$	7.73	0.026

4.1.1. Participation Factor Analysis

Participation factors are the multiplication of the corresponding elements in the right and left eigenvectors of a state matrix. It can be used for evaluating the association degree between state variables and modes. In this paper, we performed the participation factor analysis based on the state matrix A_{sys} in Equation (31).

The participation factors of state variables for the ultralow-frequency oscillation mode are shown in Figure 9. As can be seen from Figure 9, the dynamics of synchronous machines, governors, and turbines were mainly involved in the ultralow-frequency oscillation mode, and the generators with a larger T_w were more involved. The dynamics of PV hardly participate in the ultralow-frequency oscillation mode. This is mainly because PV generation uses power control modes and does not participate in the frequency regulation.

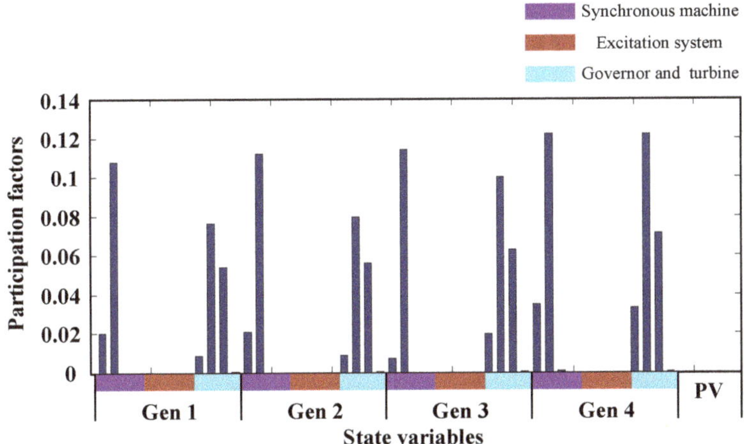

Figure 9. Participation factors. The state variables of synchronous machines contain δ, ω, E'_d, and E'_q. The state variables of excitation systems contain U_{ex1}, U_{ex2}, and U_{ex3}. The state variables of governors and turbines contain X_1, X_2, P_{GV}, and P_m. The state variables of PV generation contain U_{dc}, X_V, Y_d, Y_q, i_{gd}, and i_{gq}.

4.1.2. Different Output Powers

When the output power of PV generation increased from 100 to 600 MW, the root locus of the ultralow-frequency oscillation mode changed, as shown in Figure 10, and the corresponding damping ratio and frequency are shown in Table 3. In Figure 10, the abscissa axis correspond to the real parts of eigenvalues, and the vertical axis corresponds to the imaginary parts of eigenvalues. It can be seen from the results that the changes in PV output power had little effect on the ultralow-frequency oscillation mode.

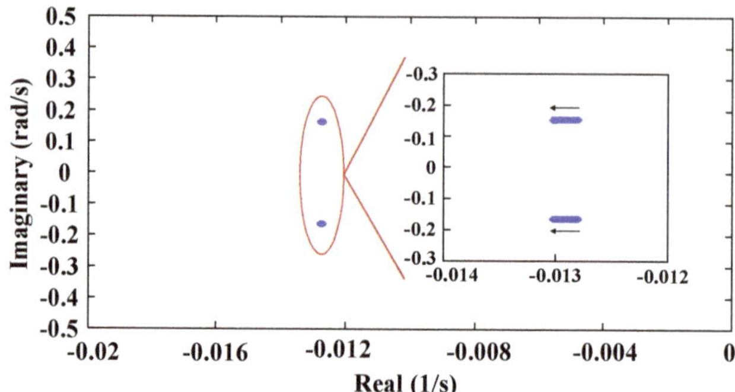

Figure 10. Root locus of the ultralow-frequency oscillation.

Table 3. Damping ratio and frequency with increasing of PV output power.

Output Power (MW)	Damping Ratio (%)	Frequency (Hz)
100	7.73	0.026
200	7.75	0.026
300	7.77	0.026
400	7.79	0.026
500	7.81	0.026
600	7.82	0.026

4.1.3. Different Locations

The ultralow-frequency oscillation modes for PV generation connected to different locations are shown in Table 4. It can be seen that the connections of PV generation with different buses had little effect on the ultralow-frequency oscillation mode.

Table 4. The ultralow-frequency oscillation modes for PV generation connected to different locations.

Location	Eigenvalues	Damping Ratio (%)	Frequency (Hz)
Bus 5	$-0.0126 \pm 0.1635i$	7.71	0.026
Bus 6	$-0.0127 \pm 0.1635i$	7.73	0.026
Bus 10	$-0.0131 \pm 0.1636i$	7.96	0.026
Bus 11	$-0.0131 \pm 0.1636i$	7.97	0.026

4.1.4. Replacing Generator

Table 5 shows the ultralow-frequency oscillation modes when a hydropower unit was replaced by PV generation. According to the results of the damping torque analysis, a larger T_w of a hydropower unit provided more negative damping. Because T_w values of Gen1 and Gen2 were small, they provided less negative damping to the system. When they were replaced by PV generation, the system damping ratio reduced. Since the T_w of Gens 3 and 4 were large, they provided more negative damping to the system. When they were replaced by PV generation, the system damping ratio was improved.

Table 5. The ultralow-frequency oscillation modes by replacing a generator.

Generator Replaced	Eigenvalues	Damping Ratio (%)	Frequency (Hz)
Gen 1	$-0.0050 \pm 0.1580i$	3.14	0.025
Gen 2	$-0.0049 \pm 0.1580i$	3.12	0.025
Gen 3	$-0.0160 \pm 0.1644i$	9.64	0.026
Gen 4	$-0.0159 \pm 0.1644i$	9.62	0.026

4.2. Actual System of a County in Sichuan Province, China

An integrated PV-hydro system in a county of Sichuan Province in China was selected as the second test system with its structure shown in Figure 11. The system was connected to an external grid through a double feeder, which could be disconnected from an outside grid and then achieve an islanded operation.

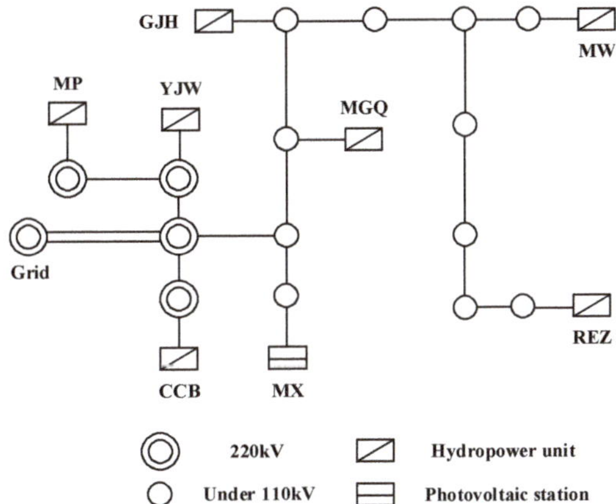

Figure 11. Structure diagram of a county town in China. MP, YJW, CCB, MGQ, GJH, MW, and REZ are the abbreviations for the names of hydropower stations, MX is the abbreviation for the name of a PV station.

The output powers of the sources are shown in Table 6. The ultralow-frequency oscillation modes of the system under different operating modes are shown in Table 7. When connected to the network, the overall damping of the system was relatively strong, since the external power grid can help stabilize the frequency. During island operation, the damping of the ultralow-frequency oscillation mode became smaller, and it was easier to excite the ultralow-frequency oscillation.

Table 6. The output powers of sources.

Name	Output Power (MW)	
	Grid-Connected Mode	**Island Mode**
MP	45	10
YJW	60	13
CCB	54	12
MHQ	36	12
GJH	44	10
MW	23	5
REZ	37	9
MX	100	20

Table 7. Ultralow-frequency oscillations under different operating modes.

Operating Mode	Eigenvalues	Damping Ratio (%)	Frequency (Hz)
On-grid	$-1.172 \pm 0.42i$	94.2	0.066
Off-grid	$-0.032 \pm 0.33i$	9.7	0.053

The participation factors are shown in Figure 12. Figure 12 indicates that the dynamics of PV hardly participated in the ultralow-frequency oscillation mode. The root locus of the PV output power increasing from 20 to 70 MW is shown in Figure 13. In Figure 13, the abscissa axis corresponds to the real parts of eigenvalues, and the vertical axis corresponds to the imaginary parts of eigenvalues. Figure 13 indicates that the root positions of the ultralow-frequency oscillation mode changed very little. The conclusion is the same as that obtained by studying the two-zone and four-machine system.

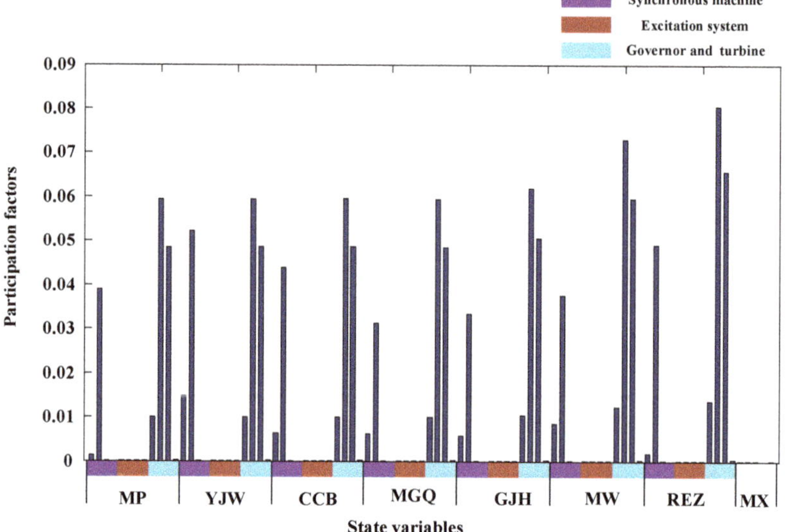

Figure 12. Participation factors. The state variables are the same as in Figure 9.

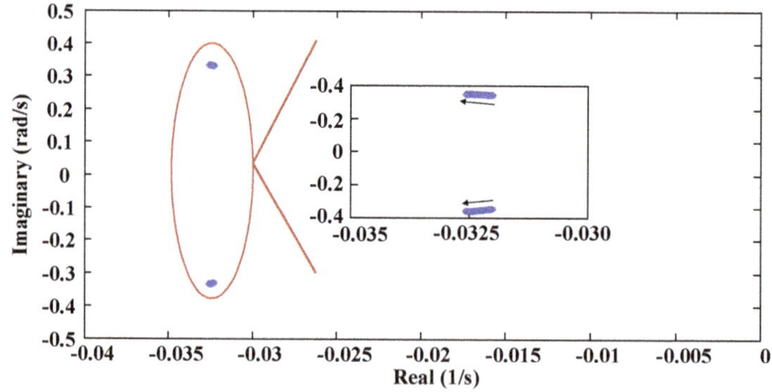

Figure 13. Root locus of the ultralow-frequency oscillation.

Remark: In order to diminish the negative influences of the ultralow-frequency oscillation, some methods have been proposed. First, by quitting the frequency regulation function of hydropwer generators with negative damping, the oscillation could be eliminated [25]. Second, some optimization methods for the PID parameters of hydropower governors were proposed, which take into account the tradeoff between the performance of primary frequency regulation and the suppression of ultralow-frequency oscillations [4]. In addition, some researchers have added a governor's power system stabilizer on the speed control side of a hydropower generator to increase its damping in the ultralow-frequency band, thereby suppressing ultralow-frequency oscillations [9]. In this paper, we mainly focused on analyzing the impact of PV generation on ultralow-frequency oscillations. The methods to suppress ultralow-frequency oscillations will be included in our future work.

5. Conclusions

In this paper, a small-signal dynamic model of an integrated PV-hydro system was established. The small-signal stability analysis method was used to analyze and study the impact of PV generation on ultralow-frequency oscillation modes. The main conclusions are summarized as follows:

(1) The dynamics of synchronizers, governors, and turbines were mainly involved in ultralow-frequency oscillation modes, while the dynamics of PV were hardly involved.

(2) Different output powers and locations of PV generation changed the distribution of the power flow but had very little effect on ultralow-frequency oscillation modes.

(3) When a synchronous machine in the system was replaced by PV generation, the ultralow-frequency oscillation mode changed significantly. In addition, when the negative damping characteristic of the replaced unit was relatively strong, the damping of the system was improved after the replacement.

Ultralow-frequency oscillation is a special phenomenon of hydropower systems. It is mainly caused by the negative damping of governors and turbines in the ultralow-frequency band. However, PV generation usually use power control modes and do not reserve power for frequency regulation. The reason for the ultralow-frequency oscillations of hydropower systems is mainly the small-signal stability problem due to frequency regulation. With the power control mode, the PV generation did not participate in frequency regulation, and thus it has little influence on ultralow-frequency oscillation. Our future research will focus on the influence of PV generation on ultralow-frequency oscillation when it is involved in the system frequency regulation.

Author Contributions: S.W. and X.W. did modeling and analysis. G.C. provided the test system data. S.W. and X.W. wrote the manuscript. Y.X. revised the manuscript. All the authors have read and approved the final manuscript. All authors have read and agreed to the published version of the manuscript.

Funding: This research was supported in part by the National Key R&D Program of China (2018YFB0905200), in part by the National Natural Science Foundation of China (51807005).

Conflicts of Interest: The authors declare no conflicts of interest.

Nomenclature

T_w	water hammer time constant of a turbine
K_P	proportional parameter of a governor
K_I	integral parameter of a governor
K_D	differential parameter of a governor
D_T	damping torque
S_T	synchronous torque
C_{dc}	DC capacitor
U_{dc}	DC-side output voltage
V_k	AC-side output voltage
L_f	AC inductor
i_g	AC-side output current
V_g	voltage of a parallel point with a power system
I_{sc}	short-circuit current of PV cells
U_{oc}	open-circuit voltage of PV cells
I_m	current of PV cells at the maximum power
U_m	voltage of PV cells at the maximum power
T	temperature of PV cells
T_{air}	temperature of the air
S	light intensity
U_{ocref}	open-circuit voltage in standard conditions
I_{scref}	short-circuit current in standard conditions
U_{mref}	voltage of the maximum power point in standard conditions
S_{ref}	light intensity in standard conditions
T_{ref}	temperature in standard conditions

k	compensation coefficient
α	compensation coefficient
β	compensation coefficient
γ	compensation coefficient
U^*_{dc}	reference value of a DC-side voltage
i^*_{gd}	reference value of a d-axis current
i^*_{gq}	reference value of a q-axis current
i_{gd}	d-axis current
i_{gq}	q-axis current
v_{gd}	d-axis voltage
v_{gq}	q-axis voltage
ω	angular frequency of a system
ω_0	base angular frequency
H	inertia constant of a synchronous machine
P_m	mechanical power
P_e	electromagnetic power
D	damping coefficient of a synchronous machine
E'_d	d-axis transient voltage
E'_q	q-axis transient voltage
X_d	d-axis unsaturated reactance
X_q	q-axis unsaturated reactance
X'_d	d-axis unsaturated transient reactance
X'_q	q-axis unsaturated transient reactance
I_d	d-axis current
I_q	q-axis current
E_{fd}	excitation voltage
T'_{d0}	d-axis unsaturated subtransient time
T'_{q0}	q-axis unsaturated subtransient time
U_m	terminal voltage
U_{ref}	reference input excitation voltage
K_a	amplifier gain
K_f	stabilizer gain
T_a	time constant of an amplifer
T_f	time constant of a stabilizer
T_r	time constant of a measurement
K_W	gain of a frequency deviation
b_p	permanent difference coefficient
K_{P1}	gain of a governor
K_{I1}	integral gain of a governor
K_{P2}	gain of a servo system
T_F	time constant of stroke feedback
T_W	time constant of water hammer

References

1. Schleif, F.R.; White, J.H. Damping for the Northwest-Southwest Tieline Oscillations—An Analog Study. *IEEE Trans. Power Appar. Syst.* **1966**, *PAS-85*, 1239–1247. [CrossRef]
2. Villegas, H.N. Electromechanical oscillations in hydro dominant power systems: An application to the Colombian power system. Master's Thesis, Iowa State University, Ames, IA, USA, 2011.
3. Fu, C.; Liu, Y.; Tu, L.; Li, P.; Hong, C.; Li, P.; Wu, C.; Xu, M.; Zhao, R. Experiment and Analysis on Asynchronously Interconnected System of Yunnan Power Grid and Main Grid of China Southern Power Grid. *South. Power Syst. Technol.* **2016**, *10*, 1–5.
4. Chen, G.; Tang, F.; Shi, H.; Yu, R.; Wang, G.; Ding, L.; Liu, B.; Lu, X. Optimization Strategy of Hydrogovernors for Eliminating Ultralow-Frequency Oscillations in Hydrodominant Power Systems. *IEEE J. Emerg. Sel. Top. Power Electron.* **2018**, *6*, 1086–1094. [CrossRef]

5. Lu, X.; Chen, L.; Chen, Y.; Min, Y.; Hou, J.; Liu, Y. Ultra-low-frequency Oscillation of Power System Primary Frequency Regulation. *Autom. Electr. Power Syst.* **2017**, *41*, 64–70.

6. Zheng, C.; Ding, G.; Liu, B.; Zhang, X.; Wang, J.; Xue, A.; Bi, T. Analysis and control to the ultra-low frequency oscillation in southwest power grid of China: A case study. In Proceedings of the 2018 Chinese Control And Decision Conference, Shenyang, China, 9–11 June 2018; pp. 5721–5724.

7. Deng, W.; Wang, D.; Wei, M.; Zhou, X.; Wu, S.; He, P.; Kang, J. Influencing Mechanism Study on Turbine Governor Parameters Upon Ultra-low Frequency Oscillation of Power System. *Power Syst. Technol.* **2019**, *43*, 1371–1377.

8. Huang, W.; Duan, R.; Jiang, C.; Zhou, J.; Gan, D. Stability Analysis of Ultra-low Frequency Oscillation and Governor Parameter Optimization for Multi-machine System. *Autom. Electr. Power Syst.* **2018**, *42*, 185–193.

9. Liu, S.; Wang, D.; Ma, N.; Deng, W.; Zhou, X.; Wu, S.; He, P. Study on Characteristics and Suppressing Countermeasures of Ultra-low Frequency Oscillation Caused by Hydropower Units. *Proc. CSEE* **2019**, *39*, 5354–5362.

10. Rashid, K.; Ellingwood, K.; Safdarnejad, S.M.; Powell, K.M. Designing Flexibility into a Hybrid Solar Thermal Power Plant by Real-Time, Adaptive Heat Integration. *Comput. Aided Chem. Eng.* **2019**, *47*, 457–462.

11. Jong, P.D.; Barreto, T.B.; Tanajura, C.A.S.; Kouloukoui, D.; Oliveira-Esquerre, K.P.; Kiperstok, A.; Torres, E.A. Estimating the impact of climate change on wind and solar energy in Brazil using a South American regional climate model. *Renew. Energy* **2019**, *141*, 390–401. [CrossRef]

12. Shah, R.; Mithulananthan, N.; Bansal, R.C. Oscillatory stability analysis with high penetrations of large-scale photovoltaic generation. *Energy Convers. Manag.* **2013**, *65*, 420–429. [CrossRef]

13. Ding, M.; Wang, W.; Wang, X.; Song, Y.; Chen, D.; Sun, M. A Review on the Effect of Large-scale PV Generation on Power Systems. *Proc. CSEE* **2014**, *34*, 1–14.

14. Quintero, J.; Vittal, V.; Heydt, G.T.; Zhang, H. The Impact of Increased Penetration of Converter Control-Based Generators on Power System Modes of Oscillation. *IEEE Trans. Power Syst.* **2014**, *29*, 2248–2256. [CrossRef]

15. Du, W.; Wang, H.; Xiao, L. Power system small-signal stability as affected by grid-connected photovoltaic generation. *Eur. Trans. Electr. Power* **2012**, *22*, 688–703. [CrossRef]

16. Eftekharnejad, S.; Vittal, V.; Heydt, G.T.; Keel, B.; Loehr, J. Impact of increased penetration of photovoltaic generation on power systems. *IEEE Trans. Power Syst.* **2013**, *28*, 893–901. [CrossRef]

17. Eftekharnejad, S.; Vittal, V.; Heydt, G.T.; Keel, B.; Loehr, J.S. Small Signal Stability Assessment of Power Systems with Increased Penetration of Photovoltaic Generation: A Case Study. *IEEE Trans. Sustain. Energy* **2013**, *4*, 960–967. [CrossRef]

18. Ge, J.; Du, H.; Zhao, D.; Ma, J.; Qian, M.; Zhu, L. Influences of Grid-connected Photovoltaic Power Plants on Low Frequency Oscillation of Multi-machine Power Systems. *Autom. Electr. Power Syst.* **2016**, *40*, 63–70.

19. Kundur, P.; Balu, N.J.; Lauby, M.G. *Power System Stability and Control*; McGraw-Hill: New York, NY, USA, 1994.

20. Rashid, K.; Mohammadi, K.; Powell, K. Dynamic simulation and techno-economic analysis of a concentrated solar power (CSP) plant hybridized with both thermal energy storage and natural gas. *J. Clean. Prod.* **2020**, *248*, 119193. [CrossRef]

21. Deng, J.; Xia, N.; Yin, J.; Jin, J.; Peng, S.; Wang, T. Small-Signal Modeling and Parameter Optimization Design for Photovoltaic Virtual Synchronous Generator. *Energies* **2020**, *13*, 398. [CrossRef]

22. Yazdani, A.; Dash, P.P. A Control Methodology and Characterization of Dynamics for a Photovoltaic (PV) System Interfaced With a Distribution Network. *IEEE Trans. Power Deliv.* **2009**, *24*, 1538–1551. [CrossRef]

23. Pogaku, N.; Prodanovic, M.; Green, T.C. Modeling, Analysis and Testing of Autonomous Operation of an Inverter-Based Microgrid. *IEEE Trans. Power Electron.* **2007**, *22*, 613–625. [CrossRef]

24. Shi, J.; Shen, C. Impact of DFIG wind power on power system small signal stability. In Proceedings of the 2013 IEEE Pes Innov. Smart Grid Technol. Conference, Washington, DC, USA, 24–27 February 2013; pp. 1–6.

25. Chen, L.; Lu, X.; Chen, Y.; Min, Y.; Mo, K.; Liu, Y. Online Analysis and Emergency Control of Ultra-low-frequency Oscillations Using Transient Energy Flow. *Autom. Electr. Power Syst.* **2017**, *41*, 9–14.

Review

Applications of Triple Active Bridge Converter for Future Grid and Integrated Energy Systems

Van-Long Pham and Keiji Wada *

Department of Electrical Engineering and Computer Science, Tokyo Metropolitan University, Tokyo 192-0397, Japan; vanlongpham206@gmail.com
* Correspondence: kj-wada@tmu.ac.jp; Tel.: +81-42-677-2751

Received: 27 February 2020; Accepted: 20 March 2020; Published: 1 April 2020

Abstract: Renewable energy systems and electric vehicles (EVs) are receiving much attention in industrial and scholarly communities owing to their roles in reducing pollutant emissions. Integrated energy systems (IES), which connect different types of renewable energies and storages, have become common in many applications, such as the grid-connected photovoltaic (PV) and battery systems, fuel cells and battery/supercapacitor in EVs. The advantages of all energy sources are maximized by utilizing connection and control strategies. Because many storage systems and household loads are mainly direct current (DC) types, the DC grid has considerable potential for increasing the efficiency of distribution grids in the future. In IES and future DC grid systems, the triple active bridge (TAB) converter is an isolated bidirectional DC-DC converter that has many advantages as a core circuit. Therefore, this paper reviews the characteristics of the TAB converter in current applications and suggests next-generation applications. First, the characteristics and operation modes of the TAB converter are introduced. An overview of all current applications of the TAB converter is then presented. The advantages and challenges of the TAB converter in each application are discussed. Thereafter, the potential future applications of the TAB converter with an adaptable power transmission design are presented.

Keywords: triple active bridge; integrated energy systems; DC grid; isolated bidirectional DC-DC converter; multiport converter

---

1. Introduction

In current years, renewable energy systems and electric vehicles (EVs) have been increasingly used [1–6]. Renewable energy systems include, but are not limited to, photovoltaic (PV), wind power, biomass, hydro, and geothermal systems. The rooftop PV system is a common and essential energy source for residential loads. Other than battery EVs and hybrid EVs (HEVs), many fuel cell EVs (FCEVs) are also developed and implemented [7–9]. However, the fuel cells have the disadvantage of a slow transient response. Therefore, the battery or supercapacitor is used to adapt to the fast response in EVs and other applications [10]. EVs may use a lithium-ion battery, supercapacitor, fuel cell, or their integration to optimize power density. By connecting different types of energy and storage, electrical systems become integrated energy systems (IES) [11]. For example, a household electrical system connects PV, EVs, and/or storage systems. The EVs use integrated fuel cells, and/or batteries, and/or supercapacitors. The IESs can then optimize the advantages of all the energy systems. Therefore, this type of IES will be developed more in the future.

Direct current (DC) grids offer many advantages over alternating current (AC) grids, such as potentially higher efficiencies, and reduced filter effort [12–14]. DC-DC converters are one of the most important technologies for future DC grids. They offer precise control ability for power flow with high reliability. Dual active bridge (DAB), an isolated bidirectional dc-dc converter, has been proposed for many applications [15,16]. The DAB converter includes two full-bridge inverters that are connected by an isolation transformer at high-frequency operation, as shown in Figure 1. It has advantages such as a bidirectional power flow with high efficiency. However, it can only connect two ports; thus, many DAB converters need to be used to connect different elements to the DC-bus in the IES. Moreover, it may require a communication bus to control power flow.

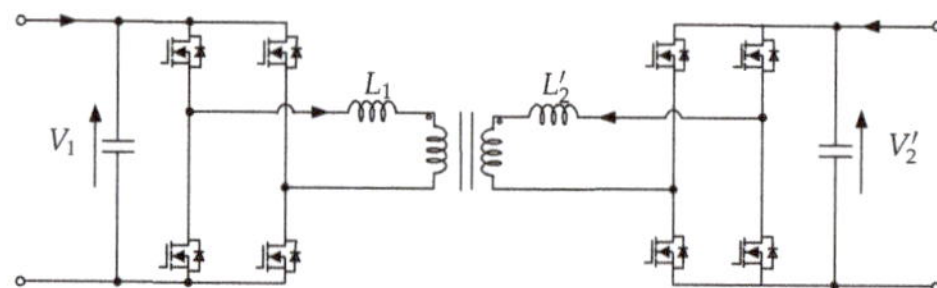

Figure 1. Dual active bridge (DAB) converter circuit.

Therefore, the triple-active-bridge (TAB) converter was proposed to connect one more element by adding one more port to the DAB converter, as shown in Figure 2 [17]. The advantages of the DAB converter can be kept in the TAB converter. Moreover, it is not only applicable to one more port but also enables flexible power transmission between three ports, as shown in Figure 3. This shows that three DAB converters are required in Figure 3 to achieve flexible power transmission between three elements, which is achieved by only one TAB converter in Figure 3b. Also, the communication between the three elements is not necessary when using the TAB converter. Therefore, the control of the total system is more straightforward. In addition, in comparison to other multiport converters, the TAB converter has the advantage by using a transformer, which not only converts the voltage ratio but also improves the safety of the system. Therefore, the TAB converter is proposed for many applications in IES and DC grid.

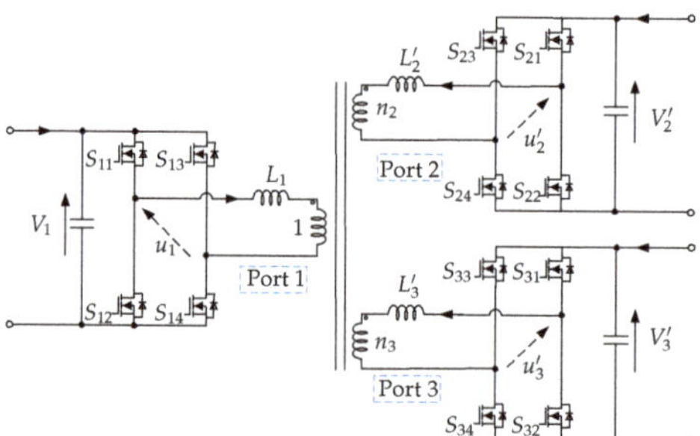

Figure 2. Triple-active-bridge (TAB) converter circuit.

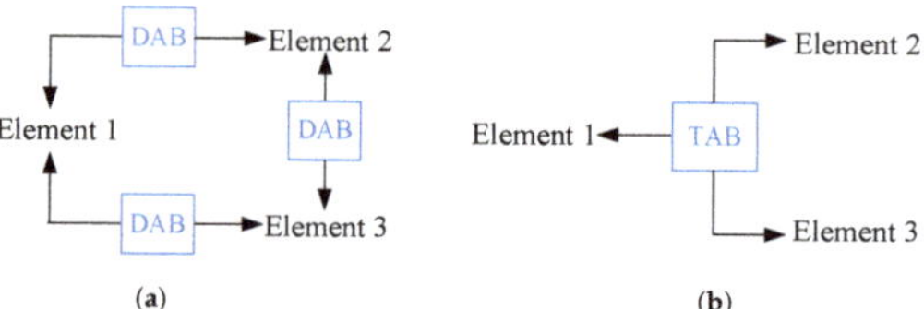

Figure 3. Comparison of a system using DAB and TAB converters. (**a**) System using DAB converter, (**b**) System using TAB converter.

Because there is an increase in the number and types of microgrids [18,19], the TAB converter is proposed to develop micro-grid systems, as shown in [20–25]. The future DC grid has potential in household applications, for which the TAB converter has an advantage [19–22]. It can be used to connect medium and low voltages in the DC distribution grid [26–30]. An uninterruptible power supply (UPS), which uses a battery to support the source in the worst-case scenario, is a suitable application for the TAB converter [31–33]. The TAB converter was discussed as an important part that can improve the reliability of the distribution system in the data center [34,35]. The proposed system in EVs applications is discussed in [36–40].

However, researches on the TAB converter focuses on each separated application and characteristic. It loses an overview of the current status, such as the advantages and challenges of the TAB converter in current applications. Also, there are many new applications such as all electrical ships, autonomous underwater vehicles, etc., that involve replacing other energy systems with electricity. Therefore, this paper presents an overview of current applications. The advantages and challenges of the TAB converter in different applications are analyzed. Then, future applications, which use storage systems, are suggested for use with the TAB converter. Also, potential applications and concepts in next-generation applications are proposed. This becomes a reference for researchers and engineers in the related topic for improving the TAB converter in current applications and extending applications.

2. Characteristics of the TAB Converter

2.1. Configuration and Model

Figure 2 shows the circuit diagram of the TAB converter. It includes a three-winding transformer connecting port-1, port-2 and port-3. Each port has an inductance connected in series, and a full bridge inverter. The series inductances, L_1, L_2', and L_3' include the leakage and external inductances on each port. The symbols n_2 and n_3 are the turn ratios of the port-2 and port-3. The phase voltages between the leg midpoints of each port, u_1, u_2', and u_3', have the amplitudes V_1, V_2', and V_3', respectively. S_{11} to S_{14}, S_{21} to S_{24}, and S_{31} to S_{34} are the control signals of ports -1, -2 and -3, respectively. Figure 4 shows a three-phase TAB converter made by adding one switching leg to each inverter [41]. It is proposed for high power applications. Each phase of the three-phase TAB converter can be modeled as a single-phase TAB converter [26,41].

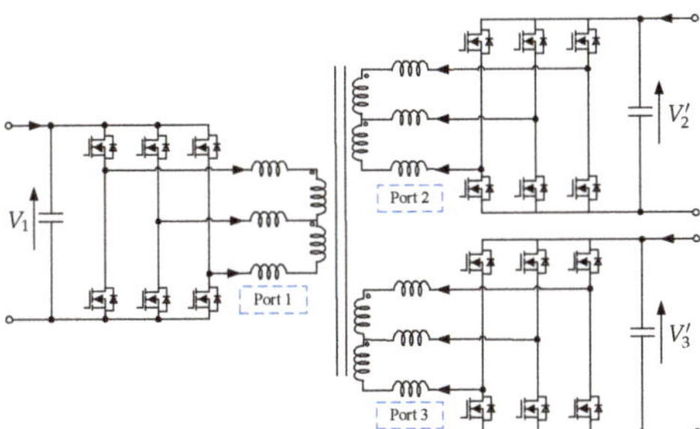

Figure 4. Three-phase TAB converter.

Figure 5 shows Y-type and Δ-type equivalent circuits of the TAB converter, respectively [40]. The voltages and currents of the port-2 and port-3 refer to the port-1 as the following equation.

$$\begin{cases} u_2 = u_2'/n_2 \\ u_3 = u_3'/n_3 \\ i_2 = i_2'n_2 \\ i_3 = i_3'n_3 \end{cases} \tag{1}$$

where u_2 and u_3 are the port-1-referred voltages, and i_2 and i_3 are the port-1-referred currents, respectively. Thus, it has the following equation.

$$\begin{cases} L_2 = L_2'/n_2^2 \\ L_3 = L_3'/n_3^2 \\ V_2 = V_2'/n_2 \\ V_3 = V_3'/n_3 \end{cases} \tag{2}$$

where L_2 and L_3 are port-1-referred inductances, and V_2 and V_3 are port-1-referred voltage amplitudes, respectively.

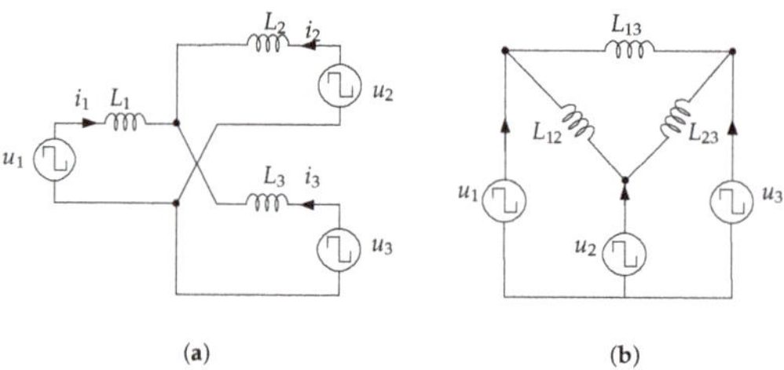

(a)   (b)

Figure 5. Equivalent circuit of the TAB converter. (**a**) Y-type, (**b**) Δ-type.

2.2. Power Transmission and Control Methods

Based on the above model, the power transmission of the TAB converter is expressed by the following equations.

$$P_1 = \frac{V_1 V_2 \varphi_2 (\pi - |\varphi_2|) L_3 + V_1 V_3 \varphi_3 (\pi - |\varphi_3|) L_2)}{2\pi^2 f(L_1 L_2 + L_2 L_3 + L_3 L_1)} \tag{3}$$

$$P_2 = \frac{V_2 V_1 (-\varphi_2)(\pi - |\varphi_2|) L_3 + V_2 V_3 (\varphi_3 - \varphi_2)(\pi - |\varphi_3 - \varphi_2|) L_1}{2\pi^2 f(L_1 L_2 + L_2 L_3 + L_3 L_1)} \tag{4}$$

$$P_3 = \frac{V_3 V_1 (-\varphi_3)(\pi - |\varphi_3|) L_2 + V_3 V_2 (\varphi_2 - \varphi_3)(\pi - |\varphi_2 - \varphi_3|) L_1}{2\pi^2 f(L_1 L_2 + L_2 L_3 + L_3 L_1)} \tag{5}$$

where P_1, P_2, and P_3 are the transmission powers of the of ports-1, -2, and -3, respectively. f is the switching frequency of the converter. φ_2 and φ_3 are phase shift angles on the port-2, and port-3 with reference to the port-1, respectively.

Therefore, the power transmission ability of each port can be controlled by two-phase shift angles, φ_2 and φ_3, as shown in Figure 6a. The phase shift angles between u_1, u_2, and u_3 decide amplitude and direction of the transmission power among the three ports [21–25]. The decoupling matrix is implemented into the control loop to remove the interference among the ports [25,40]. Besides, the combined duty cycles and two-phase shift angles methods were discussed to improve the efficiency of the TAB converter, as shown in Figure 6b [38–40,42]. δ_1, δ_2, and δ_3 are duty cycle variations of u_1, u_2, and u_3, respectively. This technique can extend the soft switching at light load but it reduces the power transmission ability of the converter [42]. Furthermore, upon adding more variables to the system, the control system and calculation of the controller become more complex. Therefore, the two-phase shift angles method is widely used for controlling the TAB converter in many applications. The combined duty cycle and two-phase shift angles method can be considered for applications that usually operate under light load condition.

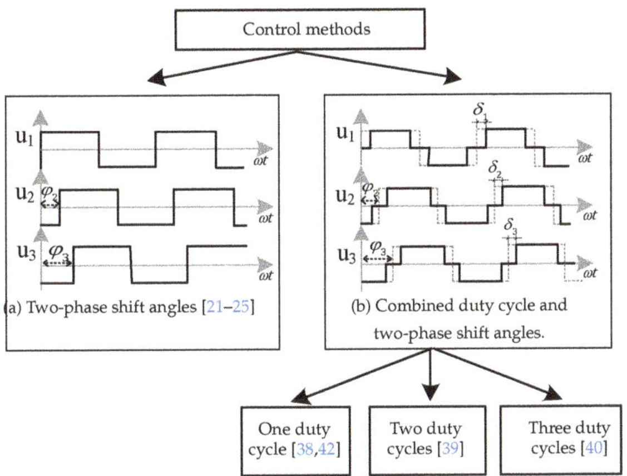

Figure 6. Control methods of the TAB converter.

2.3. Operation Mode and Characteristics

By controlling the phase shift angles, the TAB converter has many working operation modes, as shown in Figure 7.

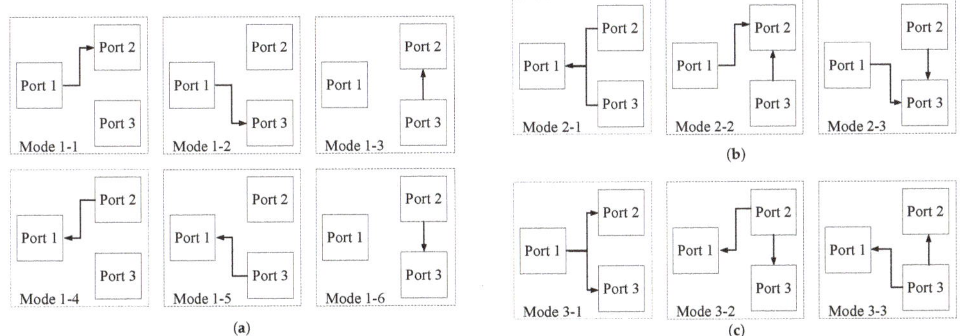

Figure 7. Operation modes of TAB converter: (**a**) Single input single output (SISO), (**b**) Dual input single output (DISO), (**c**) Single input dual output (SIDO).

It can be categorized into three groups. The single input single output (SISO) modes when the power comes from one port to another while the third port does not get power. The dual input single output (DISO) modes, which supplies power to one port from the other two ports. The single input dual output (SIDO) modes when power from one port is supplied to the two other ports. This shows the flexible power transmission ability of the TAB converter.

Figure 8 shows the phase shift operation of the TAB converter in all operation modes by using two phase-shift angles. The horizontal and vertical axes show the phase shift values of port 2 and port 3, φ_2 and φ_3, respectively. The phase shift angles are decided by the operating power, inductance values, and voltages as the relationship in (2)–(4). The series inductances are important elements of the TAB converter. A normalized design method for the inductances was discussed in [21,22]. Voltage variations, which are in many applications, need to be considered in operation modes. Figure 8a shows the phase shift operation when the voltages of the three ports are all 100%. Figure 8b shows the phase shift operation when V_1 and V_2 are both 100%, whereas V_3 is 80%. It shows that the SISO modes are the most critical in the design and control of the TAB converter, where the phase-shift angles are the largest.

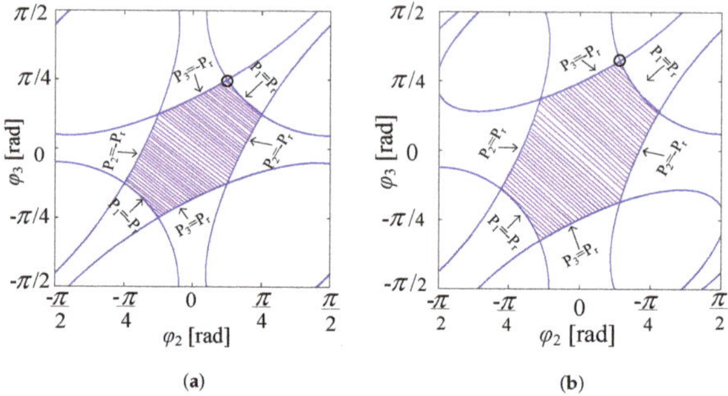

Figure 8. Phase shift operation of TAB converter in two different voltage states: (**a**) V_1, V_2, and V_3 are all 100%, (**b**) V_1 and V_2 are both 100%, V_3 is 80%.

The experimental waveform of each port in operation mode 1-2 is shown in Figure 9 to explain the characteristic in one of the most critical operation modes. The figure shows that the root mean square (RMS) and peak current of port 2, i_2', are much higher in the voltage variation condition. In this case, the combined duty cycle and phase shift control can be applied to reduce the RMS and peak currents of port 2 [38–40,42].

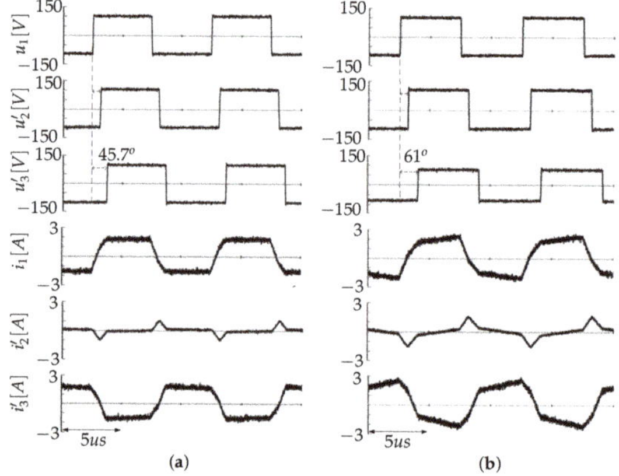

Figure 9. Experimental waveform of TAB converter in mode 1-2: (**a**) V_1, V_2, and V_3 are all 100%; (**b**) V_1 and V_2 are both 100%, V_3 is 80%.

3. Current Applications of TAB Converter

3.1. Microgrids

A microgrid is an integrated system that combines sources, storage systems, and loads [18,19]. The DC bus Microgrid systems using TAB converter were discussed in [20–22]. Figure 10 shows the comparison of a DC bus microgrid using the TAB converter and the conventional converter. It shows that the system using the TAB converter can reduce the required number of DC-DC converters and communication lines. Consequently, the cost of the system is reduced.. The rapidly increasing use of the EVs will bring much change in household electrical systems in the future. The DC household electrical system is discussed in [19]. The TAB converter can be used in household electrical systems for a power range of 10 kilowat (kW), as discussed in [21,22]. The EVs or storage can be charged from the grid or directly from the rooftop PV. In addition, EVs have a high-power battery, which can be a storage and support system for the household load or grid [5]. By using the TAB converter, the DC bus microgrid system becomes more flexible in setup and more optional in operation, which are advantages.

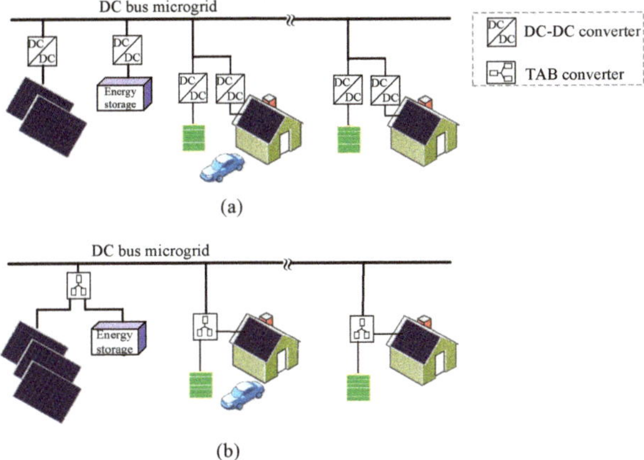

Figure 10. Comparison of DC bus microgrid using the TAB converter and the conventional DC-DC converter. (**a**) The system uses the conventional DC-DC converter. (**b**) The system uses the TAB converter.

An autonomous DC microgrid using the TAB converter is proposed in [23–25], as shown in Figure 11. One port of the TAB converter is connected to one element, and the remaining two ports are connected to the autonomous DC microgrid to control the power and voltage. The TAB converter can change the control target to keep a constant voltage and for different loads depending on the condition of the system. The transformers of the TAB converters isolate all the parts of the microgrid. Therefore, if one element has an error, other elements still work well, and the system is easy to extend at any time. The DC/AC converter can be added to a port that connects to an AC load or source. This idea can be applied to the traditional microgrid system when some elements are used to improve reliability. In the future microgrid system, the TAB converter is a promising circuit when combining two above systems.

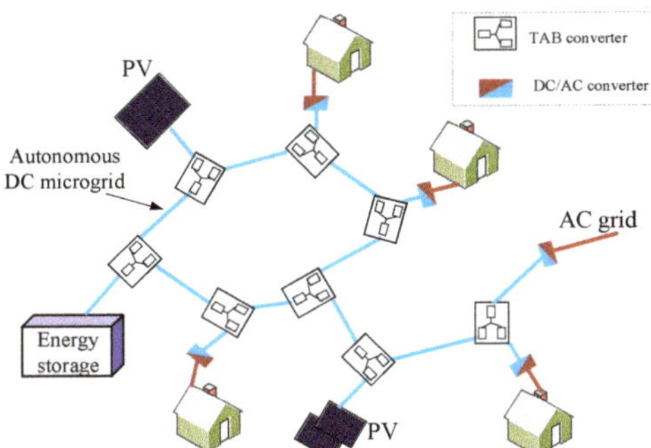

Figure 11. TAB converter for an autonomous DC microgrid system.

3.2. Connection of Medium Voltage Grid and Low Voltage Grid

The medium voltage DC (MVDC) grid has recently created a need for the investigation of the concept of the DC distribution grid [43–45]. The overall stability and efficiency of the DC grid are increased by controlling the power flow [26]. Therefore, a three-phase TAB converter is proposed for the MVDC grid in [26–30], as shown in Figure 12. The nominal voltage of the MVDC grid is 5 kV. Two LVDC grids work at 380 V and 760 V, respectively. The 760 V LVDC grid is examined as a ±380 V LVDC grid. The TAB converter can use all operation modes in Figure 7 for this application because the power is required to be transferred in any direction among the three ports. It can operate as an equal three DAB converter as shown in the comparison in Figure 3. Therefore, the system can reduce the number of components and costs by using the TAB converter. In typical operation, the power mainly comes from MVDC to LVDC as the operation modes 1-1, 1-2, and 3-1. However, in some critical situations, the remaining operation modes of the TAB converter can be used for this application. Then, three grids can support each other. This shows that the TAB converter can be optimally utilized for all operation modes in this application.

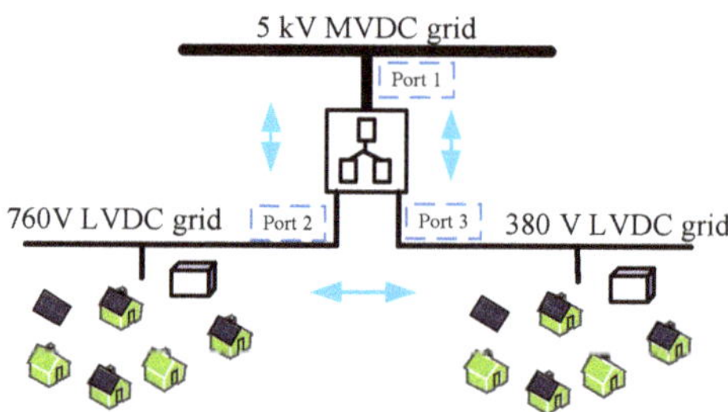

Figure 12. TAB converter connection medium-voltage direct current (MVDC) and low-voltage direct current (LVDC).

A three-phase TAB converter rated 150 kW is designed to connect 5 kV MVDC and two LVDC sources in [26–30]. The modulation strategy and transformer design are considered to operate at a rated power [26]. The soft-switching limitation in each port is analyzed. Subsequently, a parallel-phase operation (PPO) is proposed to extend the soft-switching component of the converter [27]. All ports achieve soft switching by applying the multiport duty cycle method. The current and voltage measurements require isolation to the MVDC part, as discussed in [28]. The dynamic of the current control loop is discussed in [29]. The instantaneous current control is applied to the TAB converter to achieve a high dynamic power response. The challenges of using the 1200 V SiC MOSFET for the TAB converter in MVDC and LVDC is discussed in [30].

Other configurations, which can connect multiple single-phase TAB converters, can also be applied to interconnect MVDC and LVDC, as shown in Figure 13 [46–48]. It can connect two ports in series and one port in parallel to connect two MVDC grids and one LVDC grid, as shown in Figure 13a. Figure 13b shows the system which connects one port in series and two ports in parallel to connect one MVDC grid and two LVDC grids. The number of modules is selected depending on the available semiconductor devices and operating voltage. The insulated-gate bipolar transistor (IGBT) devices rated 4.5 kV are considered for high voltage applications. More research to utilize the devices and voltage range of each module should be discussed more in the future.

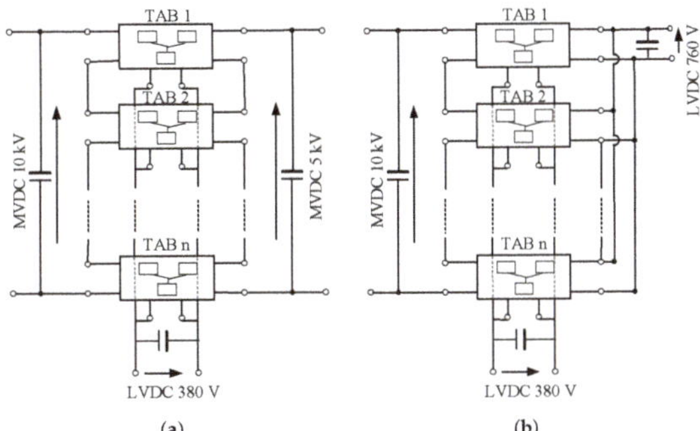

Figure 13. Multiple TAB converter for high voltage high power application: (**a**) two ports in series, one port in parallel; (**b**) one port in series, two ports in parallel.

3.3. Uninterrupted Power Supply Systems

The TAB converter was proposed for the uninterrupted power supply (UPS) systems [31,32], as shown in Figure 14. The power source, load, and battery are connected using one TAB converter as discussed in [31,32], where the power source and electric load may require a frequency isolation of 50 Hz and 60 Hz, respectively. The system operates typically in the mode 1-1. However, all remaining modes can be used in an abnormal situation. By using a TAB converter with a three-winding transformer, the safety of the system is improved. The operation of the system is flexible with simple control.

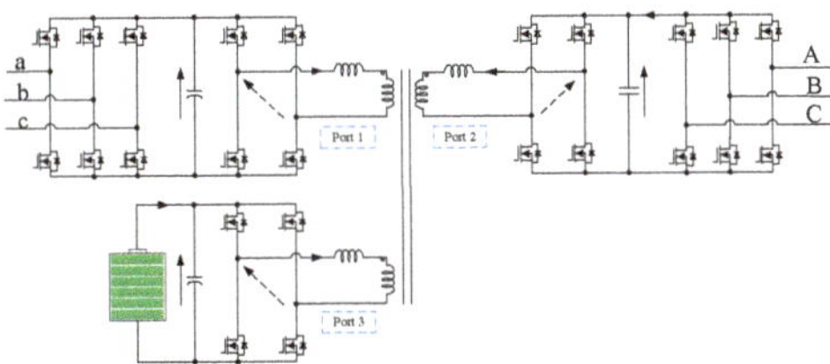

Figure 14. TAB converter in UPS application.

A UPS system powered by a fuel cell is proposed by using a TAB converter in [33,42,49], as shown in Figure 15. The system is flexible and can work on grid-connected or stand-alone modes. The fuel cells have a slow transient response, which is a drawback. The supercapacitor can be used as storage to support the fuel cells to respond quickly in transient time. Depending on the load and the status of the supercapacitor and fuel cell, it may operate in different modes. This could be discussed more in the future. For high power applications, the three-phase TAB converter can also be used. In low power applications, the half-bridge TAB converter is suitable [50]. This shows that the TAB converter is an excellent selection for these proposed systems where all operation modes are achieved by one converter.

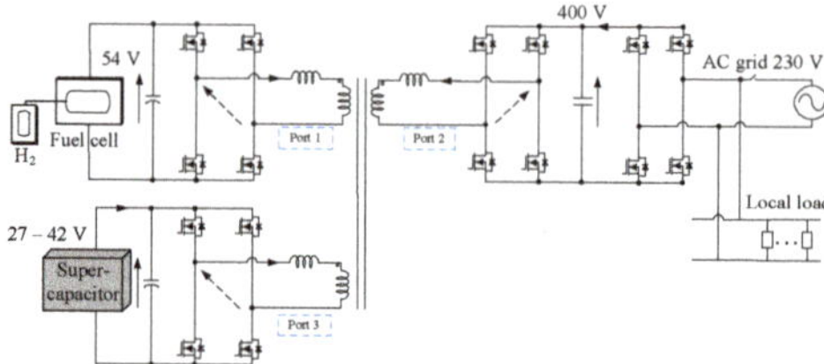

Figure 15. TAB converter in line-interactive UPS using fuel cell and supercapacitor application.

3.4. Power Distribution for Data Center

The reliability of the distribution systems for data centers is of the most importance points. Interruptions of power systems in the data center have the high cost of millions of dollars an hour [51]. Therefore, the TAB converter was proposed to improve the reliability and availability of power distribution of data centers in [34,35], as shown in Figure 16. This system can operate as a UPS system to prevent power faults in distribution systems of the data center. The number of converters can be reduced by sharing a battery between two distribution lines. Consequently, using the TAB converter can reduce the cost of the system. Furthermore, two distribution lines can support each other in a critical situation. Therefore, all operation modes of the TAB converter can be used in this application, which maximizes the advantages of the TAB converter.

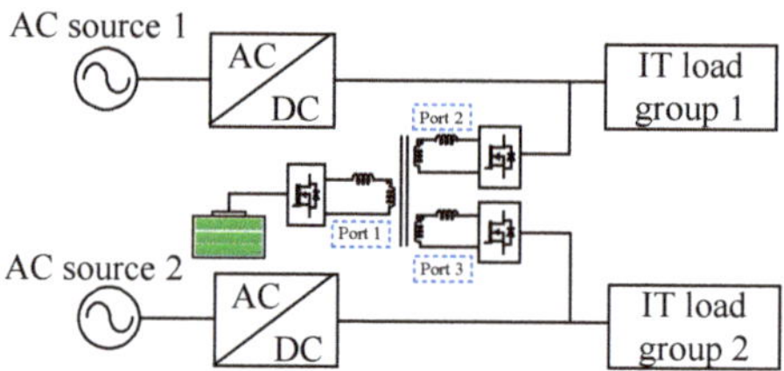

Figure 16. TAB converter in power distribution for data center application.

Typically, the data center distribution system has many independent loads. The power requirement at a time is different for each load. Some loads may require high power while others need low power. Balancing the load can improve the sustainability of the system. In order to improve the reliability of the data center distribution system, the TAB converter was proposed to share the power between the distribution lines in [52], as shown in Figure 17. The power distribution in each load group is evaluated. The TAB converter balances the power load for the data center system. This can improve the reliability of the system from 96.00% to 97.53%. Therefore, the TAB converter is a promising candidate in this application. However, the stability and accuracy will be the challenge for controlling the TAB converter in this application because of high-reliability requirements.

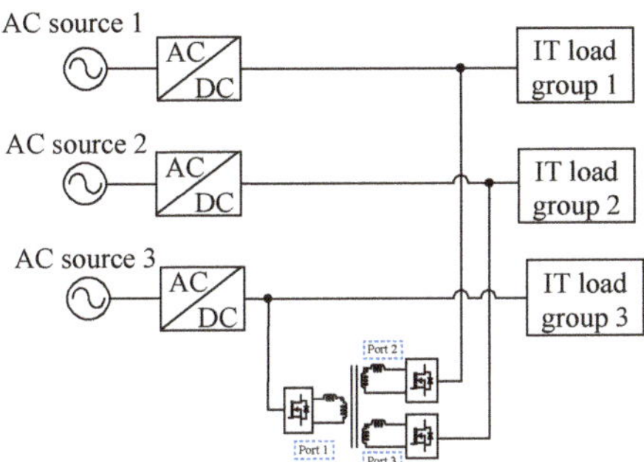

Figure 17. Improving the reliability of power distribution for data center by using TAB converter.

3.5. Electric Vehicles

Electric vehicles (EVs) normally have a high voltage (HV) battery of 300 V–400 V and a low voltage (LV) battery of 14 V [53]. The HV battery supplies the main power for the motor. The LV battery supplies power for other facilities such as the fan, light, wiper, radio, etc. The on-board charger is implemented in EVs to charge the EVs at home with a power range of 3–6 kW. Therefore, the DC-DC converter and charger are combined using the TAB converter as shown in Figure 18 [36–39]. The port-1 is connected to the main battery. The port-2 is connected to the low voltage battery. The port-3 is connected to the DC source after the rectifier from the AC grid. The two DC-DC converters in the conventional system are combined by using one TAB converter. This reduces the components, size, and cost of the system. However, the idling isolation of the charger port is a challenge in this application. Therefore, phase shift combined duty cycles are proposed for this condition to reduce the peak and RMS current [36–40]. This shows that this method can be applied in other applications that have a similar critical condition.

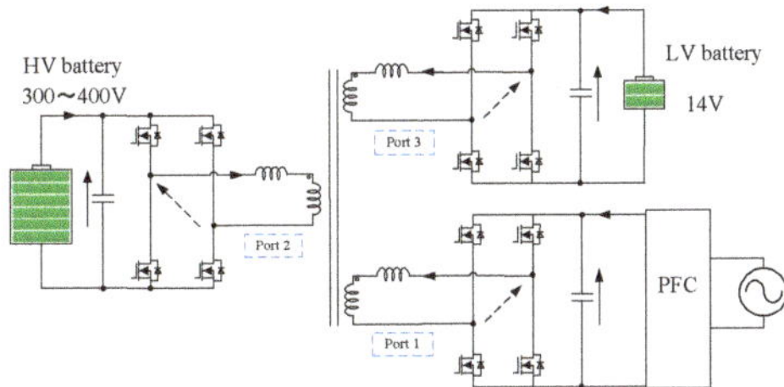

Figure 18. TAB converter for electric vehicles (EVs) application as an integrated on-board charger and DC-DC converter.

The 42 V bus for EVs was discussed in [40] to increase the power for HEVs. Therefore, the electrical system may have three voltage levels (14 V/42 V/400 V). The TAB converter has many advantages for this system because it separates voltage levels using one converter with a three-winding transformers, as shown in Figure 19. The power can be supplied from three storage systems effectively during start-up or regular running. This shows that all the operation modes of the TAB converter can be applied in this application to maximize the power and lifetime of the system. In the future, the studies on optimal power and operation modes for each condition of the system can be discussed more.

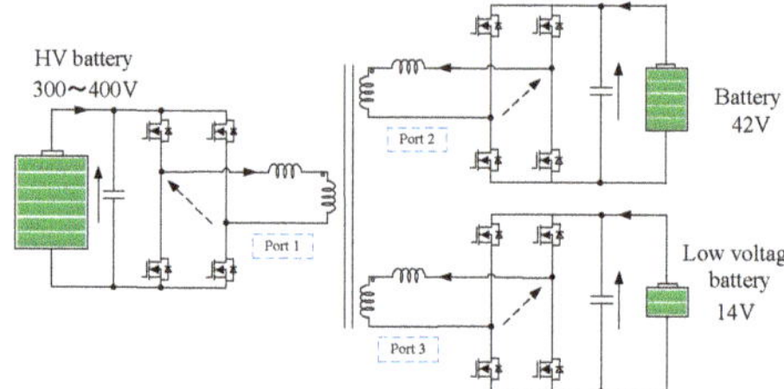

Figure 19. DC-DC converter in EVs application using the TAB converter.

3.6. More Electric Aircraft

Beside electric vehicles, the more electric aircraft (MEA) has been widely researched and discussed to reduce emissions in the world [54,55]. The electrical system in MEA requires improved efficiency, reliability, and reduced implementation costs. The electrical system usually uses a 230 V AC voltage and two DC voltages, of 270 V and 28 V, respectively. The auxiliary power unit (APU) is used to supply power when the MEA is on the ground. It can use a battery and fuel cell system of 270 V or 540 V [55–57]. The battery can also be used for emergency situations, to store energy in low load conditions, and support in heavy load conditions. Therefore, the TAB is proposed to regulate power flow in MEAs in [58], as shown in Figure 20. Two DC voltages and APU are connected and supported by one TAB converter and a centralized control system.

Using the TAB converter is an advantage because the power between three elements always works together to achieve high safety and reliability of this application. However, high current and high power design and control are the challenges of this application. More research and verification should be discussed in detail in the future.

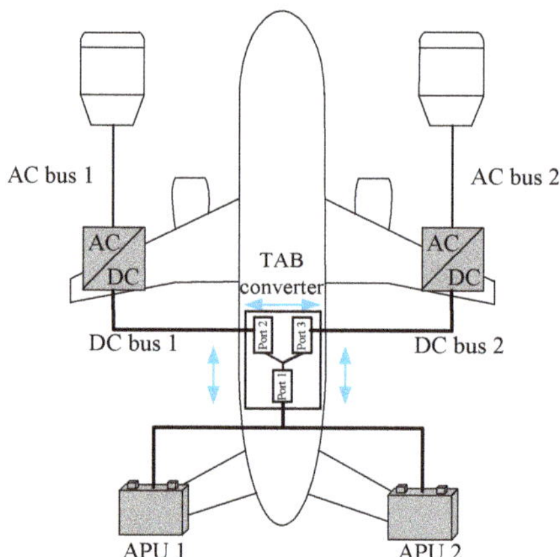

Figure 20. TAB converter in more electric aircraft application.

4. Future Applications with Battery

In recent years, there have been many new applications which aim to use more electricity as the main energy. There are many types of batteries and techniques that are developing to connect many cells in series and parallel for the high power applications [59–62]. Therefore, this part aims to propose the TAB converter for new potential applications which use the storage system.

4.1. Autonomous Underwater Vehicles

The autonomous underwater vehicles (AUVs) have spread widely in recent years [63–65]. The size and weight of the AUVs are limited, and the data can be corrected if the AUVs can work for a long time. Therefore, the power system of the AUVs is a challenge. The Lithium-ion battery is used for AUVs because of its high power density [63]. The fuel cell system for AUVs is discussed to extend their working time [64]. The integrated fuel cell and battery systems for AUVs have been discussed in recent times [65]. This can achieve a higher power density, which reduces the number of the batteries and increases the power for the AUVs. With the many advantages in the integrated power system, the TAB converter is a promising candidate for this application. A TAB converter can be connected directly between the motor driver (M), fuel cell, and battery, as shown in Figure 21.

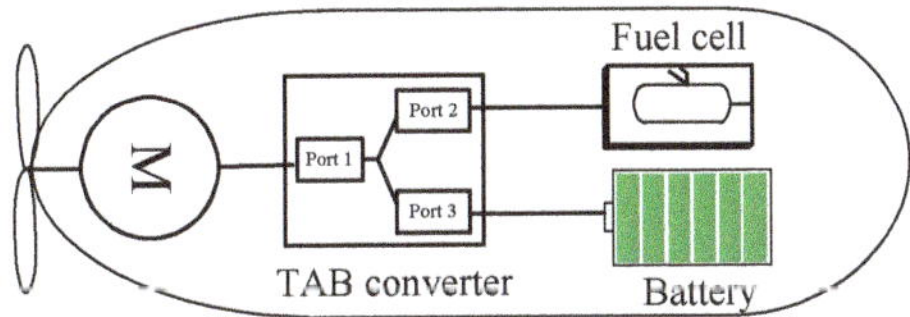

Figure 21. Proposed TAB converter for autonomous underwater vehicles (AUVs).

It can use a 48 V battery with a capacity of 34 Ah for a small AUVs [65]. A fuel cell can install a 48 V system. The converter operates at 1 to 3 kW. However, this application requires high efficiency and small size. To realize a large power AUVs system, the system can be implemented with an optimal design of the TAB converter, depending on the battery and fuel cell system.

4.2. All-Electric Ships

All-electric ships (AESs) are developed to improve the efficiency of energy in the ships [66,67]. The advantages of MVDC distribution for all-electric ships (AESs) was discussed in [68]. By using MVDC (1–8 kV) distribution, the motor is not connected to a fixed-frequency system and can be designed and operated to maximize efficiency. The advantage of the battery energy storage system (BESS) for the ship was discussed in [69]. Therefore, the electrical ship is a promising application of the TAB converter in the future, as shown in Figure 22.

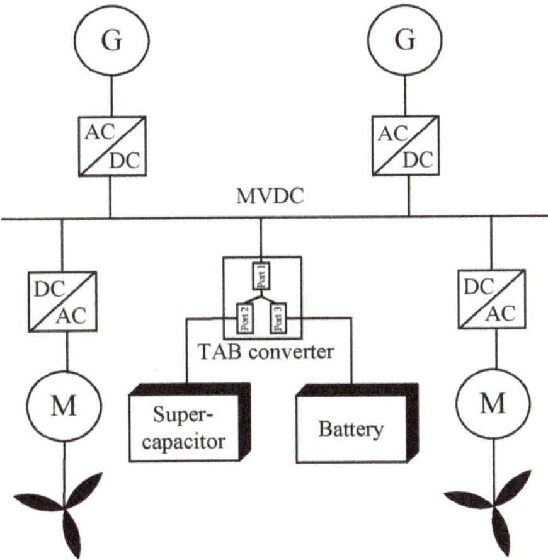

Figure 22. Proposed TAB converter for future all-electric ships.

The battery and supercapacitor can be charged by regenerative energy in the light load condition, and then used for supporting in the heavy load condition. It stabilizes the voltage and current of the electrical grid in the ship. Additionally, the battery and supercapacitor are useful auxiliary power systems for rapid-starting in booster or electric-motor modes. Depending on the size and type of ship, the power system may range from 500 kW to several megawatt. The storage capacity is selected depending on the control strategy. It can be implemented by the 400 V storage system. The TAB converter for AESs application may work at several hundred kW, which is challenged by the high voltage high power design.

5. Future Applications for Controlling Power Flow

The AC power utility system for a railway, for example, shinkansen in Japan, uses two single-phase AC voltages 25 kV or 20 kV for two direction lines [70–72]. The single-phase AC voltages are supplied from a three-phase voltage using a Scott-connected transformer, as shown in Figure 23a. The electric power load of each single-phase voltage may be unbalanced depending on the number and position of the trains in each direction. This affects the three-phase voltage side, which causes a more significant voltage fluctuation. Therefore, a railway static power conditioner (RPC) is developed to control voltage fluctuation on the three-phase voltage side in [72], as shown in Figure 23b. An RPC is constructed from two pulse width modulation (PWM) inverters, which are connected by a large DC link capacitor system. The other side of each inverter is connected to a single-phase voltage by a single-phase transformer.

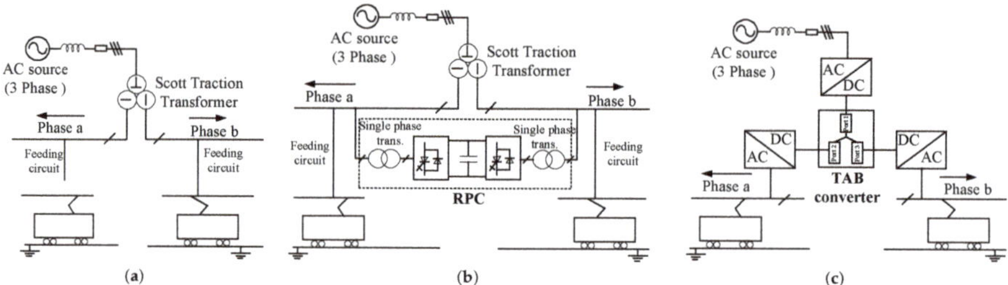

Figure 23. Proposed future concept for shinkansen power supply using the TAB converter: (**a**) conventional railway power utility system, (**b**) railway static power conditioner to balance voltage, (**c**) future concept using TAB converter.

In this way, the two inverters can work as a static synchronous compensator (STATCOM), and the reactive power is compensated by feeding an active power from one bus to another. The power utility system then needs a scott traction transformer, two inverters, two single-phase transformers, and a DC link capacitor system. Both the Scott transformer and RPC have large-capacity, high weight, and big size.

In this paper, a future concept is suggested for application in the future railway utility by using the TAB converter, as shown in Figure 23c. A TAB converter, which connected 160 kV DC voltage from the rectifier and two 44 kV DC voltages outputs for the train system. It can replace both the Scott transformer and RPC to achieve isolation and power flow control targets. The size, components, and cost are reduced. The reactive power from the load also is compensated. With this new concept, the power of each phase of the AC source is balanced, which improves the stability and reliability of the grid. The connection of many TAB converters in the module can be suggested to solve the challenges of very high voltage and power.

6. Summary and Discussion

The above sections reviewed current applications and proposed new applications of the TAB converter. This section gives a summary all applications of the TAB converter. The advantage and challenges of each application are summarized, as shown in Table 1. For each application, the TAB converter contributes to the different targets to develop conventional systems. There are also different challenges in each application, which also need more development in the future. The normal and abnormal operation modes are listed. It shows that some applications use all operation modes such as distribution systems for data center and more electric aircraft, where the TAB converter is a good selection because of saving cost and improving the reliability. Some applications do not use all operation modes. However, by using the TAB converter, it not only reduces the component but also extends the functionality of the system.

In the future, research into the TAB converter can go in two directions. Firstly, new studies can focus on solving the general issues of the TAB converter. This can have an impact on one or many applications. Secondary, the new investigation can solve a challenge in each application also is the contribution. This suggests that also the new trend of semiconductors and the magnetic components can be discussed to improve the TAB converter. It is also necessary to undertake further study of the TAB converter in these applications, which are high power and high voltage/current.

Table 1. Summary the characteristics of applications of TAB converter.

Part	Application	Meaning of the TAB Converter	Challenges	Operation Modes			Suggest/Note
				Normal	Abnormal	Do Not Use	
III.A	DC-bus microgrid (Figure 10)	Reducing the components and cost; centralizing control in each household;	Variation voltage from variation connection to each house hold system;	1-1; 1-2; 1-3; 1-6; 3-1;	2-2; 2-3;	1-4; 1-5; 1-6; 2-1; 3-2; 3-3;	By combining two ideas, the TAB can develop the future micro grid system.
	Autonomous microgrid (Figure 11)	Easy to be extended or maintained any time;	Still large number of converter complex control of the system level;	All	N/A	No	
III.B	Connection of MVDC and LVDC grid (Figure 12)	Connecting three voltage systems by only one converter with central control;	High voltage (5 kV) high power (150 kW) design and measurement (transformer, inductor, voltage, and current)	1-1; 1-2; 3-1;	Remain modes	No	N/A
III.C	UPS with AC source (Figure 14)	Improving safety and reliability; flexible operation modes;	Performance of AC grid;	1-1;	Remain modes;	No;	N/A
	UPS system using fuel cell (Figure 15)		Fast transient response strategy;	1-1;	1-2; 1-3; 2-2; 3-1;	1-4; 1-5; 1-6; 2-1; 2-3; 3-2; 3-3;	N/A
III.D	Data center 1 (Figure 16)	Reducing the components, cost, and size; balancing two loads;	Stability and accuracy of the system;	No	All	No	Break time of data center has high cost as millions dollar an hour;
	Data center 2 (Figure 17)	Balancing three loads, increasing reliability of the system from 96.00% to 97.53%;		No	All	No	
III.E	EVs 1 (Figure 18)	Reduce the components, cost, and size of the system;	Virtually isolating the AC charger side;	1-1; 1-3; 1-5; 3-3;	1-4; 2-1; 2-2;	1-2; 1-6; 2-3; 3-1; 3-2;	N/A
	EVs 2 (Figure 19)		Optimal transmission power of each element;	1-1; 1-2; 1-6; 3-1;	1-3; 1-4; 1-5; 2-1; 2-2;	No	N/A

Table 1. *Cont.*

Part	Application	Meaning of the TAB Converter	Challenges	Operation Modes			Suggest/Note
				Normal	Abnormal	Do Not Use	
III.F	More Electric Aircraft (Figure 20)	Improving the reliability of the system;	High current and power design;	All	No	No	Increase voltage of the system to reduce operating current;
IV.A	AUV [proposed] (Figure 21)	Central power control by only one converter;	High efficiency and high power density;	1-4; 1-5; 1-6;	1-2; 2-1; 2-3; 3-2;	1-1; 1-3; 2-2; 3-1; 3-3;	N/A
IV.B	AES [proposed] (Figure 22)	DC grid for ship to maximize the advantages; central power control by only one converter;	High voltage (1–8 kV) and high power 500 kW-MW design and measurement;	1-1; 1-2; 1-4; 1-5; 2-1; 3-1;	1-3; 1-6; 2-2; 2-3; 3-2; 3-3	No	Connecting many TAB converter in module for very high power and voltage application;
V	Power supply for train [proposed] (Figure 23)	Reducing the component, size and cost;	Very high voltage (160 kV) and very high power (20 MW) design and measurement;	1-1; 1-2; 3-1;	remain modes	No	

7. Conclusions

The TAB converter is considered a promising circuit for next-generation DC grid and integrated energy systems, which have vast market prospects. The advantages of the TAB converter include multiple interfacing ports with isolation, achievable implementation of centralized controls, and improved flexibility of electric systems. This paper reviewed the characteristics of the TAB converter and the research status of the current applications. The paper also showed that the TAB converter could be an upgraded solution for conventional systems. Additionally, some potential future applications with the battery were presented. A future concept for AC voltage railway systems was also proposed. The design and performance optimization of the TAB converter for high power applications with high reliability will be the trend in the future.

Author Contributions: V.-L.P. analyzed the operation and characteristics of the TAB converter; V.-L.P. summarized the current applications of the TAB converter; two authors actively proposed new applications of the TAB converter. The paper was written by V.-L.P.; two authors actively participated in revising the paper; the paper was supervised by K.W.; two authors have read and agreed to the published version of the manuscript.

Funding: This research received no external funding.

Conflicts of Interest: The authors declare no conflict of interest.

Abbreviations

The following abbreviations are used in this manuscript:

MDPI	Multidisciplinary Digital Publishing Institute
DOAJ	Directory of open access journals
TLA	Three letter acronym
LD	linear dichroism

References

1. Chan, C.C. The State of the Art of Electric, Hybrid, and Fuel Cell Vehicles. *Proc. IEEE* **2007**, *95*, 704–718. [CrossRef]
2. Huang, A.Q.; Crow, M.L.; Heydt, G.T.; Zheng, J.P.; Dale, S.J. The Future Renewable Electric Energy Delivery and Management (FREEDM) System: The Energy Internet. *Proc. IEEE* **2011**, *9*, 133–148. [CrossRef]
3. Blaabjerg, F.; Teodorescu, R.; Liserre, M.; Timbus, A.V. Overview of Control and Grid Synchronization for Distributed Power Generation Systems. *IEEE Trans. Ind. Electron.* **2006**, *53*, 1398–1409. [CrossRef]
4. Kim, J.; Yu, C.J.; Khammuang, M.; Lui, J.; Almujahid, A.; Daim, T. Forecasting Battery Electric Vehicles. In Proceedings of the 2017 IEEE Technology & Engineering Management Conference (TEMSCON), Santa Clara, CA, USA, 8–10 June 2017.
5. Yilmaz, M.; Krein, P.T. Review of the Impact of Vehicle-to-Grid Technologies on Distribution Systems and Utility Interfaces *IEEE Trans. Power Electron.* **2013**, *28*, 5673–5689. [CrossRef]
6. Longo, M.; Yaïci, W.; Foiadelli, F. Electric Vehicle Charge with Residential's Roof Solar Photovoltaic System: A Case Study in Ottawa. In Proceedings of the 2017 IEEE 6th International Conference on Renewable Energy Research and Applications (ICRERA), San Diego, CA, USA, 5–8 November 2017; pp. 121–125.
7. Sharaf, O.Z.; Orhan, M.F. An overview of fuel cell technology: Fundamentals and applications. *Renew. Sustain. Energy Rev.* **2014**, *32*, 810–853. [CrossRef]
8. He, H.; Zhang, Y.; Wan, F. Control strategies design for a fuel cell hybrid electric vehicle. In Proceedings of the IEEE Vehicle Power and Propulsion Conference, Harbin, China, 3–5 September 2008; pp. 1–6.
9. US Department of Energy. *The Department of Energy Hydrogen and Fuel Cells Program Plan*; An integrated strategic plan for the research, development, and demonstration of hydrogen and fuel cell technologies; US Department of Energy: Washington, DC, USA, 2011.
10. Gao, W. Performance comparison of a fuel cell-battery hybrid powertrain and a fuel cell-ultracapacitor hybrid powertrain. In Proceedings of the Power Electronics in Transportation (IEEE Cat. No.04TH8756), Novi, MI, USA, 21–22 October 2004; pp. 143–150.

11. Shivarama Krishna, K.; Sathish Kumar, K. A review on hybrid renewable energy systems. *Renew. Sustain. Energy Rev.* **2015**, *52*, 907–916. [CrossRef]

12. Planas, E.; Andreu, J.; Gárate, J.I.; De Alegría, I.M.; Ibarra, E. AC and DC Technology in Microgrids: A Review. *Renew. Sustain. Energy Rev.* **2015**, *43*, 726–749. [CrossRef]

13. Wunder, B.; Ott, L.; Szpek, M.; Boeke, U.; Weis, R. Energy efficient dc-grids for commercial buildings. In Proceedings of the Telecommunications Energy Conference (INTELEC), Vancouver, BC, Canada, 28 September–2 October 2014; pp. 1–8.

14. Boeke, U.; Wei, R.; Mauder, A.; Hamilton, L.; Ott, L. Efficiency Advantages of ±380 v dc Grids in Comparison with 230 v/400 v ac Grids. Available online: http://dcgrid.tue.nl/files/DCC_G_D6_1_Public_summary_V1_0.pdf (accessed on 27 February 2020).

15. Inoue, S.; Akagi, H. A Bidirectional Isolated DC–DC Converter as a Core Circuit of the Next-Generation Medium-Voltage Power Conversion System. *IEEE Trans. Power Electron.* **2007**, *22*, 535–542. [CrossRef]

16. Zhao, B.; Song, Q.; Liu, W.; Sun, Y. Overview of Dual-Active-Bridge Isolated Bidirectional DC–DC Converter for High-Frequency-Link Power-Conversion System. *IEEE Trans. Power Electron.*, **2014**, *29*, 4091–4106. [CrossRef]

17. Duarte, J.L.; Hendrix, M.; Simoes, M.G. Three-Port Bidirectional Converter for Hybrid Fuel Cell Systems. *IEEE Trans. Power Electron.* **2007**, *22*, 480–487. [CrossRef]

18. Chowdhury, S.; Chowdhury, S.P.; Crossley, P. *Microgrids and Active Distribution Networks*, 1st ed.; IET: London, UK, 2009.

19. Prabhala, V.A.; Baddipadiga, B.P.; Fajri, P.; Ferdowsi, M. An Overview of Direct Current Distribution System Architectures & Benefits. *Energies* **2018**, *11*, 2463.

20. Kado, Y.; Shichijo, D.; Deguchi, I.; Iwama, N.; Kasashima, R.; Wada, K. Power flow control of three-way isolated DC/DC converter for Y-configuration power router. In Proceedings of the 2015 IEEE 2nd International Future Energy Electronics Conference (IFEEC), Taipei, Taiwan, 1–4 November 2015; pp. 1–5.

21. Pham, V.-L.; Wada, K. Normalization Design of Inductances in Triple Active Bridge Converter for Household Renewable Energy System. *IEEJ J. Ind. Appl.* **2020**, *9*, 3.

22. Pham, V.-L.; Wada, K. Design of Series Inductances in Triple Active Bridge Converter Using Normalization Procedure for Integrated EV and PV System. In Proceedings of the 2019 10th International Conference on Power Electronics and ECCE Asia (ICPE 2019—ECCE Asia), Busan, Korea, 27–30 May 2019.

23. Kado, Y.; Shichijo, D.; Wada, K.; Iwatsuki, K. Multiport power router and its impact on future smart grids. *Radio Sci.* **2016**, *51*, 1234–1246. [CrossRef]

24. Nishimoto, K.; Kado, Y.; Wada, K. Implementation of Decoupling Power Flow Control System in Triple Active Bridge Converter Rate 400 V, 10 kW, and 20 kHz. *IEEJ J. Ind. Appl.* **2018**, *7*, 410–415.

25. Nakagawa, S.; Arai, J.; Kasashima, R.; Nishimoto, K.; Kado, Y.; Wada, K. Dynamic performance of triple-active bridge converter rated at 400 V, 10 kW, and 20 kHz. In Proceedings of the IEEE International Future Energy Electronics Conference and ECCE Asia, Kaohsiung, Taiwan, 3–7 June 2017; pp. 1090–1094.

26. Neubert, M.; Gorodnichev, A.; Gottschlich, J.; De Doncker, R.W. Performance analysis of a triple-active bridge converter for interconnection of future dc-grids. In Proceedings of the 2016 IEEE Energy Conversion Congress and Exposition (ECCE), Milwaukee, WI, USA, 18–22 September 2016; pp. 1–8.

27. Neubert, M.; Van Hoek, H.; Gottschlich, J.; De Doncker, R.W. Soft-switching operation strategy for three-phase multiport-active bridge DC-DC converters. In Proceedings of the IEEE 12th International Conference on Power Electronics and Drive Systems (PEDS), Honolulu, Hawaii, USA, 12–15 December 2017; pp. 1207–1213.

28. Gottschlich, J.; Weiler, P.; Neubert, M.; De Doncker, R.W. Delta-sigma modulated voltage and current measurement for medium-voltage DC applications. In Proceedings of the 2017 19th European Conference on Power Electronics and Applications (EPE'17 ECCE Europe), Warsaw, Poland, 11–14 September 2017; pp. 1–9.

29. Neubert, M.; Engel, S.P.; Gottschlich, J.; De Doncker, R.W. Dynamic power control of three-phase multiport active bridge DC-DC converters for interconnection of future DC-grids. In Proceedings of the IEEE 12th International Conference on Power Electronics and Drive Systems (PEDS), Honolulu, HI, USA, 12–15 December 2017; pp. 639–646.

30. Engelmann, G.; Sewergin, A.; Neubert, M.; De Doncker, R.W. Design Challenges of SiC Devices for Low- and Medium-Voltage DC-DC Converters. In Proceedings of the 2018 International Power Electronics Conference (IPEC-Niigata 2018-ECCE Asia), Niigata, Japan, 20–24 May 2018; pp. 3979–3984.

31. Zhao, C.; Kolar, J.W. A novel three-phase three-port UPS employing a single high-frequency isolation transformer. In Proceedings of the 2004 IEEE 35th Annual Power Electronics Specialists Conference (IEEE Cat. No.04CH37551), Aachen, Germany, 20–25 June 2004; pp. 4135–4141.

32. Ahmed Adam, A.H.; Hou, S.; Chen, J. Analysis, Design, and Performance of Isolated Three-Port UPS Converter for High-Power Applications. In Proceedings of the 2019 IEEE International Conference on Environment and Electrical Engineering and 2019 IEEE Industrial and Commercial Power Systems Europe (EEEIC / I&CPS Europe), Genova, Italy, 10–14 June 2019; pp. 1–7.

33. Tao, H.; Duarte, J.L.; Hendrix, M.A.M. Line-Interactive UPS Using a Fuel Cell as the Primary Source. *IEEE Trans. Ind. Electron.***2008**, *55*, 3012–3021.

34. Yu, Y.; Masumoto, K.; Wada, K.; Kado, Y. A DC Power Distribution System in a Data Center Using a Triple Active Bridge DC-DC Converter. *IEEJ J. Ind. Appl.* **2018**, *7*, 202–209. [CrossRef]

35. Yu, Y.; Masumoto, K.; Wada, K.; Kado, Y. Power Flow Control of a Triple Active Bridge DC-DC Converter Using GaN Power Devices for a Low-Voltage DC Power Distribution System. In Proceedings of the 2017 IEEE 3rd International Future Energy Electronics Conference and ECCE Asia (IFEEC 2017—ECCE Asia), Kaohsiung, Taiwan, 3–7 June 2017; pp. 772–777.

36. Kim, S.Y.; Song, H.; Nam, K. Idling Port Isolation Control of Three-Port Bidirectional Converter for EVs. *IEEE Trans. Power Electron.* **2012**, *27*, 2495–2506. [CrossRef]

37. Kim, S.Y.; Jeong, I.; Nam, K.; Song, H.S. Three-Port Full Bridge Converter Application as a Combined Charger for PHEVs. In Proceedings of the 2009 IEEE Vehicle Power and Propulsion Conference, Dearborn, MI, USA, 7–10 September 2009; pp. 461–465.

38. Ling, Z.; Wang, H.; Yan, K.; Gan, J. Optimal Isolation Control of Three-Port Active Converters as a Combined Charger for Electric Vehicles. *Energies* **2016**, *9*, 715. [CrossRef]

39. Nguyen, D.D.; Fujita, G.; Ta, M.C. New Soft-Switching Strategy for Three-Port Converter to be Applied in EV Application. In Proceedings of the 2017 IEEE 3rd International Future Energy Electronics Conference and ECCE Asia, Kaohsiung, Taiwan, 3–7 June 2017.

40. Zhao, C.; Round, S.D.; Kolar, J.W. An Isolated Three-Port Bidirectional DC-DC Converter with Decoupled Power Flow Management. *IEEE Trans. Power Electron.* **2008**, *23*, 2443–2453. [CrossRef]

41. Tao, H.; Duarte, J.L.; Hendrix, M.A.M. High-Power Three-Port Three-Phase Bidirectional DC-DC Converter. In Proceedings of the IEEE Industry Applications Annual Meeting, New Orleans, LA, USA, 23–27 Septemper 2007; pp. 2022–2029.

42. Tao, H.; Kotsopoulos, A.; Duarte, J.L.; Hendrix, M.A. Transformer-Coupled Multiport ZVS Bidirectional DC-DC Converter with Wide Input Range. *IEEE Trans. Power Electron.* **2008**, *23*, 771–781. [CrossRef]

43. Mura, F.; De Doncker, R. Design aspects of a medium-voltage direct current (MVDC) grid for a university campus. In Proceedings of the 8th International Conference on Power Electronics—ECCE Asia, Jeju, Korea, 30 May–3 June 2011; pp. 2359–2366.

44. Priebe, J.; Wehbring, N.; Moser, A. Design of Medium Voltage DC Grids– Impact of Power Flow Control on Grid Structure. In Proceedings of the 2018 53rd International Universities Power Engineering Conference (UPEC), Glasgow, Scotland, 4–7 September 2018; pp. 1–6.

45. Stieneker, M.; Mortimer, B.J.; Hinz, A.; Müller-Hellmann, A.; De Doncker, R.W. MVDC Distribution Grids for Electric Vehicle Fast-Charging Infrastructure. In Proceedings of the 2018 International Power Electronics Conference (IPEC-Niigata 2018-ECCE Asia), Niigata, Japan, 20–24 May 2018; pp. 598–606.

46. Tran, Y.; Dujic, D. A multiport medium voltage isolated DC-DC converter. In Proceedings of the IECON 2016—42nd Annual Conference of the IEEE Industrial Electronics Society, Florence, Italy, 23–26 October 2016; pp. 6983–6988.

47. Ilango, S.; Viju Nair, R.; Chattopadhyay, R.; Bhattacharya, S. Photovoltaic and Energy Storage Grid Integration with Fully Modular Architecture using Triple Port Active Bridges and Cascaded H-Bridge Inverter. In Proceedings of the IECON 2018—44th Annual Conf. of the IEEE Industrial Electronics Society, Washington, DC, USA, 21–23 October 2018; pp. 1400–1405.

48. Schäfer, J.; Bortis, D.; Kolar, J.W. Multi-port multi-cell DC/DC converter topology for electric vehicle's power distribution networks. In Proceedings of the 2017 IEEE 18th Workshop on Control and Modeling for Power Electronics (COMPEL), Stanford, CA, USA, 9–12 July 2017; pp. 1–9.

49. Michon, M.M.J.A.; Duarte, J.L.; Hendrix, M.; Simoes, M.G. A three-port bi-directional converter for hybrid fuel cell systems. In Proceedings of the 2004 IEEE 35th Annual Power Electronics Specialists Conference, Aachen, Germany, 20–25 June 2004; pp. 4736–4742.

50. Tao, H.; Duarte, J.L.; Hendrix, M.A.M. Three-Port Triple-Half-Bridge Bidirectional Converter with Zero-Voltage Switching. *IEEE Trans. Power Electron.* **2008**, *23*, 782–792.

51. Wiboonrat, M. An empirical study on data center system failure diagnosis. In Proceedings of the 3rd International Conference on Internet Monitoring and Protection, Bucharest, Romania, 29 June–5 July 2008.

52. Yu, Y.; Wada, K. Simulation Study of Power Management for a Highly Reliable Distribution System using a Triple Active Bridge Converter in a DC Microgrid. *Energies* **2018**, *11*, 3178. [CrossRef]

53. Yilmaz, M.; Krein, P.T. Review of battery charger topologies, charging power levels, and infrastructure for plug-in electric and hybrid vehicles. *IEEE Trans. Power Electron.* **2013**, *28*, 2151–2169. [CrossRef]

54. Sarlioglu, B.; Morris, C.T. More Electric Aircraft: Review, Challenges, and Opportunities for Commercial Transport Aircraft. *IEEE Trans. Transp. Electrif.* **2015**, *1*, 54–64. [CrossRef]

55. Wheeler, P.W.; Clare, J.C.; Trentin, A.; Bozhko, S. An overview of the more electrical aircraft. *J. Aerosp. Eng.* **2012**, *227*, 578–585. [CrossRef]

56. Roboam, X.; Langlois, O.; Piquet, H.; Morin, B.; Turpin, C. Hybrid power generation system for aircraft electrical emergency network. *IET Electr. Syst. Transp.* **2011**, *1*, 148–155. [CrossRef]

57. Wu, S.; Li, Y. Fuel cell applications on more electrical aircraft. In Proceedings of the 17th International Conference on Electrical Machines and Systems (ICEMS), Hangzhou, China, 24–25 October 2014; pp. 198–201.

58. Giuliani, F.; Buticchi, G.; Liserre, M.; Dehnonte, N.; Cova, P.; Pignoloni, N. GaN-based triple active bridge for avionic application. In Proceedings of the 2017 IEEE 26th International Symposium on Industrial Electronics (ISIE), Edinburgh, Edinburgh, Scotland, 19–21 June 2017; pp. 1856–1860.

59. Li, J.; Zhou, S.; Han, Y. Review of structures ans control of battery- supercapacitor hybrid energy storage system for electric vehicles. *Advances in Battery Manufacturing, Service, and Management Systems*, Wiley-IEEE Press: Hoboken, NJ, USA, 2017; pp. 303–318.

60. Pham, V.L.; Duong, V.T.; Choi, W. A Low Cost and Fast Cell-to-Cell Balancing Circuit for Lithium-Ion Battery Strings. *Electronics* **2020**, *9*, 248. [CrossRef]

61. Pham, V.L.; Khan, A.B.; Nguyen, T.-T.; Choi, W. A Low Cost, Small Ripple, and Fast Balancing Circuit for Lithium-Ion Battery String, In Proceedings of IEEE Transportation Electrification Conference and Expo Asia-Pacific, ITEC 2016, Busan, Korea, 1–4 June 2016; pp. 861–865.

62. Pham, V.L.; Nguyen, T.T.; Tran, D.H.; Vu, V.B.; Choi, W. A New Cell-to-Cell Fast Balancing Circuit for Lithium-Ion Battery in Electric Vehicles and Energy Storage System. In Proceedings of IEEE 8th International Power Electronic and Motion Control Conference (IPEMC-ECCE Asia), Hefei, China, 22–25 May 2016; pp. 2461–2465.

63. Bradley, A.M.; Feezor, M.D.; Singh, H.; Sorrell, F.Y. Power systems for autonomous underwater vehicles. *IEEE J. Ocean. Eng.* **2001**, *26*, 526–538. [CrossRef]

64. Hyakudome, T.; Nakatani, T.; Yoshida, H.; Tani, T.; Ito, H.; Sugihara, K. Development of fuel cell system for long cruising lange Autonomous Underwater Vehicle. In Proceedings of the 2016 IEEE/OES Autonomous Underwater Vehicles (AUV), Tokyo, Japan, 6–9 November 2016; pp. 165–170.

65. Albarghot, M.M.; Iqbal, M.T.; Pope, K.; Rolland, L. Sizing and dynamic modeling of a power system for the MUN explorer autonomous underwater vehicle using a fuel cell and batteries. *J. Energy* **2019**, *2019*, 4531497. [CrossRef]

66. Thongam, J.S.; Tarbouchi, M.; Okou, A.F.; Bouchard, D.; Beguenane, R. All-electric ships—A review of the present state of the art. In Proceedings of the 2013 Eighth International Conference and Exhibition on Ecological Vehicles and Renewable Energies (EVER), Monte Carlo, Monaco, 27–30 March 2013; pp. 1–8.

67. Seenumani, G.; Sun, J.; Peng, H. Real-Time Power Management of Integrated Power Systems in All Electric Ships Leveraging Multi Time Scale Property. *IEEE Trans. Control Syst. Technol.* **2012**, *20*, 232–240. [CrossRef]

68. Tessarolo, A.; Castellan, S.; Menis, R.; Sulligoi, G. Electric generation technologies for all-electric ships with Medium-Voltage DC power distribution systems. In Proceedings of the 2013 IEEE Electric Ship Technologies Symposium (ESTS), Arlington, VA, USA, 22–24 April 2013; pp. 275–281.

69. Kim, K.; Park, K.; Ahn, J.; Roh, G.; Chun, K. A study on applicability of Battery Energy Storage System (BESS) for electric propulsion ships. In Proceedings of the 2016 IEEE Transportation Electrification Conference and Expo, Asia-Pacific (ITEC Asia-Pacific), Busan, Korea, 1–4 June 2016; pp. 203–207.

70. Morimoto, H.; Ando, M.; Mochinaga, Y.; Kato, T. Development of railway static power conditioner used at substation for shinkansen. In Proceedings of the Power Conversion Conference-Osaka 2002 (Cat. No.02TH8579), Osaka, Japan, 2–5 April 2002; pp. 1108–1111.
71. Uzuka, T.; Hase, S.; Mochinaga, Y.; Takeda, M.; Miyashita, T.; Ueda, T. A static voltage fluctuation compensator for AC electric railway. In Proceedings of the 2004 IEEE 35th Annual Power Electronics Specialists Conference (IEEE Cat. No.04CH37551), Aachen, Germany, 20–25 June 2004; pp. 1869–1873.
72. Uzuka, T.; Ikedo, S.; Ueda, K.; Mochinaga, Y.; Funahashi, S.; Ide, K. Voltage fluctuation compensator for Shinkansen. *Electr. Eng. Jpn.* **2008**, *162*, 25-33. [CrossRef]

Article

Flexibility Assessment of Multi-Energy Residential and Commercial Buildings

António Coelho [1,*], Filipe Soares [1] and João Peças Lopes [2]

[1]   Centre for Power and Energy Systems, INESC TEC, 4200-465 Porto, Portugal; filipe.j.soares@inesctec.pt
[2]   FIEEE, Faculty of Engineering, University of Porto, 4200-465 Porto, Portugal; jpl@fe.up.pt
*   Correspondence: antonio.m.coelho@inesctec.pt

Received: 15 April 2020; Accepted: 20 May 2020; Published: 28 May 2020

Abstract: With the growing concern about decreasing CO_2 emissions, renewable energy sources are being vastly integrated in the energy systems worldwide. This will bring new challenges to the network operators, which will need to find sources of flexibility to cope with the variable-output nature of these technologies. Demand response and multi-energy systems are being widely studied and considered as a promising solution to mitigate possible problems that may occur in the energy systems due to the large-scale integration of renewables. In this work, an optimal model to manage the resources and loads within residential and commercial buildings was developed, considering consumers preferences, electrical network restrictions and CO_2 emissions. The flexibility that these buildings can provide was analyzed and quantified. Additionally, it was shown how this model can be used to solve technical problems in electrical networks, comparing the performance of two scenarios of flexibility provision: flexibility obtained only from electrical loads vs. flexibility obtained from multi-energy loads. It was proved that multi-energy systems bring more options of flexibility, as they can rely on non-electrical resources to supply the same energy needs and thus relieve the electrical network. It was also found that commercial buildings can offer more flexibility during the day, while residential buildings can offer more during the morning and evening. Nonetheless, Multi-Energy System (MES) buildings end up having higher CO_2 emissions due to a higher consumption of natural gas.

Keywords: CO_2 emissions; commercial buildings; flexibility quantification; flexibility optimization; HVAC systems; multi-energy systems; network operation; residential buildings

---

1. Introduction

1.1. Motivation and Aim

The growing worldwide concern with global warming is leading to the implementation of crosscutting policies to reduce greenhouse gas emissions, in particular in the transport and electricity and heat generation sectors, which produced two-thirds of the world's CO_2 emissions in 2016 [1].

The increasing pressure to reduce CO_2 emissions in these sectors is leading policy-makers to incentivize the adoption of emissions-free technologies, such as electrical vehicles (EVs) and renewable energy sources [2]. Despite the evident environmental benefits that these technologies yield, they also pose new challenges to network operators, which are facing higher uncertainties due to the volatile behavior of renewable energy technologies.

As a consequence, the network operation paradigm has been changing in the last decades, relying increasingly in flexibility provided by generation and demand side assets. As the integration of renewables is expected to boost in the coming years, it is of utmost importance to find new ways of increasing energy systems flexibility.

The increased flexibility needs of future energy systems can primarily be satisfied at the demand side, where the integration of different energy systems (electricity, heat and gas) will be one of the most promising options to deliver additional flexibility while keeping the user comfort level and avoiding service disruption. Multi-Energy Systems (MES) can help in increasing the penetration of renewable energy sources (RES) due to the possibility of switching between energy sources and large-scale heat and gas storage systems, which are, in general, cheaper than electricity storage technologies, such as Li-ion batteries [3].

1.2. Literature Review

MES are able to integrate different energy vectors through the use of resources that can transform one energy vector to another in order to meet the energy requirements of the system. The concept of Energy Hub (EH) is being vastly used in the literature to model this kind of systems [4]. Nonetheless, the EH can be used to model any type of building, even those that only use electricity. EHs are able to model the interactions between production, delivery and consumption of different energy vectors through a conversion matrix. Electricity, heat, cooling, gas or hydrogen are some of the vectors that can be integrated within the EH concept. It also considers the efficiency and other characteristics of the resources present in the hub. A standardized matrix model for the MES was developed in [5] in order to facilitate the modeling, integration and optimization of these systems. Later, a linear method was presented in [6], which does not use dispatch factors and thus reduces the complexity of the problem.

The main advantages foreseen for MES is the flexibility that these systems are able to provide to the operation of energy systems. Several works focus on the use of different type of resources like virtual storage by heating/cooling sources, technologies like combined heat and power (CHP) [7], fuel cells (FC) [8], power-to-gas (P2G) [9], microgrids [10], fuel stations [11] or parking lots [12] in order to study the advantages that their operation can bring to the energy systems. This flexibility can be used to help increasing renewables integration [3,13], solving technical problems in electricity networks [14] and decreasing CO_2 emissions [15].

Other studies on flexibility focus on the characteristics that different types of buildings, like commercial or residential, can offer to system operators or to aggregators. This is due to the fact that this type of buildings have different demand profiles, with non-coincident peak consumption periods, and use different types of heating and cooling resources. Commercial buildings usually have heating, ventilation and air conditioning (HVAC) systems to increase workers and visitors' comfort levels and thus are able to provide both heating and cooling energy. This is an important source of flexibility since it encompass different energy vectors and has a high degree of controllability, namely through the temperature set-point and fans speed. These systems can be controlled through demand response (DR) programs as presented in [16] and can participate in electricity markets, as studied in [17]. Although the majority of the residential buildings do not have centralized HVAC systems to control space temperature evenly, they usually have individual thermostats to define temperature set-points for each room. This way, these systems are also able to provide flexibility to system operators or aggregators, as show in [18,19].

In [18,19] it is also described an home energy management system. This system is responsible for acquiring and processing the required data from appliances and managing flexible loads, as well as establishing communication between the aggregator, users and appliances. The main functionalities of the home energy management system will be used in this work under the name of energy hub management system (EHMS), which will be responsible for controlling the resources of the EH.

System operators and aggregators are also able to use more flexibility from energy system by controlling other loads, like water heating [20], lighting [21] or even refrigeration systems [22] present in some type of commercial and residential buildings, through DR programs. The flexibility provided by EVs and its benefits to the electrical network is also vastly studied in the literature under the concept of V2G, as in [23].

1.3. Contributions and Advantages of the Proposed Model

This paper proposes an innovative approach to optimize the flexibility provided by multi-energy residential and commercial buildings with the aim of solving technical problems in electricity networks operation. The model generally contributes to increase networks robustness and may be used by system operators to increase the quality of service indexes. It aims at minimizing costs, considering the prices of energy, the costs of using DR programs and the costs of CO_2 emissions, while keeping the network operating within the specified technical limits.

The proposed model considers different sources of flexibility from residential and commercial buildings. This flexibility comes from changing the flexible load patterns or by managing the resources within the buildings. The flexible loads include thermal loads, like space heating and cooling, water heating or refrigeration systems, and also other loads, like EVs or lighting. Space heating and cooling is assumed to be controlled through temperature set-points defined for centralized HVAC systems and thermostats. Although only some type of resources are used in this work, such as electric boilers (EB), gas boilers (GB), CHPs, air conditioners (AC) and heat pumps (HP), the proposed model is prepared to include other resources like P2G or combined cooling, heat and power (CCHP). The studies performed look forward to a future where these technologies are widely used, since although these resources and their combinations are not extensively used nowadays, the situation may change in the coming years as the costs of these technologies have been showing a clear decreasing trend.

The optimization problem used assumes that consumers provide the parameters necessary for its resolution, like temperature ranges, lighting levels or EVs charging needs. The optimization problem differs from [7–10,14,16–23] by allowing the participation of different sources of load flexibility (e.g., PVs, EVs, HVAC systems, thermal loads, lighting) in DR programs and by considering the constraints imposed by the consumers needs and by the technical characteristics of the electrical network.

In the literature, the models developed for HVAC systems considering thermal models are usually non-linear models [24,25] or quadratic [26]. In order to linearize these models, some works use a sequential linearization approach [17] or assume a linear model with parameters calculated using regression methods [16]. A new linear model for HVAC systems was developed in this work. This linear model differs from the other linearized models by discretizing the air flow rate supply variable. This way, the approach for the linearization of the models is simpler as it does not require to be done individually for each building.

With this model, the authors studied and compared the flexibility that can be provided by buildings to electrical network operators considering two scenarios: the buildings only have electrical loads (only-electrical buildings) and the buildings have loads that consume different types of energy (MES buildings). In the literature there are several studies regarding the advantages of MES, but this study specifically compares the same buildings with two different structures: one purely electrified and other with MES. Electrification and MES are two different perspectives of looking into future scenarios of the power systems development. They have aroused great interest in the scientific community and this study offers a new level of comparison of both of them.

Furthermore, the flexibility offered by commercial and residential buildings is studied and compared. There are several works in the literature that study the flexibility that commercial [16,17] and residential [18,19] buildings are able to offer to network operators or aggregators but these works only focus on each type of buildings individually. Commercial and residential buildings have different patterns of consumption which may reflect the availability of flexibility in different times of the day. In this work, it was considered the operation of both type of buildings and the flexibility that each building is able to provide was compared.

The proposed approach was applied to a case study that includes an electrical network with technical problems. The results achieved show how the model manages the flexibility from the resources available to overcome the identified technical problems. In addition to the global flexibility quantification, the contribution of different flexible resources was also analyzed. The importance of a detailed model to explore every source of flexibility within buildings in order to mitigate any problem present in power networks is thus highlighted.

Lastly, an annual analysis of the different cases was also performed with emphasis on the costs, energy consumed and CO_2 emissions.

1.4. Paper Organization

The rest of the paper is organized as follows: Section 2 explains the problem and the model developed in this study; in Section 3, the mathematical model of the proposed approach is presented; Section 4 provides the case studies and numerical results; finally, Section 5 presents the main findings of the work developed.

2. Problem and Model Description

This work focus on the flexibility that can be provided by residential and commercial buildings to a system operator. This flexibility is obtained by changing the resources' point of operation. The buildings' flexibility will be quantified and analyzed for two scenarios: when they have resources that only use electricity and when they have multi-energy resources. The model developed allows the system operator to optimize the flexibility of the buildings' resources through a DR-based program in order to solve network technical problems. Therefore, the model assumes that the system operator is able to control the resources and loads of the EHs by communicating with their EH management system.

2.1. System Operator

The system operator is responsible to satisfy the energy needs of the consumers, while maintaining the network operating within the predefined technical limits. When the network is operating in normal conditions, systems operators may use the clients' flexibility to minimize operating costs (OPEX) or to maximize RES integration. When technical problems occur, flexibility may be used to solve them. Of course that consumers should be paid for the flexibility provided. The cost of this DR-based service is assumed to be established in a contract set between the system operator and the consumer.

The consumers present in the network are commercial and residential buildings. They are represented by EHs and it is assumed that they have an EHMS integrated. The EHMS is responsible for the control of all the assets within the EH, to fulfill the consumers' load requirements and optimize the EH. When needed, the system operator communicates with the EHMSs to request the delivery of flexibility.

EHs are connected to the electrical network in a certain point and they consume or inject electricity in that same point. This way, the system operator is able to manage the consumption or injection of electricity in that point by by using the flexibility of the EHs. When needed, the system operator orders the resources to decrease or increase the consumption of one energy vector or change the load patterns through the DR program, always taking into account the limits predefined by the client.

2.2. Energy Hub

EHs represent commercial and residential buildings and they are formed by input and output energy points, conversion technologies, storage systems and different types of loads (Figure 1).

The input energy points considered are the electricity consumed from the electrical network and from local PVs and the gas consumed from the gas network. The output point is the electricity injected into the electrical network.

Conversion technologies can transform one type of energy to another and storage systems can store energy to be used in another time.

The loads to be satisfied include electricity, gas, heating and cooling loads. They can be inflexible (e.g., electrical equipment and cookers) or flexible (lighting, space heating and cooling, water heating, refrigeration and EVs). Inflexible loads have to be completely satisfied while flexible loads can be modified, according to the flexible limits imposed by the consumers, in order to optimize the EH.

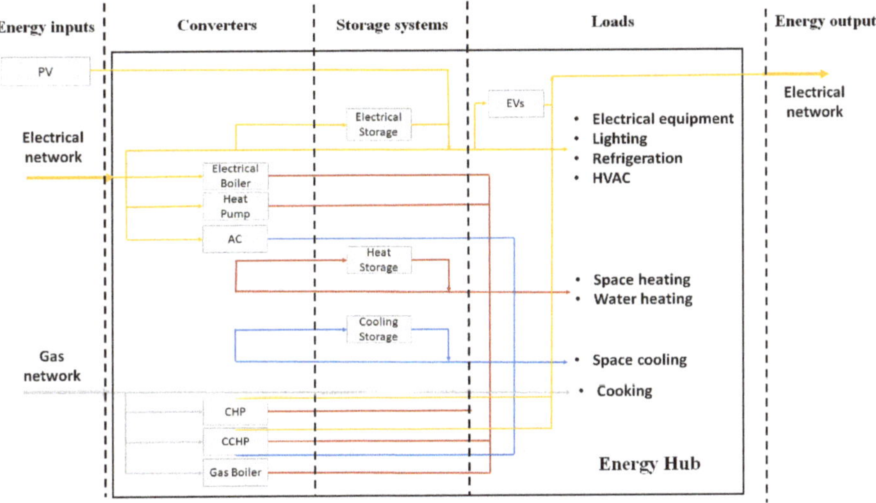

Figure 1. Example of an energy hub scheme.

2.3. Energy Hub Management System

As previously explained, an EHMS is responsible for controlling the assets of the EH in order to fulfill the requirements of the consumers. They are also responsible to communicate with the system operator to inform their energy needs and their availability to participate in the DR program. These requirements are initially set in the EHMS and they involve defining different restrictions and flexibility levels for each type of load. The definition of requirements for each type of load, resources and storage system is explained in the following paragraphs. A scheme of the EHMS is presented in Figure 2. This concept was adapted from [27] and extended for the application in an EH.

The space heating and cooling temperature requirements for each space are initially communicated by consumers to the EHMS and the actual temperature of each room is calculated by the EHMS using a thermal model. The HVAC systems control the heating and cooling power of commercial buildings and thermostats control the heating and cooling power of residential buildings. These systems are responsible for activating the necessary resources to provide the cooling and heating power according to the temperature requirements of the consumers and the actual room temperature calculated by the EHMS.

The consumers communicate the temperature for water heating and refrigeration and the range of acceptable temperatures (i.e., their temperature flexibility) to the EHMS. Water heating and refrigeration loads are also defined by thermodynamic models and calculated by the EHMS. Electric equipment, lighting and gas cooking loads are defined according to the profiles sent by the consumers. Lighting in commercial buildings has a flexibility range that is also defined by the clients.

The restrictions of EVs are defined by the consumers and they involve defining the hours when the EVs will connect and disconnect from the EH and the state-of-charge (SOC) desired at the moment of disconnection.

Storage systems, PVs and other conversion resources have no requirements and can be managed as better fits the EHMS. The only limits imposed are related with their technical characteristics. In summary, the inputs that the consumers need to provide to configure the EHMS are the following:

- Spaces temperature set points and flexibility bands;
- Hot water temperature and flexibility band;
- Refrigeration systems temperatures and flexibility bands;
- Profiles of electrical equipment, lighting and cooking;
- Lighting flexibility bands;
- EVs connection and disconnection hours;
- Final SOC of the EVs.

These inputs need to be provided only once. After that, the consumer may provide new inputs if he wants to reconfigure the EHMS.

With all this information, the EHMS will calculate the energy required by the EH and manage the resources, storage systems and flexible loads in order to satisfy all the requirements imposed by the consumers.

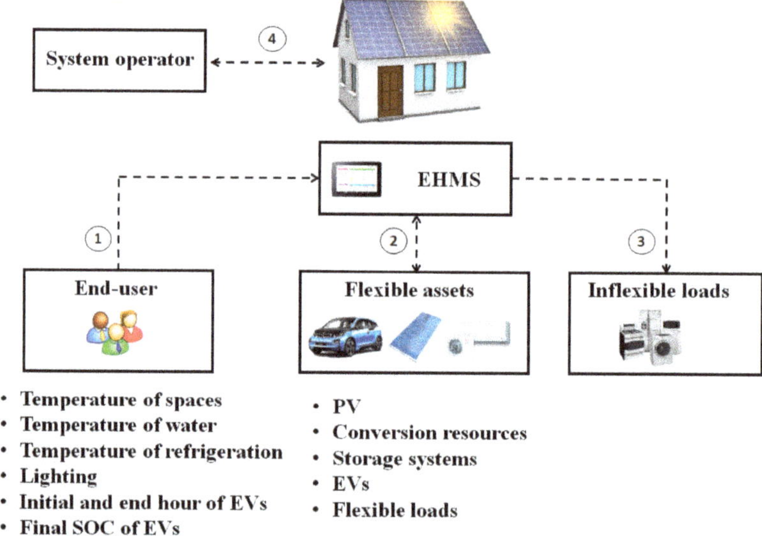

Figure 2. Energy hub management system scheme, adapted from [27].

3. Model Constraints

The model developed in this work and presented in this section has the goal of optimizing the costs of the resources present in the network in order to fulfill all technical requirements from the network and from the resources.

3.1. Objective Function

The objective function in Equation (1) considers the costs of buying electricity as seen in Equation (2) and gas in Equation (3), the costs of DR in Equation (4), the profit of selling electricity in Equation (5), the costs of CO_2 emissions (6) and penalization costs for voltage violations in Equation (7). The costs of DR consider the modification of the profile of lighting and temperature of rooms, water and refrigeration systems.

$$Cost = Cost^{electricity} + Cost^{gas} + Cost^{DR} - Cost^{sold} + Cost^{CO^2} + Cost^{viol} \qquad (1)$$

$$Cost^{electricity} = \sum_j \sum_t I_{j,t}^w . \pi_t^{elec}, \quad \forall j \in \{EB\} \qquad (2)$$

$$Cost^{gas} = \sum_j \sum_t I^g_{j,t}.\pi^{gas}_t, \quad \forall j \in \{EB\} \tag{3}$$

$$Cost^{DR} = \sum_j \sum_t (\Theta^{DR,room,-}_{i,t+1} + \Theta^{DR,room,+}_{i,t+1} + \Theta^{DR,water,-}_{i,t+1} + \Theta^{DR,water,+}_{i,t+1}$$
$$+ \Theta^{DR,refr,-}_{i,t+1} + \Theta^{DR,refr,+}_{i,t+1} + P^{DR,light,-}_{i,t+1} + P^{DR,light,+}_{i,t+1}).\pi^{DR}_t, \quad \forall j \in \{EB\} \tag{4}$$

$$Cost^{sold} = \sum_j \sum_t O^w_{j,t}.\pi^{sold}_t, \quad \forall j \in \{EB\} \tag{5}$$

$$Cost^{CO_2} = \sum_j \sum_t (I^w_{j,t}.\pi^{CO_2,elec}_t + I^g_{j,t}.\pi^{CO_2,gas}_t), \quad \forall j \in \{EB\} \tag{6}$$

$$Cost^{viol} = \sum_j \sum_t V^{viol}_{j,t}.\pi^{viol}_t, \quad \forall j \in \{EB\} \tag{7}$$

The objective function is subjected to operational network constraints which are presented in the following subsections.

3.2. Network Constraints

In this section, the operation of an electrical network is presented. Some of the variables presented are linked with the variables of the EH modeling in Section 3.3, as it will be discussed.

3.2.1. Power Flow Constraints

The considered network is a radial distribution network. A branch power flow model (DistFlow) [28,29] was considered and modeled through Equations (8)–(13).

$$P_{i,j} = P_j + \sum_{kj-k} P_{j,k} + r_{i,j}.I^2_{i,j}, \quad \forall i,j \in \{NL\} \tag{8}$$

$$Q_{i,j} = Q_j + \sum_{kj-k} Q_{j,k} + x_{i,j}.I^2_{i,j}, \quad \forall i,j \in \{NL\} \tag{9}$$

$$V^2_j = V^2_i - 2.(r_{i,j}.P_{i,j} + x_{i,j}.Q_{i,j}) + (r^2_{i,j} + x^2_{i,j}).I^2_{i,j}, \quad \forall i,j \in \{NL\} \tag{10}$$

$$\underline{V_j} <= V_j <= \overline{V_j}, \quad \forall j \in \{NB\} \tag{11}$$

$$\underline{I_{i,j}} <= I_{i,j} <= \overline{I_{i,j}}, \quad \forall i,j \in \{NB\} \tag{12}$$

$$I^2_{i,j}.V^2_i = P^2_{i,j} + Q^2_{i,j}, \quad \forall i,j \in \{NL\} \tag{13}$$

The mathematical formulation presented above is non-linear and non-convex. This formulation was simplified using a linear version of the DistFlow, the LinDistFlow [30,31] in Equations (14)–(17). In this linear approximation, the branch loss terms $(r^2_{i,j} + x^2_{i,j}).I^2_{i,j}$ were neglected. This is possible if it is assumed that the line losses are much smaller than the branch power terms $P_{i,j}$ and $Q_{i,j}$. Voltages are also nearly balanced and $(V_i - V_0)^2 \approx 0$. These assumptions will lead to the expression $V^2_i \approx -V^2_0 + 2.V_0.V_i$. The substitution of this term in Equation (10) leads to Equation (16). Differently from the DC-power flow, this model considers lines resistance and models active and reactive power flow. Variable V^{viol}_j represents a voltage violation and it is used to allow the problem to converge even when a solution with all the voltages inside the allowable limits (lower limit—$\underline{V_j}$; upper limit—$\overline{V_j}$) does not exist. This value has a high penalization in the objective function.

$$P_{i,j} = P_j + \sum_{kj-k} P_{j,k}, \quad \forall i,j \in \{NL\} \tag{14}$$

$$Q_{i,j} = Q_j + \sum_{kj-k} Q_{j,k}, \quad \forall i,j \in \{NL\} \tag{15}$$

$$V_j = V_i - \frac{r_{i,j}.P_{i,j} + x_{i,j}.Q_{i,j}}{V_0}, \quad \forall i,j \in \{NL\} \tag{16}$$

$$\underline{V_j} - V_j^{viol} <= V_j <= \overline{V_j}, \quad \forall j \in \{NB\} \tag{17}$$

The power flow was calculated for each instant t. It should be mentioned that this model can only be applied in radial networks.

Equation (18) defines that the power P_j in each bus is related with the power of all EHs in that bus. The calculation of the power $I_{n,t}^w$ for each EH is presented in Section 3.3.

$$P_j = \sum_n I_{n,t}^w \quad \forall n \in \{EH^j\}, \quad \forall j \in \{NB\}, \tag{18}$$

3.2.2. Generator Units Constraints

Equations (19) and (20) define the limits of PV and wind generators.

$$\underline{P_{t,n}^{wind}} <= P_{t,n}^{wind} <= \overline{P_{t,n}^{wind}}, \quad \forall n \in \{NW\} \tag{19}$$

$$\underline{P_{t,n}^{PV}} <= P_{t,n}^{PV} <= \overline{P_{t,n}^{PV}}, \quad \forall n \in \{NPV\} \tag{20}$$

3.3. Energy Hub Modelling Constraints

In this section, the operation for an EH is presented. Some of the variables are linked with the variables of the network modelling in Section 3.2 and of the resources modelling in Section 3.4, as it will be explained.

3.3.1. Load

Equations (21)–(24) represent the consumption of each type of load. The load of each energy vector is equal to the sum of the energy coming from the energy network, storage systems or the converters i generating that energy vector.

$$L_t^w = \sum_i E_t^{i,load}, \quad \forall i \in \{N^w, S^w, C^w, EV\} \tag{21}$$

$$L_t^h = \sum_i E_t^{i,load}, \quad \forall i \in \{S^h, C^h\} \tag{22}$$

$$L_t^c = \sum_i E_t^{i,load}, \quad \forall i \in \{S^c, C^c\} \tag{23}$$

$$L_t^g = \sum_i E_t^{i,load}, \quad \forall i \in \{N^g\} \tag{24}$$

The electrical load L_t^w is equal to the sum of the electricity needed for EVs, HVAC systems, refrigeration systems, electrical equipment and lighting, as seen in Equation (25). The heating load L_t^h is equal to the sum of the heat needed to supply the heating for spaces and water as seen in Equation (26). The cooling load L_t^c is equal to the sum of the cooling needs for space cooling as seen in Equation (27). The gas load L_t^g is equal to the sum of gas needed for cooking as seen in Equation (28).

$$L_t^w = \sum_n P_{n,t}^{EV,cha} + \sum_i P_{i,t}^{fan} + P_{i,t}^{refr} + P_{i,t}^{equip} + P_{i,t}^{light}, \quad \forall i \in \{R\}, \quad \forall n \in \{EV\} \tag{25}$$

$$L_t^h = \sum_i P_{i,t}^{h,resources} + P_{i,t}^{water}, \quad \forall i \in \{R\} \tag{26}$$

$$L_t^c = \sum_i P_{i,t}^{c,resources}, \quad \forall i \in \{R\} \tag{27}$$

$$L_t^g = \sum_i P_{i,t}^{cook}, \quad \forall i \in \{R\} \tag{28}$$

3.3.2. Energy Hub Inputs

Equation (29) and (30) represents the energy inputs of each EH. Each energy input is equal to the sum of the energy that goes from the energy network to the storage systems, loads and converters i that consume that energy vector.

$$I_t^w = \sum_i E_t^{net,i}, \quad \forall i \in \{L^w, S^w, C^w, EV\} \tag{29}$$

$$I_t^g = \sum_i E_t^{net,i}, \quad \forall i \in \{L^g, C^g\} \tag{30}$$

3.3.3. Energy Hub Outputs

Equations (31) represents the electrical output of each EH, i.e., the electricity sold to the network. The energy output is equal to the sum of the energy that goes from the EVs, storage systems or converters i generating electricity to the electrical network.

$$O_t^w = \sum_i E_t^{i,net}, \quad \forall i \in \{S^w, C^w, EV\} \tag{31}$$

3.3.4. Energy Converters and Storage Systems Input

Equations (32)–(35) represent the energy inputs of each type of storage system or converter inside the EH. The energy input of each system j is equal to the sum of the energy that goes from the network, storage systems or converter i generating that energy vector to that system.

$$E_t^{j,in} = \sum_i E_t^{i,j}, \quad \forall i \in \{N^w, S^c, G^w\}, j \in \{S^c, C^w\} \tag{32}$$

$$E_t^{j,in} = \sum_i E_t^{i,j}, \quad \forall i \in \{N^h, S^c, G^h\}, j \in \{S^c, C^h\} \tag{33}$$

$$E_t^{j,in} = \sum_i E_t^{i,j}, \quad \forall i \in \{S^c, G^c\}, j \in \{S^c, C^c\} \tag{34}$$

$$E_t^{j,in} = \sum_i E_t^{i,j}, \quad \forall i \in \{N^g\}, j \in \{C^g\} \tag{35}$$

Equation (36) represent the limits of the energy inputs of each storage system or converter j.

$$0 <= E_t^{j,input} <= \overline{P}^j, \quad \forall j \in \{S, C\} \tag{36}$$

3.3.5. Energy Resources Output

Equations (37) to (39) represent the output of the storage systems or resources present in the EH. The output of each system is the sum of energy that goes from that system i to the network, storage system, load or resource j that consumes that energy vector. It should be noted that the electricity sold to the electrical network is considered in Equation (37).

$$E_t^{i,out} = \sum_i E_t^{i,j}, \quad \forall i \in \{S^w, G^w\}, j \in \{N^w, S^w, L^w, C^w\} \tag{37}$$

$$E_t^{i,out} = \sum_i E_t^{i,j}, \quad \forall i \in \{S^h, G^h\}, j \in \{S^h, L^h, C^h\} \tag{38}$$

$$E_t^{i,out} = \sum_i E_t^{i,j}, \quad \forall i \in \{S^c, G^c\}, j \in \{S^c, L^c, C^c\} \tag{39}$$

3.4. Load, EVs, Storage and PV Contraints

In this section the constraints of the different loads, EVs, storage systems and PVs are presented. As mentioned, some of the variables are linked with the variables from the energy huh modelling in Section 3.3. In Table A1 of Appendix A, it is synthesized the values of the main parameters used in the models presented in this section.

3.4.1. Thermal Model

The temperature in a room is calculated by Equation (40) [19]. Equation (41) set the temperature at which the room has to be, taking into account the temperature set by the consumer $\hat{\Theta}_{i,t}^{room}$ and the temperature changes due to the participation in the DR program. Equations (42) and (43) define the limits imposed by the consumers for the increase $\widehat{\Theta}_{i,t+1}^{DR,room,+}$ and decrease $\widehat{\Theta}_{i,t+1}^{DR,room,-}$ of the temperature in a room for the participation in the DR program.

$$\Theta_{i,t+1}^{room} = \beta_t . \Theta_{i,t}^{room} + (1 - \beta_t)(\Theta_t^O + R_i . (P_{i,t}^h - P_{i,j,t}^c + P_{i,t}^{Inf})) + l_{i,t}, \quad \forall i \in \{R\} \tag{40}$$

$$\Theta_{i,t}^{room} = \hat{\Theta}_{i,t}^{room} - \Theta_{i,t+1}^{DR,room,-} + \Theta_{i,t+1}^{DR,room,+}, \quad \forall i \in \{R\} \tag{41}$$

$$0 <= \Theta_{i,t+1}^{DR,room,-}, <= \widehat{\Theta}_{i,t+1}^{DR,room,-}, \quad \forall i \in \{R\} \tag{42}$$

$$0 <= \Theta_{i,t+1}^{DR,room,+} <= \widehat{\Theta}_{i,t+1}^{DR,room,+} \quad \forall i \in \{R\} \tag{43}$$

Equation (44) defines the infiltration rate present in the rooms [32].

$$P_{i,t}^{Inf} = \frac{ACH_i . v_i . CV^{air} . (\Theta_i^O - \Theta_{i,t+1}^{room})}{3600}, \quad \forall i \in \{R\} \tag{44}$$

In order to calculate the energy necessary for the cooling and heating needs, it is necessary to define first the thermal parameters for each room. Equation (45) defines the thermal resistance [33,34].

$$R_i = \frac{X_i}{k_i . A_i^{wall}}, \quad \forall i \in \{R\} \tag{45}$$

The thermal reactance is calculated by using the heat capacity from the air in the room and from the walls. For the sake of simplicity, the outside walls and windows parameters were not considered in this study. The heat capacity contribution of the materials inside the room were not considered as well. Equation (46) defines the thermal reactance [35].

$$C_i = CV^{air} . v_i + c^{wall} . A_i^{wall} . X_i . d_i, \quad \forall i \in \{R\} \tag{46}$$

Parameters $ACH_i, v_i, X_i, k_i, A_i^{wall}, c_i^{wall}$ and d_i were assumed according to the information from [36].

3.4.2. HVAC System

Commercial buildings have an HVAC system installed. Their typical configuration is represented in Figure 3. The cooling coil provides the cooling power and the heating coils in each room provide the heating power. There is a supply fan which moves the cooling or heating air throughout the entire system. For each room there is also an individual variable air volume (VAV) box connected to the HVAC system, which regulates the temperature inside a room.

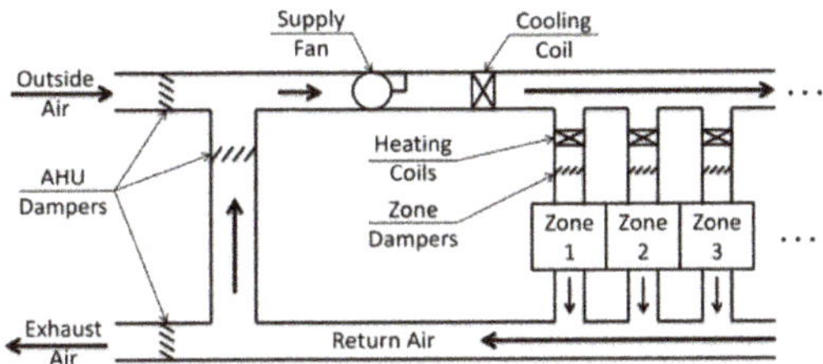

Figure 3. Scheme of a heating, ventilation and air conditioning (HVAC) system [37].

Equations (47)–(50) define the cooling and heating power provided by the respective coils [16].

$$P_{i,t}^{c,coil} = P_{i,t}^{c}, \quad \forall i \in \{R\} \tag{47}$$

$$P_{i,t}^{h,coil} = P_{i,t}^{h}, \quad \forall i \in \{R\} \tag{48}$$

$$P_{i,t}^{c,coil} = c^{air}.m_{i,t}.(\Theta_{i,t}^{room} - \Theta_{t}^{C,c}), \quad \forall i \in \{R\} \tag{49}$$

$$P_{i,t}^{h,coil} = c^{air}.m_{i,t}.(\Theta_{t}^{C,h} - \Theta_{i,t}^{room}), \quad \forall i \in \{R\} \tag{50}$$

The coils are air-water heat exchangers and Equations (51) and (52) define the power needed to power the cooling and heating coils [26], respectively, from the heating and cooling resources of the EH.

$$P_{i,t}^{c,resources} = \frac{P_{i,t}^{c,coil}}{\eta^{c,coil}.COP^{c,coil}}, \quad \forall i \in \{R\} \tag{51}$$

$$P_{i,t}^{h,resources} = \frac{P_{i,t}^{h,coil}}{\eta^{h,coil}.COP^{h,coil}}, \quad \forall i \in \{R\} \tag{52}$$

The relationship between the fan power and the airflow is between a quadratic and a cubic function, depending on the fan system [33]. This behavior is explained by the fan needing more power to compensate the leakage in the system, when an increase in airflow is necessary. In this work, it was assumed that the fan system model used is the same as the one presented in [16], as it is a model for commercial buildings. This model is defined in Equation (53) and it is applied to the commercial buildings considered in this work. The values of b_1, b_2, b_3 and b_4 are taken from the same reference. Nonetheless, if the required data was available, they could be identified using several techniques, including the ordinary least squares technique [25].

$$P_t^{fan} = b_1.m_t^3 + b_2.m_t^2 + b_3.m_t + b_4 \tag{53}$$

The total airflow supplied by the fan system is equal to the summation of the individual airflow of each room as seen in Equation (54).

$$m_t = \sum_i m_{i,t}, \quad \forall i \in \{R\} \tag{54}$$

The value of η^h and η^c is 0.98 and the value of COP^h and COP^c is 5.92. The discharge air cooling temperature $\Theta_t^{C,c}$ is set to 12.5 °C and the discharge air heating temperature $\Theta_t^{C,h}$ is set to 50 °C. The values to calculate the fan power are the following: $b_1 = 2.57 \times 10^{-12}$, $b_2 = -4.45 \times 10^{-9}$, $b_3 = 1.46 \times 10^{-4}$ and $b_4 = 4.71 \times 10^{-3}$. Each fan has a limit of 20,000 cfm.

Equation (49) present a non-linearity between two continuous variables $m_{i,t}$ and $\Theta_{i,t}^{room}$. Variable $m_{i,t}$ was discretized by assuming binary variables $b_{j,i,t}^{fan}$ that are multiplied by a small value of 100 cfm. This way, the problem can optimize the supply air flow rate with a rate of 100 cfm. By assuming a maximum of 20,000 cfm for a fan, there would exist 200 new binary variables. This would increase the complexity of the problem. Thus, instead of increasing the values by 100 each time, the supply air flow rate can be increase by 100, 200, 300, 400, 2000, 3000, 4000 and 10,000 cfm. This way, the problem can still look for solutions between 0 and 20000 cfm while needing only 8 new binary variables. This linearity is present in Equation (55).

$$m_{i,t} = \sum_n b_{i,t}^n . m^n, \quad \forall n \in \{M\}, \quad \forall i \in \{R\} \tag{55}$$

Nonetheless, with the discretization of $m_{i,t}$ there is still a non-linearity due to the multiplication of the binary variables $b_{i,t}^n$ and $\Theta_{i,t}^{room}$. This non-linearity can be solved by replacing that multiplication by a new continuous variable $y_{i,t}$, as represented in Equation (56), and by adding the restrictions represented by Equations (57)–(60).

$$y_{i,t} = b_{i,t}^n . \Theta_{i,t}^{room}, \quad \forall i \in \{R\}, \quad \forall n \in \{M\} \tag{56}$$

$$y_{i,t} <= b_{i,t}^n . \overline{\Theta}_{i,t}^{room}, \quad \forall i \in \{R\}, \quad \forall n \in \{M\} \tag{57}$$

$$y_{i,t} <= \Theta_{i,t}^{room}, \quad \forall i \in \{R\} \tag{58}$$

$$y_{i,t} >= \Theta_{i,t}^{room} - \overline{\Theta}_{i,t}^{room} . (1 - b_{i,t}^n), \quad \forall i \in \{R\}, \quad \forall n \in \{M\} \tag{59}$$

$$y_{i,t} >= 0, \quad \forall i \in \{R\} \tag{60}$$

With the discretization of $m_{i,t}$, Equation (53) also presents several non-linearities due to the multiplication of several binary variables. It was considered an additional binary variable y that replaces the multiplication of the two binary variables. In Equations (61)–(63) it is presented a general approach for the linearization of all these non-linearities.

$$y <= b_1 \tag{61}$$

$$y <= b_2 \tag{62}$$

$$y >= b_1 + b_2 - 1 \tag{63}$$

3.4.3. Thermostats

Contrary to the HVAC systems, thermostats used in residential buildings control directly the heating and cooling resources according to the thermal model defined in Equation (40). This way, $P_{i,t}^c$ and $P_{i,t}^h$ are the power needs that are requested from the heating and cooling converters as seen in Equations (64) and (65).

$$P_{i,t}^{c,resources} = P_{i,t}^c, \quad \forall i \in \{R\} \tag{64}$$

$$P_{i,t}^{h,resources} = P_{i,t}^h, \quad \forall i \in \{R\} \tag{65}$$

3.4.4. Water Heating

The power necessary to heat the water to the required temperature is presented in Equation (66) [20]. The required temperature of the water is calculated according to the temperature defined by the consumer $\widehat{\Theta}_{i,t}^{water}$ and the changes from the DR program as seen in Equation (67). The consumers also define the temperature limits of the DR program $\widehat{\Theta}_{i,t+1}^{DR,water,-}$ and $\widehat{\Theta}_{i,t+1}^{DR,water,+}$, as present in Equations (68) and (69).

$$P_{i,t}^{water} = c^w . m_i . (\Theta_{i,t}^{water} - \Theta_j^{water,i}), \quad \forall i \in \{R\} \tag{66}$$

$$\Theta_{i,t}^{water} = \widehat{\Theta}_{i,t}^{water} - \Theta_{i,t}^{DR,water,-} + \Theta_{i,t}^{DR,water,+}, \quad \forall i \in \{R\} \tag{67}$$

$$0 <= \Theta_{i,t+1}^{DR,water,-}, <= \widehat{\Theta}_{i,t+1}^{DR,water,-}, \quad \forall i \in \{R\} \tag{68}$$

$$0 <= \Theta_{i,t+1}^{DR,water,+} <= \widehat{\Theta}_{i,t+1}^{DR,water,+}, \quad \forall i \in \{R\} \tag{69}$$

The value of c^w is 4.18 J/g·°C (considering water at 18 C°) and initial temperature $\Theta_j^{water,i}$ is set to 15 °C in Winter. The mass of water m_i is defined according to the water consumption profile.

3.4.5. Refrigeration

The model used in this work to represent a refrigerator is represented in Equations (70) and (71) [22]. The required temperature of the refrigeration system is calculated according to the temperature defined by the consumer $\widehat{\Theta}_{i,t}^{refr}$ and the changes from the DR program (72). The consumers also define the temperature limits of the DR program $\widehat{\Theta}_{i,t+1}^{DR,refr,-}$ and $\widehat{\Theta}_{i,t+1}^{DR,refr,+}$ as present in Equations (73) and (74).

$$\Theta_{t+1}^{refr} = \varepsilon_i \Theta_{i,t}^{refr} + (1 - \varepsilon_i).(\Theta_{i,t}^{room} - COP^{refr}.\frac{P_{i,t}^{refr}}{a}) + l_t, \quad \forall i \in \{R\} \tag{70}$$

$$\varepsilon_i = e^{\frac{-\Delta t.a}{m^{cap}}}, \quad \forall i \in \{R\} \tag{71}$$

$$\Theta_{i,t}^{refr} = \widehat{\Theta}_{i,t}^{refr} - \Theta_{i,t}^{DR,refr,-} + \Theta_{i,t}^{DR,refr,+}, \quad \forall i \in \{R\} \tag{72}$$

$$0 <= \Theta_{i,t+1}^{DR,refr,-}, <= \widehat{\Theta}_{i,t+1}^{DR,refr,-}, \quad \forall i \in \{R\} \tag{73}$$

$$0 <= \Theta_{i,t+1}^{DR,refr,+} <= \widehat{\Theta}_{i,t+1}^{DR,refr,+}, \quad \forall i \in \{R\} \tag{74}$$

The coefficient of performance COP^{refr} was assumed to be 3 and the insulation a was assumed to be 15 kJ/°C for residential systems and 850 kJ/°C for supermarket systems.

3.4.6. Other Loads

The power necessary to supply energy to cooking and other electrical equipment, according to their load profiles, is defined in Equations (75) and (76).

$$P_{i,t}^{equip} = L_{i,t}^{equip}, \quad \forall i \in \{R\} \tag{75}$$

$$P_{i,t}^{cook} = L_{i,t}^{cook}, \quad \forall i \in \{R\} \tag{76}$$

The power necessary to supply energy for lighting is defined in Equation (77), which already takes into account the changes due to the participation in the DR program. This participation is constrained by the limits imposed by the consumers $\widehat{P}_{i,t+1}^{DR,light,-}$ and $\widehat{P}_{i,t+1}^{DR,light,+}$, as seen in Equations (78) and (79).

$$P_{i,t}^{light} = L_{i,t}^{light} - P_{i,t}^{DR,light,-} + P_{i,t}^{DR,light,+}, \quad \forall i \in \{R\} \tag{77}$$

$$0 <= P_{i,t+1}^{DR,light,-}, <= \widehat{P}_{i,t+1}^{DR,light,-}, \quad \forall i \in \{R\} \tag{78}$$

$$0 <= P_{i,t+1}^{DR,light,+} <= \widehat{P}_{i,t+1}^{DR,light,+}, \quad \forall i \in \{R\} \tag{79}$$

3.4.7. Electric Vehicles

The operation of EVs is defined by constraints (80)–(85). Equation (80) defines the state-of-charge (SOC) of the EV. Equation (81) represents the limit of the EV SOC. Equations (82) and (83) set the range for charging and discharging power.

$$SOC_{n,t+1}^{EV} = SOC_{n,t}^{EV} + P_{n,t}^{EV,cha} \cdot \eta_n^{EV,cha} - \frac{P_{n,t}^{EV,dis}}{\eta_n^{EV,dis}}, \quad \forall i \in \{EV\} \tag{80}$$

$$\underline{SOC_{n,t}^{EV}} <= SOC_{n,t}^{EV} <= \overline{SOC_{n,t}^{EV}}, \quad \forall i \in \{EV\} \tag{81}$$

$$0 <= P_{n,t}^{EV,cha} <= b_{n,t}^{EV,cha} \cdot \overline{P_{n,t}^{EV,cha}}, \quad \forall i \in \{EV\} \tag{82}$$

$$0 <= P_{n,t}^{EV,dis} <= b_{n,t}^{EV,dis} \cdot \overline{P_{n,t}^{EV,dis}}, \quad \forall i \in \{EV\} \tag{83}$$

Equation (84) assures that EVs do not charge and discharge at the same time.

$$b_{n,t}^{EV,cha} + b_{n,t}^{EV,dis} <= 1, \quad \forall i \in \{EV\} \tag{84}$$

EVs have to be charged according to the consumer preferences. Consumers should define the hour of day (HD) that an EV should be charged and the minimum state-of-charge required $\widehat{SOC}^{EV}$, taking into account the limits of charging. This is defined in Equation (85).

$$SOC_{n,HD}^{EV} >= \widehat{SOC}^{EV}, \quad \forall i \in \{EV\} \tag{85}$$

It should be noted that the sum of the charging power of all EVs is the same as the variable $E_t^{EV,input}$ used to model the EH in Section 3.3.4. Similarly, the sum of the discharging power of all EVs is the same variable as $E_t^{EV,output}$ used in Section 3.3.5. This is presented in Equations (86) and (87).

$$E_t^{EV,in} = \sum_n P_{n,t}^{EV,cha}, \quad \forall i \in \{EV\} \tag{86}$$

$$E_t^{EV,out} = \sum_n P_{n,t}^{EV,dis}, \quad \forall i \in \{EV\} \tag{87}$$

3.4.8. Storage Systems

The operation of storage units installed in the EH is defined by constraints (88)–(92). Equation (88) defines the SOC of the storage system. Equation (89) represents the storage limit capacity. Equations (90) and (91) set the range for charging and discharging power.

$$SOC_{n,t+1}^{sto} = SOC_{n,t}^{sto} + P_{n,t}^{sto,cha} \cdot \eta_n^{sto,cha} - \frac{P_{n,t}^{sto,dis}}{\eta_n^{sto,dis}}, \quad \forall n \in \{S\} \tag{88}$$

$$\underline{SOC_{n,t}^{sto}} <= SOC_{n,t}^{sto} <= \overline{SOC_{n,t}^{sto}}, \quad \forall n \in \{S\} \tag{89}$$

$$0 <= P_{n,t}^{sto,cha} <= b_{n,t}^{sto,cha} \cdot \overline{P_{n,t}^{sto,cha}}, \quad \forall n \in \{S\} \tag{90}$$

$$0 <= P_{n,t}^{sto,dis} <= b_{n,t}^{sto,dis} \cdot \overline{P_{n,t}^{sto,dis}}, \quad \forall n \in \{NS\} \tag{91}$$

As storage system cannot charge or discharge at the same time, Equation (92) defines this constraint.

$$b_{n,t}^{sto,cha} + b_{n,t}^{sto,dis} <= 1, \quad \forall n \in \{S\} \tag{92}$$

Again, it should be noted that the sum of the charging power of all storage devices is the same as the variable $E_t^{i,input}$, $\quad \forall i \in \{\mathcal{S}\}$ used to model the EH in Section 3.3.4. Similarly, the sum of the discharging power of all storage devices is the same variable as $E_t^{i,output}$, $\quad \forall i \in \{\mathcal{S}\}$ used in Section 3.3.5. This is presented in Equations (93) and (94).

$$E_t^{i,in} = P_{n,t}^{sto,cha}, \quad \forall i, n \in \{\mathcal{S}\} \tag{93}$$

$$E_t^{i,out} = P_{n,t}^{sto,dis}, \quad \forall i, n \in \{\mathcal{S}\} \tag{94}$$

3.4.9. PVs

Equation (95) defines the power generated by a PV.

$$\underline{P_t^{PV}} <= P_t^{PV} <= \overline{P_t^{PV}} \tag{95}$$

4. Case Study and Results

4.1. Case Study Description

Figure 4 presents the network used as a test case. It is a modified IEEE 37 node test feeder with penetration of wind and solar energy. There is 2.85 MW of wind generation at bus 3, 9, 10, 14 and 20, and 1 MW of solar generation installed at bus 4 and 20.

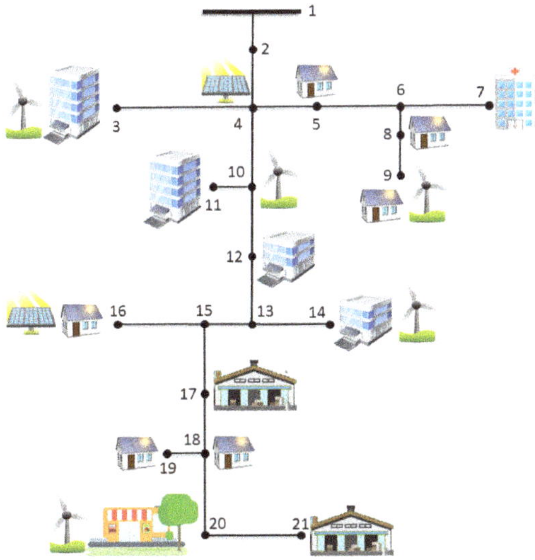

Figure 4. Adapted IEEE 37 node test feeder.

This network has several loads, including residential buildings, commercial buildings, an hospital, warehouses, medium and small offices and a supermarket. The location and number of buildings in the network is presented in Table 1. This table also presents the type of loads of these buildings, besides space heating and cooling, lighting and electrical equipment. The supermarket has 10 refrigeration systems, while the residential buildings only have 1 system each. Refrigeration systems have a maximum power of 5 kW for supermarkets and 1 kW for residential buildings.

These buildings are represented by the models shown in Sections 2 and 3. They are supplied by the electrical network, the gas grid and by local PVs. The technical restrictions of the gas grid were not considered in this study.

Table 1. Information of the buildings in the network.

Buildings	Bus	Number of Buildings	Other Loads
Residential	5, 8, 9, 16, 18, 19	10, 5, 5, 5, 5, 5	Refrigeration, Water
Small Office	12, 14	5, 5	Water
Medium Office	3, 11	1, 1	Water
Warehouse	17, 21	1, 1	-
Supermarket	20	1	Refrigeration, Water
Hospital	7	1	Water

The resources assumed for each case, i.e., only electrical buildings and MES buildings, are presented in Table 2. All the buildings also have heat storage systems installed. The efficiency of EB is 0.95, GB is 0.9 and the efficiency of CHP for electricity is 0.35 and for heat is 0.45. The efficiency of heating storage systems is 0.88. The COP of both HP and AC systems is 3.45.

Table 2. Maximum power of resources: only electrical buildings vs. Multi-Energy System (MES) buildings.

Buildings	Resources (kWh)						
	Electrical			MES			
	EB	HP	AC	GB	CHP	HP	AC
Residential	1	1	1	1	-	1	1
Small Office	10	-	5	5	10	-	5
Medium Office	100	-	75	50	100	-	75
Warehouse	50	50	50	50	-	50	50
Supermarket	-	100	50	-	50	50	50
Hospital	200	-	100	100	300	-	100

In every building, 10% of the occupants have an EV. EVs have a charging and discharging power limit of 10 kW and an efficiency of 0.9. The battery capacity of these vehicles varies between 35 kWh and 100 kWh.

PVs are installed in every building: 10 kW in small offices, 50 kW in medium offices, 20 kW in warehouses, 20 kW in the supermarket and 200 kW in the hospital. The systems' efficiency is 0.227.

The parameters and profiles of each building are taken from [36]. The assumptions for the parameters of the models were already presented in Section 3.

The wind and PV generation profiles are presented in Figure 5 and they were taken from [38], for the day 7 January 2019. The outside temperature is presented in Figure 6 and was taken from [38] for the same day. The average monthly wind and PV generation profiles are presented in Figures A1 and A2 in the Appendix B and they were calculated with the information taken from [38] for the years 2017 and 2018. The average monthly outside temperature is presented in Figure A3 in the Appendix B and was calculated from the information taken from [38] for the years 2017 and 2018.

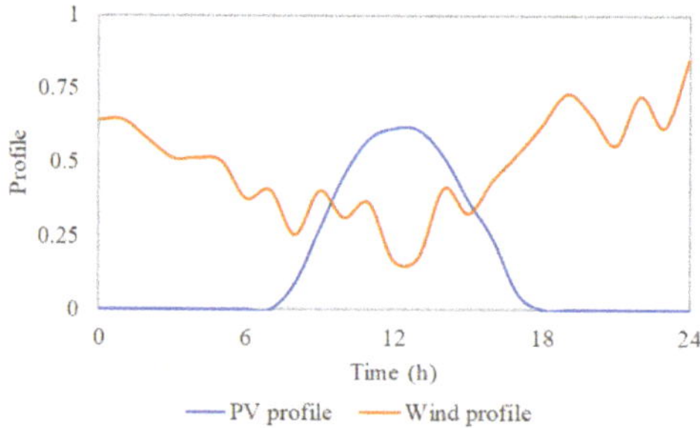

Figure 5. Profiles of solar and wind generation.

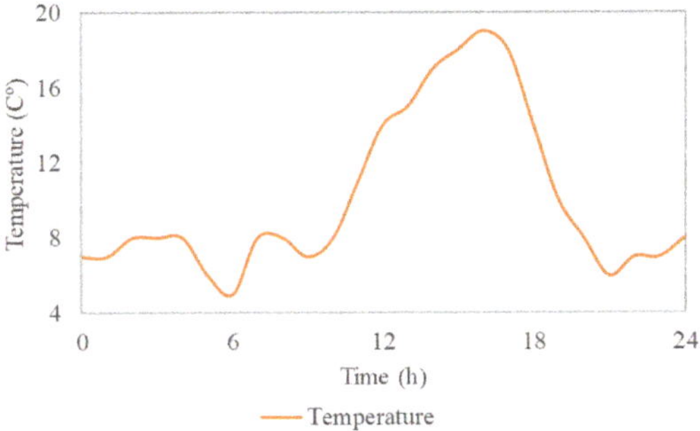

Figure 6. Profiles of outside temperature.

In Figure 7, the prices of electricity and gas are shown. They were taken from ENTSO-E [39] and from the MIBGAS platform [40], respectively, for the day 7 January 2019. The price of the electricity sold to the grid was assumed to be 0.5 of the electricity price in order to enforce self-consumption and avoid injection into the grid. The average monthly electricity and gas prices are presented in Figures A4 and A5 in the Appendix B and they were calculated with the information from [39] and [40], respectively, for the years 2017 and 2018.

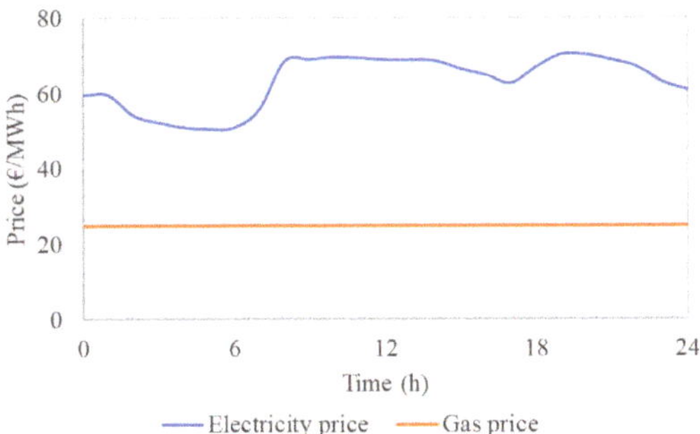

Figure 7. Electricity and gas prices.

The price of DR services related with temperatures flexibility was assumed to be 0.5 €/°C. This figure was assumed for the purpose of this work, as no supporting reference was found in the literature. Although it is outside the scope of this work, it is a very interesting topic that should be further studied in a dedicated work.

The price of each voltage violations was assumed to be very high to force the optimisation model to give priority to the resolution of these technical problems. This value was set to 10,000,000 €.

The cost of the CO_2 emissions from the natural gas is calculated by considering the average market price of CO_2 emissions and a factor representing the total emissions per MWh of natural gas consumed. The CO_2 emissions market price considered was 25 €/$tonCO^2$ for every month and it was taken from [41], whereas the emission factor was set to 0.2 $tonCO^2$/MWh and it was taken from [42]. This means that the CO_2 emissions cost for natural gas was set to 5 €/MWh for the entire year.

The cost of the CO_2 emissions from electricity is calculated as follows:

- First, the percentage of coal and natural gas in the electricity generation mix is determined, defining a daily profile for each of these resources;
- Then, the daily profiles are multiplied by the CO_2 emissions market price for that day and by a factor representing the total emissions of each resource per MWh of generated electricity.

The daily profiles of coal and natural gas in the electricity generation mix were calculated with the information from [43] for the years 2016, 2017 and 2018 and are presented in Figures A6 and A7 of Appendix B. The emission factors per MWh of electricity generated for coal and natural gas were set to 1 $tonCO^2$/MWh and 0.41 $tonCO^2$/MWh and they were taken from [44]. The average monthly cost of the CO_2 emissions from the electricity generated is presented in Figure A8 of Appendix B.

The cost of the CO_2 emissions, for both electricity and natural gas, will not have a significant impact in the results as they are much lower than electricity and gas prices.

The optimization problem described in the previous sections is a mixed-integer linear programming problem (MILP) and was solved with pycharm [45], using the CPLEX optimizer in a core i7, 3.5 GHz PC.

4.2. Results

In this section it is studied how the flexibility provided by the different buildings can be used to solve voltage problems that may appear in the network. To this end, the following cases were created:

- Only electrical buildings without optimization (no DR);
- Only electrical buildings with optimization (with DR);
- MES buildings without optimization (no DR);
- MES buildings with optimization (with DR).

With these cases it will be possible to study and compare the flexibility provided by electrical and MES buildings and by residential and commercial buildings, as well as their annual costs, energy consumption and CO_2 emissions.

The organization of this section is as follows:

- In Section 4.2.1, grid technical problems occurring in 7 January 2019 are analyzed;
- In Section 4.2.2, the daily load profiles of the proposed cases are studied;
- In Sections 4.2.3 and 4.2.4, the flexibility that only-electrical vs. MES buildings and commercial vs. residential buildings are able to offer for that day is analyzed;
- In Section 4.2.5, it is studied how the flexibility provided by the different resources and loads are used to mitigate voltage problems;
- Finally, Section 4.2.6 presents the annual results for energy consumption, CO_2 emissions and costs.

4.2.1. Grid Technical Problems

In this section it is presented the technical results of the network on the 7 January 2019. The voltages in all buses at peak hour and the voltages at bus 20 during all day are presented and analyzed.

In Figure 8 it is presented the voltages in every bus of the network at 6h. It is possible to observe that buses 18, 19, 20 and 21 present undervoltage problems in the scenario with "electrical buildings without optimization", while in the scenario with "electrical buildings with optimization" only bus 20 presents a slight undervoltage problem. The other scenarios do not present this type of problems. These problems occur due to the fact that these buses have high consumption of electricity at this hour, as it will be analyzed in Section 4.2.2, and also because they are the farthest from the feeding point, as it is possible to observe in Figure 4.

Figure 9 shows the daily profile of the voltages in bus 20. It is possible to observe that severe voltage problems occur in the "electrical buildings without optimization" scenario. Buses 18, 19 and 21 also reveal similar undervoltage problems. The importance of DR programs is easily perceptible in the "electrical buildings with optimization" scenario, as almost all undervoltage problems cease to exist. Nevertheless, there are still some minor problems in buses 20 and 21. In both MES scenarios, with and without DR, no voltage problems occur. Even without using the DR program, the system was never with technical problems as part of the MES buildings consumption relies on other energy vectors besides electricity.

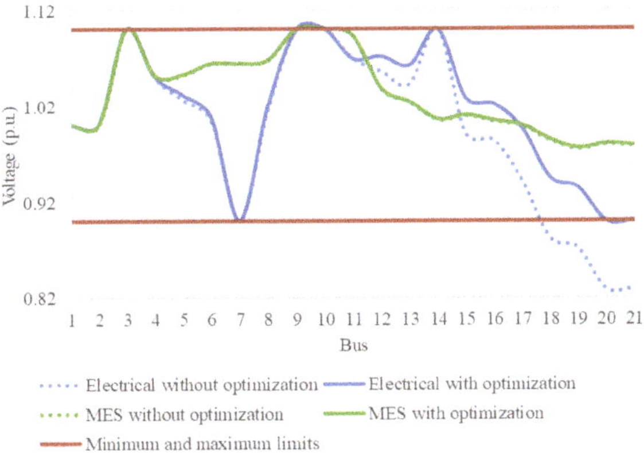

Figure 8. Voltages in the buses at 6 h (peak hour).

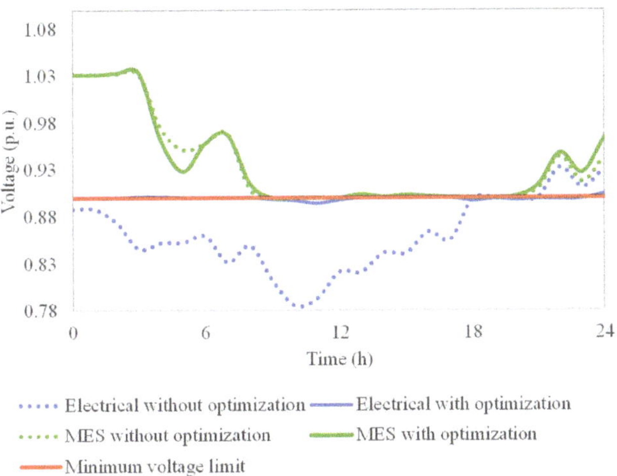

Figure 9. Voltages during the day at bus 20.

In total, in the scenario with electrical buildings without optimization, there were 54 cases of undervoltage affecting 4 buses, while in the scenario with electrical buildings with optimization there were 14 cases affecting 2 buses. This means that flexibility activation through DR programs reduce the number of undervoltage problems by 74%.

4.2.2. Daily Load Diagrams

In this section it is presented the electricity and gas consumed in the network during the day of 7 January 2019. The changes in the load due to the participation in DR programs are analyzed.

In Figure 10 it is presented the electrical consumption in the network. It is possible to observe that without optimization, the case with "only electrical" buildings consumes more electricity than the case with MES buildings. The peak consumption at 6h in the scenario with electrical buildings without optimization is due to the fact that electricity prices are cheaper at this hour and clients start heating the spaces, especially in commercial buildings. This higher consumption of electricity causes several voltage problems in the network, as it was seen in the previous section.

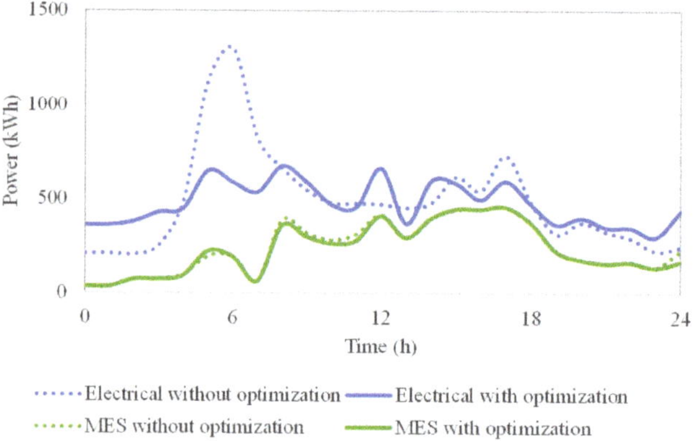

Figure 10. Power consumed in the network.

The utilization of flexibility to solve the referred problems impose changes to the daily load profiles. It is possible to observe that the main changes occur in the scenario with only electrical buildings with optimization as this case presents several voltage problems during the entire day. This way, the electrical load was decreased in this case.

In the cases without optimization, the peak load is 85% lower in the scenario MES buildings in comparison with the scenario only electrical buildings; comparing the same cases with optimization, it is observable that the peak load is reduced by 67%.

Figure 11 shows the overall gas consumption in the network. The peak consumption is at 7 h as most of the buildings with gas also start heating their spaces at this hour.

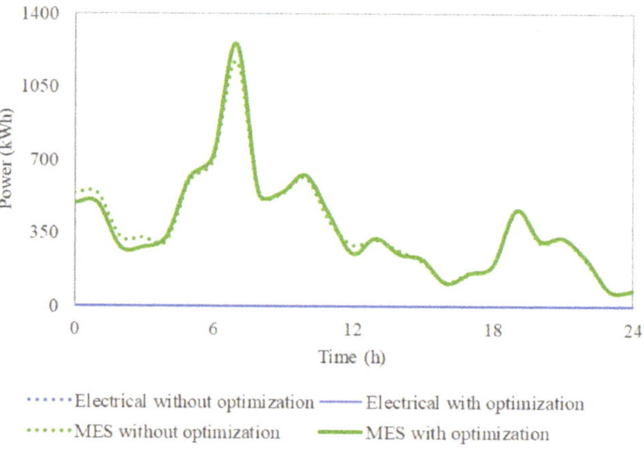

Figure 11. Gas consumed in the network.

4.2.3. Flexibility of Only Electrical and MES Buildings

In this section, the flexibility that electrical and MES buildings are able to offer is analyzed and quantified. The minimum and maximum flexibility represents the minimum and maximum consumption that each building is able to consume at each hour.

Figure 12 presents the minimum and maximum flexibility (red lines) that the buildings are capable of providing in the only electrical and MES scenarios. Eventually, this flexibility may be activated if proper DR-based programs are implemented.

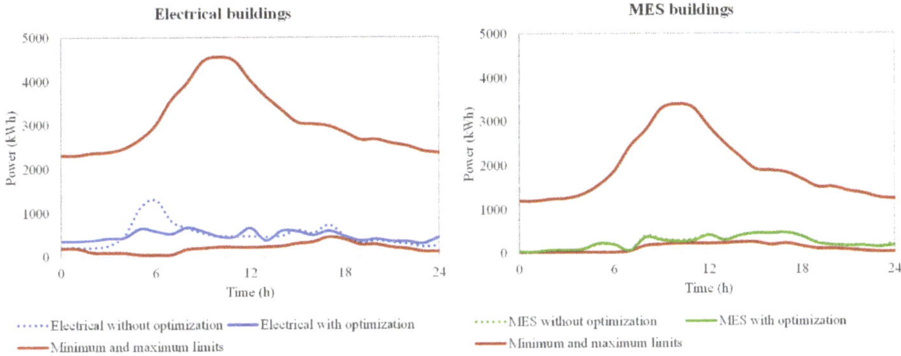

Figure 12. Total flexibility offered by electrical and MES buildings.

The most important conclusion to retain is that buildings have the capability to provide a significant band of flexibility that may be very important to system operators, as they can use it to solve low voltage and/or branches' overload problems in distribution networks, as it will be seen in Section 4.2.1.

Although the electrical buildings scenario have a larger flexibility band, their minimum limit is slightly higher as all the loads rely entirely on electricity. In the MES scenario, the flexibility band is generally straighter since there are less loads that use electricity. However, in this case, the overall buildings' consumption is lower than in the only electrical scenario, what may be an interesting situation for system operators since networks utilization will be lower.

It is possible to observe that only electrical buildings have major modifications in their profiles during the first 7 h of the day. This happens to avoid higher peaks of consumption at 7 h, when buildings start heating their spaces. Buildings accumulate heat in the previous hours decreasing their energy needs at 7 h. This is also shown in Section 4.2.5.

In average, the total minimum consumption of MES buildings is 40% lower than electrical buildings while the maximum consumption is 37% lower.

4.2.4. Flexibility of Residential and Commercial Buildings

In this section, the flexibility that residential and commercial buildings are able to offer is analyzed and quantified.

In Figure 13 it is presented the range of flexibility of residential and commercial buildings. It is possible to observe that commercial buildings have a higher range of flexibility than residential buildings. They have a higher flexibility during the day as they are intensively used during these periods, with a higher number of appliances and loads connected and capable of increasing or decreasing their energy consumption. Air heating/cooling and water heating are good examples of loads that are a great source of flexibility in these periods.

Residential buildings have higher flexibility after 16h and during the night. This happens since residential buildings do not usually have EVs connected between 8 h–16 h, which are a significant source of flexibility. Additionally, the utilization of other residential equipment (e.g., heating and cooling) is also reduced during this period, contributing to decrease the flexibility margin.

It is possible to observe that the main modifications in the load profiles with the activation of flexibility occurred in the case with "only electrical" buildings, to both residential and commercial buildings. Nonetheless, the flexibility utilization has a higher impact in residential buildings during the first 7 h of the day.

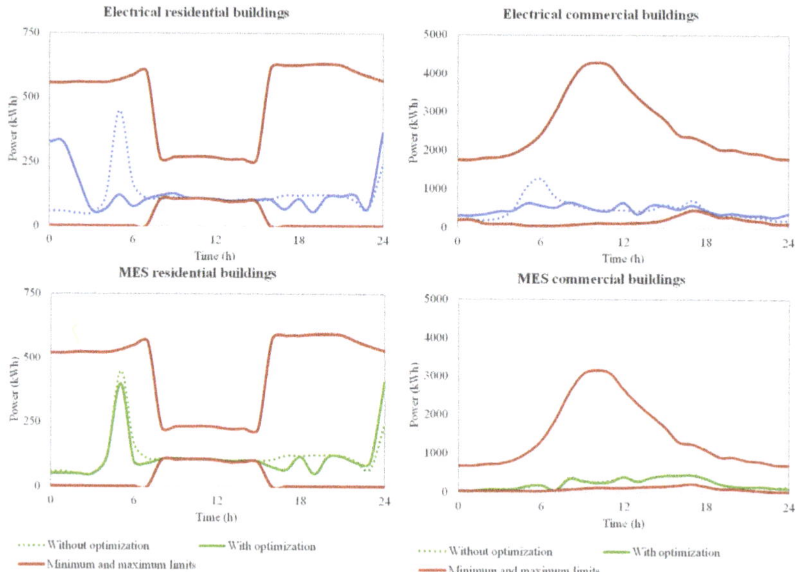

Figure 13. Total flexibility offered by residential and commercial buildings.

4.2.5. Impact on the Daily Load Profiles of Flexible Loads

The flexibility that buildings are able to offer comes essentially from using resources that consume gas instead of electricity and from DR programs. DR programs are capable of changing the load patterns of lighting, refrigeration, water heating and space heating. In this section, it is analyzed the flexibility that each source of flexibility is able to provide to the system operator. This highlights the importance of considering every possible source of flexibility within a building.

Figure 14 shows the lighting consumption in a supermarket. The lighting systems can be decreased by 10% in relation to their initial value.

It is possible to observe a decrease in the lighting consumption in the scenario electrical buildings with optimization. This is the only scenario where the DR program was activated for lighting. In the MES scenarios, changes in the lighting consumption patterns did not happen (i.e., DR programs were not activated) since all the network problems detected were solved by replacing part of the electricity consumption by gas. This is the reason why the series of the scenarios electrical buildings without optimization, MES buildings without optimization and MES buildings with optimization overlap.

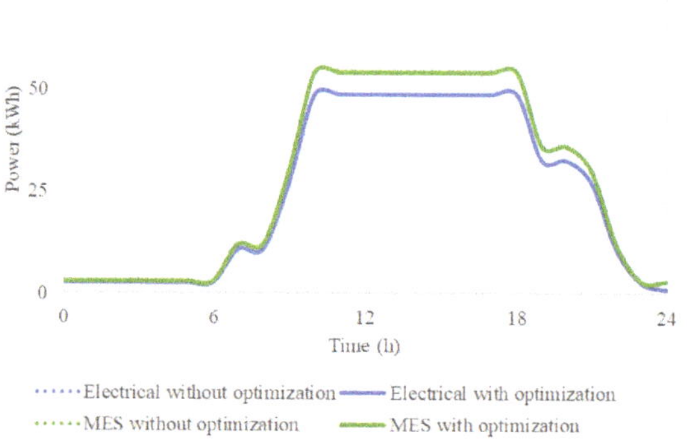

Figure 14. Power consumed for lighting in the supermarket.

In Figure 15 it is presented the temperature and the energy consumed for space heating in a room of the supermarket. Each type of building and each room has its own temperature requirements and different schedules of the HVAC system. In order to simplify the analysis of the results, only the requirements for the temperature of a room in a supermarket are presented, although the requirements for the other buildings are similar:

- HVAC system is always "on";
- Between 6 h and 23 h:

    – The temperature set-point is 21 °C;
    – The temperature can vary between 15 °C and 25 °C.

- Between 0 h and 6 h and after 23 h:

    – There is no set-point;
    – The temperature can vary between 15 °C and 25 °C.

As Figure 15 shows, the temperature requirements were met in all the scenarios. It is possible to observe that the case electrical buildings with optimization was the only case where flexibility of the space temperature was used. In this case, the DR program was activated and the temperature in the room decreased below the set-point but remained inside the allowed interval. This happened since the DR program was activated and the energy consumed for heating this room was decreased to zero. In the other scenarios, the temperature strictly followed the set-point defined.

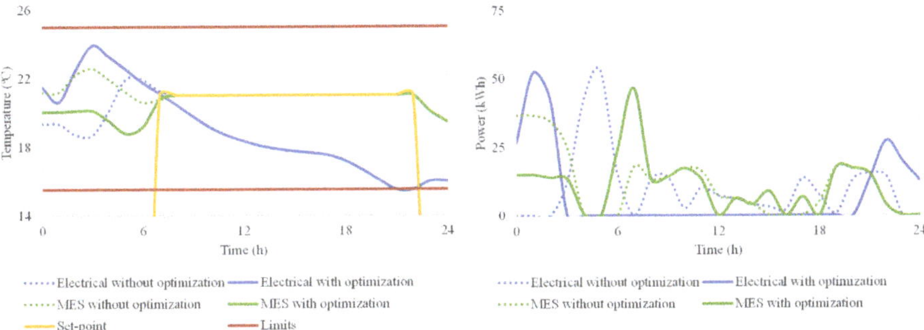

Figure 15. Temperature and power consumed for heating in the main room of the supermarket.

It is also possible to observe that between 0 h and 6 h and after 23 h, in all scenarios, the temperature was different from 21 °C due to the absence of set-point.

Figure 16 shows the temperature profile of the refrigeration systems of the supermarket as well as the respective power profile (the series of the scenarios without optimization are overlapped). The temperature requirements of all refrigeration systems present in this network are the following:

- The temperature is set to −5°C for the entire day;
- The temperature can vary between −6 °C and −3 °C by using the DR program.

More flexibility was used in the scenario electrical buildings with optimization, as the temperature increases due to the reduction of the power consumption to provide flexibility to the grid.

In the scenario MES buildings with optimization, the system operator did not request any flexibility. However, some flexibility was used to decrease energy costs at the expense of a slightly higher temperature in the refrigeration system. An example of the flexibility utilization for this purpose is shown in Figure 16, where more power was consumed between 5–7 h to decrease temperature, as the energy cost in this period was lower.

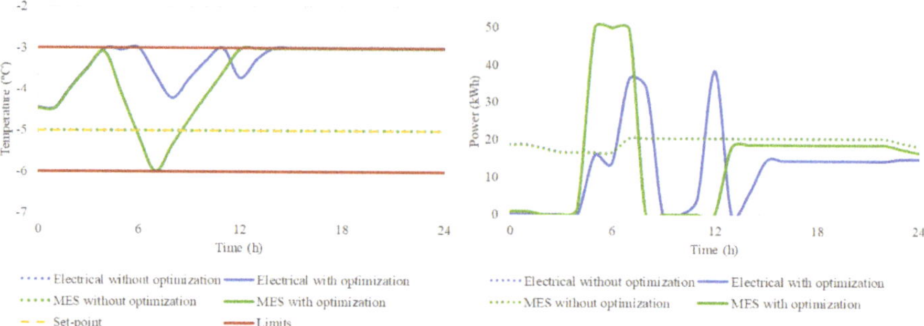

Figure 16. Temperature and power consumed by the refrigeration systems of the supermarket.

In the other scenarios the temperature is the same as the set-point as the DR program is not available and their series are overlapped.

Figure 17 presents the water temperature in the hospital at bus 13 and the respective power consumption. The water temperature requirements for all buildings present in the network are the following:

- The temperature is set to 49 °C for the entire day;
- The temperature can vary between 44 °C and 54 °C by using the DR program.

As it is possible to observe, the scenario electrical buildings with optimization was the only case where there was a decrease in the water temperature, with the correspondent power decrease to provide flexibility to the grid.

It is important to notice that the power necessary to heat the water depends on the set-point of the water temperature, which is a flexible value, and the water volume consumed, which is not flexible and it is set by the consumer. This way, the flexibility provided by heating the water is provided only by changing the water temperature and not the used volume.

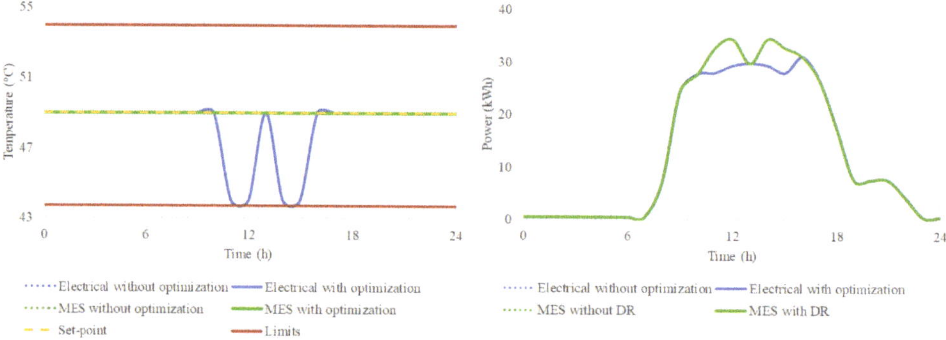

Figure 17. Temperature and power consumed for heating the water in a hospital.

In Figures 18 and 19 it is presented the charging and discharging of EVs in a residence and a medium size office. It is possible to observe that the charging and discharging patterns in the scenario with electrical buildings with optimization differ from both MES buildings scenarios and the scenario with electrical buildings without optimization. This reflects the utilization of EVs flexibility to mitigate the undervoltage problems in the network.

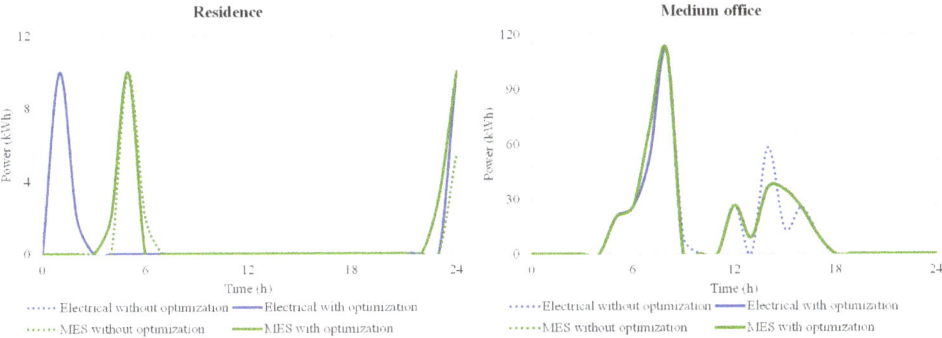

Figure 18. Electrical vehicles (EVs) charging power in a medium size office.

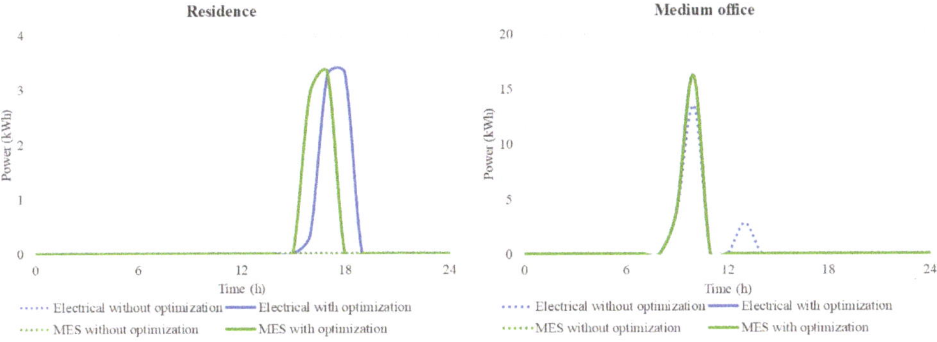

Figure 19. EVs discharging power in a medium size office.

Figure 20 presents the electricity and heat production profiles of the CHP unit in the supermarket. As it is possible to observe, there is a peak at 7 h. Comparing these figures with Figure 10, it is possible to observe that the supermarket consumes much less electricity from the network at this hour and relies more on the gas network and the CHP unit for the provision of electricity and heat. This is because gas prices are lower than electricity prices and by consuming less electricity the power network is not so overload and its limits are more relaxed.

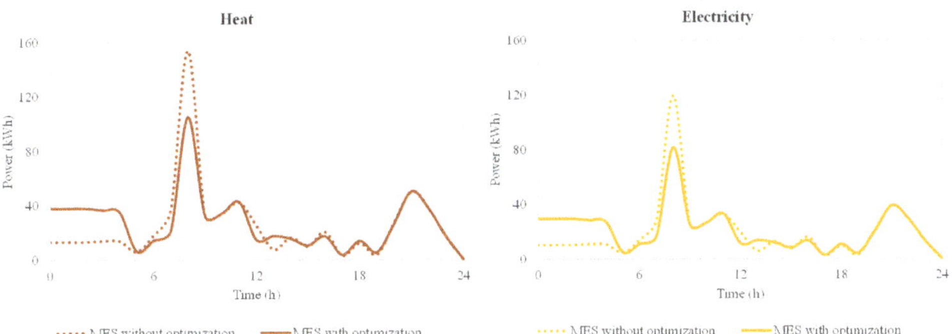

Figure 20. Provision of heat and electricity by the combined heat and power (CHP) unit of the supermarket.

4.2.6. Annual Results

In this section, it is first presented the annual consumption of electricity and gas as well as the annual emissions of CO_2. After, the costs of buying electricity and gas, the costs of the emitted CO_2 and the costs of using DR programs are presented. The results obtained in this section allow the generalization of the results presented in the previous sections.

From Figure 21 and Table 3 it is possible to observe that the cases with only electrical buildings consume naturally more electricity than the cases with MES buildings. In total, the electricity consumed in the case with MES buildings without optimization is 33% lower than the case with only electrical buildings, while in the cases with optimization it is 30% lower. The utilization of flexibility (in the scenario with optimization) decreases the consumption of electricity by 6% in the only-electrical buildings cases, while in the MES buildings it only decreases 1%. This happens since in the cases with only electrical buildings, DR program are activated more often due to the technical problems in the network.

The gas consumption is very low in only electrical buildings cases, as the gas is only used for cooking in residential buildings. In the cases with MES buildings, more gas is used during the winter as it is the main source of heating.

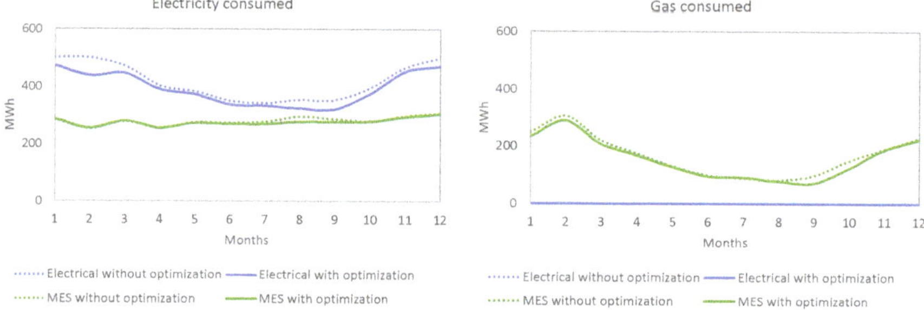

Figure 21. Annual consumption of electricity and gas.

Table 3. Annual consumption of electricity and gas in each scenario.

Case	Electricity (MWh)	Gas (MWh)
Electrical without optimization	5035	3
Electrical with optimization	4750	3
MES without optimization	3372	2028
MES with optimization	3328	1904

In Figure 22 the annual emissions of CO_2 for each case are presented. It is possible to observe that MES buildings have naturally higher emissions than only electrical buildings. Although they consume less electricity, the gas consumption is much higher which makes CO_2 emissions higher. It is also possible to observe that the cases with optimization have lower emissions as they also consume less energy due to the activation of downward flexibility. These results show that DR programs can have an impact on the CO_2 emissions from buildings.

In total, only electrical buildings have 40% and 44% lower emissions of CO_2 than MES buildings in the cases without optimization and with optimization, respectively. DR programs (i.e., optimization scenarios) lowered the emissions of CO_2 by 6% in the case with only electrical buildings and 4% in the case with MES buildings.

The costs of electricity and gas consumption, emissions of CO_2 and DR programs are presented in In Figure 23. As expected from the previous results, the costs of electricity are higher in the cases with only electrical buildings as more electricity is consumed than in the MES buildings cases. The costs of gas and CO_2 emissions are higher in the MES buildings scenarios. The case with only electrical buildings with optimization has the higher costs related with flexibility activation (DR programs), as this is the case that presents more technical problems in the electrical network.

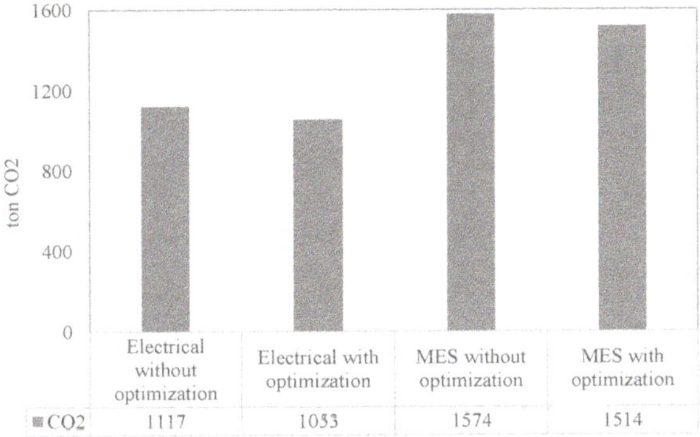

	Electrical without optimization	Electrical with optimization	MES without optimization	MES with optimization
■CO2	1117	1053	1574	1514

Figure 22. Annual emission of CO_2.

It is also possible to observe that the total costs at the end-of the year are lower in the case with MES buildings with optimization, as they consume less electricity that have higher prices when compared to natural gas.

In total, without optimization, the annual costs of electricity in the cases with only electrical buildings are 32% higher than the cases with MES buildings, the CO_2 costs are 40% lower and the total annual costs are 10% higher. With optimization, the annual electricity costs are 29% higher, the costs of CO_2 emissions are 44% lower and the total annual costs are 29% higher.

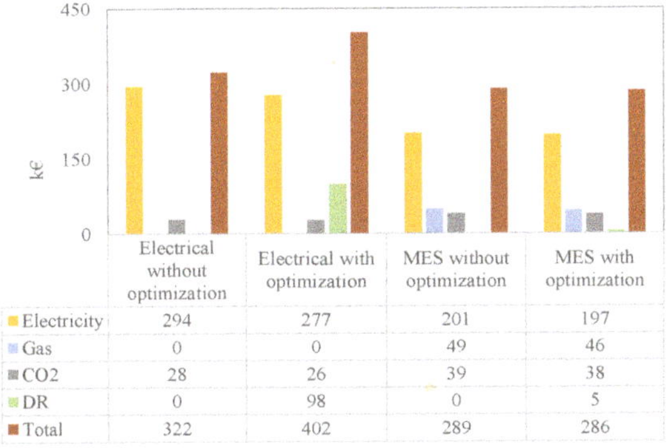

	Electrical without optimization	Electrical with optimization	MES without optimization	MES with optimization
■Electricity	294	277	201	197
■Gas	0	0	49	46
■CO2	28	26	39	38
■DR	0	98	0	5
■Total	322	402	289	286

Figure 23. Annual costs of electricity, gas, CO_2 emissions and DR programs.

5. Conclusions

This work presented a model to optimize energy consumption in buildings, aiming at minimizing costs while satisfying the technical constraints of the power network. The model developed is capable of controlling a wide variety of loads (thermal loads, lighting, EVs, PVs, etc.) taking into account the flexibility that their owners are willing to provide. This flexibility can be used for technical matters, i.e., to improve networks operation, or for economic purposes, i.e., decrease overall energy cost, including CO_2 costs. A novel linearized model for the HVAC system was also developed in this work, decreasing the complexity of the problem. The model was used with a test network to quantify and compare the capability of only electrical buildings and MES buildings, both commercial and residential, to offer flexibility. Annual costs, energy consumption and CO_2 emissions were also studied.

Analyzing the grid technical problems for the day 7 January 2019, in the scenarios with only electrical buildings, there are a number of undervoltage problems detected in the network and they cannot be entirely solved even when flexibility is activated through DR programs. These stressful operating conditions are not verified in the MES scenarios, where no technical problems were detected. In the only electrical buildings scenarios, the voltage problems detected were greatly reduced after the activation of the DR program—the number of buses with voltages below 0.9 p.u. was reduced from 54 to 14, representing a 74% reduction. The flexibility provided in the DR scenario was obtained from electrical and thermal loads, which can offer flexibility through thermal storage, lightings, EVs or CHP units and thus contribute to reduce the power absorbed from the grid. It is thus possible to conclude that DR programs are an important tool for system operators as they can enable important changes in the load profiles to avoid voltage problems or overloaded branches.

As expected, it is in the MES scenarios that electricity consumption is lower. One of the outcomes of this study is the quantification of the network load reduction that can be achieved in the MES scenarios in comparison with the only electrical buildings scenarios. In average, in the scenarios without DR, the load is 85% lower during the peak hour while in the scenarios with DR it is 67% lower. In the only electrical scenarios, the DR program was capable of reducing the peak load by 55%, although it was not enough to mitigate all voltage problems.

With lower electricity consumption, it is not surprising that buildings can reach lower absolute values of consumed power in MES buildings scenarios. In the comparison between commercial and residential buildings, the former revealed overall wider ranges of flexibility. Commercial buildings can generally provide more flexibility during the day, especially in the MES scenarios, while residential buildings are more flexible during the night and late afternoon.

The last outcome from this work is related with the annual costs, energy consumption and CO_2 emissions. It was concluded that MES consume ca. 33% less electricity, although they consume more gas. This way, annual energy costs are lower when considering MES buildings, presenting a reduction of ca. 32% in electricity costs and ca. 10% in total costs. In the cases with only electrical buildings, system operators need to solve technical problems through the activation of flexibility. This will increase total energy costs by ca. 25%, although the number of technical problems will be decreased by 74%. It was also concluded that only electrical buildings have ca. 44% and 40% lower emissions of CO_2, with and without optimization, respectively. This is due to the fact that MES buildings consume more natural gas, while only electrical buildings depend strictly on electricity. If the mix of electricity generation is highly based on renewable sources, the emissions of CO_2 by consuming electricity will be lower than if consuming natural gas. The annual results follow the trend of the results obtained for the day 7 January 2019.

Future work should address the integration of the restrictions of the gas network and the models of shiftable loads, such as washing machines and dryers. Other energy vectors could also be integrated, like hydrogen or district heating. The utilization of the flexibility provided by MES buildings in energy markets, trough multi-energy aggregators, should also be studied. Other points to be taken in consideration and that should be further studied are the fixed costs in the consumer's tariffs and the costs of DR programs. For planning purposes, investment costs should also be a matter of study.

Author Contributions: A.C.: conceptualization, methodology, software, writing—original draft, visualization. F.S.: conceptualization, methodology, writing—original draft, supervision. J.P.L.: writing—review and editing, supervision. All authors have read and agree to the published version of the manuscript.

Funding: This work is financed by the FCT—Fundação para a Ciência e a Tecnologia (Portuguese Foundation for Science and Technology) within the grant PD/BD/142811/2018.

Conflicts of Interest: The authors declare no conflict of interest. The funders had no role in the design of the study; in the collection, analyses, or interpretation of data; in the writing of the manuscript, or in the decision to publish the results.

Abbreviations

Parameters and Variables

P	Active power [kWh]
Q	Reactive power [kVar]
I	Current [A]
V	Voltage [V]
v	Volume [m^3]
r	Resistance [Ω]
R	Thermal resistance [$^\circ$C/kW]
x	Reactance [Ω]
X	Thickness of the walls [m]
Θ	Temperature of the room [$^\circ$C]
β	Thermal constant ($\beta = e^{\frac{-\Delta t}{CR}}$)
ACH	Infiltration rate [air changes per hour—ACH]
l	Heat gain and losses [$^\circ$C]
C	Thermal capacitance [kWh/$^\circ$C]
c	Specific heat capacity [J/g·$^\circ$C]
CV	Volumetric heat capacity [kJ/m^3·$^\circ$C]
k	Thermal conductivity of the material [kW/m·$^\circ$C]
A	Area [m^2]
a	Insulation [kWh/$^\circ$C]
d	Density of the materials of the walls [g/m^3]
m	Supply air flow rate [cfm, cubic feet per minute]
η	Efficiency [%]
COP	Coefficient of performance
ε	System inertia
L	Load
b	Binary variable
SOC	State-of-charge
E	Energy flow between resources [kWh]
I	Energy input of the energy hub [kWh]
O	Energy output of the energy hub [kWh]
π	Price [€/kWh] or [€/$^\circ$C]
$Cost$	Cost [€]

Subscripts

t	Time interval
i,j,k	Item belonging to a set
t	Time interval
0	Initial time
$wind$	Wind generator
PV	PV generator
O	Outside
h	Heating vector
c	Cooling vector
g	Gas vector
w	Electricity vector
C	Discharge
$consumer$	Consumer
DR	Demand response
$+,-$	Increase, decrease
air	Air

Superscripts

Inf	Infiltration
$wall$	Wall
$coil$	Coil
fan	Fan
$water$	Water
$refr$	Refrigeration system
$equip$	Electrical equipment
$cook$	Cooking
$light$	Lighting
in	Input
out	Output
EV	Electrical vehicle
$stot$	Storage system
net	Network
ch	Charging
dis	Discharging
net	Network
$viol$	Voltage violation
CO_2	Carbon dioxide

Sets

NL	Electrical network load
NB	Electrical network bus
NW	Electrical network wind generator
NPV	Electrical network PV
R	Energy hub room
N	Energy hub network
C	Energy hub converter
S	Energy hub storage
L	Energy hub load
EV	Energy hub EV
PV	Energy hub PV
M	Fan supply air flow values ε

Symbols

$\widehat{}$	Parameter defined by consumer
$\overline{}, \underline{}$	Maximum, minimum

Appendix A. Parameters of the Models

Table A1. Summary of the parameters used in the models presented in Section 3.4.

Parameter	Description	Value
c^{air}	Specific heat capacity of air	1.09 J/g·°C
$\eta^{h,coil}$	Efficiency of heating coil	0.98
$\eta^{c,coil}$	Efficiency of cooling coil	0.98
$COP^{h,coil}$	COP of heating coil	3.45
$COP^{c,coil}$	COP of cooling coil	4.45
$\Theta_t^{C,c}$	Discharging cooling air temperature	12.5 °C
$\Theta_t^{C,h}$	Discharging heating air temperature	50 °C
b_1	Parameter 1 of the fan model	2.57×10^{-12}
b_2	Parameter 2 of the fan model	-4.45×10^{-9}
b_3	Parameter 3 of the fan model	1.46×10^{-4}
b_4	Parameter 4 of the fan model	4.71×10^{-3}
c^w	Specific heat capacity of water	4.18 J/g·°C
$\Theta_j^{water,i}$	Initial temperature of water	15 °C
COP^{refr}	COP of refrigeration systems	3
a (Residential)	Insulation of residential refrigeration systems	15 kJ/°C
a (Supermarket)	Insulation of supermarket refrigeration systems	850 kJ/°C

Appendix B. Average Values Per Month

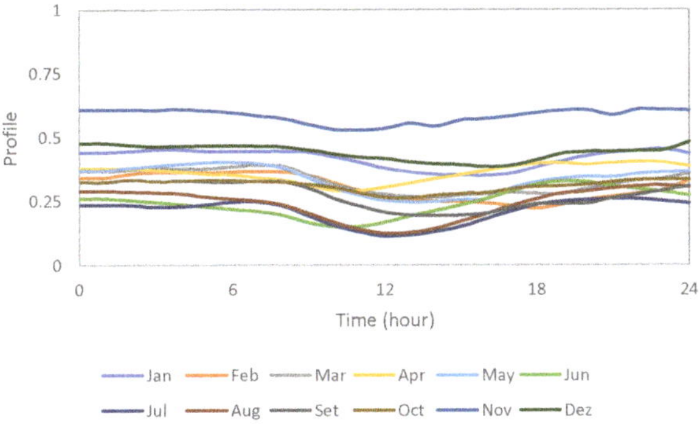

Figure A1. Average daily wind generation profiles, per month.

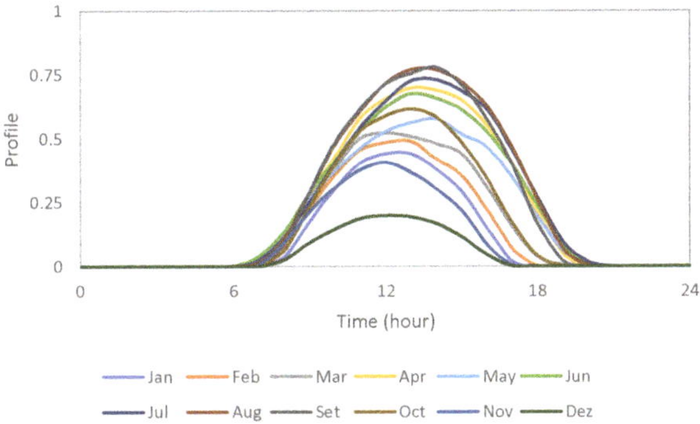

Figure A2. Average daily PV generation profiles, per month.

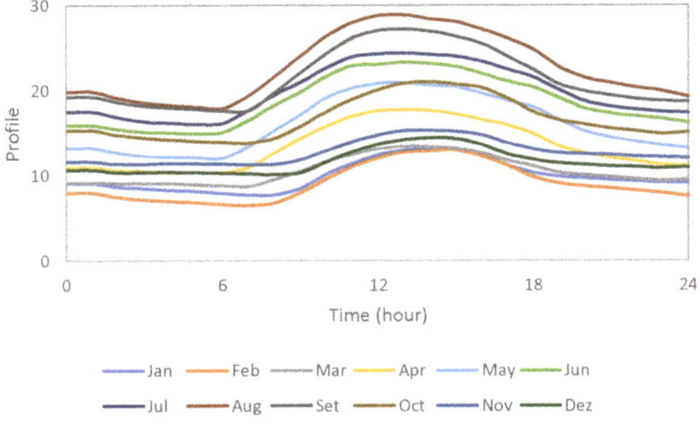

Figure A3. Average daily temperature profiles, per month.

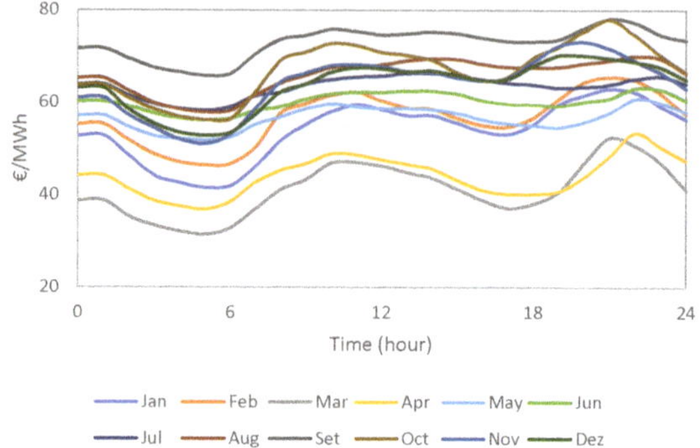

Figure A4. Average daily electricity prices, per month.

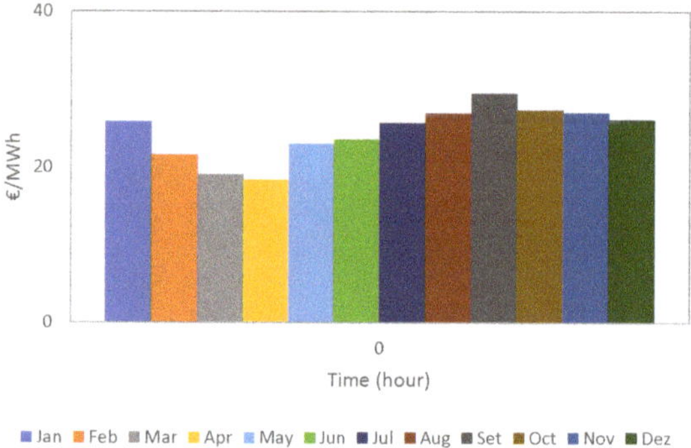

Figure A5. Average gas prices, per month.

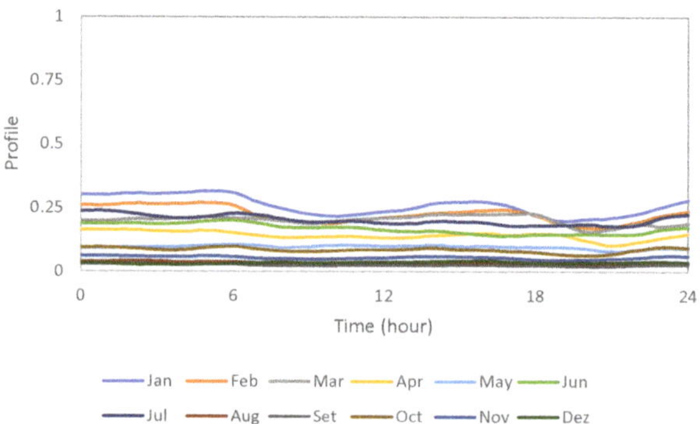

Figure A6. Average daily profile of electricity generation from coal, per month.

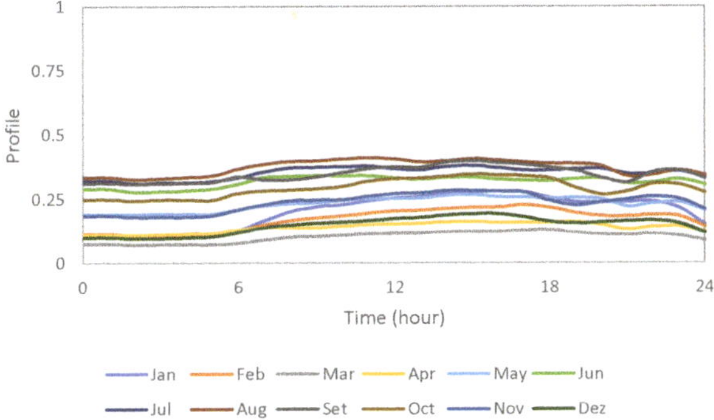

Figure A7. Average daily profile of electricity generation from natural gas, per month.

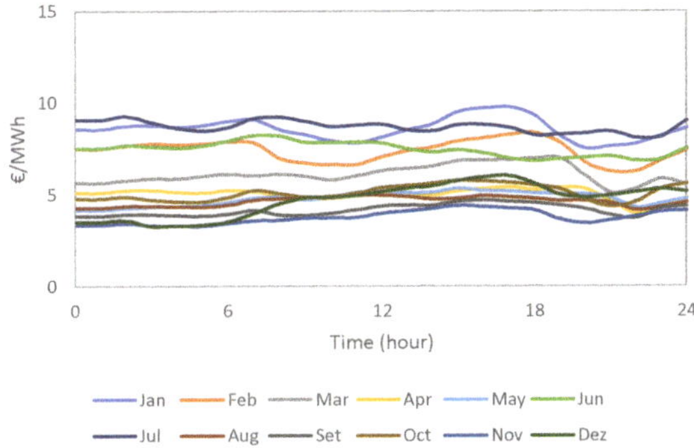

Figure A8. Average daily CO$_2$ prices for electricity consumed, per month.

References

1. Agency, I.E. *CO$_2$ Emissions from Fuel Combustion: Overview*; International Energy Agency (IEA): Paris, France, 2018.
2. Agency, I.E. *Renewables 2017: Analysis and Forecasts to 2022*; Technical Report; International Energy Agency (IEA): Paris, France, 2017.
3. Coelho, A.; Neyestani, N.; Soares, F.; Lopes, J.P. Wind variability mitigation using multi-energy systems. *Int. J. Electr. Power Energy Syst.* **2020**, *118*, 105755. [CrossRef]
4. Geidl, M.; Koeppel, G.; Favre-Perrod, P.; Klöckl, B.; Andersson, G.; Fröhlich, K. The Energy Hub—A Powerful Concept for Future Energy Systems. In Proceedings of the Third Annual Carnegie Mellon Conference on the Electricity Industry, Pittsburgh, PA, USA, 13–14 March 2007; pp. 13–14.
5. Wang, Y.; Zhang, N.; Kang, C.; Kirschen, D.S.; Yang, J.; Xia, Q. Standardized Matrix Modeling of Multiple Energy Systems. *IEEE Trans. Smart Grid* **2017**, *10*, 257–270. [CrossRef]
6. Wang, Y.; Cheng, J.; Zhang, N.; Kang, C. Automatic and linearized modeling of energy hub and its flexibility analysis. *Appl. Energy* **2017**, *211*, 705–714. [CrossRef]
7. Gimelli, A.; Muccillo, M.; Sannino, R. Optimal design of modular cogeneration plants for hospital facilities and robustness evaluation of the results. *Energy Convers. Manag.* **2017**, *134*, 20–31. [CrossRef]
8. Mohiuddin, S.; Mahmud, M.; Haruni, A.; Pota, H. Design and implementation of partial feedback linearizing controller for grid-connected fuel cell systems. *Int. J. Electr. Power Energy Syst.* **2017**, *93*, 414–425. [CrossRef]
9. Mukherjee, U.; Walker, S.; Maroufmashat, A.; Fowler, M.; Elkamel, A. Development of a pricing mechanism for valuing ancillary, transportation and environmental services offered by a power to gas energy system. *Energy* **2017**, *128*, 447–462. [CrossRef]

10. Wasilewski, J. Integrated modeling of microgrid for steady-state analysis using modified concept of multi-carrier energy hub. *Int. J. Electr. Power Energy Syst.* **2015**, *73*, 891–898. [CrossRef]

11. Moeini-Aghtaie, M.; Abbaspour, A.; Fotuhi-Firuzabad, M.; Dehghanian, P. Optimized Probabilistic PHEVs Demand Management in the Context of Energy Hubs. *IEEE Trans. Power Deliv.* **2015**, *30*, 996–1006. [CrossRef]

12. Yazdani-Damavandi, M.; Moghaddam, M.P.; Haghifam, M.R.; Shafie-khah, M.; Catalao, J.P.S. Modeling Operational Behavior of Plug-in Electric Vehicles' Parking Lot in Multienergy Systems. *IEEE Trans. Smart Grid* **2016**, *7*, 124–135. [CrossRef]

13. Li, Z.; Wu, W.; Wang, J.; Zhang, B.; Zheng, T. Transmission-Constrained Unit Commitment Considering Combined Electricity and District Heating Networks. *IEEE Trans. Sustain. Energy* **2016**, *7*, 480–492. [CrossRef]

14. Moeini-Aghtaie, M.; Abbaspour, A.; Fotuhi-Firuzabad, M.; Hajipour, E. A Decomposed Solution to Multiple-Energy Carriers Optimal Power Flow. *IEEE Trans. Power Syst.* **2013**, *29*, 707–716. [CrossRef]

15. Dzobo, O.; Xia, X. Optimal operation of smart multi-energy hub systems incorporating energy hub coordination and demand response strategy. *J. Renew. Sustain. Energy* **2017**, *9*, 045501. [CrossRef]

16. Hao, H.; Corbin, C.D.; Kalsi, K.; Pratt, R.G. Transactive Control of Commercial Buildings for Demand Response. *IEEE Trans. Power Syst.* **2017**, *32*, 774–783. [CrossRef]

17. Vrettos, E.; Andersson, G. Scheduling and Provision of Secondary Frequency Reserves by Aggregations of Commercial Buildings. *IEEE Trans. Sustain. Energy* **2016**, *7*, 850–864. [CrossRef]

18. Iria, J.; Soares, F.; Matos, M. Optimal bidding strategy for an aggregator of prosumers in energy and secondary reserve markets. *Appl. Energy* **2019**, *238*, 1361–1372. [CrossRef]

19. Iria, J.P.; Soares, F.J.; Matos, M.A. Trading Small Prosumers Flexibility in the Energy and Tertiary Reserve Markets. *IEEE Trans. Smart Grid* **2018**, *10*, 2371–2382. [CrossRef]

20. Lemmer, E.F.; Delport, G.J. The influence of a variable volume water heater on the domestic load profile. *IEEE Trans. Energy Convers.* **1999**, *14*, 1558–1563. [CrossRef]

21. Mendes, L.A.; Freire, R.Z.; Coelho, L.d.S.; Moraes, A.S. Minimizing computational cost and energy demand of building lighting systems: A real time experiment using a modified competition over resources algorithm. *Energy Build.* **2017**, *139*, 108–123. [CrossRef]

22. Stadler, M.; Krause, W.; Sonnenschein, M.; Vogel, U. Modelling and evaluation of control schemes for enhancing load shift of electricity demand for cooling devices. *Environ. Model. Softw.* **2009**, *24*, 285–295. [CrossRef]

23. Lopes, J.A.; Soares, F.J.; Almeida, P.M. Integration of electric vehicles in the electric power system. *Proc. IEEE* **2011**, *99*, 168–183. [CrossRef]

24. Vrettos, E.; Kara, E.C.; MacDonald, J.; Andersson, G.; Callaway, D.S. Experimental Demonstration of Frequency Regulation by Commercial Buildings—Part I: Modeling and Hierarchical Control Design. *IEEE Trans. Smart Grid* **2018**, *9*, 3213–3223. [CrossRef]

25. Mohammad, N.; Rahman, A. Transactive control of industrial heating–ventilation–air-conditioning units in cold-storage warehouses for demand response. *Sustain. Energy Grids Netw.* **2019**, *18*, 100201. [CrossRef]

26. Yudong Ma.; Borrelli, F. Fast stochastic predictive control for building temperature regulation. In Proceedings of the 2012 American Control Conference (ACC), Montreal, QC, Canada, 27–29 June 2012; pp. 3075–3080. [CrossRef]

27. Iria, J.; Soares, F. A cluster-based optimization approach to support the participation of an aggregator of a larger number of prosumers in the day-ahead energy market. *Electr. Power Syst. Res.* **2019**, *168*, 324–335. [CrossRef]

28. Baran, M.; Wu, F. Optimal sizing of capacitors placed on a radial distribution system. *IEEE Trans. Power Deliv.* **1989**, *4*, 735–743. [CrossRef]

29. Baran, M.; Wu, F. Optimal capacitor placement on radial distribution systems. *IEEE Trans. Power Deliv.* **1989**, *4*, 725–734. [CrossRef]

30. Gan, L.; Low, S.H. Convex relaxations and linear approximation for optimal power flow in multiphase radial networks. In Proceedings of the 2014 Power Systems Computation Conference, PSCC 2014, Wroclaw, Poland, 18–22 August 2014. [CrossRef]

31. Yeh, H.G.; Gayme, D.F.; Low, S.H. Adaptive VAR control for distribution circuits with photovoltaic generators. *IEEE Trans. Power Syst.* **2012**, *27*, 1656–1663. [CrossRef]

32. Melby, P.; Cathcart, T. *Regenerative Design Techniques: Practical Applications in Landscape Design*; Wiley: Hoboken, NJ, USA, 2002; p. 410.

33. Nicklas, S. *ASHRAE Handbook HVAC Systems and Equipment*; Technical Report 28; American Society of Heating, Refrigerating and Air-Conditioning Engineers (ASHRAE): Atlanta, GE, USA, 2016.
34. Park, S.Y.; Cho, S.; Ahn, J. Improving the quality of building spaces that are planned mainly on loads rather than residents: Human comfort and energy savings for warehouses. *Energy Build.* **2018**, *178*, 38–48. [CrossRef]
35. Ryder-Cook, D.; Mackay, D. Thermal Modelling of Buildings. Available online: https://inference.org.uk/is/papers/DanThermalModellingBuildings.pdf (accessed on 14 February 2019).
36. Commercial Reference Buildings | Department of Energy. Available online: https://www.energy.gov/eere/buildings/new-construction-commercial-reference-buildings (accessed on 11 April 19).
37. Kelman, A.; Borrelli, F. Bilinear Model Predictive Control of a HVAC System Using Sequential Quadratic Programming. *IFAC Proc. Vol.* **2011**, *44*, 9869–9874. [CrossRef]
38. Meteogalicia. Datos Históricos das redes de Estacións. Available online: https://www.meteogalicia.gal/web/inicio.action (accessed on 11 April 2019).
39. ENTSOE. Day-Ahead Prices. Available online: https://transparency.entsoe.eu/transmission-domain/r2/dayAheadPrices/show (accessed on 27 August 2019).
40. MIBGAS (Mercado Ibérico del Gas). Datos Gas. Available online: http://www.mibgas.es/pt/mercados-gas (accessed on 27 August 2019).
41. Insider, M. CO_2 European Emission Allowances. Available online: https://markets.businessinsider.com/commodities/co2-european-emission-allowances (accessed on 16 May 2020).
42. Quaschning, V. *Regenerative Energiesysteme Technologie, Berechnung, Klimaschutz*; Hanser: Munich, Germany, 2019.
43. REN. Produção Verificada—Sistema de Informação de Mercados de Energia. Available online: http://www.mercado.ren.pt/PT/Electr/InfoMercado/Prod/Paginas/Verificada.aspx (accessed on 16 May 2020).
44. U.S. Energy Information Administration. *Monthly Energy Review*; Technical Report; U.S. Energy Information Administration (EIA): Washington, DC, USA, 2020.
45. PyCharm: The Python IDE for Professional Developers by JetBrains. Available online: https://www.jetbrains.com/pycharm/ (accessed on 19 March 2019).

 *energies*

Article

Integration of Hydrogen into Multi-Energy Systems Optimisation

Peng Fu *, Danny Pudjianto, Xi Zhang and Goran Strbac

Department of Electrical and Electronic Engineering, Imperial College London, London SW7 2AZ, UK; d.pudjianto@imperial.ac.uk (D.P.); x.zhang14@imperial.ac.uk (X.Z.); g.strbac@imperial.ac.uk (G.S.)
* Correspondence: p.fu16@imperial.ac.uk; Tel.: +447456959559

Received: 14 March 2020; Accepted: 31 March 2020; Published: 1 April 2020

Abstract: Hydrogen presents an attractive option to decarbonise the present energy system. Hydrogen can extend the usage of the existing gas infrastructure with low-cost energy storability and flexibility. Excess electricity generated by renewables can be converted into hydrogen. In this paper, a novel multi-energy systems optimisation model was proposed to maximise investment and operating synergy in the electricity, heating, and transport sectors, considering the integration of a hydrogen system to minimise the overall costs. The model considers two hydrogen production processes: (i) gas-to-gas (G2G) with carbon capture and storage (CCS), and (ii) power-to-gas (P2G). The proposed model was applied in a future Great Britain (GB) system. Through a comparison with the system without hydrogen, the results showed that the G2G process could reduce £3.9 bn/year, and that the P2G process could bring £2.1 bn/year in cost-savings under a 30 Mt carbon target. The results also demonstrate the system implications of the two hydrogen production processes on the investment and operation of other energy sectors. The G2G process can reduce the total power generation capacity from 71 GW to 53 GW, and the P2G process can promote the integration of wind power from 83 GW to 130 GW under a 30 Mt carbon target. The results also demonstrate the changes in the heating strategies driven by the different hydrogen production processes.

Keywords: hydrogen; multi-energy systems; power system economics; renewable energy generation; whole system modelling

---

1. Introduction

The new-found interest in hydrogen from both industry and academia has stimulated research exploring the application of hydrogen as a potential option for decarbonizing major parts of the energy system. Hydrogen can play a key role alongside electricity in the low carbon economy due to its low-cost storability, flexibility, low-carbon hydrogen production technologies, and the opportunity to re-energise the gas distribution network. Considering the expensive investment for large scale deployment of electric storage to facilitate the mass integration of renewable energy sources (RES), hydrogen storage can potentially serve as an alternative, when coordinated with other hydrogen-related technologies, to fulfil the same functionality at a lower cost due to its reduced capital cost [1,2]. It is also a flexible energy vector that can be produced from various primary energy sources. Two main sources were considered in this paper: natural gas and electricity via electrolysers. Hydrogen can be used as fuel for electricity generation, fuel for a hydrogen boiler for heating, and it can also be used to power fuel cells for transport and co-generation for electricity and heat. This raises important questions on how the hydrogen should be integrated with other energy systems and the importance of whole-energy system optimisation, compared to the traditional silo planning approach.

The need to address these questions has triggered the development of multi-energy systems modelling approaches to assess the technical and cost implications of integrating hydrogen into the overall energy system. Compared to the energy system planning approach without considering synergies between different sectors, multi-energy systems, whereby different energy vectors (e.g., electricity, heat and gas, etc.) can operate in a coordinated fashion at various levels, introduces an important opportunity to improve the system planning technically, economically, and environmentally [3]. Few applications of hydrogen include hydrogen boilers, hydrogen fuel-cells for vehicles, and micro-CHP (combined heat and power), which demonstrate that hydrogen brings interaction to all sectors across the energy landscape. However, the impacts of this cross-energy vector interaction on the energy system capacity requirement and operation, primary energy demand, values, and options that hydrogen create have not been thoroughly investigated, especially in the context of multi-energy systems. The benefits and the system implications of integrating a hydrogen supply chain should be identified through a whole-system approach to capture complex interactions across different technologies, different energy vectors, and coordination between investment in energy infrastructure and operating decisions.

Authors have proposed the use of a holistic optimisation model for electricity system investment and operation decisions to assess the value of bulk and distributed energy storages in the future low-carbon electricity systems [4]. Enhancing the model in [4], the authors proposed the integrated electricity and heat energy model in [5]. The model was used to analyse the system implications and cost performance of alternative heating decarbonisation strategies including the use of hydrogen, electrification, and hybrid heat pumps. The analyses considered the interactions between electricity sectors and heat sectors. Similarly, Zhang et al. [6] quantified the benefits of the integration of heat, particularly district heating, and the electricity system. Through integrated planning, the flexibility that exists in the heating system can be utilised to support the electricity system, which otherwise has to count on the flexibility measures within the electricity system itself. The series of case studies demonstrated that district heating network and application of thermal energy storage would enhance the flexibility of the overall energy system, thus delivering substantial cost savings to meet the carbon target. Chaudry et al. [7] proposed a combined gas and electricity system optimisation model to solve short-term operation problems. The model links two energy sectors through gas turbine generation. The proposed combined gas and electricity system model has demonstrated its value for assessing the consequences of the failure of vital facilities.

Previous research on the integration of hydrogen in the overall energy system has mostly focused on its industrial production, transmission, and distribution [8]. Authors designed a future hydrogen supply chain which covered production, storage, and distribution for the UK transport demand. A few previous studies have considered the interactions of hydrogen with other energy sectors. Almansoori et al. [9] adopted optimisation techniques to develop the hydrogen supply chain for the transport sector; the optimal infrastructure structure and operation costs were determined through the proposed model. The mixed-integer linear programming (MILP) model was proposed to optimise the design and operation of integrated wind–hydrogen–electricity networks [10]. The optimisation only considered the transport demand supplied by hydrogen. However, the design and operation of the wider power sector and other hydrogen production processes were not considered. Samsatli et al. [11] proposed a comprehensive spatiotemporal MILP model to optimise the integrated electricity–hydrogen value chains to supply the space and water heating demand in GB. The impacts of P2G facilities in the integrated electricity and gas system were analysed in [12–14], which proved the flexibility and effectiveness of P2G facilities. In [15], a power-to-hydrogen-and-heat scheme was proposed, in which the power-to-heat and power-to-hydrogen processes were coupled through adopting the heat recovery from the P2G process. The synergy among the electricity sector, transport sector, and hydrogen sector was analysed in [16], but the heat sector was not considered.

This paper proposes an electricity–heat-transport–hydrogen economical optimisation with environmental constraints at the national level, which simultaneously considers the infrastructure capital expenditures (CAPEX) and whole system operative expenditures (OPEX), thus meeting the specific carbon target at a lower whole-system cost. The key contributions of this paper can be summarised as follows:

1)   It incorporates modelling of the hydrogen system into a combined optimization multi-energy systems model considering both investment and operation at the system level.
2)   It assesses the system implications, economic, and environmental impact of different hydrogen production infrastructures across the whole system level.
3)   It also investigates the impacts of hydrogen integration on each individual energy sector under different carbon targets.

The remainder of this paper is organized as follows. Section 2 presents the integrated multi-energy systems model. In Section 3, a series of case studies are performed to compare the advantages and disadvantages of adopting G2G and P2G individually with the system without hydrogen integration in different carbon scenarios. The conclusions are provided in Section 4.

2. Integrated Multi-Energy Systems Model

2.1. Interactions in the Multi-Energy Systems

Interactions take place through the energy conversion between different energy carriers to supply energy demand and to ensure optimal and secured operation. There are numerous interactions between different energy sectors in the proposed model, as shown in Figure 1. In this model, gas-heated reformers combined with carbon capture storage (GHR-CCS) and electrolysers are the technologies for the G2G and P2G processes, respectively. GHR and electrolysis link hydrogen with the gas and electricity system together respectively. The electricity generation can also use hydrogen as a fuel to make electricity and hydrogen systems interactive. Meanwhile, the transportation demand can be supplied by electricity or hydrogen through electric vehicles (EV) and hydrogen fuel-cell vehicles (HFCV), respectively, which means that electricity and hydrogen systems are also coupled in the transportation sector. The hydrogen boiler can function in the same way as existing gas boilers in the heating system, but brings zero-emissions, which maintains resilience for householders through promoting the diversity of energy carriers. The other components like electricity storage and thermal energy storage or other heating devices (resistance) can also be added to the model, which were omitted here for brevity.

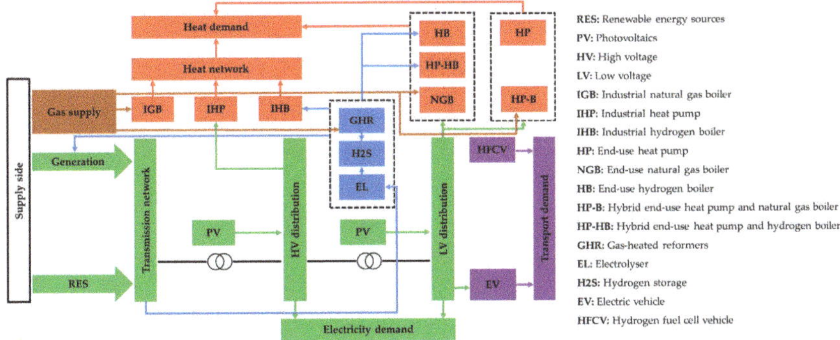

Figure 1. Interaction of integrated electricity–heat–transport–hydrogen system.

2.2. Objective Function

The model was formulated as a MILP problem with a 1-year time horizon and hourly time resolution to capture the interactions across investment and operating decisions. The objective function

(Equation (5)) is to minimize the overall annuitized investment and operation cost. The fixed and variable operating and maintenance costs of all the technologies were also considered, which were omitted in the equations for brevity. The total cost in the electricity system is formulated in Equation (1).

$$C_e = \sum_{g=1}^{G} \pi_g \cdot n_g \cdot \overline{P}_g + \sum_{r=1}^{R} \left(\pi_{hvs_r} \cdot n_{hvs_r} + \pi_{lvs_r} \cdot n_{lvs_r} \right) + \sum_{l=1}^{L} \pi_{f_l} \cdot f_l + \sum_{g=1}^{G} \sum_{t=1}^{T} Z\left(\gamma_g, P_{g,t}, \pi_{nl}, \pi_{st}, \mu_{g,t} \right) \tag{1}$$

The electricity system investment cost includes the annuitized capital cost of new-built generation, reinforcement cost of transmission, and distribution networks. The operational cost of the electricity system consists of the operation cost of the conventional generation, which is fossil fuel based as well as the no-load cost that is a function of the number of synchronized units and start-up cost and the fixed and variable O&M cost of all the units including RES. The total cost in the heating system, which includes the cost in the DHN and end-use heating appliances are formulated in Equations (2) and (3), respectively. The operational cost in the heat sector is mainly from the natural gas consumption of industrial and end-use natural gas boiler.

$$C_{h,hn} = \sum_{r=1}^{R} \left(\pi_{hp_r} \cdot n_{hp_r} + \pi_{ngb_r} \cdot n_{ngb_r} + \pi_{hb_r} \cdot n_{hb_r} + D(DH_r) \right) + \sum_{r=1}^{R} \sum_{t=1}^{T} \gamma_{ngb_r} \cdot H_{ngb_r,t} \tag{2}$$

$$C_{h,ed} = \sum_{r=1}^{R} \left(\pi_{ehp_r} \cdot n_{ehp_r} + \pi_{engb_r} \cdot n_{engb_r} + \pi_{ehb_r} \cdot n_{ehb_r} \right) + \sum_{r=1}^{R} \sum_{t=1}^{T} \gamma_{engb_i} \cdot H_{engb_i,t} \tag{3}$$

District heating is supplied by industrial heat pumps (HPs), natural gas boilers (NGBs), and hydrogen boilers (HBs) and end-use heating appliances include air source heat pumps (ASHPs), end-use NGBs, and HBs. The total cost in the hydrogen system is formulated in Equation (4). The operational cost in the hydrogen system refers to the natural gas consumption of GHR.

$$C_{h2} = \sum_{i=1}^{I} \left(\pi_{el_i} \cdot n_{el_i} + \pi_{smr_i} \cdot n_{smr_i} + \pi_{hs_i} \cdot n_{hs_i} \right) + \sum_{i=1}^{I} \sum_{t=1}^{T} \gamma_{smr_i} \cdot Q_{smr_i,t} + \sum_{l=1}^{L} \pi_{f_{ht_l}} \cdot f_{ht_l} \tag{4}$$

The investment of the hydrogen system cost includes the annuitized capital cost of the electrolyser, GHR-CCS, hydrogen storage, and the hydrogen transmission pipelines. The study assumes the cost of EV and HFCV is the same, and therefore, their costs can be omitted from the optimisation problem; the portfolio of EV and HFCV is optimised based on their system integration costs rather than by the vehicle's capital cost. It is trivial to include different EV and HFCV costs in the objective function; at present, it is the interest of this paper to evaluate the competitiveness of these technologies on the basis of their system integration costs.

$$Min \; \varphi = C_e + C_{h,hn} + C_{h,ed} + C_{h2} \tag{5}$$

2.3. Constraints

The proposed model is subject to several constraints in each energy system. Constraints associated with the electricity system include those in Equations (6)–(22). All constraints are applied to each time interval within the optimisation time horizon (1-year) ($\forall t \in T$) for all regions and locations ($\forall r \in R$, $\forall i \in I$). Electricity supply and demand are balanced in each time interval (Equation (6)). The electricity demand includes non-heat demand, the demand of HPs, and the electricity consumption of electrolysis and EV. DC power flow model Equation (7) is applied to determine power flows in the electricity network. The production of all the generating units is within their installed capacity including renewable energy units (Equations (8) and (9)). The change in the generation of a thermal unit within a single time interval is limited by its ramping capacity (Equations (10) and (11)), and the number of units being synchronised is constrained in Equations (12)–(14).

$$\sum_{g=1}^{G} P_{g,t} = \sum_{r=1}^{R} \left(DE_{r,t} + \frac{H_{hpr,t}}{\eta_{hp}} + \frac{H_{ehpr,t}}{\eta_{ehpr,t}} + \frac{Q_{elr,t}}{\eta_{el}} + \frac{V_{evr,t}}{\eta_{ev}} \right) \tag{6}$$

$$-\left(\overline{f_l} + f_l\right) \leq F(G, D, \theta) \leq \overline{f_l} + f_l \tag{7}$$

$$\mu_{g,t} \cdot \underline{P}_g \leq P_{g,t} \leq \mu_{g,t} \cdot \overline{P}_g \tag{8}$$

$$\mu_{g,t} \leq n_g \tag{9}$$

$$P_{g,t} - P_{g,t-1} \leq \mu_{g,t} \cdot R_g^{up} \cdot \Delta t \tag{10}$$

$$P_{g,t-1} - P_{g,t} \leq \mu_{g,t-1} \cdot R_g^{down} \cdot \Delta t \tag{11}$$

$$\mu_{g,t} - \mu_{g,t-1} = st_{g,t} - dst_{g,t} \tag{12}$$

$$st_{g,t} \leq n_g - \mu_{g,t-1} \tag{13}$$

$$dst_{g,t} \leq \mu_{g,t-1} \tag{14}$$

The reinforcement cost of the electricity distribution network is expressed as a function of peak flow in the local distribution system [4], which are formulated in Equations (15) and (16). The concept of reverse power flow, meaning that due to high PV penetration, the net energy may flow in the opposite direction was also considered in this model (Equations (17) and (18)).

$$DE_{r,t} + \frac{H_{hpr,t}}{\eta_{hp}} + \frac{V_{evr,t}}{\eta_{ev}} + \frac{H_{ehpr,t}}{\eta_{ehpr,t}} - P_{hvpv_r,t} - P_{lvpv_r,t} \leq \overline{hvs_r} + hvs_r \tag{15}$$

$$\kappa \cdot DE_{r,t} + \frac{V_{evr,t}}{\eta_{ev}} + \frac{H_{ehpr,t}}{\eta_{ehpr,t}} - P_{lvpv_r,t} \leq \overline{lvs_r} + lvs_r \tag{16}$$

$$P_{hvpv_r,t} + P_{lvpv_r,t} - DE_{r,t} - \frac{H_{hpr,t}}{\eta_{hp}} - \frac{V_{evr,t}}{\eta_{ev}} - \frac{H_{ehpr,t}}{\eta_{ehpr,t}} \leq \sigma_{hv} \cdot \left(\overline{hvs_r} + hvs_r\right) \tag{17}$$

$$P_{lvpv_r,t} - \kappa \cdot DE_{r,t} - \frac{V_{evr,t}}{\eta_{ev}} - \frac{H_{ehpr,t}}{\eta_{ehpr,t}} \leq \sigma_{lv} \cdot \left(\overline{lvs_r} + lvs_r\right) \tag{18}$$

Frequency response and operating reserve are two balancing services considered in this model. The supplementary frequency response and operating reserve can also be provided by the heating sector (HP) and hydrogen sector (electrolyser) as well as the transport sector (EV). The system frequency response requirement is directly related to the level of system inertia [17], which can be treated as linear to the online synchronous capacity in each time period (Equations (19) and (20)). The operating reserve requirement is determined by the forecasting errors of electricity load and renewable energy generation (Equations (21) and (22)).

$$rsp_{g,t} \leq \mu_{g,t} \cdot \overline{rsp}_g \tag{19}$$

$$\sum_{r=1}^{R} \left(rsp_{hpr,t} + rsp_{ehpr,t} + rsp_{elr,t} + rsp_{evr,t} \right) + \sum_{g=1}^{G} rsp_{g,t} \geq \overline{SF}_t \tag{20}$$

$$res_{g,t} \leq \mu_{g,t} \cdot \overline{res}_g \tag{21}$$

$$\sum_{r=1}^{R} \left(res_{hpr,t} + res_{ehpr,t} + res_{elr,t} \right) + \sum_{g=1}^{G} res_{g,t} \geq \overline{SR}_t \tag{22}$$

Heating demand is supplied by either DHN or end-use appliances (Equations (23)–(28)). The industrial-sized HPs, NGBs, and HBs are deployed on the heat network for district heating. All the appliances were also used as end-use heating appliances. Hybrid electric HPs and natural gas boilers (Hybrid HP-NGBs) have been proven to have a significant overall economic advantage over other individual heat devices like HP-only and DHN [5]. With the presence of HBs, the hybrid HPs and HBs (Hybrid HP-HBs) can be another heating strategy.

$$H_{dhn_r,t} + H_{edu_r,t} = DH_{r,t} \tag{23}$$

$$H_{dhn_r,t} = H_{hp_r,t} + H_{ngb_r,t} + H_{hb_r,t} \tag{24}$$

$$H_{edu_r,t} = H_{ehp_r,t} + H_{engb_r,t} + H_{ehb_r,t} \tag{25}$$

$$H_{dhn_r,t} = \lambda_{dhn} \cdot DH_{r,t} \tag{26}$$

$$H_{edu_r,t} = \lambda_{edu} \cdot DH_{r,t} \tag{27}$$

$$\lambda_{dhn} + \lambda_{edu} = 1 \tag{28}$$

Similarly, the transport demand is supplied by electric vehicles (EVs) and hydrogen fuel cell vehicles (HFCVs) (Equations (29)–(31)). EVs are modelled as flexible loads that can provide a demand-side response (DSR). Constraints (32) and (33) describe the demand reduction and the energy balance for demand shifting.

$$V_{ev_r,t} = \lambda_{ev_r} \cdot DT_{r,t} + d_{r,t}^+ - d_{r,t}^- \tag{29}$$

$$V_{hfcv_r,t} = \lambda_{hfcv_r} \cdot DT_{r,t} \tag{30}$$

$$\lambda_{ev_r} + \lambda_{hfcv_r} = 1 \tag{31}$$

$$d_{r,t}^- \leq \varepsilon \cdot DT_{r,t} \tag{32}$$

$$\sum_{t \in D} d_{r,t}^- = \sum_{t \in D} d_{r,t}^+ \tag{33}$$

The hydrogen system in this model considers the hydrogen supply chain from production to end-users via hydrogen storage, transmission, and distribution. Currently, the national gas transmission system and the local gas distribution system in the UK are constructed primarily from carbon steel. The unprotected iron and carbon steel pipelines suffer from embrittlement due to the diffusion of hydrogen into the material, which results in a reduction of structural integrity and can potentially cause the fracture. Therefore, these materials are not suitable for hydrogen networks [18]. The model assumes new hydrogen transmission through pipelines is needed in addition to the existing transmission gas pipelines. At the distribution level, the government-sponsored Iron Mains Replacement Programme is expected to convert a majority of the current natural gas distribution network to polyethylene pipes that are hydrogen tolerant by 2030 [19,20]. As the upgrade of the gas distribution network will occur in all scenarios, we assumed in this study that the local natural gas distribution systems will be compatible with hydrogen use by 2050 and the upgrade cost of gas distribution was excluded from the model. The hydrogen energy balancing is formulated in Equation (34). The hydrogen storage model is formulated by Equations (35)–(38). The hydrogen transmission capacity is optimised by using the transportation model [21]. The output of all the technologies is limited by its own capacity. To save the length of the article, these are not listed in the paper.

$$\sum_{i=1}^{I} \left(Q_{el_i,t} + Q_{smr_i,t} + SH_{i,t}^- - SH_{i,t}^+ \right) = \sum_{g \in I_i} \frac{P_{h2g,t}}{\eta_{h2g}} + \sum_{r=1}^{R} \left(\frac{H_{hb_r,t}}{\eta_{hb}} + \frac{H_{ehb_r,t}}{\eta_{ehb}} + \frac{V_{hfcv_r,t}}{\eta_{hfcv}} \right) \tag{34}$$

$$SH_{i,t}^+ \leq n_{hs_i} \tag{35}$$

$$SH_{i,t}^- \leq n_{hs_i} \tag{36}$$

$$SHS_{i,t} \leq n_{hs_i} \cdot SHST_i \tag{37}$$

$$SHS_{i,t} = SHS_{i,t-1} - SH_{i,t}^- + \eta_{hs} \cdot SH_{i,t}^+ \tag{38}$$

The total direct carbon emissions of the whole system do not exceed the regulated amount of carbon emissions including electricity, heat, and hydrogen sectors. The carbon emission constraints are formulated as Equation (39).

$$\sum_{t=1}^{T}\sum_{g=1}^{G} P_{g,t}{\cdot}C_g + \sum_{t=1}^{T}\sum_{r=1}^{R}\left(H_{ngb_r,t}{\cdot}C_{ngb} + H_{engb_r,t}{\cdot}C_{engb}\right) + \sum_{t=1}^{T}\sum_{i=1}^{I} Q_{smr_i,t}{\cdot}C_{smr} \leq CT \tag{39}$$

The MILP problem defined in this section was implemented in the FICO Xpress optimisation tool [22] and solved by the Newton Barrier method.

3. Case Studies

3.1. System Description and Assumptions

The proposed integrated multi-energy systems model in Section 2 was applied to analyse and optimise the GB 2050 energy systems, with 30 Mt and 10 Mt carbon targets. The benefits of integrating hydrogen into energy systems were analysed. The study used a simplified GB transmission system representing five main regions: (1) Scotland (SCOT); (2) North England and Wales (EW-N); (3) Middle England and Wales (EW-M); (4) South England and Wales (EW-S); and (5) London (LON). Three neighbouring systems, Ireland (IE), Continental Europe (CE), and Norway (NOR) are connected with GB through finite interconnectors. The model also considers the investment and operation in IE and CE; this will enable the cross-border interactions to be modelled more accurately. The topology of the simplified network together with the length and transmission capacity is illustrated in Figure 2.

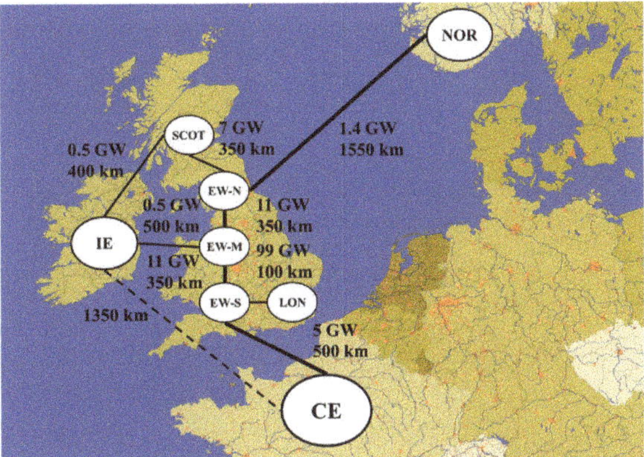

Figure 2. The simplified topology of the interconnected GB system.

The cost and operation data for different types of electricity generation are listed in Table 1 [2,23]. We assumed that the cost of hydrogen-fuelled generation (H2-CCGT, H2-OCGT) is 1.2 times that of the conventional CCGT and OCGT based on the parameters in [24]. The data on heating and hydrogen technologies are listed in Tables 2 and 3, respectively [2,25]. Other assumptions on the costs or operational parameters are omitted here for brevity.

Table 1. Economic and operational parameters of the generation units.

Generation	Capital Cost (£/kW)	Fixed O&M (£/kW/year)	Discount Rate (%)	Lifetime (Years)	Marginal Cost (£/MWh)	Carbon Emissions (kg/MWh)
Nuclear	5191	83.4	9.5%	40	5.0	0
CCGT	581	16.6	7.5%	25	37.7	318.8
OCGT	312	8.2	7.5%	30	54.2	520.6
Gas-CCS	2361	41.6	13.8%	25	33.1	31.9
Coal-CCS	3403	82.0	13.5%	25	35.4	80.5
H2-CCGT	697	17.0	7.5%	25	0	0
H2-OCGT	374	8.5	7.5%	30	0	0
Wind	1642	30.9	8.9%	23	0	0
PV	452	6.2	5.8%	25	0	0

Table 2. Economic and operational parameters of the heating technologies.

Heating Technology	Capital Cost (£/kW)	Fixed O&M (£/kW/year)	COP (%)	Discount Rate (%)	Lifetime (Years)
End-use HP	600	22.0	200%–300%	3.5%	12
End-use NGB	75	6.0	95%	3.5%	12
End-use HB	75	6.0	95%	3.5%	12
Industrial HP	480	17.6	380%	3.5%	12
Industrial NGB	35	2.8	98%	3.5%	12
Industrial HB	35	2.8	98%	3.5%	12

Table 3. Economic and operational parameters of the hydrogen production technologies.

Hydrogen production	Capital Cost (£/kW)	Fixed O&M (£/kW/year)	Discount Rate (%)	Lifetime (Years)	Carbon Emission (kg/MWh)	Efficiency (%)
Electrolyser	465	48.5	10%	30	0	74%
GHR-CCS	384	24.4	10%	40	21.9	84%

In order to quantify the economic benefit of integrating a hydrogen system with different hydrogen production technologies, the whole system costs of four different scenarios were compared under different carbon targets. The four scenarios are described as follows:

1) REF: This is the counterfactual scenario assuming that there is no hydrogen integration across the whole energy system.

2) P2G: Hydrogen is integrated into the energy system, which is produced only by the P2G process (i.e., electrolyser).

3) G2G: Similar to Equation (2), but hydrogen is produced only by G2G process (i.e., GHR-CCS).

4) OPT: The model was used to optimise the capacity of different hydrogen production processes (G2G and P2G).

3.2. The Economic Benefit of Hydrogen Integration

This section compares the economic performance of P2G, G2G, and OPT by comparing the costs against the costs of the counterfactual scenario (REF). Figure 3 shows the whole system cost savings for each scenario under two different carbon targets. The G2G process was identified as the most cost-effective hydrogen production technology under the carbon target 30 Mt, which can reduce the cost by £3.9 bn/year (6.5%). The P2G scenario can bring £2.1 bn/year (3.8%) cost-savings. The OPT scenario brings further cost savings up to £6.6 bn/year (11.2%) by optimally combining the portfolio of hydrogen production technologies. In the heating sector, the use of hybrid heating technology (hybrid HP-HBs), which is based on the use of high COP (coefficient of performance) HPs, to supply the baseload of heat demand while providing the flexibility to use hydrogen to supply peak demand or when there is scarcity in the low-carbon electricity generation output.

The flexibility provided by the hydrogen system can reduce the total power generation capacity requirement from 71 GW to 53 GW in the G2G scenario and reduce electricity operation costs by £5.1 bn/year under a 30 Mt carbon target. Integration of hydrogen also allows hydrogen-fuelled power generation to displace higher cost low-carbon technologies such as nuclear and CCS while supporting better integration of renewables by providing flexibility and balancing fluctuating renewable energy in the system. It is worth noting that under the P2G scenarios, the investment in renewable energy, especially wind power, has increased significantly due to the P2G facilities, which can help integrate renewable energy, as its power consumption can be adjusted to follow the renewable generation. The excess of renewable energy can be stored cost-effectively and used when the output of the renewable is low.

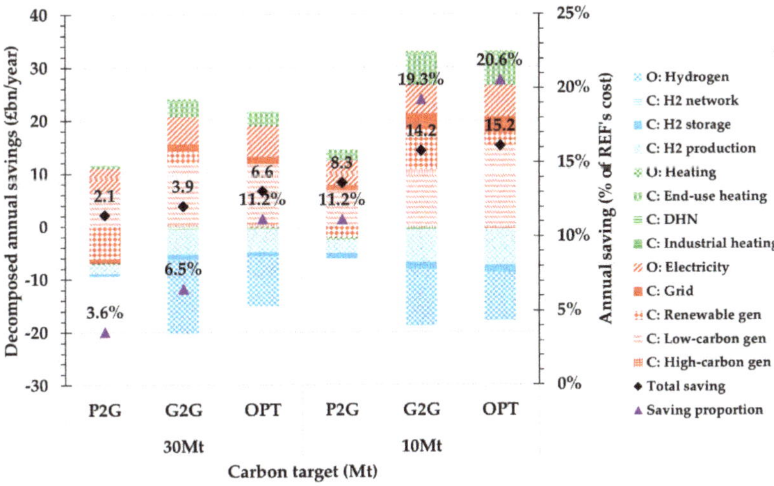

Figure 3. Saving from hydrogen integration under different carbon targets.

Most of the benefits gained in the heat sector through the deployment of HB is driven by the reduced investment in the end-use heating appliances and industrial heating appliances under 30 Mt and 10 Mt carbon targets, respectively, which also further achieved the cost savings in the distribution network reinforcement due to the peak demand reduction that was compensated by hydrogen-based heat.

The increased cost of hydrogen integration is mainly from the hydrogen system, which is dominated by the operation cost of the hydrogen system in the G2G and OPT scenarios. The increased cost in the P2G scenarios mainly comes from the hydrogen production investment. The economic benefit of the integration of the hydrogen system is influenced by the carbon target. In the 10 Mt case, due to the zero-emission characteristics of the P2G process, it plays a more important role in the low-carbon scheme, and its economic benefits become stronger than the G2G process. However, the G2G process still has an economic advantage, especially saving on the investment of electricity infrastructure (e.g., generation and grid network). The annual saving of the OPT scenario under a carbon target of 10 Mt increased to 15.2 bn/year (20.6% of total cost in the REF scenario).

3.3. Impact of Hydrogen Integration on the Electricity System

The integration of hydrogen into the system makes the application of hydrogen fuel power generation (such as H2-CCGT and H2-OCGT) advantageous, thus reshaping the power system potentially, and using HB as the main low-carbon heat source reduces the electricity peak demand and the need for a new power system capacity compared with the system capacity needed if the heat is decarbonised through electrification only. This section compares the capacity and annual electricity generation mix between different cases under given carbon targets.

Figure 4 shows the portfolio of electricity generation capacity in each scenario. It can be observed that the G2G process can reduce the capacity requirement of electricity generation significantly (and other electricity infrastructure, e.g., network) compared to P2G. The choice of hydrogen production pathway will have significant implications for the electricity system. A large part of the low-carbon generation is replaced by the hydrogen-fuelled generation because sufficient flexibility can be provided from the hydrogen-fuelled generation without carbon emission, a more expensive source of flexibility like gas-ccs is not necessary. The high-carbon generation capacity reduction is driven by the enhanced flexibility and presence of hydrogen-fuelled generation. P2G can significantly promote the integration of wind power as the electrolyser can absorb excess wind power, which improves the wind power utilisation. The availability of firm low carbon generation such as nuclear is more critical for energy system decarbonisation under a more demanding carbon target. It is worth emphasising that the carbon emissions from the G2G process limit the large-scale deployment of hydrogen-fuelled generation in the electricity system.

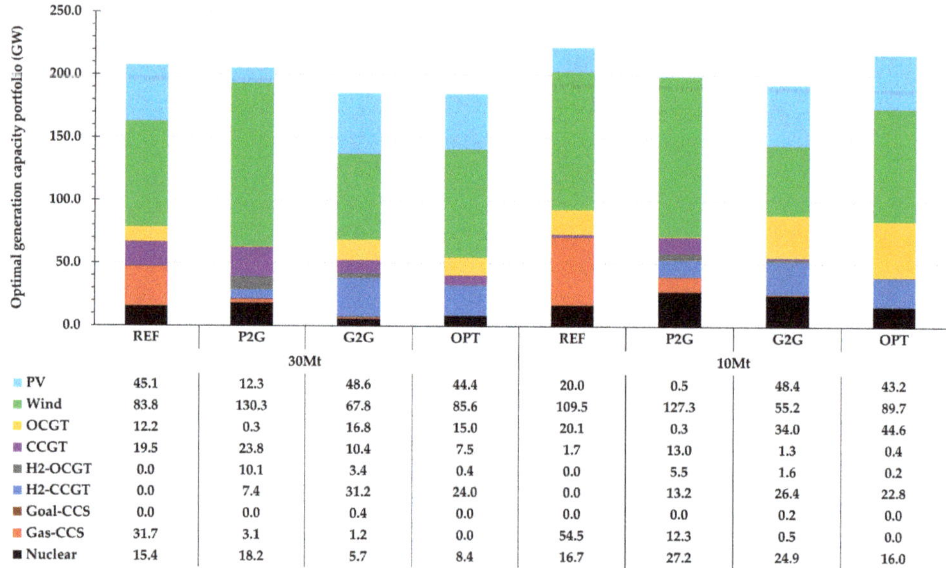

	REF	P2G	G2G	OPT	REF	P2G	G2G	OPT
		30Mt				10Mt		
PV	45.1	12.3	48.6	44.4	20.0	0.5	48.4	43.2
Wind	83.8	130.3	67.8	85.6	109.5	127.3	55.2	89.7
OCGT	12.2	0.3	16.8	15.0	20.1	0.3	34.0	44.6
CCGT	19.5	23.8	10.4	7.5	1.7	13.0	1.3	0.4
H2-OCGT	0.0	10.1	3.4	0.4	0.0	5.5	1.6	0.2
H2-CCGT	0.0	7.4	31.2	24.0	0.0	13.2	26.4	22.8
Goal-CCS	0.0	0.0	0.4	0.0	0.0	0.0	0.2	0.0
Gas-CCS	31.7	3.1	1.2	0.0	54.5	12.3	0.5	0.0
Nuclear	15.4	18.2	5.7	8.4	16.7	27.2	24.9	16.0

Figure 4. Optimal generation capacity portfolio in different scenarios.

Figure 5 shows the portfolio of annual electricity production in each scenario, where the annual wind power generation in the REF cases were 240 TWh and 278 TWh under 30 Mt and 10 Mt carbon targets, respectively. It can be observed that the annual wind power generation in the P2G cases increased to 435 TWh and 421 TWh, which were 67% and 61% of total generation under the 30 Mt and 10 Mt carbon targets, respectively. The increased system ability to integrate wind is driven by the use of the P2G facility, which allows the excess renewable energy to be stored cost-effectively via hydrogen storage, thus reducing the curtailment rate of wind and solar power. The difference in the annual generation mix is further reflected by the increased generation of nuclear power in the G2G cases compared with the P2G cases, where the nuclear power generation in the G2G cases is notably higher than in the case of P2G as well as its capacity. The main reason is that the G2G facility cannot help integrating more renewable energy, and it will give priority to nuclear power as low-carbon power generation. Meanwhile, the relatively high carbon emission of GHR increases the integration costs of a hydrogen system under a 10 Mt carbon target. Thus, the installed capacity of hydrogen-fuelled generation and its production decrease.

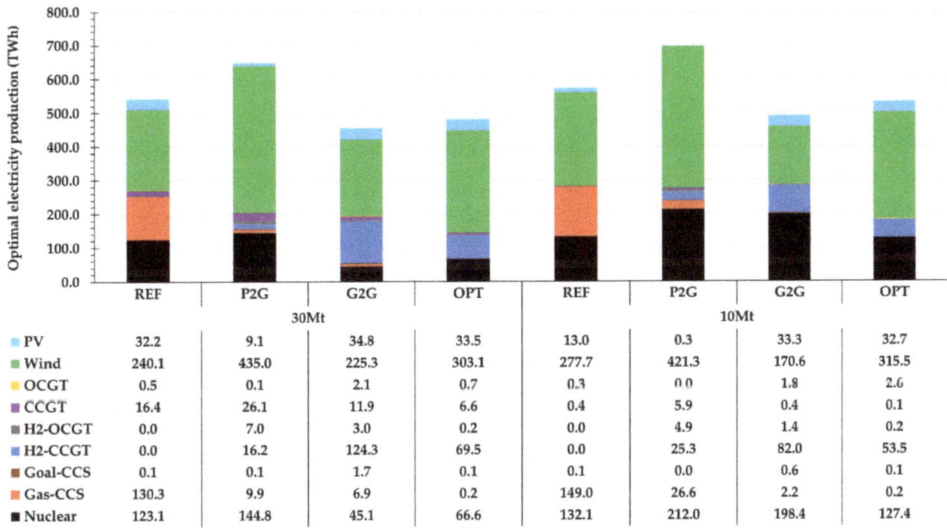

	REF	P2G	G2G	OPT	REF	P2G	G2G	OPT
			30Mt				10Mt	
PV	32.2	9.1	34.8	33.5	13.0	0.3	33.3	32.7
Wind	240.1	435.0	225.3	303.1	277.7	421.3	170.6	315.5
OCGT	0.5	0.1	2.1	0.7	0.3	0.0	1.8	2.6
CCGT	16.4	26.1	11.9	6.6	0.4	5.9	0.4	0.1
H2-OCGT	0.0	7.0	3.0	0.2	0.0	4.9	1.4	0.2
H2-CCGT	0.0	16.2	124.3	69.5	0.0	25.3	82.0	53.5
Goal-CCS	0.1	0.1	1.7	0.1	0.1	0.0	0.6	0.1
Gas-CCS	130.3	9.9	6.9	0.2	149.0	26.6	2.2	0.2
Nuclear	123.1	144.8	45.1	66.6	132.1	212.0	198.4	127.4

Figure 5. Optimal electricity production in different scenarios.

3.4. Impact of Hydrogen Integration on the Heating System

The mix of heating technology and annual heat production under different carbon targets are shown in Figure 5. The national DHN pathway only contributes a small part of heat demand in each scenario under both carbon targets due to the expenditure associated with the deployment of heat networks.

As shown in Figure 6, in a system with hydrogen, the heating pathway is shifted from end-use HP-NGB to end-use HP-HB, which drives less investment in the electricity sector as can be derived from Figure 4. It is worth noting that hybrid HP-HB can dominate the heating market, which is up to 73% in the OPT scenario when the carbon target becomes tighter (10 Mt).

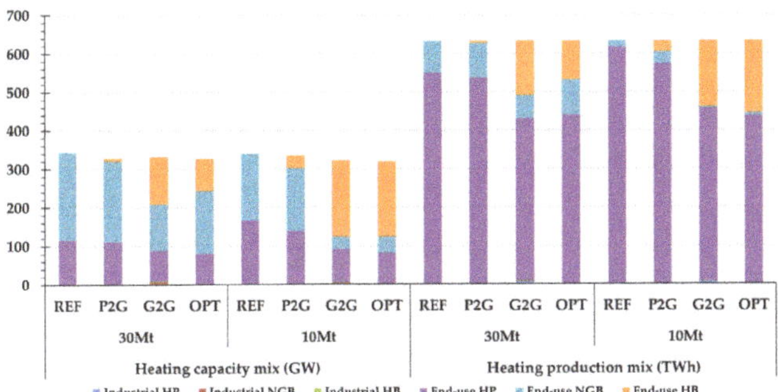

Figure 6. Heating technology capacity and production portfolio under different carbon targets.

Hydrogen integration also has a notable impact on the annual heat production mix. It can be observed that HP supplies the baseload while NGB only provides a little part of heat demand during the peak load due to its emissions and less flexibility when the hydrogen integration is not enabled, or its integration is not cost-effective (e.g., the P2G pathway). In the G2G case, the heat provided by HB increases to 23% and 27% under the 30 Mt and 10 Mt carbon targets, respectively. In the OPT scenario, the P2G process brings zero-emission hydrogen production, which offsets the carbon emissions from the G2G process and makes HB production increase further to 29% under a 10 Mt carbon target. Generally, the P2G process is necessary to offset the carbon emissions from the hydrogen system under a demanding carbon target.

3.5. Impact of Hydrogen Integration on the Transport Sector

In this model, assume the efficiency of HFCV is 40% lower than EV [26]. Figure 7 shows the proportion of each transport technology in different scenarios under different carbon targets. When the carbon target is 30 Mt, in the P2G scenario, EV still accounts for the most market. If hydrogen production shifts from P2G to G2G or a combined pathway, the cost of hydrogen system integration will be reduced so that the hydrogen will become more competitive in the transport sector, which will allow HFCV to dominate the transport market. However, when the carbon target is tightened to 10 Mt, EV takes back the domination position due to the hydrogen production process, which is less competitive. Only through least-cost hydrogen production (OPT), can the HFCV occupy 35% of the market share. In summary, from the whole-system point of view, the deployment of HFCV is sensitive to the costs of hydrogen and lower costs of hydrogen drive investment in HFCV. Emission is another factor affecting the deployment of HFCV, and the P2G process is necessary for the development of HFCV due to its zero-emission feature. However, the capital cost of HFCV is still not competitive compared to the current EVs [27].

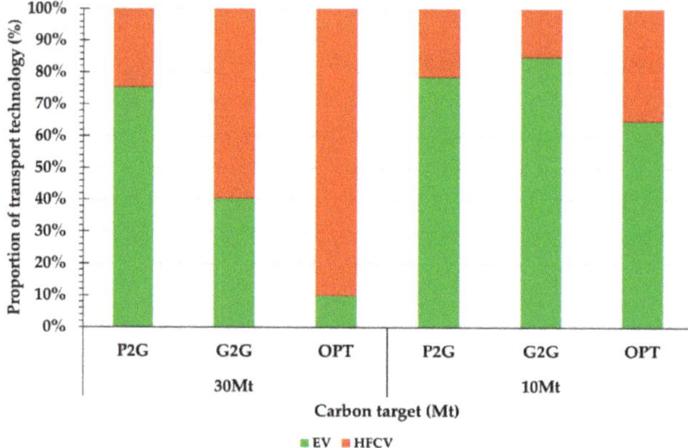

Figure 7. The proportion of each transport technology.

3.6. Impact of Hydrogen Integration on Carbon Emission

The integration of the hydrogen system will influence the decarbonisation strategies of other energy sectors. Figure 8 shows the total carbon emissions of each energy system in each scenario. The integration of the hydrogen system shifts the carbon emissions from the electricity system to other energy sectors in all the P2G, G2G, and OPT scenarios. Decarbonisation of the heat sector under a 30 Mt carbon target will require a higher integration of low-carbon heat supply technologies (HP and HB) and increased investment cost in the electricity and hydrogen sectors. When the carbon target is set strictly to 10 Mt, the high hydrogen integration cost of P2G also increases the cost of the decarbonisation in the electricity and heat sectors. When more economical hydrogen production methods (G2G and OPT) are adopted to produce hydrogen, the penetration of hydrogen in the electricity and heat sectors increases further, and the carbon emissions of the whole system mainly come from the hydrogen system due to a higher share of hydrogen-fuelled power generation, and HB replaces most of the NGB, as shown in

Figures 4 and 6, respectively. This case study indicates that hydrogen integration may shift the carbon emissions from the electricity and heat sector to the hydrogen system since it is more cost-effective to decarbonise the energy through hydrogen. It is worth noting that the carbon emissions of the heat sector are far lower than the other energy sectors in the G2G and OPT scenario with the 10 Mt carbon target due to the massive deployment of HB, indicating that the hydrogen integration will play an important role in the cost-effective transition towards a zero-carbon future energy system.

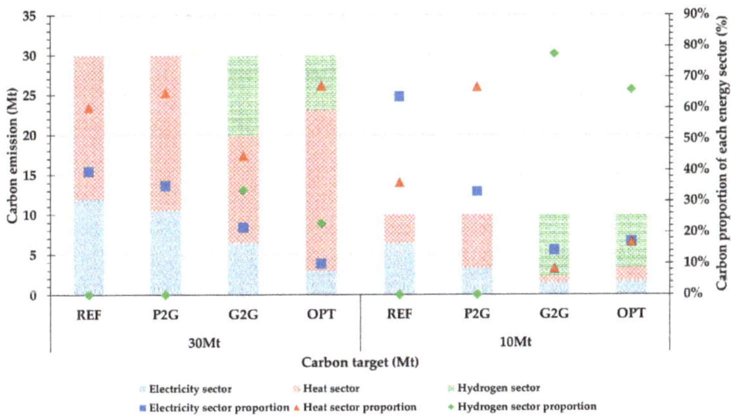

Figure 8. Carbon emission mix under different carbon targets.

3.7. Hydrogen Production Technologies Deployment

The deployment of an electrolyser requires more electricity system investment, which increases the 34.4 GW and 52.1 GW peak demand under 30 Mt and 10 Mt carbon targets in the P2G scenario, respectively. From Table 4, it can be observed that GHR-CCS dominated the hydrogen production technology due to the economic advantages of GHR as mentioned in the above case. However, the capacity of the electrolyser increases when the carbon target becomes stricter because of its zero-carbon emissions feature. In terms of the annual hydrogen production, when the carbon target changes from 30 Mt to 10 MT, the annual hydrogen production of the electrolyser increases in all scenarios. In contrast, the annual hydrogen production of GHR-CCS decreases in the G2G scenario and the share of GHR-CCS decreases in the OPT scenario, mainly due to the need to meet a stricter carbon target.

Table 4. Capacity and the annual output of different hydrogen production technologies.

Scenarios		Capacity (GW)		Proportion (%)		Production (TWh)		Proportion (%)	
		EL	GHR	EL	GHR	EL	GHR	EL	GHR
30 Mt	**P2G**	19.3	0	100%	0%	93.4	0	100%	0%
	G2G	0	79.7	0%	100%	0	461.4	0%	100%
	OPT	9.1	60.1	13.1%	86.9%	25.5	313.5	7.5%	92.5%
10 Mt	**P2G**	27.2	0	100%	0%	128.3	0	100%	0%
	G2G	0	103.0	0%	100%	0	353.5	0%	100%
	OPT	11.4	95.5	10.7%	89.3%	39.6	301.3	11.6%	88.4%

3.8. The Relation between the P2G Facility and Wind Power

In this system, aside from P2G being able to absorb the excess wind power to integrate more wind power into the system, the P2G facility can also offer a flexible load, thus providing ancillary services like frequency response through the interrupted operation, which also increases the wind power integration potentially. So, as shown in Figure 9, P2G can promote the integration of wind power, and G2G plays the opposite role.

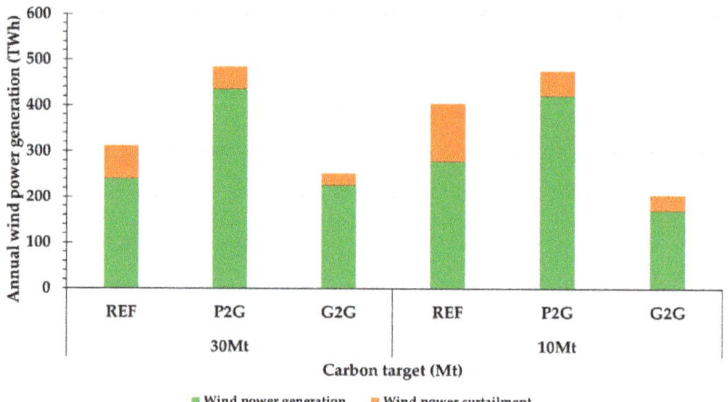

Figure 9. The wind power capacity and its load factor under different carbon targets.

3.9. Sensitivity Studies

This section conducts sensitivity studies investigating the drivers for different hydrogen production technologies. The impact of two important drivers (i.e., the capital cost of wind power and natural gas price on the hydrogen production mixes) are investigated below.

3.9.1. Sensitivity Analysis of Wind Power Capital Cost

Figure 10 illustrates the wind power capacity and hydrogen production technology capacity mix in a series of wind power capital cost scenarios under the carbon targets of 30 Mt and 10 Mt. It can be observed that the competitiveness of the P2G facility is highly sensitive to the variation of wind power capital cost, while the penetration of the P2G facility is much more robust under the stricter carbon targets. In terms of G2G capacity, with the increase in capital costs of wind power, less wind power will be installed; consequently, the G2G capacity will need to increase to achieve the overall carbon target.

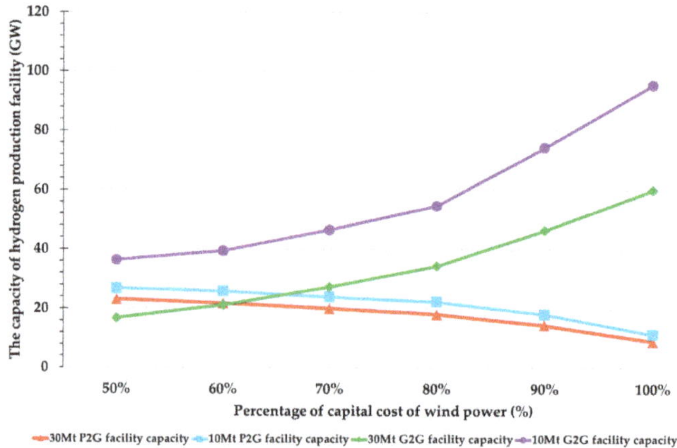

Figure 10. The sensitivity study on the capital cost of wind power.

3.9.2. Sensitivity Analysis of Natural Gas Price

As mentioned before, the cost of large-scale hydrogen integration is dominated by the operation cost of the hydrogen system, which is highly sensitive to the natural gas price. The above case studies are based on the natural gas price of 67 p/therm. The G2G process has a dominating role in the integration of hydrogen in the OPT scenario. As shown in Figure 11, when the natural gas price drops by 50% to 33.5 p/therm, the overall integration cost of the G2G facilities will be reduced due to the decrease of the natural gas price, which will weaken the integration of P2G facilities. On the contrary, if the price of natural gas increases by 50% to 100.5 p/therm, the P2G's zero-emission advantage will enable it to integrate more capacity than the G2G process under the 30 Mt carbon target. However, the capacity of G2G is still higher than that of P2G due to the higher demand for hydrogen integration under the 10 Mt carbon target. The G2G still has an economic advantage compared to P2G. In terms of hydrogen production, the rise of natural gas prices has significantly reduced the production of G2G, in contrast, the production of P2G is slowly rising, and exceeds that of G2G under the 30 Mt carbon target when the natural gas price is 100 p/therm.

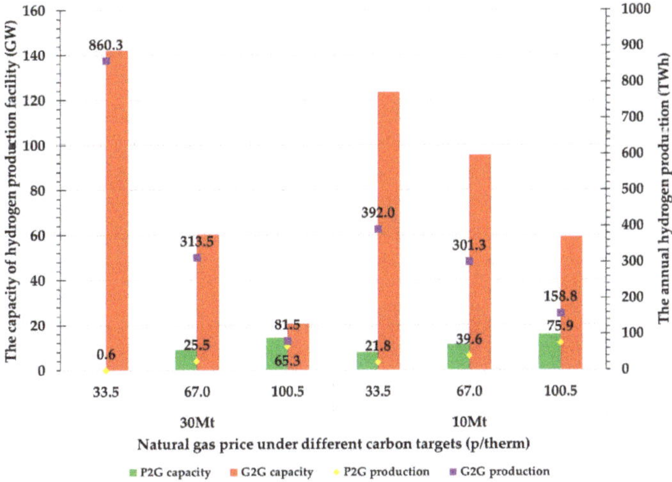

Figure 11. The sensitivity study on the natural gas price.

4. Conclusions

This paper proposes an integrated electricity, heat, transport, and hydrogen energy systems model to investigate the impact of different hydrogen production technologies on multi-energy systems. The model was applied to optimise the decarbonisation strategies of the whole energy system while assessing the values of hydrogen integration. The studies demonstrate that hydrogen integration through the G2G process brings more economic benefits when compared to the P2G process, which can deliver £3.9 bn/year and £14.2 bn/year cost savings under the 30 Mt and 10 Mt carbon targets, respectively. The OPT pathway can offset the carbon emissions from the G2G process and achieve further cost savings. The results also clearly demonstrate the changes in the electricity side driven by the different hydrogen integration strategies. The G2G process can reduce the total power generation capacity requirement from 71 GW to 53 GW, and the P2G can increase the integration of wind power capacity from 83 GW to 130 GW under the 30 Mt carbon target. The integration of hydrogen will promote the deployment of HB, which, combined with HP, will dominate the heating market, which is up to 73% in the OPT scenario under the 10 Mt carbon target. From the perspective of the transport sector, the development of HFCV is highly related to the integration cost of the hydrogen system, especially in a demanding carbon scenario. The HFCVs can occupy 90% market share in the OPT scenario under a 30 Mt carbon target, however, when the carbon target becomes tighter (10 Mt), the integration cost of the hydrogen system increases, and the market share of HFCV will decrease to 35% in the OPT scenario. Finally, the series of sensitivity studies indicate that the integration of the P2G facility is highly sensitive to the wind power capital cost, and the G2G facility is highly sensitive to the natural gas price. The uncertainty in various costs will need to be taken into consideration when deciding the hydrogen production technologies.

It is worth mentioning that we only considered the direct carbon emissions of all the technologies during their energy production in this study. The indirect emissions generated during the equipment construction or energy transmission were ignored. The carbon emissions along the life cycle of the energy supply were investigated in [28,29]. Our future work will address this limitation, making carbon emission analysis more integrated.

Author Contributions: P.F. and D.P. developed the models used in the paper and conducted quantitative case studies and interpreted the results. X.Z. provided the technical data regarding the heat network and other parameters in the heating system and also gave valuable suggestions on the design of the case studies and analysis of the results. G.S. provided research guidance and led the drafting of the main conclusions. All authors have read and agreed to the published version of the manuscript.

Funding: The authors are grateful for the valuable support and funding received from the UK EPSRC IDLES project (reference: EP/R045518/1) and from the UK EPSRC The Active Building Centre (reference: EP/S016627/1). All contents and views expressed in this paper are the sole responsibility of the authors and do not necessarily express the views of the project consortia.

Conflicts of Interest: The authors declare no conflicts of interest.

Abbreviations

G2G	Gas-to-gas
CCS	Carbon capture and storage
P2G	Power-to-gas
GB	Great Britain
RES	Renewable energy sources
CHP	Combined heat and power
MILP	Mixed-integer linear programming
CAPEX	Capital expenditures
OPEX	Operative expenditures
GHR-CCS	Gas-heated reformers combined with carbon capture storage (GHR-CCS)
EL	Electrolyser
H2S	Hydrogen storage
EV	Electric vehicle
HFCV	Hydrogen fuel-cell vehicle
PV	Photovoltaics
HV	High voltage
LV	Low voltage
DHN	District heating network
IGB	Industrial natural gas boiler
IHP	Industrial heat pump
IHB	Industrial hydrogen boiler
HP	End-use heat pump
NGB	End-use natural gas boiler
HB	End-use hydrogen boiler
HP-B	Hybrid end-use heat pump and natural gas boiler
HP-HB	Hybrid end-use heat pump and hydrogen boiler
HB	End-use hydrogen boiler
COP	Coefficient of performance

Notation

Sets

I	Set of locations
R	Set of regions
T	Set of operating time intervals
D	Set of operating days
G	Set of conventional generators
PV	Set of PV generation units
HV	High voltage distribution network
LV	Low voltage distribution network
L	Set of transmission/interconnection corridors
DHN	Set of district heating network

Functions

$Z(\cdot)$	Generation operating cost function
$F(\cdot)$	Power flow function
$D(\cdot)$	District heating network cost function

Parameters

Δt	Time interval (h)
π_g	Generation investment cost (£/GW/year)
π_f	Transmission network cost (£/GW/year)
π_{hv}	Electricity high-voltage distribution network cost (£/GW/year)
π_{lv}	Electricity low-voltage distribution network cost (£/GW/year)
π_{hp}	Industrial heat pump cost (£/GW/year)
π_{ehp}	End-use heat pump cost (£/GW/year)
π_{ngb}	Industrial natural gas boiler cost (£/GW/year)
π_{engb}	End-use natural gas boiler cost (£/GW/year)
π_{hb}	Industrial hydrogen boiler cost (£/GW/year)
π_{ehb}	End-use hydrogen boiler cost (£/GW/year)
π_{el}	Electrolyser investment cost (£/GW/year)
π_{smr}	GHR-CCS investment cost (£/GW/year)
π_{hs}	Hydrogen storage investment cost (£/GW/year)
π_{ht}	Hydrogen pipeline investment cost (£/GW/km/year)
π_{nl}	Generation no-load cost (£/h)
π_{st}	Generation start-up cost (£/start)
γ	The operation cost of each system (£/GWh)
DE	Electricity demand (GW)
DH	Heat demand (GW)
DT	Transport demand (GW)
$\overline{f}$	Existing electricity transmission capacity (GW)
$\overline{hvs}$	Existing high-voltage distribution network capacity (GW)
$\overline{lvs}$	Existing low-voltage distribution network capacity (GW)
$\overline{P}$	The power rating of a generation unit (GW)
$\underline{P}$	Minimum stable generation (GW)
R^{up}	Ramping up limit (GW/h)
R^{down}	Ramping down limit (GW/h)
$\overline{rsp}$	Frequency response limit (GW)
$\overline{res}$	Spinning reserve limit (GW)
$\overline{SF}$	System frequency response requirement (GW)
$\overline{SR}$	System operation reserve requirement (GW)
η	Energy conversion efficiency (%)
ε	The ratio of flexible transport demand (%)
κ	The ratio of electricity demand at high-voltage distribution level (%)
σ	Reverse power flow coefficient (%)

SHST	Hydrogen storage duration (h)
CT	Carbon target (tCO_2/year)
C	Direct carbon emission of each technology (tCO2/GWh)

Variables

n	The additional capacity of technologies (GW)
f	Additional transmission network capacity (GW)
d^-	Reduction in transport load due to DSR (GW)
d^+	Increased transport load due to DSR (GW)
P	Electricity generation (GW)
st	Number of generating units being synchronized
dst	Number of generating units being de-synchronized
μ	Number of units in operation
H	Heating production (GW)
Q	Hydrogen production (GW)
V	Transportation supply (GW)
θ	Voltage angle
λ	Penetration of production technologies (%)
rsp	Frequency response (GW)
res	Spinning reserve (GW)
SH^+	Hydrogen production by storage (GW)
SH^-	Hydrogen consumed by storage (GW)
SHS	The energy content of hydrogen storage (GWh)

References

1. Sanders, D.; Hart, A.; Ravishankar, M.; Brunert, J.; Strbac, G.; Aunedi, M. *An Analysis of Electricity System Flexibility for Great Britain*; Carbon Trust/Imperial College: London, UK, 2016.
2. Strbac, G.; Pudjianto, D.; Sansom, R.; Djapic, P.; Ameli, H.; Shah, N.; Brandon, N.; Hawkes, A.; Qadrdan, M. *Analysis of Alternative UK Heat Decarbonisation Pathways*; Imperial College London: London, UK, 2018.
3. Mancarella, P. MES (multi-energy systems): An overview of concepts and evaluation models. *Energy* **2014**, *65*, 1–17. [CrossRef]
4. Pudjianto, D.; Aunedi, M.; Djapic, P.; Strbac, G. Whole-systems assessment of the value of energy storage in low-carbon electricity systems. *IEEE Trans. Smart Grid* **2013**, *5*, 1098–1109. [CrossRef]
5. Zhang, X.; Strbac, G.; Teng, F.; Djapic, P. Economic assessment of alternative heat decarbonisation strategies through coordinated operation with electricity system—UK case study. *Appl. Energy* **2018**, *222*, 79–91. [CrossRef]
6. Zhang, X.; Strbac, G.; Shah, N.; Teng, F.; Pudjianto, D. Whole-system assessment of the benefits of integrated electricity and heat system. *IEEE Trans. Smart Grid* **2018**, *10*, 1132–1145. [CrossRef]
7. Chaudry, M.; Jenkins, N.; Strbac, G. Multi-time period combined gas and electricity network optimisation. *Electr. Power Syst. Res.* **2008**, *78*, 1265–1279. [CrossRef]
8. Damen, K.; van Troost, M.; Faaij, A.; Turkenburg, W. A comparison of electricity and hydrogen production systems with CO2 capture and storage—Part B: Chain analysis of promising CCS options. *Prog. Energy Combust. Sci.* **2007**, *33*, 580–609. [CrossRef]
9. Almansoori, A.; Shah, N. Design and operation of a future hydrogen supply chain: Snapshot model. *Chem. Eng. Res. Des.* **2006**, *84*, 423–438. [CrossRef]
10. Samsatli, S.; Staffell, I.; Samsatli, N.J. Optimal design and operation of integrated wind-hydrogen-electricity networks for decarbonising the domestic transport sector in Great Britain. *Int. J. Hydrog. Energy* **2016**, *41*, 447–475. [CrossRef]
11. Samsatli, S.; Samsatli, N.J. The role of renewable hydrogen and inter-seasonal storage in decarbonising heat–Comprehensive optimisation of future renewable energy value chains. *Appl. Energy* **2019**, *233*, 854–893. [CrossRef]
12. Li, G.; Zhang, R.; Jiang, T.; Chen, H.; Bai, L.; Li, X. Security-constrained bi-level economic dispatch model for integrated natural gas and electricity systems considering wind power and power-to-gas process. *Appl. Energy* **2017**, *194*, 696–704. [CrossRef]

13. Clegg, S.; Mancarella, P. Integrated modeling and assessment of the operational impact of power-to-gas (P2G) on electrical and gas transmission networks. *IEEE Trans. Sustain. Energy* **2015**, *6*, 1234–1244. [CrossRef]

14. Qadrdan, M.; Ameli, H.; Strbac, G.; Jenkins, N. Efficacy of options to address balancing challenges: Integrated gas and electricity perspectives. *Appl. Energy* **2017**, *190*, 181–190. [CrossRef]

15. Li, J.; Lin, J.; Song, Y.; Xing, X.; Fu, C. Operation optimization of power to hydrogen and heat (P2HH) in ADN coordinated with the district heating network. *IEEE Trans. Sustain. Energy* **2019**, *10*, 1672–1683. [CrossRef]

16. Fu, P.; Pudjianto, D.; Zhang, X.; Strbac, G. Evaluating strategies for decarbonising the transport sector in Great Britain. In Proceedings of the IEEE Milan PowerTech, Milan, Italy, 23–27 June 2019; pp. 1–6.

17. Teng, F.; Strbac, G. Full stochastic scheduling for low-carbon electricity systems. *IEEE Trans. Autom. Sci. Eng.* **2017**, *14*, 461–470. [CrossRef]

18. Brandon, N.; Kurban, Z. Clean energy and the hydrogen economy. *Philos. Trans. R. Soc. A Math. Phys. Eng. Sci.* **2017**, *375*, 20160400. [CrossRef] [PubMed]

19. Speirs, J.; Balcombe, P.; Johnson, E.; Martin, J.; Brandon, N.; Hawkes, A. A greener gas grid: What are the options. *Energy Policy* **2018**, *118*, 291–297. [CrossRef]

20. Heap, R. *Potential Role of Hydrogen in the UK Energy System*; Energy Research Partnership: Birmingham, UK, 2016.

21. Villasana, R.; Garver, L.; Salon, S. Transmission network planning using linear programming. *IEEE Trans. Power App. Syst.* **1985**, *PAS-104*, 349–356. [CrossRef]

22. FICO Xpress Optimization. Available online: https://www.fico.com/en/products/fico-xpress-optimization (accessed on 28 March 2020).

23. BEIS. *Electricity Generation Costs*; The Stationery Office: London, UK, 2016.

24. Element Energy. *Hy-Impact Series Study 3: Hydrogen for Power Generation Opportunities for hydrogen and CCS in the UK Power Mix*; Element Energy: Cambridge, UK, 2019.

25. Walker, I.; Madden, B.; Tahir, F. *Hydrogen Supply Chain Evidence Base*; Element Energy Ltd.: Cambridge, UK, 2018.

26. Efficiency Compared: Battery-Electric 73%, Hydrogen 22%, ICE 13%. Available online: https://insideevs.com/news/332584/efficiency-compared-battery-electric-73-hydrogen-22-ice-13/ (accessed on 24 October 2019).

27. Offer, G.; Howey, D.; Contestabile, M.; Clague, R.; Brandon, N. Comparative analysis of battery electric, hydrogen fuel cell and hybrid vehicles in a future sustainable road transport system. *Energy Policy* **2010**, *38*, 24–29. [CrossRef]

28. Chen, G.; Chen, B.; Zhou, H.; Dai, P. Life cycle carbon emission flow analysis for electricity supply system: A case study of China. *Energy Policy* **2013**, *61*, 1276–1284. [CrossRef]

29. Kannan, R.; Leong, K.; Osman, R.; Ho, H. Life cycle energy, emissions and cost inventory of power generation technologies in Singapore. *Renew. Sustain. Energy Rev.* **2007**, *11*, 702–715. [CrossRef]

Article

Medium- and Long-Term Integrated Demand Response of Integrated Energy System Based on System Dynamics

Shuhui Ren [1], Xun Dou [1,*], Zhen Wang [1], Jun Wang [1] and Xiangyan Wang [2]

[1]   College of Electrical Engineering and Control Science, Nanjing TECH University, Nanjing 211816, China; rsh360167160@163.com (S.R.); wzwqlx@gmail.com (Z.W.); wjnjut@163.com (J.W.)
[2]   China Electric Power Research Institute, Nanjing 210003, China; wangxiangyan@epri.sgcc.com.cn
*   Correspondence: dxnjut@njtech.edu.cn; Tel.: +86-139-1471-9418

Received: 5 January 2020; Accepted: 22 January 2020; Published: 6 February 2020

Abstract: For the integrated energy system of coupling electrical, cool and heat energy and gas and other forms of energy, the medium- and long-term integrated demand response of flexible load, energy storage and electric vehicles and other demand side resources is studied. It is helpful to mine the potentials of demand response of various energy sources in the medium- and long-term, stimulate the flexibility of integrated energy system, and improve the efficiency of energy utilization. Firstly, based on system dynamics, the response mode of demand response resources is analyzed from different time dimensions, and the long-term, medium-term and short-term behaviors of users participating in integrated demand response are considered comprehensively. An integrated demand response model based on medium-and long-term time dimension is established. Then the integrated demand response model of integrated energy system scheduling and flexible load, energy storage and electric vehicles as the main participants is established to simulate the response income of users participating in the integrated demand response project, and to provide data sources for the medium- and long-term integrated demand response system dynamics model. Finally, an example is given to analyze the differences in response behaviors of flexible load, energy storage and electric vehicle users in different time dimensions under the conditions of policy subsidy, regional location and user energy preferences in different stages of the integrated energy system.

Keywords: integrated energy system; integrated demand response; medium- and long-term; system dynamics; user decision

---

1. Introduction

In recent years, in order to improve energy use efficiency and deal with problems such as the deterioration of the ecological environment, an integrated energy system [1–3] for the energy interconnected network has been established, and the mutual conversion between different energy sources [4,5] has become a hot topic for research in various countries around the world [6–8]. Integrated demand response (IDR) provides an important entry direction for the two-way interaction between supply and demand sides in an integrated energy system. At the end of consumption, users can achieve the same effect by selecting different types of energy. The user's short-term, medium-term, and even long-term energy use has been broadened, and the impact of integrated demand response on the medium- and long-term load curve and integrated energy system planning is increasing. Therefore, it is of great significance to study the integrated demand response for integrated energy systems.

For integrated demand response, existing studies can consider the response type, participation equipment, and participation method [9–14] of the IDR from the system operation level. Based on the reduction of load, transfer of load, and alternative load [15], analyze the cost of the IDR resource to the overall system operation or the impact of integrated energy utilization efficiency; at the system planning level, there have been studies that can comprehensively consider the peak-cutting and valley-filling effect of the coordinated operation of combined heat and power units and electricity to gas, but focus more on the impact of IDR on equipment capacity and typical loads. Most are still from the perspective of the power system, and analyze the impact of other forms of energy supply on the power load [10,16]. At present, most of them are based on the typical load day or random scenarios to establish the IDR model and analyze its impact on planning and operation. It is a short-term integrated demand response, and there is not much research on the medium- and long-term integrated demand response.

Generally speaking, the user's long-term investment in integrated energy-consuming equipment behavior will affect the maximum potential and flexibility of its IDR, and this maximum potential will affect the user's recognition and participation in the IDR, which ultimately determines the user of an IDR event. The actual participation effect [17,18], system dynamics can integrate the analysis of the long-term potential, medium-term potential, and short-term response of demand response resources into the same model [19], but currently analyzes the factors affecting integrated demand response resources from different time dimensions. Research is not integrated, and it is not integrated in terms of the coupling between long-, medium-, and short-term user behavior and the time-varying nature of demand response resources.

In response to the above problems, based on the existing research, based on system dynamics, this paper analyzes the response methods of demand response resources in different time dimensions, comprehensively considers the user's long-term investment, updates and behaviors of integrated energy equipment, whether to sign integrated demand response contracts in the medium- term. A dynamic model of integrated demand response system based on the medium and long-term time dimension is established according to system dynamics. In the short-term time scale, an integrated demand response model of integrated energy system scheduling and flexible loads, energy storage, and electric vehicles were established as participants. Based on the long-term incentives of policy subsidy, the example analyzes the policy subsidy at different stages of the integrated energy system, regional location, and user energy preferences; flexible load, energy storage, and response behavior of electric vehicle users in different time dimensions.

2. Multi-Energy Load-Oriented Integrated Demand Response Model for Integrated Energy Systems

2.1. Modeling Principles

The medium- and long-term integrated demand response model for integrated energy systems includes three parts: a long-term integrated demand response decision model, a medium-term integrated demand response decision model, and a short-term integrated demand response decision model. The modeling principle is shown in Figure 1.

In the long-term integrated demand response decision model, users compare the utility value that can be provided during the life cycle of integrated energy equipment with their own expectations, decide whether to invest in integrated energy equipment, and invest in integrated energy equipment can enhance users' long-term potential. That is, the maximum capacity that users can respond to. In the medium-term integrated demand response decision-making model, users decide whether to sign an integrated demand response contract based on the integrated demand response revenue accumulated during the validity period of the contract, and users who sign an integrated demand response contract have medium-term potential, and only users with medium-term integrated demand response potential can make short-term response decisions. In the short-term integrated demand response decision model, a simulated response is run based on the integrated energy system scheduling and integrated demand response to determine the response capacity and response revenue, and whether the decision is to respond.

For a user, if an integrated demand response contract is not signed, whether it responds and the response capacity is not considered. If the user signs an integrated demand response contract, the response decision is made. If the user has also invested in an integrated energy equipment, contact the compared with other users who have not made long-term decisions, the long-term integrated demand response potential and medium-term integrated demand response potential are higher, and this user has more respond capacity.

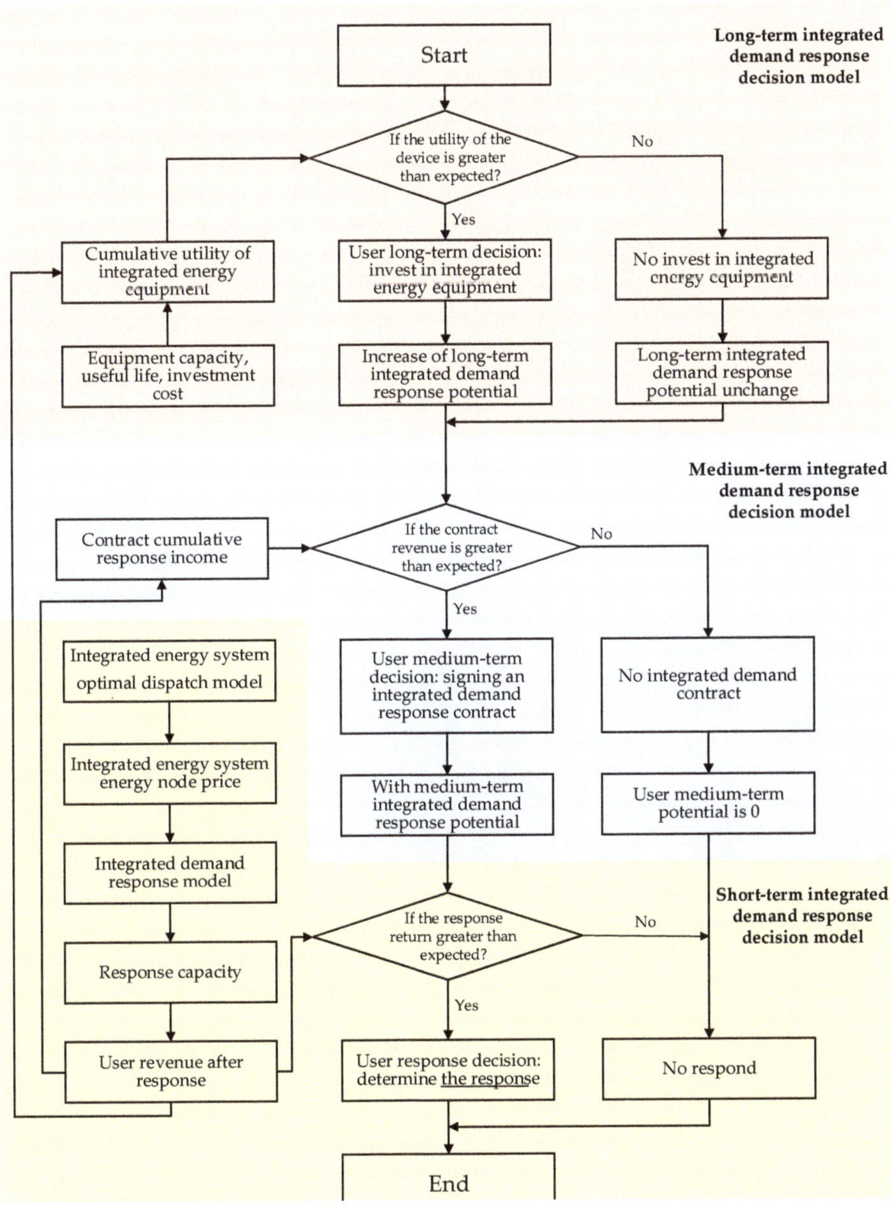

Figure 1. Medium- and long-term integrated demand response modeling process for integrated energy systems.

2.2. Dynamic Model of Integrated Demand Response System Based on Medium- and Long-Term Time Dimension

A system dynamics model that can effectively characterize different time dimensions and step sizes is used to study the integrated demand response behavior at different time scales. The causal circuit diagram is shown in Figure 2. Where, "→" represents the main circuit, which reflects the long-term integrated user response. The interrelationships among decision-making, medium-term decision-making and short-term decision-making, "+" indicates positive correlation between variables, and "-" indicates negative correlation between variables.

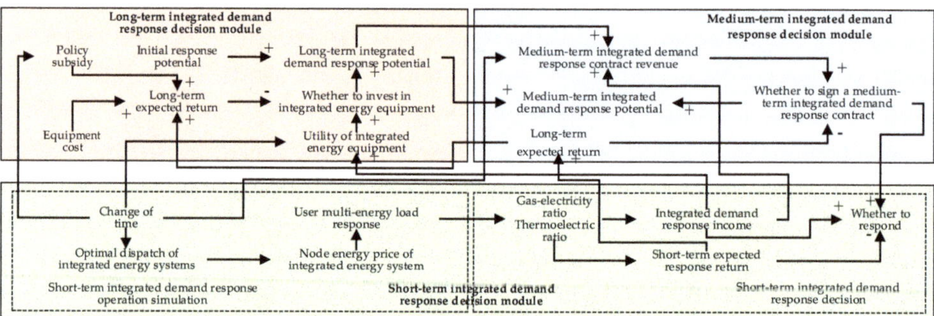

Figure 2. Causal loop diagram.

The dynamic cause-effect circuit diagram of the integrated demand response system based on the medium- and long-term time dimension is divided into three major modules: long-term integrated demand response decision module, medium-term integrated demand response decision module, and short-term integrated demand response decision module. The short-term integrated demand response decision module is divided into Run simulation and short-term integrated demand response decision-making for short-term integrated demand response. The user's gas-to-electricity ratio and heat-to-electricity ratio change as the user's load increases. The integrated demand response income is given by the integrated demand response model based on electricity, gas, and heat loads. Demand response contract revenue and integrated energy equipment performance are provided by the integrated demand response operation simulation based on the cumulative demand response contract time and equipment usage time cumulatively. The user's expected response revenue, medium-term expected revenue, and long-term expected revenue are affected by the user's own characteristics. Users with different electricity, gas, and heat load ratios expect different response returns, and the expected response returns will affect the medium-term expected returns, which in turn affects long-term expected returns with equipment costs and policy subsidy in different periods.

2.2.1. Long-Term Integrated Demand Response Decision Model

The user's long-term integrated demand response decision, considers whether to replace high-efficiency integrated energy equipment, increase inventory or modify production lines, and other technological transformation projects that increase the potential of integrated demand response. The user's long-term decision model considers the capacity of integrated energy equipment, equipment costs, and policy subsidy. Medium-term expected returns, long-term expected returns and other variables, among which policy subsidy as long-term incentives will have a greater impact on users' decisions. Users comprehensively consider the effectiveness of integrated energy-use equipment and long-term expected returns, and make the long-term decision-making that whether to invest in integrated energy-use equipment. The specific stock flow diagram is shown in Figure 3, and the mathematical model is shown in formula.

$$S = \begin{cases} 1, U_{ie} > R_{le} \\ 0, other \end{cases} \tag{1}$$

where S represents whether to invest in integrated energy equipment. U_{ie} represents integrated energy equipment utility. R_{le} represents long-term expected return.

$$U_{ie} = I_{el} \tag{2}$$

where I_{el} represents integration of integrated demand response income with equipment life.

$$R_{le} = R_{me} \times (1 + R_{lu}) + P_s + C_e \tag{3}$$

where R_{me} represents medium-term expected return. R_{lu} represents long-term user expectation improvement rate. P_s represents policy subsidy. C_e represents equipment cost.

$$P_{lr} = P_{ir} \times (1 + R_{ai} \times S) \tag{4}$$

where P_{lr} represents long-term integrated demand response potential. P_{ir} represents initial response potential. R_{ai} represents response improvement rate after investing in integrated energy equipment.

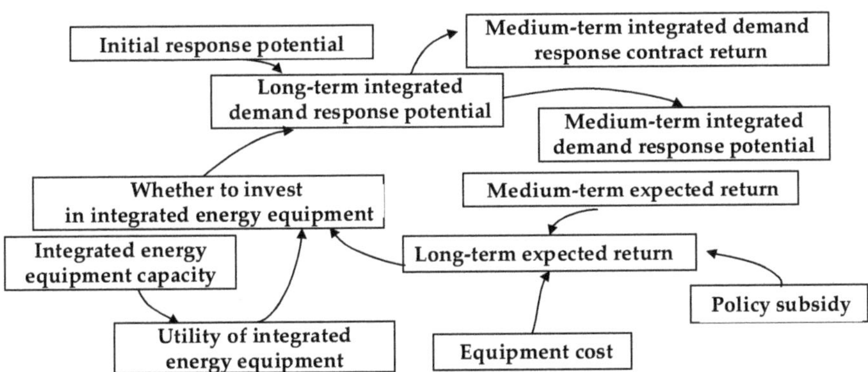

Figure 3. Long-term integrated demand response decision module stock flow chart.

2.2.2. Medium-Term Integrated Demand Response Decision Model

The user's medium-term decision considers whether to sign a medium-term integrated demand response contract. The user's medium-term decision model considers variables such as contract duration, long-term integrated demand response potential, integrated demand response income, and expected response income, among which the long-term integrated demand response potential determines the medium-term integrated demand In response to the maximum capacity of the contract, the user comprehensively considers the medium-term integrated demand response contract revenue and medium-term expected revenue to make a medium-term decision on whether to sign an integrated demand response. The specific inventory flow chart is shown in Figure 4, and the mathematical model is shown in formula.

$$C_{mi} = \begin{cases} 1, R_{mc} > R_{me} \\ 0, other \end{cases} \tag{5}$$

where, C_{mi} represents whether to sign a medium-term integrated demand response contract. R_{mc} represents medium-term integrated demand response contract revenue.

$$R_{mc} = P_{ic} + P_{lr} \times R_c \tag{6}$$

where, P_{ic} represents points for integrated demand response income over the duration of the contract. R_c represents revenue coefficient.

$$R_{me} = E_r \times (1 + R_{mu}) \tag{7}$$

where, E_r represents expected response revenue. R_{mu} represents long-term user expectation improvement rate.

$$P_{mr} = \begin{cases} P_{lr}, C_{mu} \\ 0, other \end{cases} \tag{8}$$

where, P_{mr} represents medium-term integrated demand response potential. C_{mu} represents user sign medium-term integrated demand response contracts.

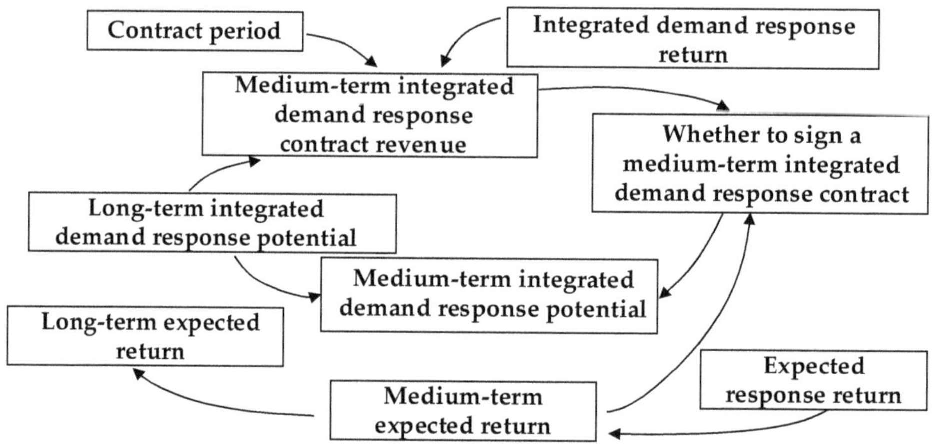

Figure 4. Medium-term integrated demand response decision module stock flow chart.

2.2.3. Short-Term Integrated Demand Response Decision Model

The user's integrated demand response decision model considers whether to participate in responding to IDR events on a daily scheduling time scale.

$$R = \begin{cases} 1, C_{mu} \&\& R_{id} > E_r \\ 0, other \end{cases} \tag{9}$$

where, R represents whether to respond. R_{id} represents integrated demand response revenue.

$$R_{si} = u(C_r, P_e), V_{sr} \tag{10}$$

where, R_{si} represents short-term integrated demand response revenue. C_r represents response capacity. P_e represents energy price. V_{sr} represents short-term integrated demand response simulation value.

$$R_{se} = u(C_{er}, P_{ex}, U_p) \tag{11}$$

where, R_{se} represents short-term expected response revenue. C_{er} represents expected response capacity P_{ex} represents expected energy price. U_p represents user preference.

The short-term integrated demand response decision model considers whether to sign a medium-term integrated demand response contract, the expected energy price, the user's own preferences, the expected response capacity, the gas-electricity ratio and the thermoelectric ratio of the self-load, response capacity, energy prices and other variables. Among them, different users have different sensitivity to gas-electricity ratio and thermoelectricity ratio and users consider short-term integrated demand response income and expected response income to make a short-term decision on whether to respond. The specific inventory flow chart is shown in Figure 5, and the mathematical model is shown in formula.

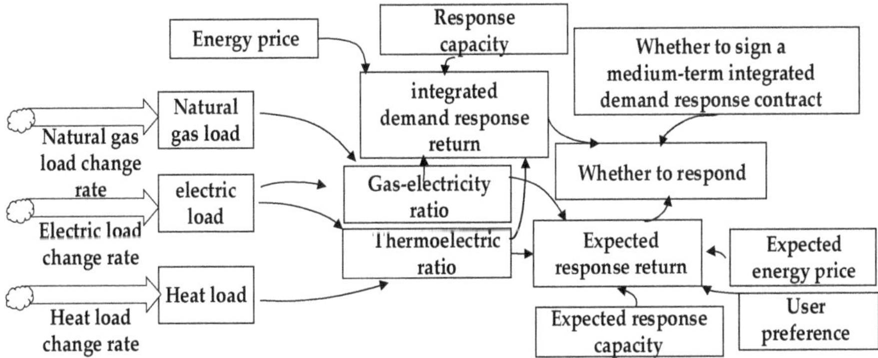

Figure 5. Short-term integrated demand response simulation module stock flow chart.

3. Short-Term Integrated Demand Response Operation Simulation

The short-term integrated demand response operation simulation in the short-term integrated demand response decision model is a data source for the dynamic model of the integrated demand response system based on the medium- and long-term time dimension.

Based on the regional electric-pneumatic interconnected integrated energy system network, the coupling nodes connect the park-level heat network through combined cooling, heat and power (CCHP) to build a park-level cool-heat-electric-gas integrated energy system. Regional-level electricity-gas interconnection network is used as the input/output of CCHP in the park-level integrated energy system to provide a network architecture for connecting the park-level integrated energy system. Establish an integrated energy system and establish a scheduling model to obtain node energy prices and then use the energy network Node multi-energy load users are targeted, considering gas-electricity/thermal-electricity replaceable loads, establishing an integrated demand response model, simulating the short-term response capacity and short-term response benefits of users participating in integrated demand response, and passing them to users for response decisions. The simulated short-term integrated demand response income of the user is integrated according to the integrated demand response contract cycle and the integrated energy equipment usage cycle, and is used as an indicator for the user's medium-term and long-term decisions. The linkage principle is shown in Figure 6.

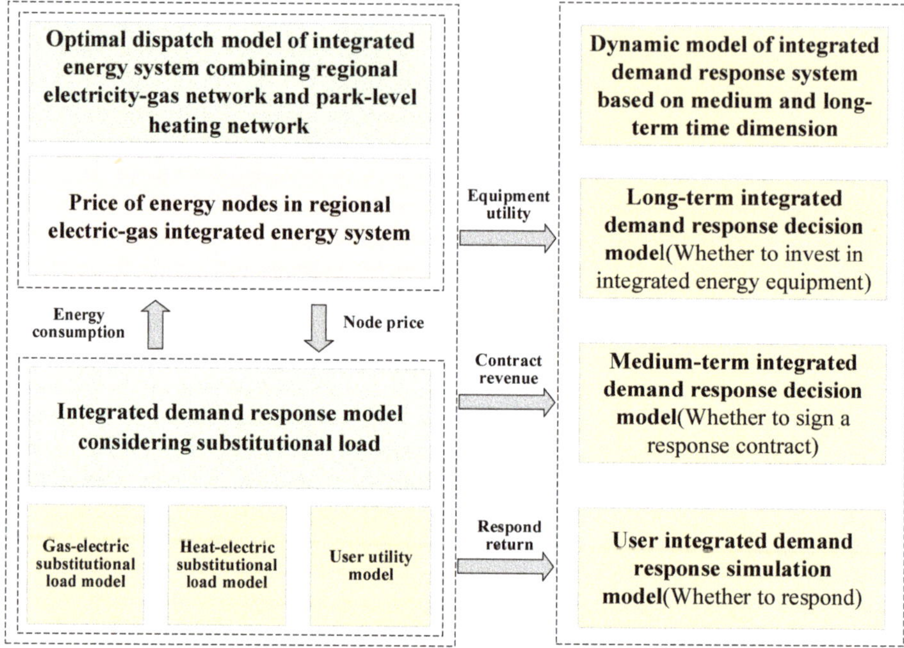

Figure 6. Modeling schematic.

3.1. Integrated Energy System Scheduling Model

The CCHP equipment of the coupling node in this paper mainly includes the use of combined heat and power, gas boilers, electric refrigerators, and absorption refrigerators. There are many related models [20,21], which are not repeated here. The electric-pneumatic network is connected to the park-level heating network to supply power and cooling while heating users. The structure of the integrated energy system is shown in Figure 7.

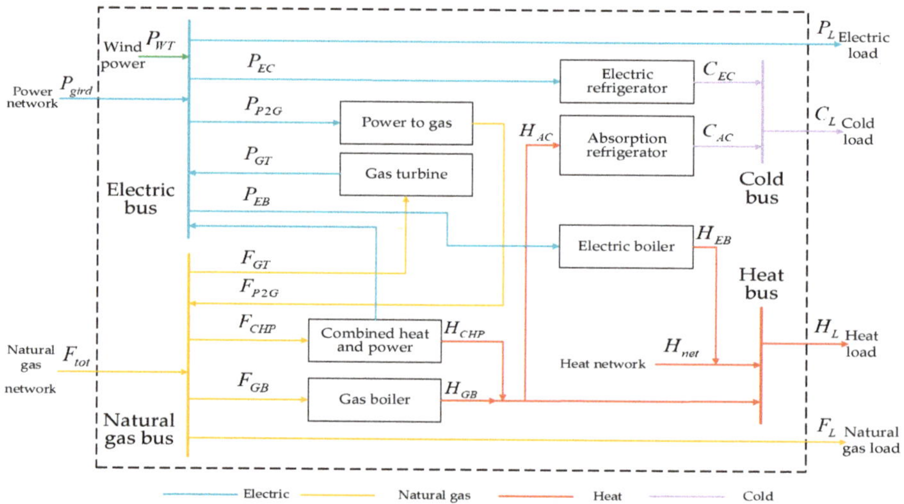

Figure 7. Integrated energy system structure.

3.1.1. Heat Network Model

The main considerations are node flow balance, node power fusion, load taking characteristics, temperature constraints of water supply and return water, and heat transfer characteristics of the pipe section [22].

1. Node traffic balance

For any node in the heat network, the sum of the incoming hot water flow is equal to the sum of the outgoing flow, that is:

$$\sum_{j \in S_i^{pipe+}} Q_{j,t}^g = \sum_{k \in S_i^{pipe-}} Q_{k,t}^g \tag{12}$$

where, S_i^{pipe+} and S_i^{pipe-} is a set of pipes connected to node i and starting and ending from node i respectively; $Q_{j,t}^g$ is the mass flow of hot water in pipe j during period t.

2. Node temperature fusion

Hot water of different temperatures flows from different pipes to the same node and is mixed. After mixing, the hot water flowing into the different pipes from the same node has the same temperature, that is:

$$\sum_{j \in S_i^{pipe+}} T_{j,t}^O Q_{j,t}^g = T_{k,t}^I \sum_{k \in S_i^{pipe-}} Q_{k,t}^g \tag{13}$$

where, $T_{j,t}^O$ is the hot water outlet temperature of pipe j in period t; $T_{k,t}^I$ is the temperature of the hot water in the pipeline k for period t.

3. Load taking characteristics

For a heat network branch that contains hot users, the load node i consumes heat at time t as the mass flow through the load node is water temperature reduced to return water temperature, which is:

$$h_{i,t}^L = cQ_{i,t}^g\left(T_{i,t}^g - T_{i,t}^h\right) \tag{14}$$

where, C is the specific heat capacity of hot water, taking 4.2 kJ/(kg·°C).

4. Variable supply and return water temperature constraints

In order to ensure the heat supply quality of heat sources and heat users, the supply and return water temperatures of heat sources and heat users need to be limited, that is:

$$T_{min}^g \leq T_{i,t}^g \leq T_{max}^g \tag{15}$$

$$T_{min}^h \leq T_{i,t}^h \leq T_{max}^h \tag{16}$$

5. Heat transfer characteristics of pipe sections

According to the modelling of heat transfer characteristics in [23], the steady-state heat transfer characteristics are as follows:

$$T_e = (T_a - T_s)\frac{x}{Rc\rho f} + T_s \tag{17}$$

where x is the distance between a point on the pipe section and the head of the pipe section; R is the thermal resistance per unit length of the pipe section; $T\,s$, $T\,e$, $T\,a$ are the head end temperature of a pipe section, x is the temperature and outside temperature; f is the flow of hot water.

The transient heat transfer characteristics are as follows:

For places near the heat source, the transient process is short. After one adjustment and before the next adjustment, the temperature of the pipe section has reached a steady state. The transient heat transfer characteristics of the pipe section at these points can be expressed as:

$$T^i(x,t) = \begin{cases} \frac{\left(T_s^{i-1}-T_s^{i-2}\right)(1-\alpha x)}{\beta x}(t-t_{i-1}) \\ +\left(T_a - T_s^{i-2}\right)\alpha x + T_s^{i-2} \ t \in [t_{i-1}, t_{i-1}+\beta x] \\ \left(T_a - T_s^{i-1}\right)\alpha x + T_s^{i-1} \ t \in (t_{i-1}+\beta x, t_i] \end{cases} \tag{18}$$

where $T^i(x,t)$ is the temperature of the heat network pipe at time t from the heat source x during the i-th period; T_s^{i-1} and T_s^{i-2} are the temperature of the heat source at time t_{i-1} and time t_{i-2}; $i = 1,2,3, \ldots$

For places far away from the heat source, the steady state has not been reached during the adjustment process. The transient heat transfer characteristics of the pipe section at these points can be expressed as:

$$T^i(x,t) = \frac{\left(T_a - T_s^{i-1}\right)\alpha x + T_s^{i-1} - T^{i-1}(x,t_{i-1})}{\beta x} \times (t-t_{i-1}) + T^{i-1}(x,t_{i-1}) \ t \in [t_{i-1}, t_i] \tag{19}$$

where, $T^i(x,t)$ is the temperature of the heat network pipe at time t_{i-1} from the heat source x during the $i-1$ period; $i = 1,2,3, \ldots$

3.1.2. Natural Gas Network Model

It mainly includes pipeline flow constraints, gas source point constraints, flow balance constraints, compressor constraints, and node pressure constraints [22].

1. Pipeline flow constraints

For an ideal adiabatic gas pipeline, considering the two-way flow of natural gas, the flow equation can be expressed as:

$$\tilde{Q}_{ij,t}\left|\tilde{Q}_{ij,t}\right| = C_{ij}^2\left(p_{i,t}^2 - p_{j,t}^2\right) \tag{20}$$

where, $\tilde{Q}_{ij,t} = (Q_{ij,t}^{in} + Q_{ij,t}^{out})/2$, which represents the average flow through pipe ij at time t, where $Q_{ij,t}^{in}$, $Q_{ij,t}^{out}$ are the first section of natural gas injection flow and the final natural gas output flow of pipeline ij at time t; C_{ij} is the constants related to the efficiency, temperature, length, inner diameter, and compression factor of pipeline ij; $p_{i,t}$, $p_{j,t}$, respectively the pressure value of the first and last nodes i and j at time t.

2. Natural gas source point constraint

$$Q_{n,\min}^N \leq Q_{n,t}^N \leq Q_{n,\max}^N \tag{21}$$

where, $Q_{n,\max}^N$, $Q_{n,\min}^N$, $Q_{n,t}^N$ are the upper and lower limits of the natural gas supply flow at the gas source point n and the output of the gas source at time t.

3. Traffic balance constraint

$$Q_{i,t}^N + \sum_{i\in\Omega_j} Q_{ij,t} + Q_{i,t}^{P2G} - Q_{i,t}^{G2P} - Q_{i,t}^L - Q_{i,t}^{CCHP} = 0 \tag{22}$$

where, $Q_{i,t}^N$ supply gas flow for node i at time t; $Q_{i,t}^{P2G}$ is the gas-to-electricity gas supply flow at node i at time t; $Q_{i,t}^{G2P}$ is the natural gas flow consumed by the gas turbine at node i at time t; $Q_{i,t}^L$ is the natural gas load at node i at time t. $Q_{i,t}^{CCHP}$ is the natural gas flow consumed by CCHP at node i at time t.

4. Compressor constraint

   Using a simplified compressor model [24] is

   $$p_{l,t} \leq \beta_{com} p_{i,t} \tag{23}$$

where, β_{com} is the compression coefficient of the compressor, and $p_{l,t}$ and $p_{i,t}$ are the pressure values of nodes l and i.

5. Node pressure constraint

   $$p_i^{min} \leq p_{i,t} \leq p_i^{max} \tag{24}$$

where, p_i^{max}, p_i^{min} are the upper and lower limits of the pressure value at node i.

3.1.3. Grid Model

Node power balance, unit output constraints, climbing constraints, branch flow constraints, and electrical model gas and gas turbine related model constraints can be referred to in the literature [24,25], and will not be described here.

3.1.4. Objective Function

The park-level heat network is coupled to the regional-level electric-pneumatic interconnected integrated energy system network through CCHP equipment. The heat load is consumed by CCHP to supply electricity and natural gas. It only needs to consider the system's dispatching costs for electricity and natural gas, mainly including the cost of thermal power generation, the cost of natural gas supply and the cost of electricity to gas, that is:

$$minF = \sum_{t \in T} \left[\sum_{g \in \Omega_G} f_1(P_{g,t}^G) + \sum_{n \in \Omega_N} f_2(Q_{n,t}^N) + \sum_{l \in \Omega_{P2G}} f_3(P_{l,t}^{P2G}) \right] \tag{25}$$

where, F is the integrated operating cost of the system; T is the number of time sections; Ω_G is thermal power units; Ω_N is a set of natural gas source points; Ω_{G2P} is gas turbine unit; Ω_{p2G} is a unit for gas to electricity. $f_1\left(P_{g,t}^G\right)$ is the power generation cost function of thermal power unit g at time t, expressed as:

$$f_1\left(P_{g,t}^G\right) = a_g\left(P_{g,t}^G\right)^2 + b_g P_{g,t}^G + c_g \tag{26}$$

where a_g, b_g, c_g are the parameters of the g consumption curve of the thermal power unit. $P_{g,t}^G$ is the active output of the thermal power unit g at time t.

$f_2\left(Q_{n,t}^N\right)$ is the cost function of gas supply point n at time t, which is expressed as:

$$f_2\left(Q_{n,t}^N\right) = C_{n,t}^N Q_{n,t}^N \tag{27}$$

where, $C_{n,t}^N$ is the gas supply cost coefficient of gas source point n at time t, $Q_{n,t}^N$ is the natural gas supply flow at gas source point n at time t.

$f_3\left(P_{l,t}^{P2G}\right)$ is the running cost function of electric gas l at time t, which is expressed as:

$$f_3\left(P_{l,t}^{P2G}\right) = C_{l,t}^{P2G} P_{l,t}^{P2G} \tag{28}$$

where, $C_{l,t}^{P2G}$ is the operating cost coefficient of electric gas l at time t, $P_{l,t}^{P2G}$ is the active power converted from electricity to gas l at time t.

3.2. Integrated Demand Response Model Considering Flexible Loads, Energy Storage, and Electric Vehicle as Participants

The multi-energy synergy of the integrated energy system expands the form of user participation in integrated demand response. Based on the difference in energy prices, users can choose different energy consumption methods to maximize their own energy efficiency. The main body is divided into flexible loads, energy storage and electric vehicles. In the power demand response, the response methods of flexible loads mainly include interruptible, translational, and transferable types, which are not described here, mainly considering the coupling and substitution relationship between multiple energy sources According to different ways of energy replacement, establish gas-electricity replacement load and heat-electricity replacement load model. For energy storage, in addition to considering traditional electric energy storage, you also need to establish models for gas storage systems and heat storage systems. For cars, the user's electric vehicle driving characteristics are considered to establish a charge and discharge model for electric vehicles.

3.2.1. Node Energy Price

First, determine the node energy price according to the node energy balance constraints of the 3.1 integrated energy system scheduling model, as shown below:

$$P_{i,t}^{G} + P_{i,t}^{W} + P_{i,t}^{CCHP} + P_{i,t}^{G2P} - P_{i,t}^{P2G} - \sum_{j \in \Omega_i} P_{ij,t} = P_{i,t}^{L} \tag{29}$$

$$Q_{i,t}^{N} + \sum_{i \in \Omega_j} Q_{ij,t} + Q_{i,t}^{P2G} - Q_{i,t}^{G2P} - Q_{i,t}^{CCHP} = Q_{i,t}^{L} \tag{30}$$

The right side of the above formula is the node power load and the node natural gas load. $P_{i,t}^{L}$ increasing 1, the objective function value F (system operating cost) will correspondingly generate a marginal value corresponding to the power network node, and use this value to represent the node electricity price, which is recorded as:

$$p_{i,t}^{e} = \partial F / \partial P_{i,t}^{L} \tag{31}$$

Similarly, whenever the node natural gas load $Q_{i,t}^{L}$ increasing 1, the value of the objective function will also generate a marginal value corresponding to the natural gas network node. Use this value to represent the natural gas price of the node and record it as:

$$p_{i,t}^{g} = \partial F / \partial Q_{i,t}^{L} \tag{32}$$

3.2.2. Flexible Load Response Model

For user i, at time t, the electric load adjusted by the energy replacement project can be expressed as:

$$q_{i,t}^{e} = \overline{q_{i,t}^{e}} - L_{i,t}^{eg} - L_{i,t}^{eh} \tag{33}$$

The adjusted natural gas load is expressed as:

$$q_{i,t}^{g} = \overline{q_{i,t}^{g}} + \rho_{e/g} L_{i,t}^{eg} \tag{34}$$

The adjusted heat load is expressed as:

$$q_{i,t}^{h} = \overline{q_{i,t}^{h}} + \rho_{e/h} L_{i,t}^{eh} \tag{35}$$

where: $L_{i,t}^{eg}$ replace the electric load with gas for the user at time t; $L_{i,t}^{eh}$ replace the electric load with heat for the user at time t; $\overline{q_{i,t}^{e}}$ The user's power load value before participating in the energy replacement project; $\overline{q_{i,t}^{g}}$ is the user's natural gas load value before participating in the energy replacement project; $\overline{q_{i,t}^{h}}$ is the user's heat load value before participating in the energy replacement project; $\rho_{e/g}$ is gas-electricity substitution coefficient; $\rho_{e/h}$ is the thermo-electric substitution factor.

3.2.3. Multi-Type Energy Storage Response Model

The electric energy storage system, gas storage system, and heat storage system can all participate in the integrated demand response, and the response can be achieved by switching the charging and discharging mode [26]. The dynamic mathematical model of electric energy storage is expressed as:

$$E_t^{ESS} = (1 - \mu_e)E_{t-1}^{ESS} + \left(P_t^{ESS,in}\eta_{ech} - \frac{P_t^{ESS,dis}}{\eta_{edis}} \right) \times \Delta t \tag{36}$$

where E_t^{ESS} is the storage capacity of electric energy storage during t period, μ_e is the loss rate of electric energy storage, $P_t^{ESS,in}$, $P_t^{ESS,dis}$ are the storage charge and discharge power during t period, and η_{ech} and η_{edis} represent the charge and discharge efficiency.

The mathematical model of gas storage dynamics is expressed as:

$$E_t^{GS} = \left(1 - \mu_g\right)E_{t-1}^{GS} + \left(F_t^{GS,in}\eta_{gch} - \frac{F_t^{GS,dis}}{\eta_{gdis}} \right) \times \Delta t \tag{37}$$

where E_t^{ESS} is the gas storage capacity of gas storage during t period, μ_g is the loss rate of gas storage, $F_t^{GS,in}$ and $F_t^{GS,dis}$ are the injection and extraction flow of the gas storage facility during the period t, and η_{gch} and η_{gdis} represent the injection and extraction efficiency.

The mathematical model of thermal storage dynamics is expressed as:

$$H_t^{HS} = (1 - \mu_h)H_{t-1}^{HS} + \left(Q_t^{HS,in}\eta_{hch} - \frac{Q_t^{HS,dis}}{\eta_{hdis}} \right) \times \Delta t \tag{38}$$

where H_t^{HS} is the heat storage capacity of thermal energy storage during t period, μ_h is the heat dissipation loss rate of heat storage, $Q_t^{HS,in}$, $Q_t^{HS,dis}$ are the heat storage and endothermic power during the period t, and η_{hch} and η_{hdis} represent the efficiency of heat absorption and heat release.

3.2.4. Electric Vehicle Response Model

Electric vehicle can be used as mobile energy storage to participate in integrated demand response, and the response is closely related to travel rules, mileage and atmospheric conditions such as sunlight. While achieving energy storage, it can meet users' travel requirements. Frequent discharge of batteries will cause a certain amount of battery life for electric vehicles. In order to slow down the degradation of battery life, it is necessary to minimize the number of charge and discharge switching times [27] of the battery. It is assumed that the electric vehicle is only discharged once a day. The electric vehicle charge and discharge limit time t_{lim} can be expressed as:

$$t_{\text{lim}} = \frac{P_d t_n + P_c t_l - (1 - S_n)C_s}{P_c + P_d} \tag{39}$$

where, P_d, P_c, C_s are the charging and discharging power and rated capacity of the electric vehicle, t_n, t_l, and S_n are the current time and off-grid, respectively. Time and state of charge of the current time.

The state of charge when the electric vehicle leaves is satisfied by the following formula:

$$S_n \geq \frac{d_{next} \times W + Q_{\min}}{C_s} \tag{40}$$

where, d_{next} represents the next mileage of the electric vehicle, W is the power consumed per kilometer, and Q min is the minimum power of the electric vehicle.

Discharge capacity of electric vehicle P_d^{EV}. It can be expressed as:

$$P_d^{EV} = \frac{P_d}{P_c + P_d}[(t_l - t_n)P_c - (1 - S_n)C_s]$$

(41)

Charging capacity of electric vehicle P_c^{EV} can be expressed as:

$$P_c^{EV} = C_s(S_n + S_l - S_a - S_{\min})$$

(42)

where, S_l, S_a, and S_{min} are the state of charge when offline, the state of charge when connected, and the lowest state of charge, respectively.

3.2.5. Objective Function

Based on microeconomics theory, users' satisfaction with electricity, gas and heat is expressed as [28,29]:

$$v_{i,t}^e = \int_{q_{i,t}^{ne}}^{q_{i,t}^e} \left(\frac{q}{a}\right)^{\frac{1}{\varepsilon_e}} dq$$

(43)

$$v_{i,t}^g = \int_{q_{i,t}^{ng}}^{q_{i,t}^g} \left(\frac{q}{a}\right)^{\frac{1}{\varepsilon_g}} dq$$

(44)

$$v_{i,t}^h = \int_{q_{i,t}^{nh}}^{q_{i,t}^h} \left(\frac{q}{a}\right)^{\frac{1}{\varepsilon_h}} dq$$

(45)

where v_i^e, v_i^g as well as v_i^h Satisfaction of electricity, gas and heat for user i, $q_{i,t}^e$, $q_{i,t}^g$, $q_{i,t}^h$ For user i's pure electrical load, pure gas load, pure thermal load, $q_{i,t}^{ne}$, $q_{i,t}^{ng}$, $q_{i,t}^{nh}$ represent the rigid load of electricity, natural gas, and heat, respectively of user i at time interval t; ε_e, ε_g, ε_h represent the user's electricity, gas, and heat demand-price elasticity coefficient.

Maximizing node user utility is expressed as:

$$\max U_i = \sum_{t=1}^{T}\left\{\lambda_e\left[v_{i,t}^e(q_{i,t}^e) - q_{i,t}^e p_t^e\right] + \lambda_g\left[v_{i,t}^g(q_{i,t}^g) - q_{i,t}^g p_t^g\right] + \lambda_h\left[v_{i,t}^h(q_{i,t}^h) - (q_{i,t}^{he}p_t^e + q_{i,t}^{hg}p_t^g)\right]\right\}$$

(46)

where U_i is the integrated energy efficiency of node user i, λ_e, λ_g as well as λ_h. The weight coefficients for user electricity, gas and heat use, $q_{i,t}^{he}$, $q_{i,t}^{hg}$ are electric energy and natural gas consumed by CCHP for pure thermal load, p_t^e, p_t^g are the node electricity price and the node gas price obtained from the integrated energy system scheduling model. Finally, the short-term response income is classified according to the proportion of electricity, gas, and heat load. According to the short-term response income and short-term response corresponding to different gas-to-electricity ratios and thermoelectric ratios, expected returns make short-term response decisions [30–32].

4. Analysis of Examples

4.1. Study Data

In the underlying integrated energy system dispatching network, the improved IEEE24-node power grid and the Belgian 20-node gas network are coupled by electricity-to-gas, gas turbines and other equipment to form an upper-layer electric-pneumatic interconnected network. The improved campus-level heat network will pass CCHP according to [17]. It is connected to the electricity-gas network to form a park-level cool-heat-electricity-gas integrated energy system as shown in Figure 8, where, the IEEE24 node system has eight generating units; nodes 2 and 22 are gas turbines, which are respectively connected to the natural gas network. Anderlues is connected to the Mons node; nodes 13 and 18 are combined cooling and heating power units, which are connected to the Liege and Zomergem nodes of the natural gas network, eight nodes of the thermal network I and eight nodes of the thermal network II; power nodes 8, 19 and 21 each connected to a wind power unit with a rated output of 100 MW. In order to maximize the wind power and avoid the natural gas network line blockage, the input end of the electricity to gas is also connected to nodes 8, 19 and 21 of the power network, and the output end is connected to the Loenhout, Peronnes, and Voeren nodes in the natural gas network are connected. The 20-node natural gas system in Belgium includes 21 gas pipelines, two pressurized stations, and two gas source points W1 and W2. The heating networks I and II are connected to an electric boiler at 11 nodes, respectively. The upper and lower limits of output are 300 MW and 40 MW, and the efficiency is 0.95.

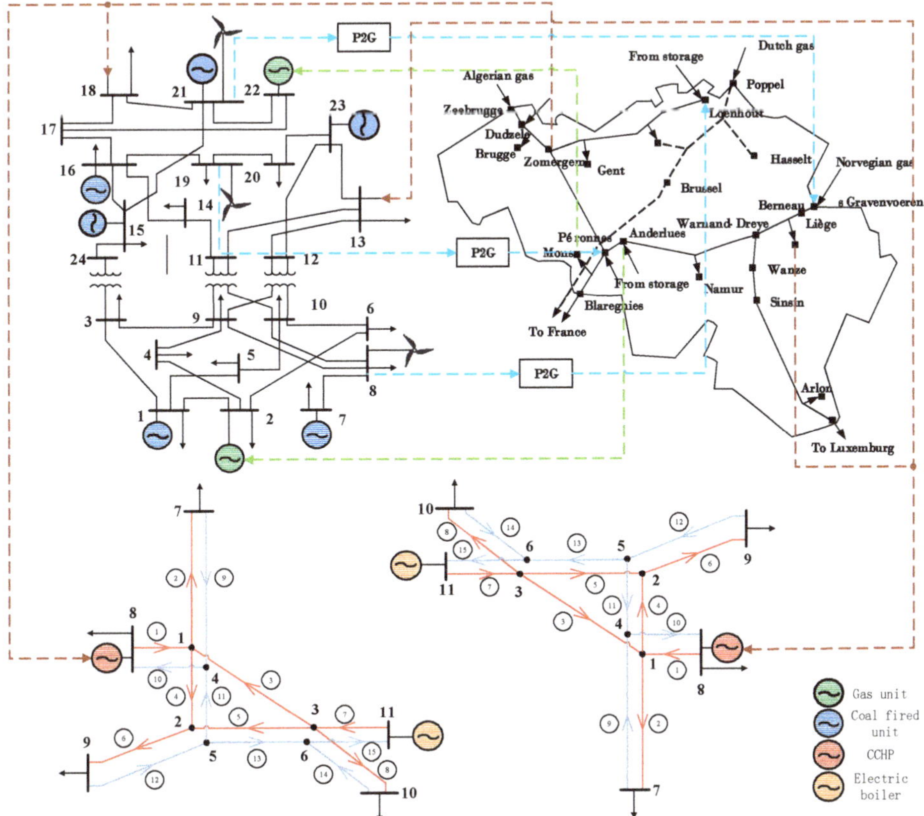

Figure 8. Integrated energy system example diagram of combined power grid, gas network and heat network.

4.1.1. Equipment Related Parameters

The relevant parameters of the gas turbine are shown in Table 1. The relevant parameters of the electric-to-gas conversion are shown in Table 2. The CCHP on the heating network I and the heating network II have the same configuration parameters, as shown in Table 3. On this basis, take coupling node 13 and coupling node 18 in the power grid for comparison and analysis. It is assumed that the two nodes have the same load before the response, corresponding to user 1 and user 2, respectively.

And their energy consumption preferences and expected response benefits to the integrated demand response are different.

Table 1. Gas turbine related parameters.

Unit	Power Network Node	Natural Gas Network Node	Active Upper Limit (MW)	Active Lower Limit (MW)	Conversion Efficiency (%)
GT1	2	Anderlues	104	0	43%
GT2	18	Liege	80	0	43%
GT3	22	Mons	80	0	43%

Table 2. P2G related parameters.

Unit	Power Network Node	Natural Gas Network Node	Enter Upper Limit (MW)	Enter Lower Limit (MW)	Methane Conversion Efficiency (%)	Run Cost ($/MW)
P2G1	8	Loenhout	88	0	60%	1.5
P2G2	19	Peronnes	88	0	60%	1.6
P2G3	21	Voeren	66	0	60%	1.5

Table 3. CCHP related parameters.

Name	Maximum Output (MW)	Lower Output Limit (MW)	Effectiveness (%)
Electric refrigerator	200	0	4
Absorption refrigeration	200	0	1.1
Cogeneration	350	0	0.75
Gas boiler	500	0	0.9
Heat Exchanger	10,000	0	0.9

4.1.2. Flexible Load Data

Take the load of four weeks in a month in the second year of winter to analyze the response of flexible load users to the integrated demand response project. The growth of gas load, heat load, and electrical load over time in one month is shown in Figure 9. The growth from Monday to Friday was relatively stable, and the load increased significantly every weekend.

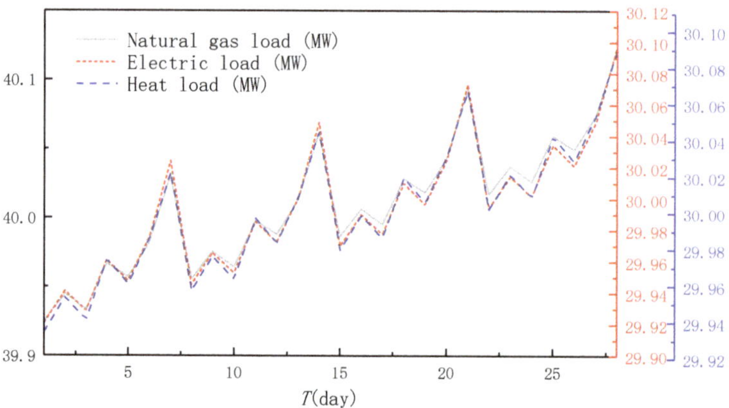

Figure 9. Flexible load growth in a month.

4.1.3. Energy Storage Load Data

Take the load of four quarters in a year to analyze the participation of energy storage users in integrated demand response projects. The change of electricity, gas and heat load over time in a year is shown in Figure 10.

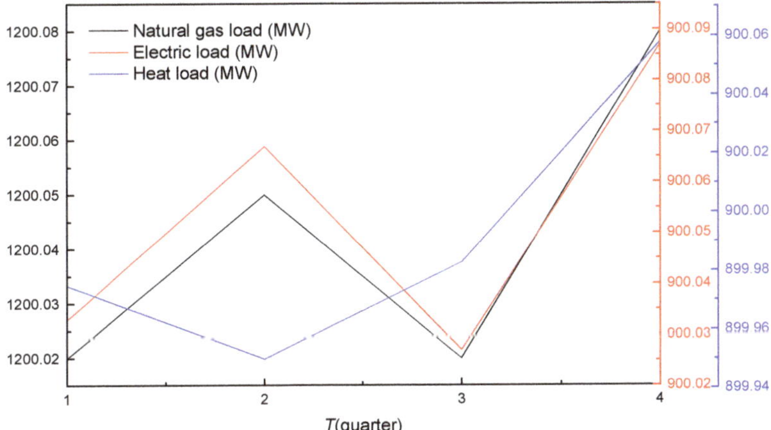

Figure 10. Quarterly load change.

4.1.4. Changes in Mileage of Electric Vehicle

The daily response of electric vehicle users is affected by the satisfaction of the mileage and the response income. The daily mileage is approximately log-normally distributed. The random distribution of the mileage of the user 1 within a month and the responses are shown in Figure 11.

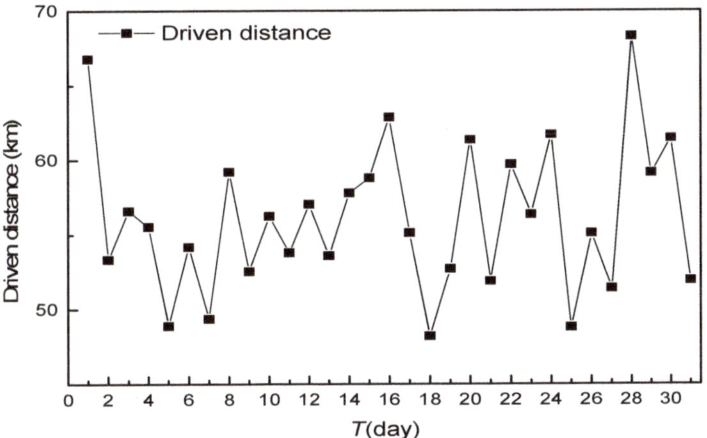

Figure 11. Electric vehicle mileage.

4.1.5. Changes in Policy Subsidy

Assume that the integrated demand response project has been implemented for five years, and the policy subsidy reflects the long-term incentive level. In order to compare and analyze the impact of different long-term incentive levels on the medium- and long-term integrated demand response decision, it is assumed that the policy subsidy changes are shown in Figure 12.

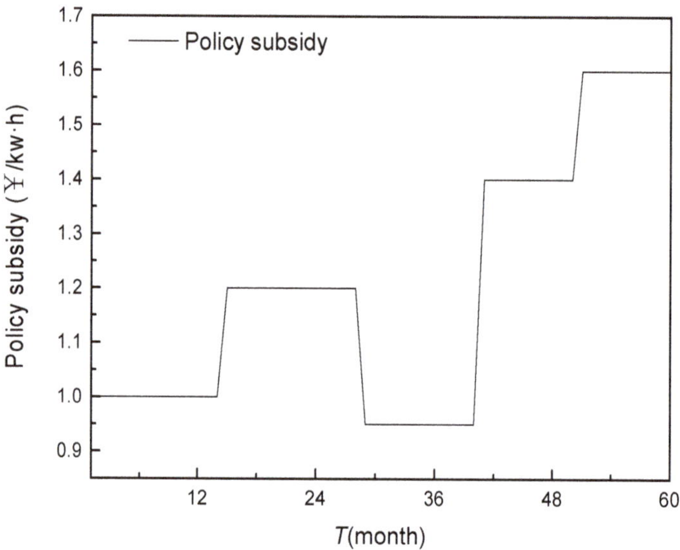

Figure 12. Policy subsidy change.

In the second year, the policy subsidy increased compared to the first year, but in the third year, the policy subsidy fell to a trough due to some reasons, and in the fourth and fifth years, it returned to the high subsidy level.

4.1.6. Changes in Policy Subsidy

According to the short-term integrated demand response operation simulation, the short-term expected response benefits are shown in Table 4, and the different parameters of different types of users' response to the long-term integrated demand are shown in Table 5.

Table 4. Short-term expected response income.

Gas-Electricity Ratio	Thermoelectric Ratio	Short-Term Expected Response Returns ($)
Greater than 1.1875 and Less than 1.33	Greater than 0.9375 and less than 1	26,931.453
Greater than 1.1875 and Less than 1.33	Greater than 1	25,435.261
Greater than 1.1875 and Less than 1.33	Less than 0.9375	25,435.261
Greater than 1.33	/	26,931.453
Less than 1.1875	/	17,954.302

Table 5. Differential parameters for medium- and long-term response.

User	Medium-Term Yield Improvement	Long-Term Yield Improvement Rate	Equipment Cost ($)
Flexible load user 1	9	0.15	835
Flexible load user 2	9	0.2	835
Energy storage user 1	9	0.15	17,167
Energy storage user 2	9	0.2	17,167
Electric vehicle users1	9	0.15	33,335
Electric vehicle users 2	9	0.2	33,335

4.2. Analysis of Short-Term Integrated Demand Response Simulation Results

Due to the continuous use characteristics of the energy storage equipment, the investment period of the energy storage equipment is set to half a year, and the investment equipment will be used. Therefore, the long-term energy storage equipment investment decision is not considered to analyze the short-term response benefits of energy storage users. It mainly analyzes the short-term response of flexible loads and electric vehicle users. The short-term response income simulation results are shown in Table 6.

Table 6. Short-term response benefit simulation results.

Gas-Electricity Ratio	Thermoelectric Ratio	Short-Term Response Income ($)
Greater than 1.1875 and Less than 1.33	Greater than 0.9375 and less than 1	28,427.645
Greater than 1.1875 and Less than 1.33	Greater than 1	26,931.453
Greater than 1.1875 and Less than 1.33	Less than 0.9375	26,931.453
Greater than 1.33	/	25,435.261
Less than 1.1875	/	13,465.727

4.2.1. Analysis of Short-Term Response of Flexible Load Users

For flexible loads, the response situation is more closely related to time. The changes in weekday and holiday loads make users' changes in expected response revenue more obvious. Therefore, this article analyzes the response of flexible load users to participate in integrated demand response from weekly load changes. The response within four weeks is shown in Figure 13. Among them, "1" on the ordinate represents the user's participation in the response, and the abscissa is the number of days. In the four weeks shown in Figure 12, users have higher responsiveness from Monday to Friday and relatively few weekends. The main reason is that users have higher requirements for energy consumption on weekends and higher expected returns on integrated demand response projects, so they have fewer response times.

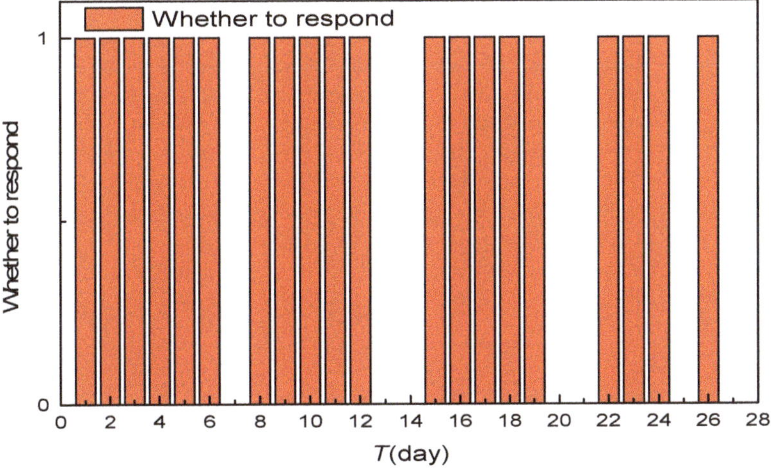

Figure 13. Short-term response of flexible load users.

4.2.2. Analysis of Short-Term Response of Electric Vehicle Users

The response of the electric vehicle is shown in Figure 14. In conjunction with Figures 11 and 14, when the electric vehicle's mileage did not exceed 58 km, the user chose to respond. This is mainly because when the mileage did not exceed 58 km, the user's travel satisfaction was within the acceptable range. However, when the mileage exceeded 58 km, the response was not enough to make up for the user's request for travel satisfaction, and it did not respond.

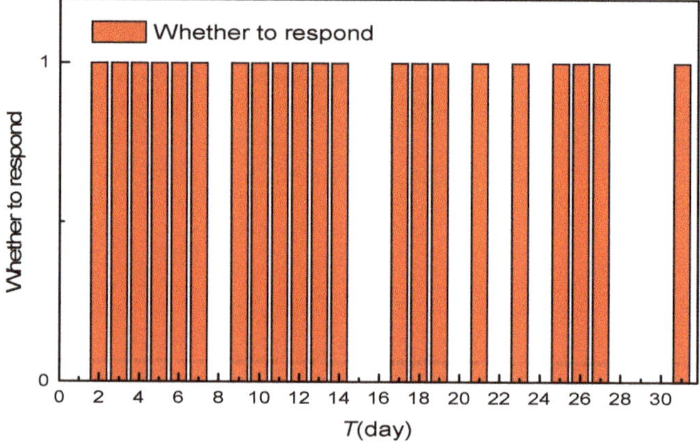

Figure 14. Short-term response of electric vehicle users.

4.3. Long-Term Integrated Demand Response Analysis of Flexible Loads, Energy Storage, and Electric Vehicle

4.3.1. Analysis of Medium- and Long-Term Decisions for Flexible Load Users

Vensim software was used to simulate the long-term integrated demand response project. The medium- and long-term integrated demand response situation of flexible load users is shown in Figure 15, where the red block is the subsidy level, and the long-term decision of users 1 and 2 is 1 for integrated investment. Energy equipment, when it was 0, did not invest. When the medium-term decision was 1, it means that it signed a medium-term integrated demand response contract, and when it was 0, it means that it did not sign a contract. User 1 started to invest in integrated energy-use equipment after the second year subsidy policy was raised, and A medium-term integrated demand response contract was signed, but its sensitivity to policy subsidy was not high, and even if the subsidy slightly declined in the third year, its investment decision on integrated energy-using equipment was unchanged, and an integrated demand response contract was also signed. User 2 first began to invest in integrated energy equipment in 2015, actively participated in integrated demand response projects and signed medium-term integrated demand response contracts, but its sensitivity to policy subsidy was high. When the subsidy fell in the third year, it was decided not to invest in integrated energy equipment. However, a medium-term contract is still signed to ensure integrated demand response participation.

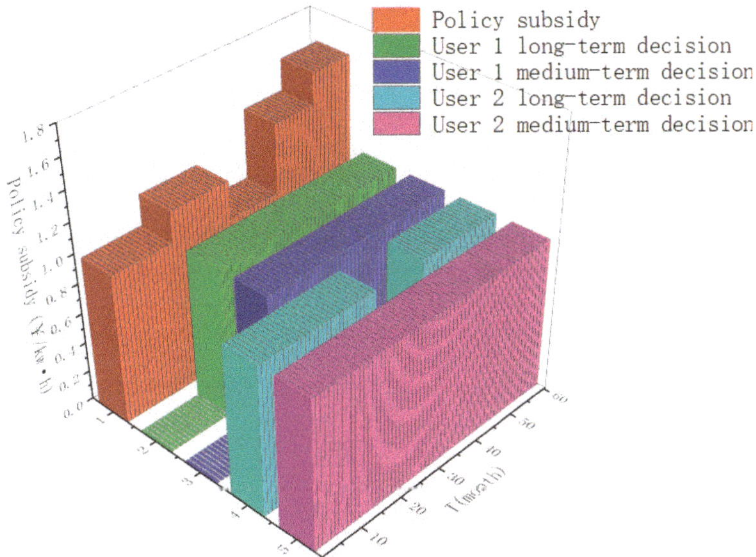

Figure 15. Medium- and long-term decision-making of flexible load users.

4.3.2. Medium- and Long-Term Decision Analysis of Energy Storage

The response of energy storage is greatly related to seasonal changes, and the changes in winter and summer make the response gains of energy storage significantly different, so this section analyzes the medium-to-long-term decisions of energy storage users from the changes in seasonal load. According to Figure 10 quarterly load changes of energy storage users, and the medium-to-long-term integrated demand response of energy storage users including electricity storage, gas storage, and heat storage systems are shown in Figure 16.

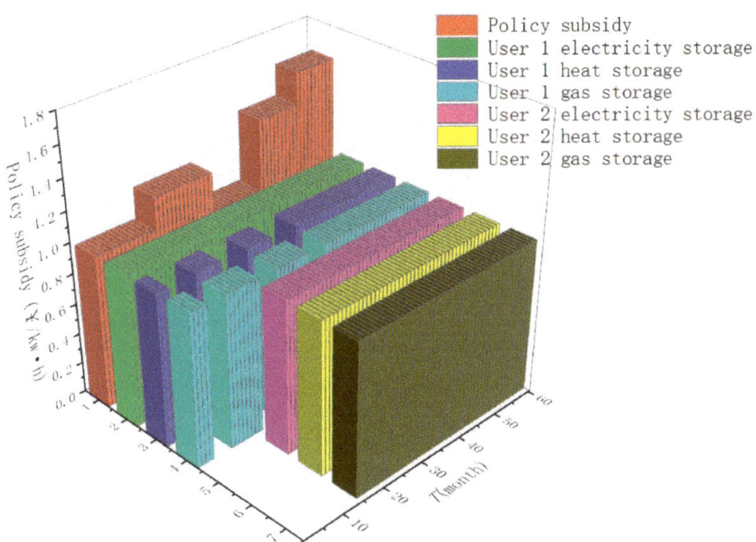

Figure 16. Long-term decision-making of energy storage users.

Combining Figures 12 and 16, User 1 had a higher preference for the energy consumption of electric loads, and has been investing in electric energy storage equipment since the first year, but did not invest in heat and gas storage in the season when the heat load and gas load were small. Equipment, after the policy subsidy rose sharply in the later period, began to invest in energy storage equipment regardless of the season. User 2's preference for electricity, heat, and gas was relatively balanced, and in the first year of policy subsidy levels, no energy storage equipment was invested. After the subsidy rose in the second year, it has been investing in energy storage equipment and responded positively.

4.3.3. Analysis of Medium- and Long-Term Integrated Decision of Electric Vehicle

According to Table 5, it can be seen that the long-term profit improvement rate of electric vehicle user 2 is higher than that of user 1, which reflects that user 2's travel satisfaction is higher than that of user 1. The medium-to-long-term integrated demand response of electric vehicle users is shown in Figure 17.

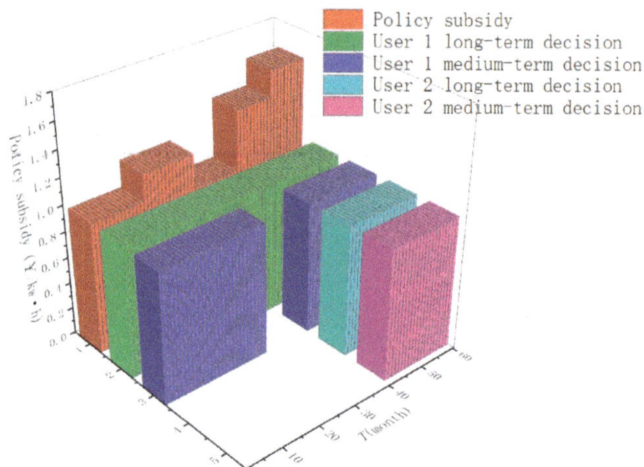

Figure 17. Medium- and long-term decision-making of electric vehicle.

Faced with the first year of policy subsidy, users with low travel satisfaction 1 began to invest in electric vehicles. Until the third year, subsidies fell, and the upgrading of electric vehicles was stopped, but investment was resumed after the subsidy rose. The travel satisfaction of 2 was high. In the face of the first to third year of subsidy, they were reluctant to invest in electric vehicles. After the sharp increase in the fourth year, they began to invest in electric vehicles.

5. Conclusions

Based on system dynamics, this paper analyzes the long-term, medium-term, and short-term integrated demand response behavior of demand response resources in different time dimensions, and establishes an integrated demand response model based on the long-term time dimension. Considering flexible loads, energy storage, and electric vehicles as IDR participation, the main body, established an integrated energy system scheduling and integrated demand response model, simulates the benefits of user participation in IDR, and provides a data source for the long-term integrated demand response model. The example is based on long-term incentives based on policy subsidy, and compares and analyzes different long-term incentive levels and regions: location, user energy preferences, and other characteristics of flexible load, energy storage, and electric vehicle users' response behavior in different time dimensions. When users of different types and locations face incentive signals from policy subsidy, their response decisions are significantly different. In the future, the uncertain impact of user response can be considered, and the user energy consumption behavior will continue to be deepened. We will analyze the multiple behaviors of different users, and do further research on a good source-charge interaction mechanism, considering issues such as economics, environmental protection, voltage quality and safety.

Author Contributions: Investigation, X.D. and Z.W.; Resources, X.W.; Writing—review and editing, S.R. and J.W. All authors have read and agreed to the published version of the manuscript.

Funding: This research was funded by the National Key R&D Program of China grant number 2018YFB0905000 and the Science and Technology Project of State Grid Corporation of China grant number SGTJDK00DWJS1800232.

Conflicts of Interest: The authors declare no conflict of interest.

References

1. National Guiding Opinions on Promoting the Development of "Internet +" Smart Energy [EB/OL]. Available online: http://rench.smesdgov.cn/ecdomain/jnrc/index/gmbgegiiejafbbodjlocigkedbdjccbo/20160308130440106.html (accessed on 8 March 2016).
2. Wang, W.; Wang, D.; Jia, H.; Chen, Z.; Guo, B.; Zhou, H.; Fan, M. Review of steady-state analysis of typical regional integrated energy system under the background of energy internet. *Proc. CSEE* **2016**, *36*, 3292–3305. (In Chinese)
3. Wang, W.; Wang, D.; Jia, H.; Chen, Z.; Guo, B.Q.; Zhou, H.M.; Fang, M.W. Steady state analysis of electricity-gas regional integrated energy system with consideration of NGS network status. *Proc. CSEE* **2017**, *37*, 1293–1304. (In Chinese)
4. Sun, H.; Guo, Q.; Pan, Z. Energy Internet: Ideas, Architecture and Future Perspectives. *Autom. Electr. Power Syst.* **2015**, *39*, 1–8. (In Chinese)
5. Sun, H.; Pan, Z.; Guo, Q. Research on Integrated Energy and Energy Management: Challenges and Prospects. *Autom. Electr. Power Syst.* **2016**, *40*, 1–8.
6. Tafreshi, S.M.M.; Lahiji, A.S. Long-term market equilibrium in smart grid paradigm with introducing demand response provider in competition. *IEEE Trans. Smart Grid* **2015**, *6*, 2794–2806. [CrossRef]
7. Favre-Perrod, P. A vision of future energy networks. In Proceedings of the 2005 IEEE Power Engineering Society Inaugural Conference and Exposition in Africa, Durban, South Africa, 11–15 July 2005; pp. 13–17.
8. Geidl, M.; Koeppel, G.; Favre-Perrod, P.; Klockl, B.; Andersson, G.; Frohlich, K. Energy hubs for the future. *Power Energy Mag. IEEE* **2006**, *5*, 24–30. [CrossRef]
9. Wu, J. Drive and Status Quo of the Development of European Integrated Energy System. *Autom. Electr. Power Syst.* **2016**, *40*, 1–7. (In Chinese)
10. Bozchalui, M.C.; Hashmi, S.A.; Hassen, H.; Canizares, C.A.; Bhattacharya, K. Optimal operation of residential energy hubs in smart grids. *IEEE Trans. Smart Grid* **2012**, *3*, 1755–1766. [CrossRef]
11. Lund, H.; Andersen, A.N. Optimal designs of small CHP plants in a market with fluctuating electricity prices. *Energy Convers. Manag.* **2005**, *46*, 893–904. [CrossRef]
12. Fragaki, A.; Andersen, A.N. Conditions for aggregation of CHP plants in the UK electricity market and exploration of plant size. *Appl. Energy* **2011**, *88*, 30–40. [CrossRef]
13. Fragaki, A.; Andersen, A.N.; Toke, D. Exploration of economical sizing of gas engine and thermal store for combined heat and power plants in the UK. *Energy* **2008**, *33*, 59–70. [CrossRef]
14. Buber, T.; von Roon, S.; Gruber, A.; Conrad, J. Demand response potential of electrical heat pumps and electric storage heaters. In Proceedings of the IECON 2013-39th Annual Conference of the IEEE Industrial Electronics Society, Vienna, Austria, 10–13 November 2013; pp. 8028–8032.
15. Gao, Y.; Wang, P.; Xue, Y. Collaborative planning of integrated electricity-gas energy systems considering demand side management. *Autom. Electr. Power Syst.* **2018**, *42*, 3–11. (In Chinese)
16. Brahman, F.; Honarmand, M.; Jadid, S. Optimal electrical and thermal energy management of a residential energy hub, integrating demand response and energy storage system. *Energy Build.* **2015**, *90*, 65–75. [CrossRef]
17. Bradley, P.; Leach, M.; Torriti, J. A review of the costs and benefits of demand response for electricity in the UK. *Energy Policy* **2013**, *52*, 312–327. [CrossRef]

18. Ruan, W.; Wang, B.; Li, Y.; Yang, S. Customer response behavior in time-of-use price. *Power Syst. Technol.* **2012**, *36*, 86–93. (In Chinese)

19. Wang, B.; Yang, X.; Yang, S. System Dynamics Analysis of Demand Response Potential and Effects Based on Medium- and Long Term Time Dimensions. *Chin. J. Electr. Eng.* **2015**, *35*, 6368–6377. (In Chinese)

20. Wang, J.; Gu, W.; Lu, S.; Zhang, C.; Wang, Z.; Tang, Y. Collaborative planning of multi-region integrated energy system combined with heat network model. *Autom. Electr. Power Syst.* **2016**, *40*, 17–24. (In Chinese)

21. Wang, C.S.; Hong, B.W.; Guo, L.; Zhang, D.J.; Liu, W.J. A general modeling method for optimal dispatching of combined cooling, heating and power microgrid. *Proc. CSEE* **2013**, *33*, 26–33. (In Chinese)

22. Dong, S.; Wang, C.; Xu, S.; Zhang, L.; Zha, H.; Liang, J. Day-ahead optimal scheduling of electricity-gas-heat integrated energy system considering dynamic characteristics of network. *Autom. Electr. Power Syst.* **2018**, *42*, 12–19. (In Chinese)

23. Yao, S.; Gu, W.; Zhang, X. Effect of Heating Network Characteristics on Ultra-short-term Scheduling of Integrated Energy System. *Autom. Electr. Power Syst.* **2018**, *42*, 89–90. (In Chinese)

24. Wei, Z.; Zhang, S.; Sun, G.; Zang, H.; Chen, S.; Chen, S. Power-to-gas considered peak load shifting research for integrated electricity and natural-gas energy systems. *Proc. CSEE* **2017**, *37*, 4601–4609. (In Chinese)

25. Correa-Posada, C.M.; Sanchez-Martin, P. Integrated power and natural gas model for energy adequacy in short-term operation. *IEEE Trans. Power Syst.* **2015**, *30*, 3347–3355. [CrossRef]

26. Yang, X.; Zhang, Y.; Lu, J.; Zhao, B.; Huang, F.T.; Qi, J.; Pan, H.W. Blockchain-based automated demand response method for energy storage system in an energy local network. *Proc. SCEE* **2017**, *37*, 3703–3716.

27. Li, Z.; Zhao, S.; Liu, Y. Control strategy and application of distributed electric vehicle energy storage. *Power Syst. Technol.* **2016**, *40*, 442–450.

28. Ren, Y.; Zhou, M.; Li, G. Bi-level model of electricity procurement and sale strategies for electricity retailers considering users' demand response. *Autom. Electr. Power Syst.* **2017**, *41*, 30–36. (In Chinese)

29. Thimmapuram, P.R.; Kim, J. Consumers' price elasticity of demand modeling with economic effects onelectricity markets using an agent-based model. *IEEE Trans. Smart Grid* **2013**, *4*, 90–397. [CrossRef]

30. Song, X.; Lin, H.; De, G.; Li, H.; Fu, X.; Tan, Z. An Energy Optimal Dispatching Model of an Integrated Energy System Based on Uncertain Bilevel Programming. *Energies* **2020**, *13*, 477. [CrossRef]

31. Liu, Z.; Yu, H.; Liu, R.; Wang, M.; Li, C. Configuration Optimization Model for Data-Center-Park-Integrated Energy Systems Under Economic, Reliability, and Environmental Considerations. *Energies* **2020**, *13*, 448. [CrossRef]

32. Lv, Z.; Wu, Z.; Dou, X.; Zhou, M.; Hu, W. Distributed Economic Dispatch Scheme for Droop-Based Autonomous DC Microgrid. *Energies* **2020**, *13*, 404. [CrossRef]

MDPI

St. Alban-Anlage 66

4052 Basel

Switzerland

Tel. +41 61 683 77 34

Fax +41 61 302 89 18

www.mdpi.com

Energies Editorial Office

E-mail: energies@mdpi.com

www.mdpi.com/journal/energies

9 783036 503424